Theory and Applications of Transport in Porous Media

Volume 29

Series editor

More information about this series at http://www.springer.com/series/6612

Auli Niemi · Jacob Bear
Jacob Bensabat
Editors

Geological Storage of CO_2 in Deep Saline Formations

Springer

Editors
Auli Niemi
Department of Earth Sciences
Uppsala University
Uppsala
Sweden

Jacob Bensabat
EWRE Ltd.
Haifa
Israel

Jacob Bear
Department Civil and Environmental Engineering
Technion-Israel Institute of Technology
Haifa
Israel

ISSN 0924-6118 ISSN 2213-6940 (electronic)
Theory and Applications of Transport in Porous Media
ISBN 978-94-024-0994-9 ISBN 978-94-024-0996-3 (eBook)
DOI 10.1007/978-94-024-0996-3

Library of Congress Control Number: 2016955068

Printed on acid-free paper

This Springer imprint is published by Springer Nature
The registered company is Springer Science+Business Media B.V.
The registered company address is: Van Godewijckstraat 30, 3311 GX Dordrecht, The Netherlands

Editorial Board

Preface

Geological storage of carbon dioxide in deep geological formations is considered a key transitional technique for reducing atmospheric emissions of greenhouse gases and thereby their inverse effects on climate. The technique as such has been in use for several decades in applications related to enhanced oil recovery, while the first industrial-scale project with purely environmental objectives of reducing atmospheric emissions of greenhouse gases was started in 1996, at Sleipner, Norway. Since then, a number of industrial, demonstration and pilot-scale projects have been and are being carried out around the world and processes and methodologies related to geological storage of carbon dioxide have been studied, theoretically and experimentally. The results have been presented in numerous papers, in scientific and professional journals and in books. The work is actively ongoing.

Deep saline formations are the geological formations evaluated to hold the highest storage potential, due to their abundance worldwide. Previously, such formations have been relatively poorly characterized, due to their low economic value and, therefore, they still need to be better quantified. Methods for characterizing, modeling, and monitoring of CO_2 storage in them are rapidly evolving in many parts of the world.

In Europe, the European Commission has included this topic in the framework of their research programs. Within these efforts, a large-scale integrating project MUSTANG,[1] with the focus on quantifying saline aquifers for geological storage of CO_2, was started in 2009. The project involved a team of 19 partners from universities, geological institutes, and companies as well as 24 affiliated parties from industries, regulators, and overseas expert organizations. The project was followed

[1]MUSTANG (2009–2014) coordinated by Auli Niemi, Uppsala Unversity, Sweden (www.co2mustang.eu).

by subsequent EU projects, PANACEA,[2] TRUST,[3] and CO2QUEST,[4] addressing various additional aspects of CO_2 geological storage. Many of the experts involved in these projects, especially the MUSTANG project, contributed to this book. Additional world-leading experts were invited to contribute material on specific topics to ensure a balanced coverage of the various subjects.

The book's objective is to present a comprehensive overview, as well as an in-depth understanding of appropriate methods for quantifying and characterizing saline aquifers that are suitable for geological storage of CO_2. It starts from a general overview of the methodology and the processes that take place when CO_2 is injected and stored in deep saline water containing formations. This is followed by presentations of mathematical and numerical models used for predicting the consequences of CO_2 injection. A description of relevant experimental methods, from laboratory experiments to field-scale site characterization, and techniques for monitoring the spreading of the injected CO_2 in the formation is presented next. Experiences from a number of important field injection projects are reviewed as are those from CO_2 natural analog sites. Finally, relevant risk management methods are presented. The book should be of interest to anyone working or planning to work or research on the topic of geological storage of CO_2.

We wish to acknowledge the European Commission for the funding of the aforementioned projects. Special thanks go also to all the project partners and external advisors in the Project Advisory Boards, for their valuable contributions and guidance. The funding for the work presented in the book and originating from projects other than those mentioned above is also greatly acknowledged. We want to express special thanks to all the reviewers and members of the Editorial Board who have helped us to improve the contents of the book. Final thanks go to the young researchers: Tian Liang, Saba Joodaki, and Maryeh Hedayati, who, with great dedication, helped in the final editing process of the book.

We hope that this book will serve all those who are working or exploring the possibility of working in the field of geological storage of CO_2 in geological formations.

Uppsala, Sweden — Auli Niemi
Haifa, Israel — Jacob Bear
Haifa, Israel — Jacob Bensabat
August 2016

---

[2]PANCEA (2012–2014) coordinated by Jacob Bensabat, EWRE, Israel (http://panacea-co2.org/).

[3]TRUST (2012–2017) coordinated by Jacob Bensabat (http://trust-co2.org/).

[4]CO2QUEST (2013–2016) coordinated by Haroun Mahgerefteh, UCL, England (http://www.co2quest.eu/).

Contents

Contributors

Jacob Bear Department of Civil and Environmental Engineering, Technion – Israel Institute of Technology, Haifa, Israel

Jacob Bensabat Environmental and Water Resources Engineering Inc., Haifa, Israel

Peter Bergmann Helmholtz Centre Potsdam, GFZ German Research Centre for Geosciences, Section Geological Storage, Potsdam, Germany

Jesus Carrera Groundwater Hydrology Group (GHS UPC-CSIC), Institute of Environmental Assessment and Water Research (IDAEA), Spanish National Research Council, Barcelona, Spain

Francois H. Cornet Institut de Physique du Globe–Strasbourg, Paris, France

Marco Dentz Institute of Environmental Assessment and Water Research (IDAEA), Spanish National Research Council (CSIC), Barcelona, Spain

Stéphanie Dias OXAND SA, Avon, France

Christine Doughty Lawrence Berkeley National Laboratory, Berkeley, CA, USA

Katriona Edlmann School of Geoscience, Edinburgh Collaborative of Subsurface Science and Engineering (ECOSSE), University of Edinburgh, Edinburgh, UK

Fritjof Fagerlund Department of Earth Sciences, Uppsala University, Uppsala, Sweden

Bruno Figueiredo Department of Earth Sciences, Uppsala University, Uppsala, Sweden

Barry Freifeld Lawrence Berkeley National Laboratory, Berkeley, CA, USA

John Gale IEA Greenhouse Gas R&D Programme (IEAGHG), Cheltenham, UK

Iulia Ghergut Angewandte Geologie, Universität Göttingen, Göttingen, Germany

Stuart M.V. Gilfillan School of Geoscience, Edinburgh Collaborative of Subsurface Science and Engineering (ECOSSE), University of Edinburgh, Edinburgh, UK

Philippe Gouze Géosciences, Université de Montpellier, CNRS, Montpellier Cedex 5, France

Yvi Le Guen OXAND SA, Avon, France

Juan Hidalgo Institute of Environmental Assessment and Water Research (IDAEA), Spanish National Research Council (CSIC), Barcelona, Spain

Christopher Juhlin Department of Earth Sciences, Uppsala University, Uppsala, Sweden

Axel Liebscher Helmholtz Centre Potsdam, GFZ German Research Centre for Geosciences, Section Geological Storage, Potsdam, Germany

Linda Luquot Spanish National Research Council (IDAEA-CSIC), Barcelona, Spain

Stefan Lüth Helmholtz Centre Potsdam, GFZ German Research Centre for Geosciences, Section Geological Storage, Potsdam, Germany

Sonja Martens Helmholtz Centre Potsdam, GFZ German Research Centre for Geosciences, Section Geological Storage, Potsdam, Germany

Franz May Bundesanstalt fuer Geowissenschaften und Rohstoffe (BGR), Hannover, Germany

Christopher Ian McDermott School of Geoscience, Edinburgh Collaborative of Subsurface Science and Engineering (ECOSSE), University of Edinburgh, Edinburgh, UK

Johannes M. Miocic School of Geoscience, Edinburgh Collaborative of Subsurface Science and Engineering (ECOSSE), University of Edinburgh, Edinburgh, UK

Fabian Möller Helmholtz Centre Potsdam, GFZ German Research Centre for Geosciences, Section Geological Storage, Potsdam, Germany

Larry Myer Lawrence Berkeley National Laboratory, Berkeley, CA, USA

Auli Niemi Department of Earth Sciences, Uppsala University, Uppsala, Sweden

Olivier Poupard OXAND SA, Avon, France

Henry Power Department of Mechanical Engineering, School of Mechanical, Materials, Manufacturing Engineering and Management, The University of Nottingham, Nottingham, UK

Dorothee Rebscher Bundesanstalt fuer Geowissenschaften und Rohstoffe (BGR), Hannover, Germany

Martin Sauter Angewandte Geologie, Universität Göttingen, Göttingen, Germany

Cornelia Schmidt-Hattenberger Helmholtz Centre Potsdam, GFZ German Research Centre for Geosciences, Section Geological Storage, Potsdam, Germany

Martin Streibel Helmholtz Centre Potsdam, GFZ German Research Centre for Geosciences, Section Geological Storage, Potsdam, Germany

Alexandru Tatomir Angewandte Geologie, Universität Göttingen, Göttingen, Germany

Chin-Fu Tsang Department of Earth Sciences, Uppsala University, Uppsala, Sweden

Victor Vilarrasa Groundwater Hydrology Group (GHS UPC-CSIC), Institute of Environmental Assessment and Water Research (IDAEA), Spanish National Research Council (CSIC), Barcelona, Spain

Jan Lennard Wolf Bundesanstalt fuer Geowissenschaften und Rohstoffe (BGR), Hannover, Germany

Zhibing Yang Department of Earth Sciences, Uppsala University, Uppsala, Sweden

Chapter 1
CO_2 Storage in Deep Geological Formations: The Concept

John Gale

Abstract The objective of this first chapter is to give an overview of the role and status of Carbon Capture and Storage (CCS) technology in a world-wide perspective. The effects of increased greenhouse gas concentrations in the atmosphere, and their related impacts on climate change, are discussed first. This is followed by a discussion on options to mitigate the emissions and the role of CCS in these efforts, in terms of both the present status at the time of writing this book and future outlook.

1.1 Introduction

The Earth's atmosphere contains gases that are collectively called the greenhouse gases. The greenhouse gases are so called because in the atmosphere they both absorb and emit radiation. This process is the fundamental cause of what is called the greenhouse effect (IPCC 1990). The main greenhouse gases in the Earth's atmosphere are water vapor, carbon dioxide (CO_2), methane, nitrous oxide, and ozone. It is a well-established scientific fact that these greenhouse gases affect significantly the Earth's temperature—without them, scientists predict that the Earth's surface would average about 33 °C colder than the present average of 14 °C (Le Treut et al. 2007).

Since the late 18th century the atmospheric concentrations of these greenhouse gases and, in particular, CO_2 has risen by some 40 %. The principal reason for the increase in global greenhouse gas emissions is the increased use of fossil fuels and global deforestation since the period we call the Industrial Revolution, the beginning of which is taken as the year 1750 (Le Treut et al. 2007). Since that time, the atmospheric concentration of CO_2 in the atmosphere has increased from 280 to 392.6 parts per million (ppm) in 2012. In May 2013 the atmospheric concentration

J. Gale (✉)
IEA Greenhouse Gas R&D Programme (IEAGHG),
Cheltenham, UK
e-mail: john.gale@ieaghg.org

A. Niemi et al. (eds.), *Geological Storage of CO_2 in Deep Saline Formations*,
Theory and Applications of Transport in Porous Media 29,
DOI 10.1007/978-94-024-0996-3_1

of CO_2 at the Mauna Loa atmospheric observatory in Hawaii, taken as the "gold standard" atmospheric monitoring laboratory in the world was recorded as 400 ppm. This is the first time that the atmospheric CO_2 concentration has reached 400 ppm in millions of years. Scientific analyses suggest that atmospheric CO_2 levels reached as much as 415 ppm during the Pliocene era, between 5 and 3 million years ago. In that period, global average temperatures have been estimated to be 3–40 °C and as much as 100 °C warmer at the poles. Sea levels have been estimated to have ranged between 5 and 40 m higher than today (Monroe 2013).

Whilst CO_2 has the highest atmospheric concentration of the greenhouse gases and also has the largest contribution to total radiative forcing of all the gases, the other gases cannot be ignored. The atmospheric concentrations of other green-house gases have also risen significantly since the late 1980s. The atmospheric concentrations of the methane (CH_4), and nitrous oxide (N_2O) in 2011 were 1803 and 324 ppb, respectively, exceeding pre-industrial levels by about 150 and 20 % (IPCC 2013).

1.2 What are the Effects of Increased Greenhouse Gas Concentrations in the Atmosphere?

The principal effects of increased greenhouse gas concentrations in the atmosphere are:

- An uptake of energy by the climate system causing average surface temperatures to rise. The IPCC 5th Assessment Report (AR5) states (IPCC 2013) that "each of the last three decades has been successively warmer at the Earth's surface than any preceding decade since 1850. In the Northern Hemisphere, 1983–2012 was likely the warmest 30-year period of the last 1400 years".
- Melting of polar ice sheets. Again, IPCC AR5 states that "Over the last two decades, the Greenland and Antarctic ice sheets have been losing mass, glaciers have continued to shrink almost worldwide, and Arctic sea ice and Northern Hemisphere spring snow cover have continued to decrease in extent".
- Increased sea levels: IPCC AR5 states that "The rate of sea level rise since the 19th century has been larger than the mean rate during the previous two millennia. Over the period 1901–2010, global mean sea levels rose by 0.19 m".

The other significant effect of increased atmospheric concentrations of CO_2 is ocean acidification. Fundamental changes in seawater chemistry are occurring throughout the world's oceans. The oceans absorb about a quarter of the CO_2 that is released into the atmosphere every year, so as atmospheric CO_2 levels increase, so do the concentrations of CO_2 in the ocean. Initially, scientists considered that there were benefits from the ocean removing CO_2 from the atmosphere. However, decades of ocean observations now show that there is also a downside in that the CO_2 absorbed by the ocean is changing the chemistry of the seawater—as more CO_2 is absorbed the pH of the oceans is increasing (IGBP, IOC, SCOR 2013).

1.3 What are the Impacts of Climate Change?

The Intergovernmental Panel on Climate Change (IPCC) was established by the United Nations Environment Programme (UNEP) and the World Meteorological Organization (WMO) in 1988 to provide the world with a clear scientific view on the current state of knowledge in climate change and its potential environmental and socio-economic impacts (IPCC 1988). The IPCC is a scientific body under the auspices of the United Nations (UN). It reviews and assesses the most recent scientific, technical and socio-economic information produced worldwide relevant to the understanding of climate change. It does not conduct any research nor does it monitor climate related data or parameters.

The IPCC publishes its results in a series of Assessment Reports, the 4th Assessment report (AR4) was in 2007 and the 5th Assessment Report (AR5) was in early 2015 (IPCC 2014).

The AR4 indicates that global climate change has already had observable effects on the environment. Points highlighted by the report include:

- Warming of the climate system is unequivocal.
- Each of the last three decades has been successively warmer than any preceding decade since 1850.
- Over the last two decades, the Greenland and Antarctic ice sheets have been losing mass, glaciers have continued to shrink almost worldwide, and Arctic sea ice and Northern Hemisphere spring snow cover have continued to decrease in extent.
- The rate of sea level rise since the mid-19th century has been larger than the mean rate during the previous two millennia.

Over the coming decades it is expected that:

- Global surface temperature change for the end of the 21st century is likely to exceed 1.5 °C relative to 1850–1900 for all modeled scenarios and may even exceed 2 °C.
- Changes in the global water cycle in response to the warming over the 21st century will not be uniform. The contrast in precipitation between wet and dry regions and between wet and dry seasons will increase.
- The global ocean will continue to warm during the 21st century. Heat will penetrate from the surface to the deep ocean and affect ocean circulation.
- The Arctic sea ice cover will continue to shrink and thin and the Northern Hemisphere spring snow cover will decrease during the 21st century as global mean surface temperature rises. Global glacier volume will further decrease.

- Global mean sea level will continue to rise during the 21st century and the rate of sea level rise will very likely exceed that observed during 1971–2010 due to increased ocean warming and increased loss of mass from glaciers and ice sheets.
- Climate change will affect carbon cycle processes in a way that will ex-acerbate the increase of CO_2 in the atmosphere. Further uptake of carbon by the ocean will increase ocean acidification.

We should also note that a recent report by The International Geosphere-Biosphere Programme (IGBP), The Intergovernmental Oceanographic Commission (IOC-UNESCO) and the Scientific Committee on Oceanic Research (SCOR) (IGBP, IOC, SCOR 2013) highlights the impacts of increased ocean acidification which are:

- Ocean acidification causes ecosystems and marine biodiversity to change. It has the potential to affect food security and it limits the capacity of the ocean to absorb CO_2 from human emissions.
- The economic impact of ocean acidification could be substantial.

The IPCC report also paints the gloomy picture that most aspects of climate change will persist for many centuries even if emissions of CO_2 are stopped now.

1.4 What Options do We Have?

The first option is to do nothing—carry on burning fossil fuels as we are now doing or even increase their use as energy demand grows in developing countries. The reality of doing this is that we will do irreparable damage to the world we live in. It is not a sensible approach. The folly of this approach was highlighted in the Stern Review (Stern 2007) in 2007, which discussed the effect of global warming on the world economy.

The Review stated that climate change is the greatest and widest-ranging market failure ever seen, presenting a unique challenge for economics. The Stern Review's main conclusion was that the benefits of strong, early action on climate change far outweigh the costs of not acting. According to the Review, without action, the overall costs of climate change will be equivalent to losing at least 5 % of global gross domestic product (GDP) each year, now and forever. Including a wider range of risks and impacts could increase this to 20 % of GDP or more, also indefinitely. Stern believed at the time that 5°–6° of temperature increase was "a real possibility."

We can take the second option which is to adapt to changing global weather patterns we are now facing. Adaptation measures may be planned in advance or put in place spontaneously in response to a local pressure (IPCC 2007a). They include large-scale infrastructure changes—such as building defenses to protect against sea-level rise or improving the quality of road surfaces to withstand hotter temperatures—as well as behavioral shifts such as individuals using less water, farmers planting different crops and more households and businesses buying flood insurance.

The IPCC describes vulnerability to climate change (IPCC 2007a) as being determined by three factors: exposure to hazards (such as reduced rainfall), sensitivity to those hazards (such as an economy dominated by rain-fed agriculture), and the capacity to adapt to those hazards (for example, whether farmers have the money or skills to grow more drought-resistant crops). Adaptation measures can help reduce vulnerability—for example by lowering sensitivity or building adaptive capacity—as well as allowing populations to benefit from opportunities of climatic changes, such as growing new crops in areas that were previously unsuitable.

Low-income countries tend to be more vulnerable to climate risks. Adaptation measures will need to address risks specifically caused by climate change, such as raising the height of sea defenses. It is still unclear how expensive these measures will be or who will pay for them, but the World Bank suggests adaptation could cost as much as the world currently spends on development assistance.

Mitigation (IPCC 2007b) on the other hand addresses the root causes, by reducing green-house gas emissions, while adaptation seeks to lower the risks posed by the consequences of climatic changes. Both approaches will be necessary, because even if emissions are dramatically decreased in the next decade, adaptation will still be needed to deal with the global changes that have already been set in motion.

1.5 What are the Options to Mitigate CO_2 Emissions?

The International Energy Agency (IEA) assesses on a regular basis the role that low carbon technological options can play in transforming the current energy system. The 2012 edition of the IEA's Energy Technology Perspectives (ETP 2012) 2 °C Scenario (2DS) identifies the technology options and policy pathways that ensure an 80 % chance of limiting long-term global temperature in-crease to 2 °C (GEA 2012). The 2DS demonstrates how energy efficiency and accelerated deployment of low-carbon technologies can help cut government expenditure, reduce energy import dependency and lower greenhouse gas emissions.

This IEA analysis shows that CCS is an integral part of any lowest-cost mitigation scenario where long-term global average temperature increases are limited to significantly less than 4 °C (the 4 °C Scenario, 4DS), particularly for 2 °C scenarios (2DS). Figure 1.1 provides a summary of the IEA's analyses of the impacts that each low carbon technology must achieve to attain the 2DS scenario.

The IEA's analysis indicates that we need all low carbon technology options to be deployed globally to meet the 2DS target. There is no individual technology that can achieve the desired levels of emissions reductions on its own. All the options on

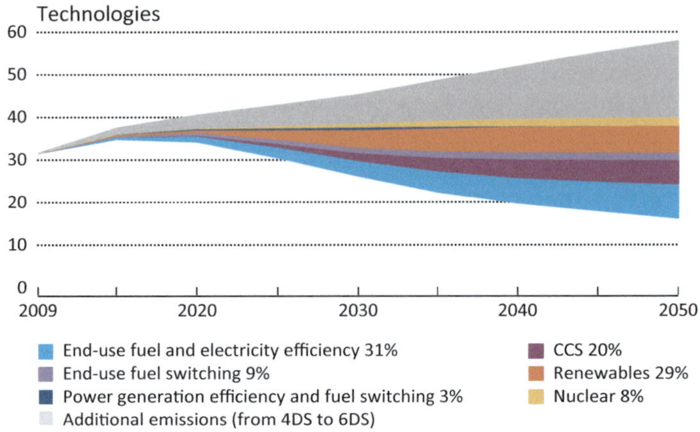

Fig. 1.1 Technology contributions needed to reach 2D scenario relative to the 4DS (ETP 2012)

the table have their own merits and demerits. Carbon capture and storage (CCS) advocates point to its ability to decarbonize the power sector without significant modification and investment in the current energy infrastructure. De-tractors point to the energy efficiency penalty associated with the capture component and uncertainties in the storage integrity of geological reservoirs. More recently it has become accepted that CCS is probably the only technology on the horizon today that would allow industrial sectors (such as iron and steel, cement and natural gas processing) to meet deep emissions reduction goals.

The IEA indicates that abandoning CCS as a mitigation option would significantly increase the cost of achieving the 2DS. The additional investment needs in electricity that are required to meet the 2DS would increase by a further 40 % if CCS is not available, with a total extra cost of USD 2 trillion over 40 years. Without CCS, the pressure on other emissions reduction options would also be higher.

More recently the IEA has considered a new scenario of 4DS (ETP 2014). The 4 °C Scenario (4DS) takes into account recent pledges made by countries to limit emissions and step up efforts to improve energy efficiency. Like the 2DS the 4DS is already an ambitious scenario that requires significant changes in policy and technologies. Moreover, capping the temperature increase at 4 °C requires significant additional cuts in emissions in the period after 2050. The need for CCS is not reduced in this scenario—rather its deployment needs to be ramped up quickly after 2020.

The IEA's analysis is not a unique one. Other example of many is the Global Energy Assessment 2012 (GEA) which uses a different assessment approach to that of the IEA which works from setting a 'business as usual' base case also. The GEA report shows that there are many combinations of energy resources, end-use, and supply technologies that can simultaneously address the multiple sustainability challenges. One of the report's key findings is that energy systems can be transformed to support a sustainable future through (a) radical improvements in energy efficiency, especially in end use, and (b) greater shares of renewable energies and advanced energy systems with carbon capture and storage (CCS) for both fossil fuels and biomass (GEA 2012).

The combination of biomass and CCS is one that is being taken up by environ-mental groups because of its potential to take emissions out of the atmosphere and reduce the atmospheric carbon budget, the so called "negative emissions" effect. Recent work to assess the global potential for bioCCS has suggested the global technical potential is large and, if deployed, could result in negative emissions up to 10 Gt of CO_2 equivalents annually. The key obstacle to the implementation of the technology is identified as the absence of a price for stored biomass based CO_2. There is therefore, a need for policy developments in this area to assist global take-up of the technology (IEAGHG 2011a).

1.6 What is Carbon Capture and Storage?

Carbon capture and storage (CCS) is the process of capturing waste carbon dioxide (CO_2) from large point sources, such as fossil fuel power plants, transporting it to a storage site by pipeline or ship, and injecting it where it will not enter the atmosphere, normally an underground geological formation. In this way release of large quantities of CO_2 into the atmosphere (from fossil fuel use in power generation and other industries) is prevented (IPCC 2005).

Geological storage of CO_2 is accomplished by injecting it in dense form into a rock formation below the Earth's surface. Porous rock formations that hold or (as in the case of depleted oil and gas reservoirs) have previously held fluids, such as natural gas, oil or brines, are potential candidates for CO_2 storage. Suitable storage formations can occur in both onshore and offshore sedimentary basins (natural large-scale depressions in the earth's crust that are filled with sediments). Coal beds also may be used for storage of CO_2 where it is unlikely that the coal will later be mined and provided that permeability is sufficient (IPCC 2005).

The IPCC Special Report on CO_2 capture and storage (IPCC SRCCS) undertook the first review of the global potential for CO_2 storage in geological formations (IPCC 2005). The IPCC SRCCS considered in detail three types of geological formations that had at that time received extensive consideration for the geological storage of CO_2. The three options were storage in: oil and gas reservoirs, deep saline formations and un-minable coal beds.

At the time of the IPCC SRCCS several other possible geological formations or structures (such as basalts, oil or gas shale's, salt caverns and abandoned mines) were considered. However it was felt at the time that these only represented niche opportunities, or had not been sufficiently studied at that time to assess their potential. This conclusion is still largely valid today.

The estimates of the technical potential for different geological storage options from the IPCC SRCCS are summarized in Table 1.1.

Table 1.1 Storage capacity for several geological storage options (IEAGHG 2011a)

Reservoir type	Lower estimate of storage capacity ($GtCO_2$)	Upper estimate of storage capacity ($GtCO_2$)
Oil and gas fields	675[a]	900[a]
Unminable coal seams (ECBM)	3–15	200
Deep saline formations	1000	Uncertain but possibly 10^4

[a]These numbers would increase by 25 % if 'undiscovered' oil and gas fields were included in this assessment. *Source* IPCC SRCCS (2005)

Whilst there have been numerous studies on the individual storage options since the IPCC SRCCS, the fact that the largest CO_2 storage potential globally lies in deep saline formations still remains a core conclusion to this day.

Since the IPCC SRCCS, our knowledge on how these storage resources can be developed, has been advanced. In the case of deep saline formations they represent a tantalizing resource for global storage of CO_2. However, they remain relatively unexplored in most regions of the world. Deep saline formations require much more extensive characterization because in general they are "virgin" formations not previously investigated. Because of this they require much longer lead times (up to 15 years of pre-exploration may be required,) that could be considered as viable for geological storage of CO_2 (IEAGHG 2011b).

Gas fields hold a greater storage potential than oil fields. Compared to deep saline formations both gas and oil fields are much better explored and have a background data set of both geological and operational/production data. On this basis they both should be more suitable for early application of CO_2 storage than deep saline formations. Storage in oil fields is typically carried out as part of enhanced oil recovery operations (EOR). In such systems the injection of CO_2 is used to maximize oil production, not for storage of the injected CO_2. However, incidental storage does occur within the reservoir and the gas recycle system which can amount to up to 90 % of the CO_2 injected (Whittaker and Perkins 2013).

The potential for geological storage in coal seams has recently been re-examined. Storage in coal seams involves replacing methane trapped in the coal structure with CO_2. This causes methane to be released, which is called CO_2 enhanced coal bed methane recovery (CO_2-ECBM). However, CO_2-ECBM is not considered as a primary extraction method but rather as a secondary technique once the coal has been dewatered and methane extracted during primary methane extraction or coal bed methane (CBM) extraction. Conventional CBM extraction may leave up to 50 % of the methane in the seam after development and production operations have been completed. It is postulated that another 20 % could potentially be recovered through the application of CO_2-ECBM in a second stage production process. Theoretically this approach therefore suggests once more that CO_2-ECBM could be a viable opportunity. However, this storage option is at an early stage of research with one project undertaken in China which has had production difficulties due to coal fines accumulation in unlined horizontal wells, and a second underway in the USA which, to date, has had limited operational success. The jury is therefore still out on CO_2-ECBM (IEAGHG 2013a).

In the last few years interest in using shale has also surfaced as a potential storage option. Again it is too early to decide whether this is a promising option for the future or not (IEAGHG 2013a).

1.7 Status of Geological Storage of CO_2

At the time of the IPCC SRCCS there were three operational commercial scale CO_2 injection projects globally, which were Sleipner in the North Sea, the IEAGHG Weyburn CO_2 Capture and Storage Project in Canada and the In-Salah Project in Algeria. Two of these, Sleipner and In-Salah, were injecting into deep saline formations, whilst Weyburn was a CO_2 flood in a depleting oil field (IEAGHG 2011a).

Since that time the number of CO_2 injection projects has grown considerably. An analysis undertaken by the IEA Greenhouse Gas R&D Programme has shown that there were, as of mid-2012, 45 small scale injection projects and 43 large scale projects (IEAGHG 2013b). Small scale projects were considered to be those injecting <100,000 tonnes, though many projects inject considerably less (<15,000 tonnes). Large scale projects were considered to be injecting >100,000 tonnes/annum.

Figure 1.2 provides an overview of the global distribution of the small or pilot CO_2 injection projects.

Whilst there are a number of individual projects in Australia, China, Europe and Japan, the majority of these projects are in North America and principally the USA. The reason for the large number of projects in the USA is that in 2003, the U.S. Department of Energy (DOE) awarded cooperative agreements to seven Regional Carbon Sequestration Partnerships (RCSPs). The seven RCSPs were tasked to determine the best geologic and terrestrial storage approaches and apply technologies to safely and permanently store CO_2 for their specific regions (RCSP 2003).

The RCSP Initiative has been implemented in three phases:

- Characterization Phase (2003–2005): Initial characterization of their region's potential to store CO_2 in different geologic formations.
- Validation Phase (2005–2011): Evaluation of promising CO_2 storage opportunities through a series of small-scale (less than 500,000 metric tons CO_2) field tests to develop understanding of injectivity, capacity, and storability of CO_2 in the various geologic formations within a wide range of depositional environments.
- Development Phase (2008–2018+): Implementation of large-scale field testing involving at least 1 million metric tons of CO_2 per project

1 BSCSP Basalt	16 MGSC Sugar Creek EOR Phase II	32 SECARB - Mississippi Saline Reservoir Test Phase II
2 Carbfix	17 MGSC Tanquary ECBM Phase II	33 South West Hub (Collie South West Hub)
3 CarbonNet	18 Mountaineer PVF	34 Surat Basin CCS Project (Previously Wandoan)
4 CIDA China	19 MRCSP Appalachian Basin (Burger) Phase II	35 SWP San Juan Basin Phase II
5 CS Energy Callide Oxyfuel Project	20 MRCSP Cincinnati Arch (East Bend) Phase II	36 Teapot Dome, Wyoming
6 CSEMP	21 MRCSP Michigan Basin Phase II	37 Lacq-Rousse
7 Fenn/Big Valley	22 Nagaoka Pilot CO2 Storage Project	38 West Pearl Queen
8 Frio, Texas	23 Otway I (Stage I)	39 WESTCARB Arizona Pilot (Cholla)
9 JCOP Yubari/Ishikari ECBM Project	24 Otway II Project (Stage 2A,B)	40 WESTCARB Northern California CO_2 Reduction Project
10 K12B	25 PCOR Lignite	41 WESTCARB Rosetta-Calpine test 1
11 Ketzin	26 PCOR Williston Basin -Phase II (NW McGregor Field)	42 WESTCARB Rosetta-Calpine test 2
12 Marshall County	27 PennWest Energy EOR Project	43 Western Kentucky
13 Masdar/ADCO Pilot project	28 Recopol	44 Zerogen Project
14 MGSC loudon Field EOR Phase II	29 SECARB - Black Warrior Basin Coal Seam Project	45 East Canton Oil Field
15 MGSC Mumford Hills EOR Phase II	30 SECARB - Central Appalachian Coal Seam Project	46 PCOR Zama Field Validation Project

Fig. 1.2 Global distribution of Pilot Scale CO_2 injection projects as of December 2013 (IEAGHG 2013b)

The RCSP program as of April 2014 has six operational one $MtCO_2$ injection projects with two more in preparation. This is the biggest geological storage program in the world (IEAGHG 2013b).

Together the pilot projects around the world have played a fundamental role in the development of the CO_2 storage component of the CCS system. Some of the key findings from these pilots is summarized as follows:

- The Otway project in Australia, was the first to specifically determine the CO_2 residual saturation in a saline aquifer. Using observed downhole pressures resulted in an average value of the CO_2 residual saturation of between 15 and 19 %.
- At the K-12 B injection offshore the Netherlands, the reservoir model was calibrated to an approximately 17 year production history and could accurately predict the pressure response for 2 short-term CO_2 injection tests.
- The Nagoka project in Japan was actually monitoring during an earthquake and physically demonstrated that seismic activity does not impact storage security. Post closure monitoring at Nagoka has also demonstrated CO_2 dissolution in the reservoir brine.

- The IEAGHG Weyburn CO_2 Monitoring and Storage Project in Canada demonstrated the importance of baseline monitoring when a supposed leak from the site was disproved based on baseline monitoring.

The pilot projects have led to the production of several best practice documents and guidelines, which vary in scope and technical detail. A number of non-site specific best practice guides have been produced, such as NETL's risk assessment and site selection manuals and WRI's CCS guidelines that outline the entire process. There are also best practice guidelines considering learnings taken from particular projects, such as the SACS best practices for the storage of CO_2 in saline aquifers, which uses, amongst others, learnings from the Sleipner storage site in the North Sea. Other examples of best practice guides are the QUALSTORE best practice guide and the EU Guidance documents. There are several documents outlining issues regarding public communication including guidelines from NETL and WRI. The Global CCS Institute commissioned CO2CRC to produce a summary of best practice guides, including a summary of the varying areas of coverage and technical detail (CO2CRC 2011).

Pilots have played a key role in helping build public confidence in geological storage (IEAGHG 2013b). It is safety/integrity of storage sites that principally gets raised in public debates on CCS. These pilot projects have assisted through:

- Establishing visitor centers at sites so the public can get first-hand experience of storage operations.
- They have enabled direct local dialogue with farmers and other key stakeholders.
- They enable the public to meet the scientists involved so they can learn and speak openly about their concerns.
- They provide the opportunity to disseminate information at a local level

Through these actions the pilot projects have helped build public confidence in CCS at a local/regional level which is important for the success of projects and CCS globally.

The pilot projects have also contributed directly to demonstration activities. For example, monitoring and operational experience from the K-12 B project, The Netherlands was used to develop permit application for ROAD project to EC. This application tested the permitting process under the EC Directive in Europe. The ROAD project permit was approved and hence was the First Injection/Monitoring permit approved by EU under the EC CCS Directive.

1.8 Outlook for CCS

Recent years have seen a considerable increase in the deployment of CCS at the demonstration scale. The Global Carbon Capture and Storage Institutes 2014 review of the status of CCS (Global CCS Institute 2014) has highlighted that:

- Four large scale CO_2-EOR projects commenced operation in 2013/4—Air Products Steam Methane Reformer Enhanced Oil Recovery (EOR) Project, Coffeyville Gasification Plant, Lost Cabin Gas Plant, all in the United States (US) and Petrobras Lula Oil Field CCS Project in Brazil.
- The world's first large-scale carbon capture and storage (CCS) project in the power sector commenced operation in October 2014 at the Boundary Dam power station in Saskatchewan, Canada.
- Two additional large-scale CCS projects in the power sector—at the Kemper County Energy Facility in Mississippi and the Petra Nova Carbon Capture Project in Texas—are planned to come into operation in late 2016 and early 2017 respectively.
- The Gorgon project in Australia will be operational in 2016. This will be the largest off shore gas processing plant capturing CO_2 for injection into a deep saline formation.
- Construction is also underway on the world's first large-scale CCS project in the iron and steel sector, the Abu Dhabi CCS Project in the United Arab Emirates (UAE).
- Construction also started in 2014 on the Archer Daniel Midland full scale bioCCS project in Illinois, USA.

The Global CCS Institute indicates that there are 22 large-scale CCS projects in operation or construction around the world—double the number at the beginning of the decade. Also there are a further 14 large-scale CCS projects in advanced planning, including nine in the power sector, many of which are anticipated to be in a position to make a final investment decision during 2015.

In the USA and Canada the CO_2-EOR projects in operation or under construction are building a pipeline infrastructure that can be used in the future for saline storage operations. In Europe and Australia there are a number of hub projects developing like Rotterdam Hub, the Teesside hub and the Collie hub in Western Australia (IEAGHG 2015). Hub and cluster projects like these will be essential to take the implementation of the technology to the next level. The challenge will be to fund these hub and cluster projects that can transport large volumes of CO_2 from multiple projects to their storage locations. In Europe it is hoped that new funding opportunities such as the NER400 might provide the financing for such project (European Council 2014).

References

CO2CRC (2011) A review of existing best practice manuals for carbon dioxide storage and regulation—a desktop study prepared for the Global CCS Institute by CO2CRC

European Council (2014) European Council (23 and 24 October 2014)—Conclusions, EUCO 169/14, Brussels, 24 Oct 2014

ETP (2012) Energy technology perspectives 2012—pathways to a clean energy system, 690 pp. ISBN: 978-92-64-17488-7, IEA (2012)

ETP (2014) Energy technology perspectives 2014—harnessing electricity's potential, 380 pp. http://www.iea.org/publications/freepublications/publication/energy-technology-perspectives-2014.html. Accessed 26 Jun 2016

GEA (2012) Global energy assessment—toward a sustainable future. International Institute for Applied Systems Analysis, Vienna, Austria and Cambridge University Press, Cambridge, UK and New York, NY, USA

Global CCS Institute (2014) The global status of CCS: 2014, Melbourne, Australia

IEAGHG (2011a) Potential for biomass and carbon dioxide capture and storage. 2011/06, July, 2011

IEAGHG (2011b) Global storage resource gap analysis for policy makers. 2011/10, Sept, 2011

IEAGHG (2013a) Potential implications on gas production from shales and coals for geological storage of CO_2. 2013/10, Aug, 2013

IEAGHG (2013b) The process of developing a CO_2 test injection: experience and best practice. 2013/13, Dec, 2013

IEAGHG (2015) Carbon capture and storage cluster projects: review and future opportunities. 2015/03, Apr 2015

IGBP, IOC, SCOR (2013) Ocean acidification summary for policymakers—third symposium on the ocean in a high-CO_2 World. International Geosphere-Biosphere Programme, Stockholm, Sweden

IPCC—Intergovernmental Panel on Climate Change (1988). http://www.ipcc.ch/organization/organization.shtml. Retrieved 28 Jun 2016

IPCC (1990) Climate change: the IPCC Scientific Assessment (1990) Report prepared for Intergovernmental Panel on Climate Change by Working Group I. In: Houghton JT, Jenkins GJ, Ephraums JJ (eds) Cambridge University Press, Cambridge, Great Britain, New York, NY, USA and Melbourne, Australia, 410 pp

IPCC (2005) IPCC Special Report on Carbon Dioxide Capture and Storage. In: Metz B, Davidson O, de Coninck, Loos M, Meyer LA (eds) Prepared by Working Group III of the Intergovernmental Panel on Climate Change. Cambridge University Press, Cambridge and NewYork, NY, 442 pp

IPCC (2007) Summary for policymakers. In: Climate change 2007: impacts, adaptation and vulnerability. In: Parry ML, Canziani OF, Palutikof JP, van der Linden PJ, Hanson CE (eds) Contribution of Working Group II to the Fourth Assessment Report of the Intergovernmental Panel on Climate Change. Cambridge University Press, Cambridge, 7–22

IPCC (2007) Summary for policymakers. In: Metz B, Davidson OR, Bosch PR, Dave R, Meyer LA (eds) Climate change 2007: mitigation. Contribution of Working Group III to the Fourth Assessment Report of the Intergovernmental Panel on Climate Change. Cambridge University Press, Cambridge and New York, NY

IPCC (2013) Summary for policymakers. In: Stocker TF, Qin D, Plattner G-K, Tignor M, Allen SK, Boschung J, Nauels A, Xia Y, Bex V, Midgley PM (eds) Climate change 2013: the physical science basis. Contribution of Working Group I to the Fifth Assessment Report of the Intergovernmental Panel on Climate Change. Cambridge University Press, Cambridge and New York, NY

IPCC (2014) Fifth Assessment Report (AR5). http://www.ipcc.ch/report/ar5/index.shtml. Accessed 28 Jun 2016

Le Treut H, Somerville R, Cubasch U, Ding Y, Mauritzen C, Mokssit A, Peterson T, Prather M (2007) Historical overview of climate change science. In: Solomon S, Qin D, Manning M, Chen Z, Marquis M, Averyt KB, Tignor M, Miller HL (eds) Climate change 2007: the physical science basis. Contribution of Working Group I to the Fourth Assessment Report of the Intergovernmental Panel on Climate Change. Cambridge University Press. Retrieved 14 Dec 2008

Monroe R (2013) What does 400 ppm look like? Original post at The Keeling Curve site. Available via Scripps Institution of Oceanography, UC San Diego https://scripps.ucsd.edu/programs/keelingcurve/. Accessed 28 Jun 2016

RCSP (2003) Regional carbon sequestration partnerships (RCSP) initiative. http://netl.doe.gov/research/coal/carbon-storage/carbon-storage-infrastructure/rcsp. Accessed 28 Jun 2016

Stern N (2007) The economics of climate change—the Stern review. Cambridge University Press, UK

Whittaker S, Perkins E (2013) Technical aspects of CO_2 enhanced oil recovery and associated carbon storage. Global CCS Institute. http://www.globalccsinstitute.com/publications/technical-aspects-co2-enhanced-oil-recovery-and-associated-carbon-storage. Accessed 28 Jun 2016

Chapter 2
Overview of Processes Occurring During CO_2 Geological Storage and Their Relevance to Key Questions of Performance

Chin-Fu Tsang and Auli Niemi

Abstract The objective of this chapter is to provide an overview and discussion of the relevancy of various physical and chemical processes to be associated with the geological storage of CO_2 at a particular site, and thereby serve as a bridge between the detailed process descriptions and modeling techniques to be presented in the following chapters and the studying and simulation of site-specific physicochemical behavior of a potential CO_2 geosequestration site. The approach adopted is to address the relevancy of a given process in terms of the specific objectives, the technical issues of concern, or the key questions associated with CO_2 geological storage, in the context of the geological settings and characteristics of the storage site. The suggested approach is exemplified by application to two field cases.

2.1 Introduction

In recent literature, many of the physical and chemical processes associated with CO_2 geological storage have been extensively and intensely investigated and reported. In Chaps. 3–5 of this book, they are reviewed and described in detail, together with modeling techniques used to simulate them. These chapters also give a substantial list of references related to these processes and their modeling. The objective of the present chapter is to provide an introductory overview and a bridge between these process descriptions and modeling techniques and the studying and simulation of physicochemical behavior of a potential CO_2 geosequestration site, and, in particular, to identify the relevant physicochemical processes to site-specific study of CO_2 injection and sequestration in the deep subsurface. The focus will be on CO_2 geosequestration in saline formations.

C.-F. Tsang (✉) · A. Niemi
Department of Earth Sciences, Uppsala University, Uppsala, Sweden
e-mail: cftsang@lbl.gov

A. Niemi
e-mail: auli.niemi@geo.uu.se

A. Niemi et al. (eds.), *Geological Storage of CO_2 in Deep Saline Formations*,
Theory and Applications of Transport in Porous Media 29,
DOI 10.1007/978-94-024-0996-3_2

The next section below gives an overview of the processes involved in a basic scenario of subsurface CO_2 injection and sequestration in a brine formation, followed by a general discussion of the geological settings and characteristics. Then the approach for discussing the relevancy of the physicochemical processes to CO_2 geosequestration at a specific site is presented. The implementation of this approach with an example of application to two field cases is described and discussed in the remaining sections of this chapter.

2.2 Overview of Processes in a Basic Scenario

For the sake of discussion in this chapter, it is useful to consider a basic scenario of injection and storage of CO_2 in brine formations as presented in Fig. 2.1, which shows a storage injection zone overlain by a caprock greater than 800 m in depth. Three main physicochemcial processes are indicated. First, there is the hydrological process of buoyancy flow of the CO_2 with its factor-of-two lower density and an order-of-magnitude lower viscosity. Thus the plume of injected CO_2 will migrate outwards from the injection well and upwards towards the caprock by buoyancy. Other hydrologic factors also come into play, which will be discussed below. Second, both injection and buoyancy provide additional pressure on the rock matrix of the formation, which may thus be deformed, with changes in matrix porosity or fracture apertures. They in turn cause changes in flow permeability and, consequently, the flow field. This is what we call the hydromechanical effect. Finally, the injected CO_2 plume will, in general, chemically react with the formation minerals. This could give rise to porosity changes near the injection well, but, positively, formation matrix minerals can react with the injected CO_2 to form new minerals in the rock matrix, thus trapping CO_2 chemically. This is the mineral trapping process for sequestration of CO_2. These main processes are discussed in more detail below.

Fig. 2.1 Schematic diagram of CO_2 geosequestration in saline formation

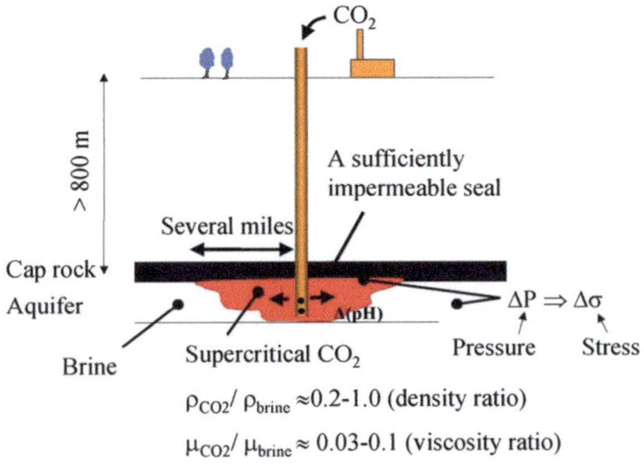

CO_2 injected into a deep brine formation will be present in three forms: a dense supercritical phase; a dissolved state in pore water, and an immobilized state through geochemical reaction with in situ minerals. The dissolved part of estimated to be from 2 % in saturated NaCl brines by weight to 7 % in typical ground water. CO_2 immobilization in formation matrix minerals is a slow process and varies considerably with rock types. The amount of CO_2 sequestered through such mineral reactions can be comparable with CO_2 dissolution in pore waters. Among all the forms that the injected CO_2 takes in a brine formation, the liquid-like supercritical phase is the main storage form and it has properties quite different from those of the pore water. For example, for storage of CO_2 at 1000 m depth, CO_2 density is about 60–75 % that of water in the formation, while its viscosity is about a factor of 15–20 times less than that of water.

The lower density of the stored supercritical CO_2 will cause buoyant flow of CO_2 to the top of the injection zone below the caprock. The flow depends on the density difference as well as the vertical and horizontal permeabilities of the formation. Because of the tendency for buoyancy flow of CO_2 to the top of the injection formation, the areal extent of the injected CO_2 will be much larger than a buoyancy-neutral fluid. For example, storage of 2.7×10^{11} kg of CO_2, injected at a rate of 350 kg/s for 30 years into a 100-m thick formation with $k_x = k_z = 10^{-13}$ m^2, has been estimated to have an increase in areal extent resulting from buoyancy flow by a factor of approximately 1.4. In this example, because of the large volume of CO_2 involved, the areal extent of the injected CO_2 can be as much as 120 km^2.

Once the injected fluid is in place, what happens if it is next to a leakage path in the caprock, such as an abandoned borehole or a fault? The density of super-critical CO_2 at a depth of 1000 m is about 600–750 kg/m^3, resulting in a significant buoyancy driving force causing an upward leakage of CO_2. However the buoyancy pressure needs to be larger enough to overcome the gas entry pressure into the caprock pores. One can estimate the thickness h of the layer of CO_2 needed to provide enough buoyancy pressure to exceed the gas entry, which turns out to be 70–170 m for a pore radius of 10^{-7} m. However, if there exists a fracture in the caprock, the effective pore radius in the fracture can be much larger and thus the thickness of CO_2 required to overcome the gas entry pressure of the fracture would be much less. In general, as CO_2 migrates upward, the flow involves evolving phases of CO_2 as well as the brine, presenting a complex phase interference effect.

The very low viscosity of supercritical CO_2 will give rise to flow instability at the CO_2-brine interface as CO_2 is being injected into the storage formation. This flow instability results in fingering. In other words, instead of piston-like flow of the CO_2 front into the injection formation, parts of the front will flow much faster in the form of fingers. This phenomenon occurs in parallel with the buoyancy flow effect discussed above. However, viscous fingering of CO_2 may not be as significant in the presence of geologic heterogeneity.

Heterogeneity of the injection formation gives rise to the fingering or channeling effect. The injected CO_2 will be channelized to follow the most permeable paths because of the spatial variation of permeability. The flow pattern will depend not only on the permeability variability and its spatial correlation range, but also on the saturation level of CO_2 in the different parts of the brine formation.

Mechanically, the main concern for liquid waste injection has been to ensure that injection pressure is safely below that which will cause hydraulic fracturing or affect well integrity (Fig. 2.2). For CO_2 injection and storage, however, both the injection and buoyancy pressures need to be considered. While injection pressure is highest around the injection well and starts to decrease after the termination of injection, buoyancy pressure extends over the entire CO_2 plume and lasts well beyond the injection period. An increase in formation fluid pressure, due to both injection and buoyant pressures, will cause local changes in the effective stress field, which, in turn, will induce mechanical deformations, possibly increasing the porosity and permeability and thus reducing the fluid pressure. However at the same time, increasing pressure may also cause irreversible mechanical failure in the caprock. This mechanical failure may involve possibly shear-slip along existing fractures and creation of new fractures (hydraulic fracturing), that reduce the sealing properties of the caprock system. In addition to these mechanical processes, replacing the native formation fluid with CO_2 may cause changes in rock mechanical properties through chemical-mechanical interactions between the CO_2 and the host rock, or through desiccation of fractures.

Chemically, at the CO_2 front where CO_2 is dissolved in water, the acidity of the groundwater is increased and many minerals comprising the host rock matrix minerals such as calcite, may dissolve readily, leading to an increase in permeability and porosity along the flow channel. This leads to a higher flow rate and increased dissolution, potentially forming what are known as wormholes. On the other hand, based on experience from enhanced oil recovery, CO_2 has been known to reduce injectivity in some cases, but to increase permeability near injection wells in others. There are also data indicating that dissolved CO_2 will cause a reduction in permeability where the carbonate minerals precipitate along the flow paths with a large pressure gradient. All these observations suggest the need for careful evaluation of the compatibility between supercritical CO_2 and geochemistry of the brine formation. Such an evaluation may also yield information useful for the design of

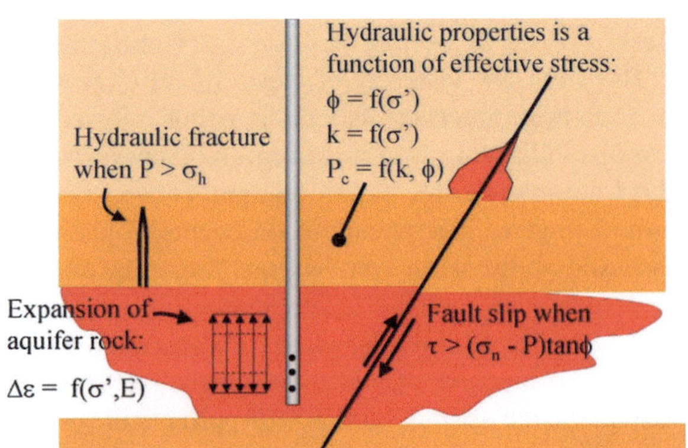

Fig. 2.2 Schematic diagram on hydromechanical changes due to CO_2 injection and storage (from Jonny Rutqvist, private communication 2012)

injection operations, such as keeping injection pressure below a certain value so that there will be no severe pressure gradients to induce precipitation or dissolution.

With CO_2 storage at or below the depth of about 1000 m, there may well be a sequence of intervening strata of confining and permeable layers separating the injection zone and the lowest underground sources of drinking water (USDW). This sequence of strata can provide a compounded margin of safety to reduce CO_2 upward leakage. Thus, each high-permeability layer serves as an injection zone for CO_2 leaking into it from below and spreading in it. The next overlying confining layer will then act as the next caprock to prevent continuing CO_2 leakage.

Because of the large volume of CO_2 being injected and stored, the displacement of in situ brine is an issue of concern. The displaced brine may migrate to neighboring formations and/or diffuse into shallower hydraulically conductive units. Potential focused migration may also occur through abandoned wells or sub-vertical faults and connected fractures.

2.3 Geological Settings and Characteristics

A number of different types of geologic formations have been proposed for CO_2 geosequestration. These include saline formations, depleted oil and gas reservoirs, coal seams, and possibly organic-rich shale. In this chapter, we focus on saline formations. For such formations, CO_2 sequestration is mainly through four different trapping mechanisms: structural or stratigraphic trapping, capillary or mobility trapping, dissolution trapping, and mineralization or chemical trapping. The first two tend to occur earlier in time, whereas the latter two are much slower.

Structural or stratigraphic trapping depends on the local geology: a low-permeability, regionally extensive caprock with high gas entry pressure, sealing faults, and anticline structures all have the potential to trap buoyant CO_2. Conversely, geologic heterogeneity that creates gaps in the low-permeability structures can promote concentrated vertical flow of CO_2 along these preferential flow paths.

Capillary or mobility trapping is due to phase interference between the immiscible CO_2 (the nonwetting gas-like phase) and the brine (the wetting phase). The mobilities of the flowing phases depend on phase distributions at the pore scale, as embodied in continuum-scale relative permeability functions. It is widely recognized that fluid distributions within the pore space differ for drainage (where CO_2 as the nonwetting phase displaces brine as the wetting phase, in the case of an advancing CO_2 plume), and imbibition (where brine as the wetting phase replaces CO_2 as the nonwetting phase, in the case of a retreating CO_2 plume). This process-dependence can be represented by using hysteretic relative permeability functions in which the nonwetting-phase residual saturation is small during drainage, but large during imbibition, leading to the trapping of significant quantities of CO_2.

Dissolution trapping occurs when CO_2 dissolves in brine. The CO_2-saturated brine density increases, making it heavier than the surrounding brine. Buoyancy forces then contribute to trapping of dissolved CO_2 due to the tendency for CO_2-saturated brine to migrate downwards, often in the form of fingering flow. Mineralization or chemical trapping occurs when CO_2 reacts with rock minerals to form carbonate compounds, effectively immobilizing the CO_2.

Among these four trapping mechanisms, structural trapping is important, especially in the early time frame. Saline formations may be embedded in several different large-scale and regional tectonic settings: (a) post-collisional, or post-orogenic, inland basins; (b) passive continental margins; e.g., paralic to shelf environment; and (c) fold and thrust belts. The inland basins contain mainly listric and normal faulting with minor periods of reverse faulting. Both inland basins and passive continental margins are assumed to have low tectonic or seismic activity. The fold and thrust belts as a whole are compressional, but they can exhibit large extensional domains, with the result that all types of faults, including large strike-slip faults, may be expected. An understanding of these large-scale CO_2-geosequestration site settings is useful because of the large amounts of CO_2 to be stored; a proper evaluation of the large CO_2 plume footprint and the even larger-scale effects of its displacement of in situ brine (see Fig. 2.1) require consideration of the characteristics of these large-scale settings.

Geological factors of importance for CO_2 geosequestration in all the different large-scale settings include faults, folds, sedimentary facies, and various caprock characteristics. Faults may offset stratigraphic layers, bring permeable formations above the caprock and the storage formation closer together, or juxtapose them. They may also be either hydraulically conductive, providing a migration path, or sealed, providing a barrier. Thus, they play a significant role in potential migration, or conversely, in flow compartmentalization and storage capacity.

Folds in large-scale reservoirs are rather a local feature, occurring in the vicinity of faults, and can be accompanied by many fractures. If a caprock layer laps on a buried fold or pre-existing inclining structure, the caprock may thin out (or "pinch" out), with potential migration of stored CO_2 or displaced brine, at that location.

For sedimentary facies, the concern is with spatially varying grain size or effective pore-size distribution, and pore structure. This spatial variability will impact estimates of storage capacity and pressure buildup. Spatially varying mineralogy also leads to different dissolution and precipitation reactions and different reaction rates of rock with the stored supercritical CO_2, in situ brine, and brine with dissolved CO_2.

With respect to the caprock, the geological properties of interest are rock type, facies distribution, thickness, and (more specifically) its integrity under different thermal and geomechanical processes, such as those involved in induced or natural seismic events.

2.4 Approach

The approach adopted in this chapter for discussing the relevancy of the physico-chemical processes to CO_2 geosequestration is to identify process relevancy in terms of specific objectives, technical issues of concern, or key questions associated with CO_2 geosequestration, in the context of the geological settings and characteristics of the geosequestration sites. We shall refer to the CO_2 geosequestration-related objectives, issues, and questions all as key questions (KQ), which will be discussed below.

In our discussion of relevant processes, we shall consider not only physical and chemical processes (P) , but also geologic features or structures (F). The latter include such features as the presence of fractures and faults, as well as rock-property heterogeneity, such as spatial variations in grain size, porosity and permeability. It is often impossible to discuss processes without also discussing the features, because features are the framework within which the processes operate. Furthermore, the separation of processes and features is, in many instances, not clear-cut. Sometimes a so-called process is defined to represent some underlying physical processes acting on "smeared-out" or averaged features, especially when detailed information on these features is not available. One example of this is dispersion, which results from solute transport through pathways of different velocities in the rock pores, with diffusion among the paths. Dispersion is then a flow and migration process in a continuum representation of this pore-scale variability. Nevertheless, on a practical level, one may separate processes (P) and features (F) by the fact that processes are represented as a term in the governing equations of a model that simulates rock physical and chemical behavior, while features are accounted for by the mesh design and material property values in the internal or boundary elements of the model.

Definitions or representations of processes and their characteristics depend also on the conceptual model used for their representations. For example, the relative permeability functions used to describe multiphase flows are used to represent the flow interference between the multiple phases in the pore or fracture aperture structure. They may display hysteretic effects, and are a function of scale. Such hysteresis and scale-dependence will also be included in our discussions as part of the processes.

As can be seen in the brief discussion above, there are no clear distinctions between features, processes, and process representations, and often this kind of grouping is somewhat arbitrary. It is probably fruitless to pursue a better definition of such distinctions, and the present chapter proposes the use of these categories only as a convenience to guide our discussion on their relevancy to CO_2 geosequestration.

Two further remarks are needed concerning our approach. First, the matter of temporal and spatial scales is critical and needs to be kept in mind in any consideration of processes and features. In fact, this applies not only to processes, but also to the key questions, objectives, or technical issues of concern related to CO_2

geosequestration. For example, the potential or risk of CO_2 migration is an issue of concern, but processes related to focused migration from a particular location and those related to average migration over a large area may be quite different.

Secondly, some of the key questions can and need to be addressed through operational strategy, which includes such activities as site selection, injection well spacing, injection schedule, monitoring plan, and associated operational re-adjustment or optimization. Furthermore, for consideration of operational strategy, there is a need of data, both generic and site-specific. Since it is impossible to have a "complete" set of data, one issue is the evaluation of an optimal set of desired data, while another issue is research towards defining and determining uncertainty due to data gaps.

In Sect. 2.5, the main objectives, technical issues of concern, or key questions associated with CO_2 geosequestration will be identified. For simplicity, they are all referred to as key questions (KQs).

Section 2.6 will provide a number of design alternatives related to operational strategy. Different alternatives involve different costs, and costing level is an important factor in any practical CO_2 geosequestration projects, but this will not be addressed in the present chapter. In Sects. 2.7 and 2.8 we shall list the main features (F), and processes (P), respectively. For this, we draw on information from Chaps. 3–5 of this book, and the current state of knowledge. Particular attention will be paid to the coupled processes, which are those processes coupling the effects of thermal, hydrologic, mechanical, and chemical processes.

These sections are then followed by Sect. 2.9 with an attempt (in the form of tables) to associate each KQ with various F and P, along with a few remarks on related operational strategy and data needs. The tables are not presumed to be complete, but may be useful as an initial step in the planning, evaluation, and development of a CO_2 geosequestration project.

Finally, in Sect. 2.10 one particular large-scale study of two potential CO_2 geosequestration sites in the literature will be reviewed, to point out the KQs considered and the features and processes included in the study. A few remarks on the application of the present work then conclude this chapter.

2.5 Key Questions

Key questions (KQs) associated with CO_2 geosequestration addressed in this chapter are those from a technical and scientific perspective. Thus, we do not discuss public acceptance, cost-benefit, or regulatory and legal aspects of the problem. We may divide the KQs in two groups: those related to the performance of CO_2 geosequestration and those related to its risk.

2.5.1 Performance-Related Key Questions

In line with current thinking in this field, we may define the performance-related KQs to be of three categories: namely, capacity of CO_2 that can be sequestrated at a given site, injectivity of CO_2 into deep saline formations, and containment of CO_2 without significant migration.

Capacity is by definition a large-scale issue, since for CO_2 geosequestration to be useful, large quantities of CO_2 will have to be stored. Capacity includes contributions from structural trapping, residual or capillary trapping, dissolution and mineralization trapping, and mechanical deformation of pore space. One controlling parameter that limits storage capacity is the pressure rise in the CO_2 storage formation. A maximum limit to the pressure rise is often required by regulatory agencies to stay safely below the level that could potentially cause hydraulic fracturing of the caprock or significant seismic events (Rutqvist et al. 2007). Another limit to pressure rise is the gas entry pressure into the overlying sealing formation, with its low permeability and high capillarity.

Injectivity, on the other hand, is probably a local issue concerning the capability of the formation to receive the injected CO_2 without unwarranted effects, such as gas entry into caprock and hydraulic fracturing leading to leakage near the injection well or above the CO_2 plume. This KQ may be partially addressed by selecting a storage saline formation with sufficiently high permeability, using horizontal injection wells, or changing the separation of injection areas over the storage formation, which are part of operational strategy.

Containment in the performance context involves the general question, how effective is the containment of CO_2 sequestration in the saline formation? In some assessments, it has been suggested that a small percentage of leakage can be tolerated for a system still considered to be effective in sequestering a significant amount of CO_2. Although the mechanisms of potential migration through faults and abandoned wells have been much studied, how large and significant the migration volume would be for a given geosequestration site remains an open issue to consider. Furthermore, separately from the containment question, a focused leakage causing environmental damage and danger at the leakage locations on the land surface is an important issue.

2.5.2 Risk-Related Key Questions

These may be identified as (a) induced seismicity, (b) focused migration, (c) diffused migration, (d) large-scale flow, brine displacement, and pressure changes, and (e) leaching and transport of minerals and chemicals to shallow groundwater systems.

With induced seismicity, we have both the potential for significant or major seismic events involving fault shear slippage (with possibly the creation of new migration paths or enhancement of existing paths) and induced seismic swarms of small magnitudes. The risk for focused migration refers specifically to migration through discrete flow paths along faults and connected fractures through the caprock. The caprock may be of a single layer or have a multilayered structure. Included in this category is also leakage through imperfectly sealed, abandoned wells or improperly constructed injection wells. Such focused migration may involve free-phase CO_2 or CO_2-rich brine. For the migration of brine with dissolved CO_2, the CO_2 may degas with reduced pressure as the fluid moves upwards. Diffused migration, on the other hand, is migration of mainly brine and CO_2-rich brine through caprock layers as a whole without discrete well-defined flow paths. It thus tends to be slow migration that covers a large horizontal area and flows vertically through a number of low-permeability layers to the shallower groundwater system, the vadose zone, and to the land surface.

The key question regarding large-scale flow, brine displacement, and pressure change arises from concerns over how the large quantity of CO_2 injected and stored underground would affect the groundwater system. The brine displaced by the stored CO_2 may enter into shallower groundwater formations, increasing their salinity, and the associated increases in pressure may also cause significant changes in the groundwater flow patterns. Finally, the leaching and transport of minerals and chemicals from CO_2-rock interactions, and from interactions between deeper brine and shallower formations during brine migration, are subjects of concern if they significantly affect the shallow aquifers.

2.6 Operational Strategy

Before we discuss features and processes involved in CO_2 geosequestration, it is useful to consider operational strategy used in CO_2 injection. Some of the key questions discussed in the last section can and should be addressed by appropriate design and planning of CO_2 geosequestration projects. Based on the operational strategy used, the impact of some of the features and processes may be reduced, while others may become more important and require careful study. Below, we identify a number of issues related to operational strategy.

Usually, the first step within operational strategy is site evaluation or characterization, based on which a site selection can be made. For example, the performance-related key question of CO_2 storage capacity can be partially addressed by selecting a site with an extensive saline formation under a good caprock layer. Similarly the risk-related key question of induced seismicity can be partially addressed by selecting a site with a low potential for seismic activity and moderate in situ stress fields. Such site selection depends on previously available data and new data that can be obtained at reasonable cost and time. Scale is also an issue: sometimes, it may be possible to select a “good” site around an injection well or even over the expected CO_2 plume footprint, but the site may not be so good over the vast area of pressure changes and brine migration induced by CO_2

injection and buoyancy flow. An added aspect of site characterization is the establishment of the baseline conditions (including dynamic conditions) of the site, such as pressure and salinity distributions, as well as natural seismic activities. Such information is needed to understand the impact of CO_2 injection and storage at the site and also the monitoring results during storage operation.

The choice of multiple vertical or horizontal CO_2 injection wells and placement of such wells is also part of operational strategy. With horizontal wells and a wide separation between injection areas, the injection pressure at the injection site may be less for the same total injection rate, but the cost may be dramatically higher. An injection strategy to target certain storage formations or to promote certain trapping mechanisms has also been explored. For example, alternating CO_2 and water injection has been suggested as a way to enhance capillary trapping of CO_2. Constant-rate injection versus variable-rate injection has also been studied, with some studies indicating an advantage for variable-rate injection in promoting CO_2 dispersion and dissolution in brine.

Concerns for pressure rise caused by CO_2 injection and buoyancy can probably be partially addressed by schemes such as brine withdrawal from the storage formation. In this case, however, treatment and disposal of the produced brine become an issue. In general, pressure-management methods can be considered to be part of the operational strategy. Finally, strategies for monitoring potential leakage and its mitigation are important steps in CO_2 geosequestration. How to best design and implement monitoring and mitigation is an important operational issue.

2.7 Features

Some of the large-scale regional features have been described in Sect. 2.3 (on regional geological setting). Of relevance to the key questions discussed in Sect. 2.5 are a number of local and large-scale geological features and structural characteristics.

First, the geometry of the saline storage formation and caprock needs to be defined and characterized. This activity includes accounting for the lateral extent of the storage formation and the boundary conditions, whether they are open, closed, or partially open. The geometry and permeability structure of the caprock has great relevance to CO_2 containment. For example, there may be an advantage in having the caprock be composed of multiple low-permeability, high-capillarity layers rather than a single low-permeability layer. The roughness of the caprock-saline formation interface has also been proposed to have a significant effect: it may provide effective traps for CO_2, since the CO_2 flows up and spreads at the caprock interface due to its buoyancy.

The CO_2 storage formation is heterogeneous with a permeability correlation structure, and this heterogeneity may subject flow to hydraulic compartmentalization, so that not all of the storage formation is utilized for CO_2 storage. Other features that affect the distribution of CO_2 flow are faults (sealed or open), folds, and fracture zones. All these complications could result in preferential flow paths, which may divert CO_2 flow to farther-away regions without fully occupying the storage formation in the near field, and which, furthermore, may also create CO_2 migration paths to shallower formations.

Not only is formation heterogeneity an issue, but also the formation property anisotropy, which is a feature that is essential for addressing some of the key questions. Not all of the geological features can be known a priori. This will give rise to uncertainty because of lack of knowledge. Some research effort has been devoted to the definition and estimation of uncertainty: it is hoped that the uncertainty can be bounded, based on site-specific data on a large scale.

2.8 Processes

Processes involved in CO_2 geosequestration are discussed in detail in Chaps. 3–5 of this book. Below, they are presented in an outline way under processes, process representations, and coupled processes.

2.8.1 Processes

Under processes, we identify below some significant hydrological, chemical, mechanical, and thermal processes relevant to CO_2 geosequestration:

- Flow of supercritical CO_2, gaseous CO_2, brine with dissolved CO_2, and in situ brine, including "impurities" injected with the CO_2.
- Pressure changes due to both injection pressure and buoyancy (gravity) pressure.
- Buoyancy-driven flow of supercritical and gaseous CO_2, and, in general, gravity-driven flow among fluids of different densities.
- Pore-scale capillarity, which contributes to entrapment of CO_2.
- Multicomponent flow-viscosity fingering.
- Heterogeneity-induced channeling or fingering and exchange between fast and neighboring slower paths.
- Macro-scale trapping of CO_2/brine due to heterogeneity and flow compartmentalization.
- Interface behavior between CO_2, brine with dissolved CO_2, in situ brine, and low-salinity fluids at shallower formations. For example, such behavior includes the capillary fringe effect at the CO_2-brine interface and potential flow instability at the interface.
- Displacement of formation brine and its interaction with shallow groundwater and rock minerals.
- Large-scale CO_2 spreading, mixing, fingering, and dissolution.

- Dissolution of CO_2 in brine and transport of dissolved CO_2 through convective flow.
- Evaporation of water into CO_2 phase and subsequent salt precipitation.
- Molecular diffusion of dissolved CO_2 between flow zones of different velocities or between fractures and the adjacent rock matrix.
- Reactions of CO_2 and supercritical CO_2 with rock minerals.
- Thermal transfer.
- Mechanical stress and displacements/deformations.
- Fracture shear slippage and propagation; induced seismicity.

2.8.2 Process Representations in Macro-Scale Models

Some processes are defined based on macro-scale conceptual models used to represent the underlying physicochemical processes occurring in a structural framework of a smaller scale. Because of this, there may be significant scale dependence in these processes, both temporally and spatially. Although we refer to them as simply "processes", it is useful to list them here so that their dependence on conceptual models of the geological system is highlighted.

- Flow and transport in the hydrogeological system with underlying heterogeneity of different scales. The representative permeability values depend on the existence of representative elementary volume. Furthermore, the representative elementary volume may be different for different processes. For example, it will be very different among pressure (a very diffusive process), thermal transport (a moderately diffusive process), and solute transport (a process much more sensitive to local-scale heterogeneity). In this respect, methods for determining parameter values for particular applications require careful consideration of the related scale involved.
- Flow anisotropy. The differences in flow permeability for different flow directions depend on the characteristics of permeability distribution in the different directions. It may also be a spatially varying quantity.
- Flow dispersion is a representation of lower-level velocity variations of CO_2 (or other fluids) in the pore structure at the scale of interest, such as the scale of the calculation element within a numerical model mesh. If the calculation element is small, a dispersion coefficient (in the context of the conventional Darcy's equation) may not be defined, and other types of governing equations involving possible non-local spatial and temporal terms may need to be used.
- Relative permeability effects on CO_2-brine flow, as represented, for example, by van Genuchten equations or Brooks–Corey curves. They may also display hysteresis effects.
- Linear sorption and higher-order kinetic effects.

2.8.3 Coupled Thermo-Hydro-Mechanical-Chemical Processes

Coupled processes are illustrated in Fig. 2.3, which shows the possible coupling between thermal effects and heat transfer (T), rock mechanical stress and deformation (M), chemical transport and reactions with rock minerals resulting in dissolution and precipitation (C), and hydrological processes (H). For hydrological processes, it is useful to consider different types of fluids; i.e., brine, brine with dissolved CO_2, and supercritical and gaseous CO_2, since they not only interfere with each other, but they also have different chemical reactions and characteristics. In Fig. 2.3, the solid arrowheads indicate a strong coupling direction, while the open arrowheads indicate a coupling direction of weaker strength. Possible coupled processes include

- Induced flow and transport of heat through thermal convection.
- Dissolution and transport of rock minerals in brine, brine with dissolved CO_2, and supercritical and gaseous CO_2.
- Flow-permeability changes due to precipitation or dissolution, changing pore size and structure.
- Chemical reactions with injected CO_2 containing impurities such as H_2S, SO_2, etc.
- Fluid-pressure-induced rock stress changes, causing porosity or fracture aperture changes, leading to changes in flow permeability.

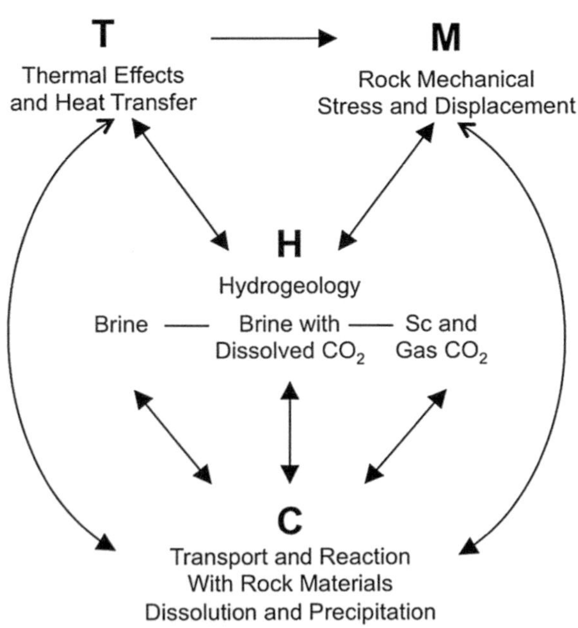

Fig. 2.3 Coupled THMC processes, coupling thermal (T), hydrological (H), rock mechanical (M) and chemical (C) effects occurring in saline formations used for CO_2 geosequestration

- Hydraulic fracturing or shear slippage.

Figure 2.3 may be useful as a framework to identify additional coupled processes, in the context of the various features presented in Sect. 2.7, which are of relevance to particular key questions associated with CO_2 geosequestration.

2.9 An Attempt to Associate Features (F) and Processes (P) with Key Questions (KQ)

This section is composed of a series of tables (Tables 2.1, 2.2, 2.3, 2.4, 2.5, 2.6, 2.7 and 2.8) that attempt to associate those F and P discussed in Sects. 2.7, and 2.8, respectively, with the KQ presented in Sect. 2.5. No claim of completeness is made here; the tables may serve as a starting point in the initial evaluation, planning, predictive modeling, design and implementation of a CO_2 geosequestration project. Details about the individual processes may be found in the other chapters of this book.

2.10 An Example Application to a Study of Large-Scale CO_2 Geosequestration at Two Potential Sites

This section provides an example of applying the above formalism on a recently published large-scale modeling of two potential sites for CO_2 geosequestration. After a summary of this study, we shall identify the key questions addressed and the important features and relevant processes included in this work.

In a series of papers (Birkholzer and Zhou 2009; Zhou et al. 2010; Zhou and Birkholzer 2011), Birkholzer, Zhou, and coworkers studied the storage capacity of

Table 2.1 KQ: Capacity

F	Porosity: lateral and vertical extent of storage formation; boundary conditions; caprock structure—structural trapping; presence of sealed or conductive faults and of migration paths; heterogeneity, and flow compartmentalization
P	Buoyancy flow, flow fingering at CO_2-water interface and at CO_2-saturated brine and formation-brine interface; CO_2 solubility in brine and solution rate; mineralization and rate; residual saturation and hysteresis in relative permeability characteristics
Coupled process	Hydromechanical effects on porosity and on fault or fracture permeability; hydromechanical effects on dissolution and mineralization
Remarks on operational strategy	Site characterization and selection are needed. The controlling parameter is pressure rise in the storage formation due to CO_2 injection and storage; it must be kept below a regulatory maximum. Pressure management methods may be applied

Table 2.2 KQ: Injectivity

F	Short-term
	Local permeability and permeability structure; potential presence of faults and abandoned wells
	Long-term
	Lateral and vertical extent of storage formation; boundary conditions; presence of sealing or nonsealing faults
P	Injection pressure and flows; two-phase flow effects
Coupled process	Hydromechanical couplings that allow porosity changes; hydrofracturing process
Remarks on operational strategy	Controlling injection peak pressure through adjustments of injection well spacing; use of horizontal wells, pressure management through brine production (where brine treatment and disposal become an issue); controlled hydrofracturing

Table 2.3 KQ: Containment

F	Boundary conditions; faults and fractures; caprock geometry and properties; anticlinal structures, presence of multiple caprock layers; abandoned wells
P	Buoyancy flow; flow channeling; multiphase flow through potential migration paths; effectiveness of structural trapping, capillary or residual trapping, and dissolution and mineralization trapping
Coupled process	Hydromechanical effects on fracture and fault permeabilities; hydrochemical effects on dissolution and precipitation, changing permeability and permeability structures
Remarks on operational strategy	Site evaluation and selection are important; development of monitoring plans and response strategies

Table 2.4 KQ: Induced seismicity

F	Faults and fracture distributions; rock mechanical conditions in the neighborhood of faults and fractures In situ stress fields
P	Changes in rock stresses and deformations; fracture dilation; shear displacements; fracturing and fracture propagation
Coupled process	Injection-pressure-induced mechanical changes; buoyancy-pressure-induced mechanical changes; effects of stress dissolution on hydraulic and mechanical properties at mechanically stressed points
Remarks on operational strategy	Analysis of the potential for induced major seismic events versus multiple minor events; monitoring plan and understanding of possible occurrences of seismic swarms needed

CO_2 at two sites with different geological characteristics, based on a model study of the pressure rise and brine migration due to an industrial-scale CO_2 injection of 5 Mt per year over 50 years. The first site is the Illinois Basin, Mount Simon

Table 2.5 KQ: Focused migration

F	Fault and fracture structure; connected fractures; heterogeneity-induced channeling; multiple caprock layers; abandoned wells and imperfectly constructed injection wells, formation pinch-out areas
P	Multiphase flow in faults, connected fractures, and heterogeneity-induced channeling paths; vertical migration through multilayer caprock; potential for accelerating migration rate, degassing
Coupled process	Coupled thermohydrological effects (e.g., cooling) on multiphase flow through vertical migration paths; coupled thermo-hydro-chemical and thermo-hydro-mechanical effects on permeability and migration paths
Remarks on operational strategy	Site evaluation and selection; Monitoring plan and response strategy needed

Table 2.6 KQ: Diffused migration

F	Permeability distribution and geometry of storage formation and caprock system; structure of shallower geologic formation and aquifers; regional variations in rock hydrologic properties
P	Flow of brine with or without dissolved CO_2 through caprock and multilayered caprock; effects of spatial variations in porosity and gas-entry pressure; long-term diffusion and retardation processes; impacts of capillary or residual trapping, dissolution and mineralization trapping
Coupled process	Effects of hydrochemical processes on flow properties and vertical diffused migration
Remarks on operational strategy	Estimate of diffused migration and its impact needed; Monitoring plan and response strategy needed

Table 2.7 KQ: Large-scale flow, brine displacement, and pressure changes

F	Regional geological structure and permeability distributions; current pressure distributions; properties and conditions at the boundaries between aquifer and aquitard systems both vertically and laterally; major faults and their permeabilities; geological features, such as folding and stratigraphic offsets; layer thinning; spatially varying pore structure; flow compartmentalization
P	Multiphase flow in near field or CO_2 injection well; flow of brine with dissolved CO_2; pressure and flow of regional brine across boundaries; interface flow behavior; effects of heterogeneity at different scales
Coupled process	Coupled thermo-hydro-chemical effects that may change permeability structure, especially at boundaries or flow constriction points
Remarks on operational strategy	Establishment of current and transient flow conditions prior to CO_2 geosequestration is important

Table 2.8 KQ: Leaching and transport of minerals and chemicals from rock matrix to shallow groundwater systems

F	Local scale
	Hydraulic structure in the storage formation and spatial rock-mineral characteristics; chemical properties of injected CO_2 stream (with or without impurities)
	Regional scale
	Regional geological structure and permeability distributions; current pressure distributions; properties and conditions at the boundaries between aquifer and aquitard systems both vertically and laterally; major faults and their permeabilities; geological features, such as folding and stratigraphic offsets; layer thinning; spatially varying pore structure; flow compartmentalization
P	CO_2-rock-water interactions, including effects of CO_2 in supercritical state, gaseous phase, and dissolved phase In the far field, chemical interactions between displaced brine into shallower or neighboring groundwater and rock systems
Coupled process	Effects of coupled thermo-hydro-chemical processes on flow patterns; mineral dissolution and precipitation
Remarks on operational strategy	Establishment of current and transient hydrochemical conditions prior to CO_2 geosequestration is important

Sandstone formation, which represents a large "open" system with continuous sealing caprock and few known fault zones. The second is the Southern San Joaquin Basin in California, with multiple sealing faults, which give rise to flow and storage volume compartmentalization. In this case, the caprock displays pinch-out in several directions. Details may be found in the references.

2.10.1 *Case of Illinois Basin, Mount Simon Sandstone Formation*

The upper left subfigure in Fig. 2.4 shows the location of the first site considered, the Mount Simon Sandstone formation. It has an area of roughly 570 km by 550 km, extending over the U.S. state of Illinois and parts of neighboring states. The depth of Mount Simon is shown in the upper right subfigure in Fig. 2.4. The formation is extensive laterally and continues to be present beyond the Illinois basin, thus allowing brine to escape into neighboring basins to the north, west, and east during a long-term CO_2 injection and migration. The southwestern model boundary is formed by the Ozark Uplift in Missouri, where the Mount Simon becomes thin or disappears.

The lower subfigure in Fig. 2.4 presents (a) the permeability distribution in a vertical cross section at roughly the middle of the Illinois Basin, showing layering in the vertical permeability with the Eau Claire serving as a caprock; and

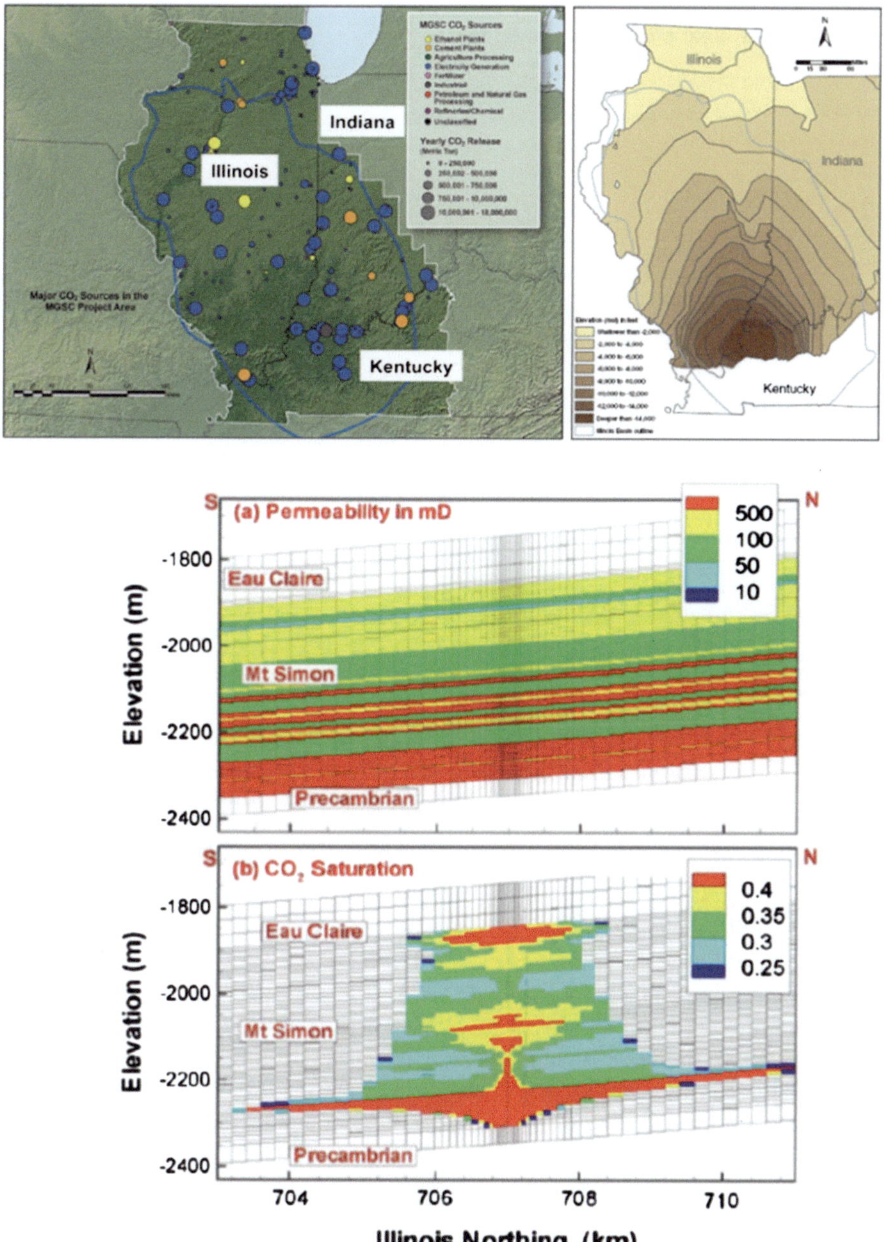

Fig. 2.4 The *upper subfigures* show the location and depth (in feet) of the Mount Simon Sandstone at Illinois Basin. The *lower subfigure* shows **a** the permeability (in mD) in a vertical cross-section at about the middle of the sandstone formation shown in the *upper figure* and **b** calculated CO_2 saturation after 50 years of CO_2 injection in the Arkosic layer into a depth interval from about 2240–2300 m. From Birkholzer and Zhou (2009)

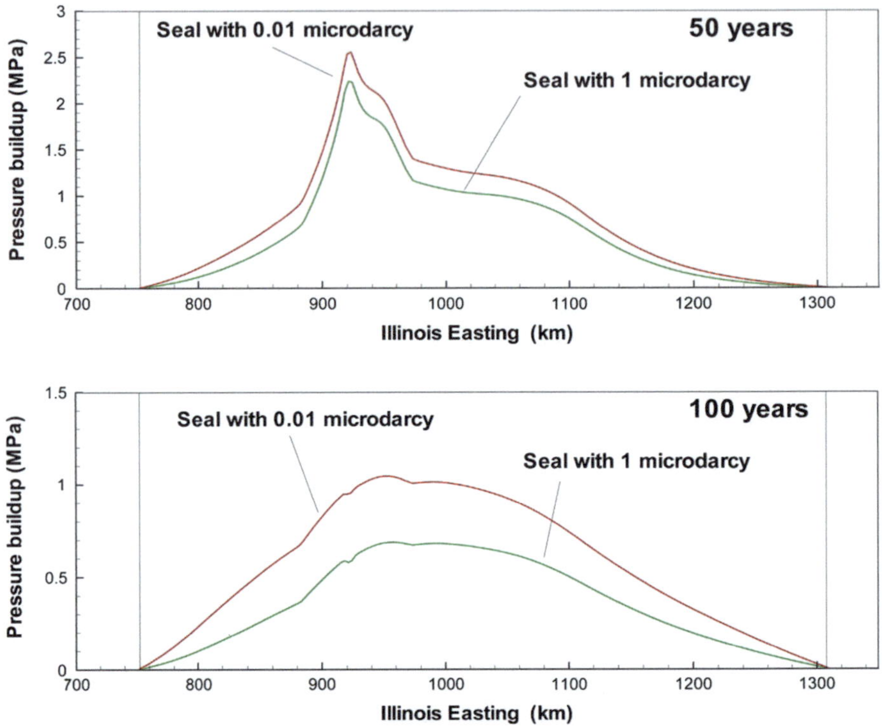

Fig. 2.5 Profiles of pressure buildup (in MPa) in the Mount Simon Sandstone formation along the east-west direction for two cases of the caprock (Eau Claire) permeability of 1 μD (*lower curve*) and 0.01 μD (*upper curve*) at 50 years and 100 years, respectively. Results are for injection of 5 Mt of CO_2 per year over the first 50 years. From Birkholzer and Zhou (2009)

(b) calculated CO_2 saturation after 50 years of CO_2 injection, showing the impact of vertical permeability variation.

Modeling was conducted using a 3D unstructured mesh, with progressive mesh refinement (down to the order of 10 m) in the core injection areas to very large grids (order of 10 km) in the far region, so that both the details of CO_2 plume multiphase flow and its spatial variability in the near field and brine migration processes in the basin scale are properly simulated. The parallel version of the TOUGH2/ECO2N simulator (Pruess et al. 1999; Zhang et al. 2008; Pruess 2005) was used to solve the multiphase flow and multicomponent transport of CO_2 and brine in response to CO_2 injection. In the simulation, full accounts are taken of changes in brine density and viscosity; changes in CO_2 density and viscosity; CO_2 solubility in brine; brine solubility in CO_2, and their dependence on fluid pressure, temperature, and salinity.

Figure 2.5 shows the calculated pressure rise along one of the transects of the model. The pressure buildup in this large regional system reaches 2.5 MPa in 50 years and will be about 1 MPa after 100 years if CO_2 injection is terminated after 50 years. Note that the pressure rise is sensitive to the permeability value of

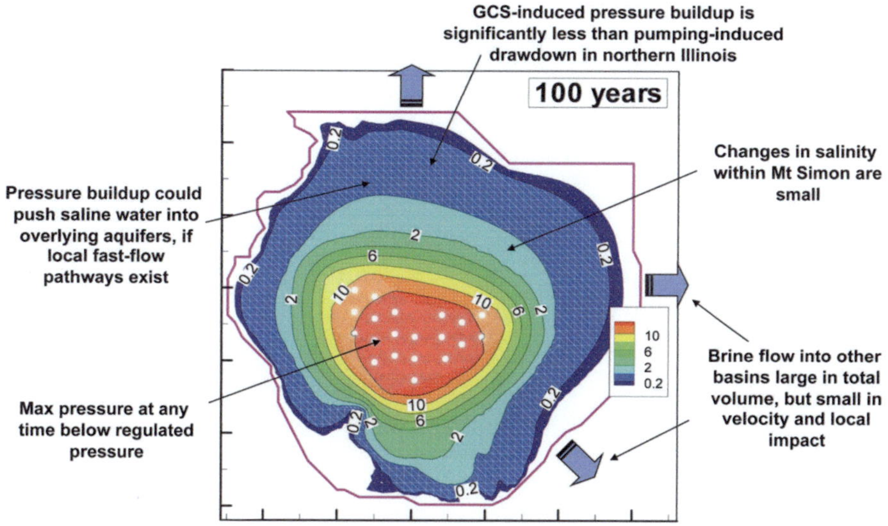

Fig. 2.6 Conclusions from study of pressure rise in Mount Simon Sandstone due to CO_2 geosequestration. Here the contours are in bars, or 0.1 MPa, and the color area is about 500 km across, with scale tick marks at 100 km intervals (from Birkholzer private communications 2012)

the sealing caprock. Even though there is a significant contrast between the permeability of the Mount Simon Sandstone (about 10–100 mD) and the caprock permeability of 1 μD, there is sufficient diffused migration of brine through the caprock to result in a 30–40 % decrease in pressure rise at 100 years, as compared with a zero-permeability caprock case, as represented by the 0.01 μD results in the figure.

The conclusions from the Mount Simon study by Birkholzer and Zhou indicate (Fig. 2.6) that the maximum pressure rise at any time in the storage formation is below the regulated limit associated with potential hydrofracturing (Rutqvist et al. 2007). Changes in salinity within the Mount Simon are small, and brine flow into neighboring basins are large in total volume but small in velocity and local impact. However, the pressure buildup could push saline water into overlying conductive formations if fast-flow pathways exist between them.

2.10.2 Case of Southern San Joaquin Basin

In contrast to Mount Simon Sandstone in the Illinois Basin, the potential CO_2 storage formation in California, Vedder Sand, in the Southern San Joaquin Basin (Fig. 2.7), has a number of major sealing faults as determined from extensive petroleum exploration studies, so that the formation is partially compartmentalized hydrologically. The formation pinches out towards the south, north, and west.

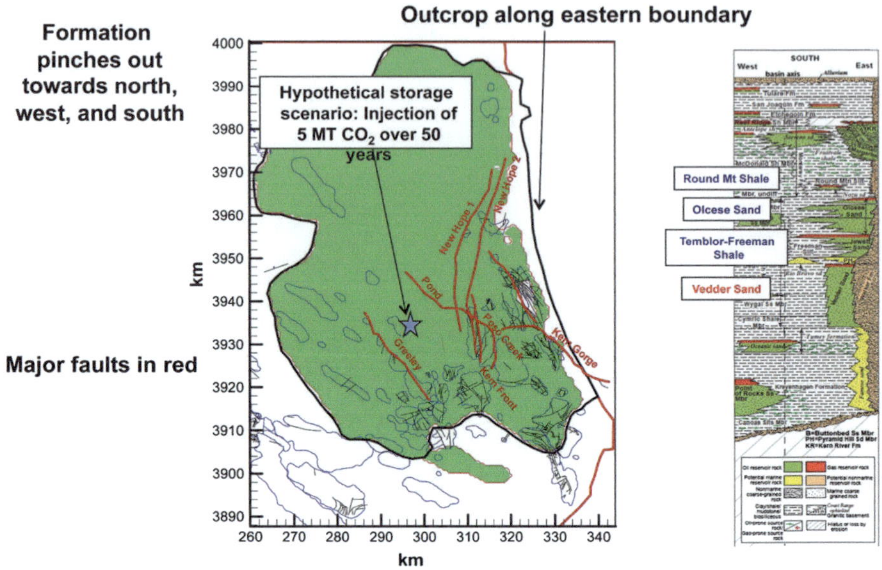

Fig. 2.7 On the *left* is the site map of the Vedder Sand formation in the San Joaquin Basin. Major faults are indicated as *red lines*. On the *right* is a vertical profile indicating the caprock, Temblor-Freeman shale separating the storage formation, Vedder Sand from an overlying conductive layer, Olcese Sand. From Zhou and Birkholzer (2011)

To the east, it outcrops along the edge of the Sierra Nevada mountain range. Vertically, the caprock is the Temblor-Freeman shale, which is overlaid by a hydraulically conductive formation called Olcese Sand. In fact, the Vedder Sand and the Olcese Sand connect in the northern area of the domain shown in Fig. 2.7.

Using the same methodology as in the modeling of CO_2 geosequestration in the Mount Simon Sandstone, the pressure rise was calculated for the Vedder Sand at the Southern San Joaquin Basin. Figure 2.8 shows the pressure rise after 50 years of CO_2 injection. These results indicate the significant effects of the sealing faults at the site, diffused water migration through the caprock, focused water migration through the caprock pinch-out and water discharge into the outcrop area.

2.10.3 Discussion of the Two Cases

In the context of the present chapter on relevant processes for CO_2 geosequestration, we may summarize the above study (with much more details in the references) as follows.

This study addresses the key question of "storage capacity" as indicated by the maximum pressure rise, which must be kept below a regulatory limit to avoid damage to caprock integrity. It also addresses the key issues of large-scale flow,

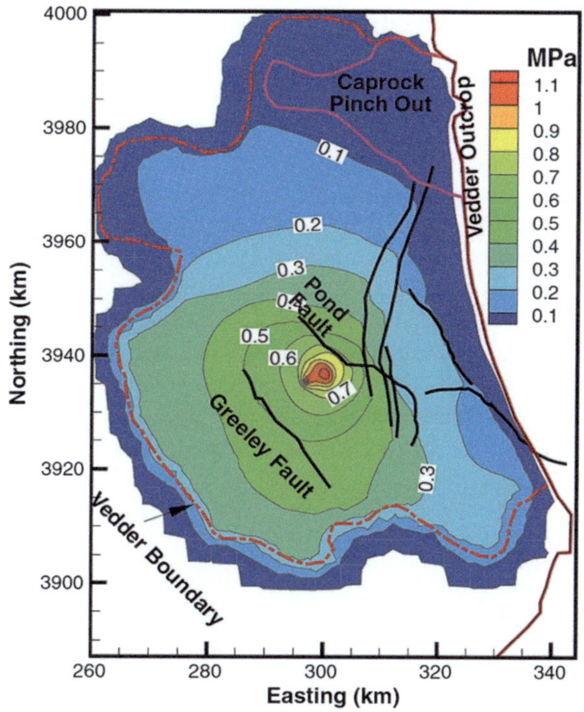

Fig. 2.8 Simulated pressure rise in MPa after 50 years of CO_2 injection into the Vedder Sand formation at the Southern San Joaquin Basin. From Zhou and Birkholzer (2011)

brine displacement, and pressure changes at large scale. Results from this study will be useful as input in addressing the key issues of induced seismicity and focused and diffused migration.

Features in the study include the geometry of the storage formation, the definition of boundary conditions, vertical permeability variations, including the effects of multi-layered caprock, the possibility of caprock pinch-out at the boundary of the storage formation, occurrence of sealing faults, and flow and CO_2 storage compartmentalization.

Processes of relevance in this study include pressure rise due to injection and buoyancy, lateral pressure propagation and brine displacement, diffused or focused vertical brine migration into overlying and underlying formations, two-phase flow and multicomponent transport of CO_2 and brine, changes in brine and CO_2 density and viscosity, CO_2 solubility in brine and (inversely) brine solubility in CO_2 and their dependence on fluid pressure, temperature, and salinity. All these processes have been incorporated into the modeling work for the study of the two sites.

2.11 Concluding Remarks

The present chapter presents a framework for discussing the relevant processes in CO_2 geosequestration in saline formations. It is suggested that relevant processes can be usefully discussed in the context of certain key questions associated with CO_2 geosequestration, including the objectives of a site-specific project and key issues of concern. A list of key questions has been identified; features within which the processes occur are also described. Tables are provided for each of the key questions to give a list of relevant features and processes. These tables, however, are not presumed to be complete, and indeed may have to be revised as additional information emerges from further research and experience. Nevertheless, it is hoped that they may serve as a starting point for considering the relevant processes, features, operational strategy, and other factors in addressing key questions related to CO_2 geosequestration at a particular site.

Acknowledgments The authors would like to gratefully acknowledge the input of Torsten Lange and Martin Sauter, on which part of Sect. 2.2 is based. Discussions with and input from Jens Birkholzer, Marco Dentz, Jesus Carrera, Tore Torp, Jacob Bear, Jacob Bensabat, and other members of the Mustang project are much appreciated. Careful review and helpful comments from Christine Doughty and Quanlin Zhou are also acknowledged with thanks. The work is funded by the European Community's Seventh Framework Programme FP7/2007-2013, under Grant Agreement No. 227286 (the MUSTANG project).

References

Birkholzer JT, Zhou Q (2009) Basin-scale hydrogeologic impacts of CO_2 storage: Capacity and regulatory implications. Int J Greenh Gas Control 3:745–756

Pruess K (2005) ECO2N: A TOUGH2 fluid property module for mixtures of water, NaCl, and CO_2 (No. LBNL-57952). Lawrence Berkeley National Laboratory, Berkeley

Pruess K, Oldenburg CM, Moridis GJ (1999) TOUGH2 user's guide version 2 (No. LBNL–43134, 751729). Berkeley National Laboratory, Berkeley

Rutqvist J, Birkholzer J, Cappa F, Tsang C-F (2007) Estimating maximum sustainable injection pressure during geological sequestration of CO_2 using coupled fluid flow and geomechanical fault-slip analysis. Energy Convers Manag Geologic Carbon Sequestration and Methane Hydrates Research from the TOUGH Symposium 2006, vol 48, pp 1798–1807

Zhang K, Wu YS, Pruess K (2008) User's guide for TOUGH2-MP—a massively parallel version of the TOUGH2 code (No LBNL-315E). Lawrence Berkeley National Laboratory, Berkeley

Zhou Q, Birkholzer JT (2011) On scale and magnitude of pressure build-up induced by large-scale geologic storage of CO_2. Greenh Gases Sci Technol 1:11–20

Zhou Q, Birkholzer JT, Mehnert E, Lin Y-F, Zhang K (2010) Modeling basin- and plume-scale processes of CO_2 storage for full-scale deployment. Ground Water 48:494–514

Chapter 3
Mathematical Modeling of CO_2 Storage in a Geological Formation

Jacob Bear and Jesus Carrera

Abstract Chapter 2 discusses, in a descriptive manner, the processes occurring during the geological storage of CO_2. In this chapter, the mathematical models describing these processes are described. The chapter starts from the basic properties of the injected CO_2 and of the native brine, proceeding to the relevant models for multiphase flow of CO_2 and brine, the related chemical and reactive transport processes, the non-isothermal effects of CO_2 injection and the mechanical deformation. The concept of degrees of freedom, facilitating the selection of a smaller number of equations to be solved, in order to obtain a complete solution for this multifaceted problem, is also discussed. The numerical and analytical approaches for solving these mathematical models are presented in the following Chap. 4.

3.1 Introduction

The general concepts underlying a GCS (Geological Carbon Storage) project have been presented in Chap. 1, together with a description of a number of actual experimental and full-scale projects. Chapter 2 introduces and discusses the processes that occur in the target formation during and following the injection of CO_2. Like in any engineering project, the planning of a GCS project involves the preparation of alternative plans and management schemes that have to be evaluated by comparing the consequences of implementing them. The *model* is the tool that can provide information on these consequences. Model results provide the

J. Bear (✉)
Department of Civil and Environmental Engineering, Technion – Israel Institute of Technology, Haifa, Israel
e-mail: cvrbear@technion.ac.il

J. Carrera
Groundwater Hydrology Group (GHS UPC-CSIC), Institute of Environmental Assessment and Water Research (IDAEA), Spanish National Research Council (CSIC), Barcelona, Spain
e-mail: jesus.carrera.ramirez@gmail.com

A. Niemi et al. (eds.), *Geological Storage of CO_2 in Deep Saline Formations*, Theory and Applications of Transport in Porous Media 29,
DOI 10.1007/978-94-024-0996-3_3

information required for risk assessment and for selecting the optimal project plan. This chapter presents and discusses the appropriate models for this purpose.

This chapter introduces the mathematical models that describe the processes that occur upon the injection of CO_2 into a deep geological formation that contains saline water. We start (Sect. 3.2) by presenting the basic properties of the injected CO_2 and of the saline water, that initially occupies the entire void space of the considered geological formation. Section 3.3 introduces and discusses the model that describes *multiphase flow* phenomena and processes that occur in the formation as the injected CO_2 displaces the indigenous saline water. In addition, the displacing CO_2 also dissolves in the native saline water, leading to a rather aggressive solution that will tend to dissolve formation minerals, while precipitating others. All these phenomena, referred to as *reactive transport*, are presented in Sect. 3.4. The temperature of the injected CO_2 is different from that of the fluid and solid matrix in the target formation. This means that the flow and reactive transport phenomena within the formation occur under *non-isothermal conditions*. A model that describes energy transport is required. This model is discussed in Sect. 3.5. The injected CO_2 is introduced into the formation at a pressure higher than that prevailing in the formation. This high pressure may affect formation properties, such as porosity and permeability and may damage the integrity of the sealing caprock. Mechanical deformation is the subject of Sect. 3.6. All these models involve a large number of variables and, consequently, a large number of equations. Section 3.7 presents and discusses the notion of degrees of freedom, which facilitates the selection of a smaller number of equations that have to be solved in order to obtain a complete solution for this multifaceted problem.

In this chapter, we are considering four extensive quantities:

- The mass of fluid phases, typically water and CO_2.
- The mass of chemical species dissolved in the fluids that occupy the void space.
- The energy of the two fluids and the solid matrix comprising the geological formation.
- The momentum of the solid matrix comprising the considered formation, combined with that of the fluids that occupy the void space.

The balance equations written for these extensive quantities constitute the core of the model, or set of mathematical models, that describe the behavior of the target formation as CO_2 is being injected.

The mathematical models are presented at the macroscopic level. This is the level at which the entire considered domain—here the geological formation—is regarded as a continuum. An important issue when dealing with real applications is that of *heterogeneity,* which is not addressed here. Most geological formations are strongly heterogeneous; this calls for the use of stochastic modeling, or the use of upscaling techniques, which is the topic of Chap. 5, and is especially relevant for solute transport models.

3.2 Properties of CO_2 and Saline Water

Two fluid phases are involved in a GCS project: the native aqueous phase, which initially occupies the entire void space of the target formation, and the CO_2 phase, which is injected into the formation through wells and spreads out in the formation, displacing the saline water. The native water is often referred to as brine, although its salinity usually lies between 10 and 100 g/L, so that it is strictly salt water. We shall use both the popular (brine) and the generic (saline water) terms. The objective of this section is to present all the major thermodynamic properties of these two phases that are relevant to their movement within the target formation during a GCS project. Specifically, we introduce the *Equations of State* (EOSs) that describe the relationships among the state variables and properties of each of the two fluid phases. The fluid phase properties discussed in this section are:

- The fluids density: ρ_{br}, ρ_{CO_2}.
- The fluids dynamic viscosity: μ_{br}, μ_{CO_2}.
- The fluids enthalpy: h_{br}, h_{CO_2}.

3.2.1 Gibbs Phase Rule

A fluid phase at a point $\mathbf{x}$ in a considered domain and at an instant of time t has the density $\rho = \rho(\mathbf{x}, t)$. This density is thus a *state variable* that, in turn, depends on two other *state variables*: pressure (p) and temperature (T). It may also depend on the concentrations of dissolved species, $(c^\gamma, \gamma = 1, \ldots, NC)$ where γ identifies each species and NC is the total number of species. Altogether, $\rho = \rho(p, T, \mathbf{c})$ where $\mathbf{c}$ is the vector of concentrations of all species. Considering for a moment a single fluid phase ($NP = 1$), with no dissolved species (i.e., one component, $NC = 1$). According to *Gibbs phase rule* (see any text on thermodynamics, e.g., Denbigh 1981), the number of independent state variables, NF (= number of *degrees of freedom*) is

$$NF = NC - NP + 2 = 1 - 1 + 2 = 2, \tag{3.2.1}$$

i.e., *two degrees of freedom*. This means that out of the 3 variables of state, ρ, p, T we have to determine (by solving the appropriate mathematical models) the value of two variables of state and the others, i.e., the third one will be determined by the thermodynamic relationship $\rho = \rho(p, T)$, referred to as an *equation of state*. This rule can be generalized to the cases of two phases (and, in GCS, two components) or cases in which chemical reactions take place. Each chemical reaction adds a constraint (i.e., NP is increased) and each species adds a "component". As we shall see in Sect. 3.4, we shall reserve the term "component" to linear combinations of chemical species that remain unaffected by chemical reactions, so that effectively the number of degrees of freedom increases with the number of such components. The number of degrees of freedom for various kinds of problems is further discussed in the Appendix.

Henceforth, for the sake of simplicity, we shall refer to density, ρ, and enthalpy, h, not as state variables, but as *properties*.

3.2.2 Properties of CO_2

The nature of a phase (its state of aggregation and properties) evolves as a function of pressure and temperature. The description of this state is made with the aid of phase diagrams and equations of state.

A *phase diagram* is a graph showing, in a compact way, the *state of aggregation (i.e., gas, liquid or solid)* of a substance (here CO_2) as a function of two state variables, typically, but not necessarily, pressure and temperature.

An *equation of state* (abbreviated EOS), expresses in the form of a mathematical equation the relationship between three variables of state of a considered substance. For example, the expression yielding density as a function of pressure and temperature is an EOS. EOS's can be divided into two groups: General EOS and Special EOS. A general EOS is valid for all substances, although it may contain substance-specific parameters. A special EOS is an equation designed specifically for the substance of interest.

A. Phase diagram for CO_2

A phase diagram of CO_2 is shown in Fig. 3.1. Two points of the phase diagram represent specific conditions of interest: (1) *the triple point (for* CO_2 *pressure of 5.11 atm and temperature of* -56.6 *°C) where all three* phases co-exist; and (2) the *critical point* (for CO_2, the temperature is 31.45 °C and the pressure is 73.1 atm) above which the transition from the liquid to the gaseous phase is smooth. The importance of this latter point lies in that below it, there is a clear difference

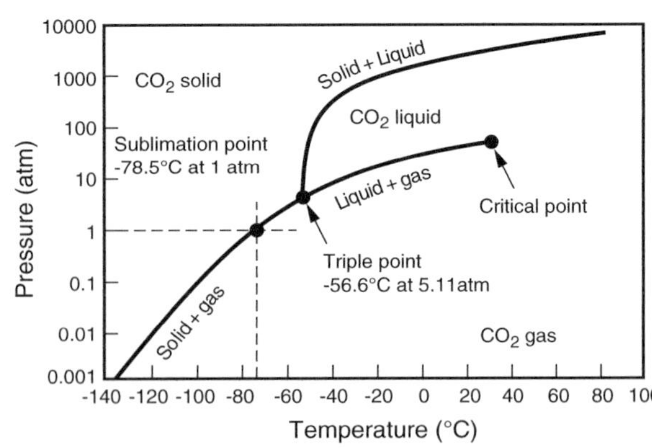

Fig. 3.1 The phase diagram for CO_2, indicating regions of different states of aggregation

between phases, and substances are either liquid or gas or solid (except at phase equilibrium lines, where two phases may co-exist, or at their intersection, triple point, where the three phases co-exist). In order to change from one phase to another, a significant amount of energy needs to be added (e.g., from liquid to gas) or taken away (e.g., from gas to liquid) from the system, even if the phase change occurs under constant pressure and temperature conditions. Above the critical point, the substance is said to be in supercritical state, and changes occur smoothly from one state to another. In deep geologic formations, suitable for CO_2 sequestration, the temperature and pressure conditions typically exceed those of the critical point of CO_2, and, therefore, the transport and storage of CO_2 occurs under supercritical conditions. This is unfortunate because the term supercritical suggests "worse than critical", which feeds all kinds of public fears. In reality, it is the opposite. Phase changes occur abruptly below the critical point. A lot of energy needs to be added at a phase equilibrium point (e.g., 100 °C for water at atmospheric pressure) to change from liquid (high viscosity and density, low compressibility) to gas (low viscosity and density, high compressibility). Under supercritical conditions, such changes occur smoothly, as pressure and temperature evolve in response to the addition of energy.

Other forms of phase diagrams are possible. In fact, it is possible to use other state variables or to take advantage of the phase diagram to display the values of other state variables. Recall, that for a single substance, Gibbs phase rule restricts to two the number of degrees of freedom. Therefore, any state variable can be expressed as a function of any other two state variables. For example, for CO_2, Fig. 3.2 displays the phase diagram in the (v, p) plane, where $v = 1/\rho$ is the specific volume. The same diagram also displays $T = T(p, v)$, that is, the temperature at which a specific volume will be found for a given pressure. For this reason, it is often called an Equation of State (as it describes the state variable T as function of the two others). A diagram like the one in Fig. 3.3 is more involved than the one in Fig. 3.1 but displays several advantages. First, it is more informative, as one can find not only the state of aggregation, but also the value of a third state variable. Secondly, it is clearer, as it expands the phase change regions (e.g., the liquid + gas region below the bubble and dew point lines in Fig. 3.2, which was lumped as a line in Fig. 3.1. This facilitates analyzing phase changes. For example, it allows identifying the specific volume when liquid CO_2 starts vaporizing (bubble point) or when gaseous CO_2 starts condensating (*dew point*). The third advantage is that it allows emphasizing some specific type of information. For example, Fig. 3.2 emphasizes that less storage volume is required per unit of storage mass (i.e., less specific volume) for high than low pressures. This is why GCS is implemented at depth, and that volume is gained by storing cold CO_2, which is one of the reasons why cold CO_2 injection is favorable, an issue we shall revisit in Chap. 7.

B. CO_2 density

As discussed earlier, the relationship between density, pressure and temperature is an Equation of State. Some options for the density EOS are discussed below.

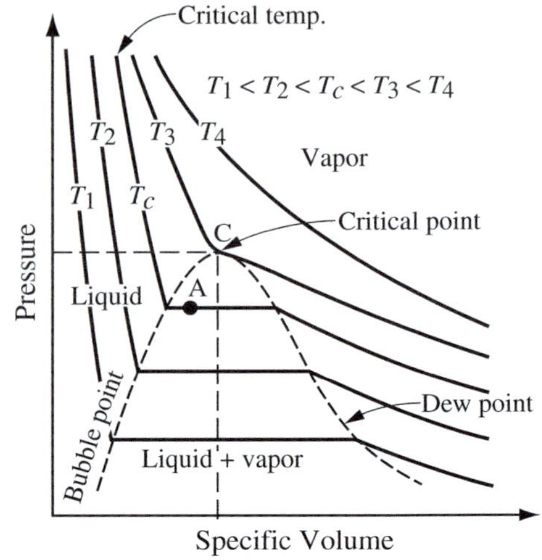

Fig. 3.2 Sketch of the equation of state $T = T(v, p)$ for a fluid

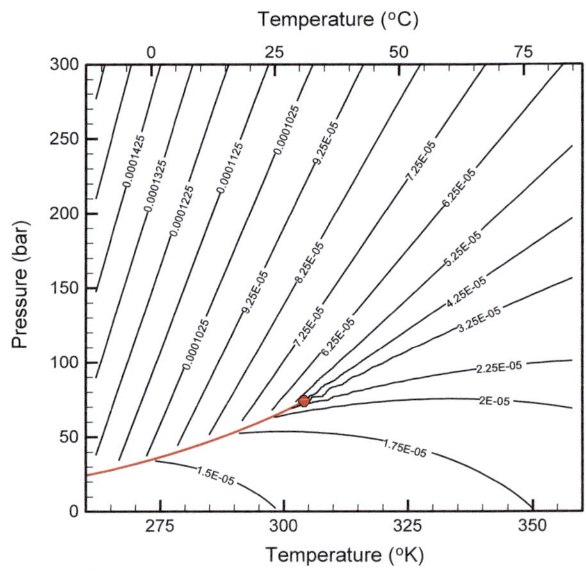

Fig. 3.3 Calculated CO_2 viscosity, using Altunin correlation (Garcia 2003)

General EOS

The best example of a general EOS is the *ideal gas law*, valid for all gas mixtures at low pressures. Other examples are the EOS's developed by Redlich and Kwong (1949), or Peng and Robinson (1976). A practical approach to describe the behavior of CO_2–H_2O is that proposed by Spycher et al. (2003). Their equation is applicable at moderate temperatures up to a pressure of 600 bar.

$$P = \frac{RT}{V - b_m} - \frac{a_m}{T^{0.5} V(V + b_m)}, \tag{3.2.2}$$

where R is the universal gas constant, V is the molar volume of the CO_2 rich phase, T is the temperature and a_m and b_m are interaction and repulsion parameters of the CO_2–H_2O mixture, respectively. These parameters can be calculated by numerous empirical equations available in the literature. The ones based on the standard mixing rules (Prausnitz et al. 1986) are:

$$a_{ij} = \left(a_i a_j\right)^{0.5}\left(1 - k_{ij}\right), \tag{3.2.3}$$

$$b_m = \sum_{i=1}^{c} y_i b_i, \tag{3.2.4}$$

where c is the number of components and y_i is the mole fraction of the i component.

For the binary H_2O–CO_2 mixture:

$$a_m = y_{H_2O}^2 a_{H_2O} + 2 y_{H_2O} y_{CO_2} a_{H_2O-CO_2} + y_{CO_2}^2 a_{CO_2}, \tag{3.2.5}$$

$$b_m = y_{H_2O} b_{H_2O} + y_{CO_2} b_{CO_2}, \tag{3.2.6}$$

where the coefficients proposed by Spycher et al. (2003) are:

$$a_{CO_2} = 7.54 \times 10^7 - 4.13 \times 10^4 T\left[\text{bar cm}^6 \text{K}^{0.5} \text{mol}^{-2}\right], \tag{3.2.7}$$

$$a_{H_2O\text{-}CO_2} = 7.89 \times 10^7 \ (\text{bar cm}^6 \text{K}^{0.5} \text{mol}^{-2}), \tag{3.2.8}$$

$$b_{CO_2} = 27.8 \ (\text{cm}^3 \ \text{mol}^{-1}), \tag{3.2.9}$$

$$b_{H_2O} = 18.18 \ (\text{cm}^3 \ \text{mol}^{-1}). \tag{3.2.10}$$

Since the amount of H_2O in the gas phase is significantly low, the value of the interaction parameter a_{H_2O} of H_2O is insignificant.

Special EOS

The Span-Wagner equation is currently the preferred method for the properties of CO_2 and liquid-gas mixtures at equilibrium. The equation covers the range from the triple point temperature to 1100 K at pressures up to 800 MPa. This EOS is an empirical representation of the *Helmholtz free energy*, A, related to two independent variables *p* and *T*. The dimensionless *Helmholtz energy function* is split into two parts: one part describes the behavior of an ideal gas at specified temperature and density. The second part describes the residual behavior due to the real fluid behavior:

$$\frac{A_{(T,p)}}{RT} = f_{(\tau,\delta)} = f^0_{(\tau,\delta)} + f^r_{(\tau,\delta)}. \tag{3.2.11}$$

Cooper (1982) developed the ideal gas part of the dimensionless Helmholtz free energy from the equation for the *specific isobaric heat capacity* of an ideal gas:

$$f^0_{(\tau,\delta)} = \ln(\delta) + a^0_1 + a^0_2\tau + a^0_3\ln(\tau) + \sum_{i=4}^{8} a^0_i \ln\left[1 - e^{-\tau\theta^0_i}\right], \tag{3.2.12}$$

where $\tau = T_C/T$ is a dimensionless temperature, $\delta = \rho/\rho_C$ is dimensionless density, and both a^0_i and θ^0_i are parameters, i = 1, … , 8. For pure CO_2, critical values of temperature and density, respectively, are $T_C = 304.1282$ K and $\rho_C = 467.6$ kg/m^3.

The form of the residual part of the dimensionless Helmholtz energy equation is (Span and Wagner 1996):

$$f^r_{(\tau,\delta)} = \sum_{i=1}^{I_1} n_i\delta^{d_i}\ln(\delta) + \sum_{i=I_1}^{I_2} n_i\delta^{d_i}\tau^{t_i}e^{-\delta^{C_i}} + \sum_{i=I_2+1}^{I_3} n_i\delta^{d_i}\tau^{t_i}e^{\left[-\alpha_i(\delta-\varepsilon_i)^2-\beta_i(\tau-\gamma_i)^2\right]} + \sum_{i=I_3+1}^{I_4} n_i\Delta^{b_i}\delta\psi e, \tag{3.2.13}$$

with

$$\begin{aligned} \Delta &= \theta + B_i\left[(\delta-\varepsilon_i)^2\right]^{a_i}, \\ \theta &= (1-\tau) + A_i\left[(\delta-1)^2\right]^{\frac{1}{2\beta_i}}, \\ \psi &= e^{\left[-C_i(\delta-1)^2-D_i(\tau-1)^2\right]}, \end{aligned} \tag{3.2.14}$$

where $d_i, t_i, c_i, a_i, \beta_i, \varepsilon_i, \gamma_i, A_i, B_i, C_i, D_i$ are parameters (Span and Wagner 1996).

The thermodynamic properties of pure CO_2 can be obtained from the Helmholtz energy equation. For example, from the thermodynamic definition of vapor pressure $p_{(T,\rho)} = (-\partial A/\partial v)_T$, the equation for vapor pressure, temperature and CO_2 density was expressed in terms of the derivative of the dimensionless Helmholtz energy function *f*,

$$\frac{p_{(T,\rho)}}{\rho RT} = 1 + \delta f^r_\delta, \tag{3.2.15}$$

where the expression for $f^r_\delta = (\partial f^r/\partial\delta)_\tau$ is given in in IAPWS (2009).

C. CO_2 viscosity

A practical expression for calculating the viscosity of supercritical CO_2, including the effect of pressure, temperature and composition, is the empirical equation proposed by Altunin and Sakhabetdinov (1972):

$$\mu_g = \mu_g^0 \cdot e^{\left(\sum_{i=1}^{4}\sum_{j=0}^{1} a_{ij}\frac{\rho_r^i}{T_r^j}\right)}, \quad (3.2.16)$$

where $\rho_r = \rho_g/\rho_c$ and $T_r = T/T_C$ are the reduced density and temperature respectively, μ_g^0 and a_{ij} coefficients are displays in Eqs. 3.2.17 and 3.2.18:

$$\mu_g^0 = \sqrt{T_r}\left(27.2246461 - 16.6346068/T_r + 4.66920556/T_r^2\right). \quad (3.2.17)$$

$$\begin{aligned} a_{10} &= 0.24856612, a_{11} = 0.004894942, a_{20} = -0.373300660 \\ a_{21} &= 1.22753488, a_{30} = 0.363854523, a_{31} = -0.774229021 \\ a_{40} &= -0.0639070755, a_{41} = 0.142507049 \end{aligned} \quad (3.2.18)$$

This correlation is valid for temperature range of 200–1300 K and pressures up to 1200 bar. Figure 3.3 displays the viscosity contours for CO_2. Under supercritical conditions, the viscosity of CO_2 is considerably lower than that of water.

D. CO_2 enthalpy

The enthalpy of a pure component in a single CO_2-rich phase may be calculated from the integration of Maxwell's relations:

$$dh_{CO_2} = c_p dT + \left[v - T\left(\frac{\partial v}{\partial T}\right)_p\right]dp, \quad (3.2.19)$$

where, h_{CO2} is the specific enthalpy of the CO_2-rich phase, v its molar volume, and c_p is the specific heat at constant pressure. At constant temperature, Eq. 3.2.19 reduces to:

$$dh_{CO_2} = \left[v - T\left(\frac{\partial v}{\partial T}\right)_p\right]dp. \quad (3.2.20)$$

Integrating (3.2.19) from pressure 0 to p, we obtain:

$$h_{CO_2} - h_{CO_2}^0 = \int_0^p \left[v - T\left(\frac{\partial v}{\partial T}\right)_p\right]dp, \quad (3.2.21)$$

where $h_{CO_2}^0$ is the ideal gas enthalpy at zero pressure and $h_{CO_2} - h_{CO_2}^0$ is the enthalpy departure function. In order to obtain the form of the departure function of enthalpy we first find an expression for the departure function for the Helmholtz energy. The variation in the Helmholtz energy with molar volume V is given by:

$$dA = -PdV. \tag{3.2.22}$$

Integrating at constant temperature from the reference volume V^0 to the system volume V gives:

$$A - A^0 = \int_{\infty}^{V} \left(P - \frac{RT}{V}\right) dV - RT\ln\frac{V}{V^0}. \tag{3.2.23}$$

The entropy function is readily obtained:

$$S - S^0 = \frac{-\partial}{\partial T}\left(A - A^0\right)_V = \int_{\infty}^{V} \left[\left(\frac{\partial P}{\partial T}\right)_V - \frac{R}{V}\right] dV + R\ln\frac{V}{V^0}, \tag{3.2.24}$$

and the enthalpy function is therefore:

$$h - h^0 = \left(A - A^0\right) + T\left(S - S^0\right) + RT(Z - 1) \tag{3.2.25}$$

Estimation of enthalpy can be obtained by using cubic EOS. For more complex equations such as the ones by Span and Wagner (1996) and Altunin (1975). Garcia (2003) describes the Altunin's correlation to compute the enthalpy of pure CO_2 from a nonlinear correlation of Z:

$$Z^0 = 1 + \rho_r \sum_{i=0}^{9} \sum_{j=0}^{J_i} b_{ij} (T_r - 1)^j (\rho_r - 1)^i, \tag{3.2.26}$$

where Z = PV/RT, $\rho_r = \rho/\rho_C$, and $T_r = T_C/T$. Figure 3.3 shows the contours of CO_2 enthalpy as calculated using the correlations of Altunin.

Another relationship between thermodynamic state variables are displayed in Mollier diagram. This diagram is commonly utilized by chemical engineers to quantify physicochemical phenomena. This plot can help us to understand what occurs during CO_2 compression, a process that is relevant for the geological sequestration of CO_2 in deep geological formations. It is generally assumed that gaseous substances heat up during compression and cool during expansion. However, this is not always true and, indeed, each gaseous substance may cool down upon compression for high pressures. For isenthalpic processes, these two intervals are divided by the Joule-Thompson inversion curve, which can be identified by the points where isotherms in Fig. 3.4 become vertical (e.g. 600 bar, -170 kJ mol^{-1}, for 80 °C). While this inversion curve has been a source of puzzlement, it is not much of an issue for CO_2 storage, because inversion occurs for pressures well above typical operation pressures. It is evident that if CO_2 were pressurized without refrigeration from 1 bar, it would enter the target formation at a relatively high temperature, possibly higher than that of the formation itself, which

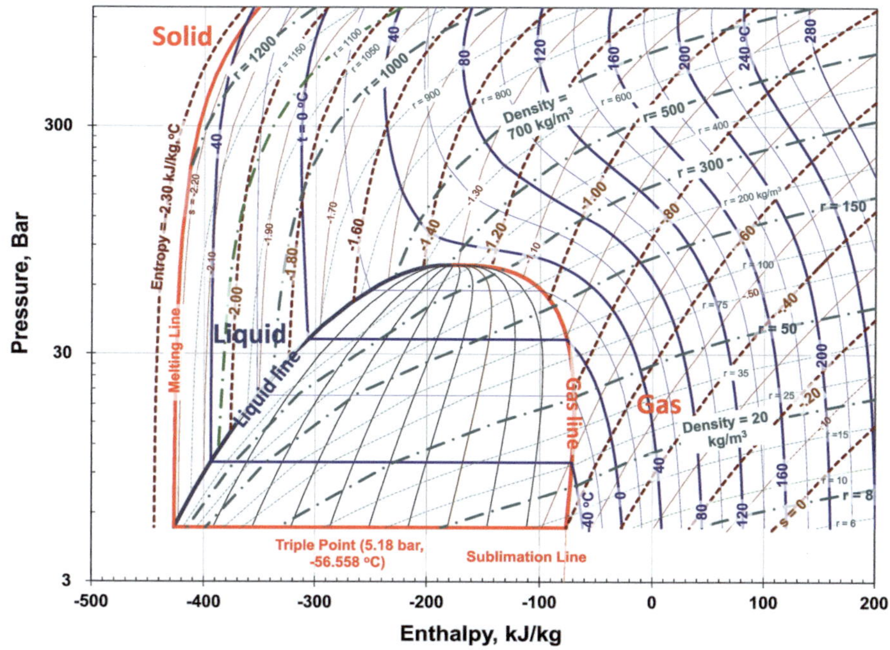

Fig. 3.4 Mollier's diagram (Pressure-Enthalpy-Density-Temperature-Entropy diagram) for pure CO_2 created from the data provided by ChemicaLogic (ChemicaLogic 1999). This diagram shows pressure (*bar*, log scale) versus specific enthalpy (kJ kg^{-1}, normal scale) for the different states. The corresponding values of temperature (°C, *blue lines*), density (kg m^{-3}, *dot-dashed green lines*), and specific entropy (kJ kg^{-1}, *brown dashed lines*) are represented by iso-lines

might influence the heat balance of the aquifer, and, consequently, the dissolution of CO_2 in the brine, and the dissolution/precipitation of minerals, in the surroundings of the injection point.

3.2.3 Properties of Aqueous Phase

The considered aqueous phase here is water that initially occupies the entire void space of the target formation. The difference between categories of water is the dissolved salt content. Any water that contains over than 10,000 ppm of dissolved solids is generally referred as salt water, while water containing between 1000 and 10,000 ppm of dissolved solids is referred to as brackish water. Water with less than 1000 ppm is often referred as freshwater. Salt saturated water is generally referred as brine (typically, more than 200,000 ppm). However, in the GCS literature, the term brine is used for referring to the aqueous phase, even though usually it is just salt water.

A. Brine density

The density of brine that contains various dissolved salts, as well as dissolved CO_2 can be expressed as

$$\rho_{br} = \rho_{br0} \cdot \exp\left[\beta_p(p_{br} - p_o) + \sum_i \beta_{w^i}\left(\omega_i^{br} - \omega_i^o\right) - \beta_T(T - T_O)\right], \quad (3.2.27)$$

where ω_i denotes the mass fraction of i species; superscript 0 denotes a reference value, β_p denotes the fluid's compressibility $\left[\beta_p = -\frac{1}{\rho}\frac{\partial\rho}{\partial p}\right]$, and β_ω, β_T denote coefficients that introduce the effects of concentration and temperature, respectively. Increasing salinity increases the density of brine.

Rowe and Chou (1970) presented a polynomial to calculate specific volume of various salt solutions at pressure above a limited temperature range.

$$\begin{aligned} v_{br} = {} & A_{(T)} - P \cdot B_{(T)} - P^2 \cdot C_{(T)} + x \cdot D_{(T)} + x^2 \cdot E_{(T)} \\ & - xP \cdot F_{(T)} - x^2P \cdot G_{(T)} - 0.5xP^2 \cdot H_{(T)}. \end{aligned} \quad (3.2.28)$$

where A, B, and C are coefficient functions for pure water and E, F, G and H are coefficient functions related to saline water, and x is the weight fraction of NaCl in water. Numerous other expressions are available. Additional data on sodium chloride solutions were provided by Zarembo and Fedorov (1975), and Potter and Brown (1977), which led to the expressions for the density of sodium chloride solutions by Batzle and Wang (1992).

Several methods for calculating brine density functions of temperature, pressure, and salinity are also reported in Adams and Bachu (2002).

IAPWS (2009) recommends calculating thermodynamic properties of water using the fundamental equation for specific Helmholtz free energy.

B. Brine viscosity

Brine viscosity is strongly dependent on temperature (decreases with increasing temperature), less dependent on salinity, and almost negligibly dependent on pressures (increases with increasing salinity or pressure). The most common viscosity expressions are shown in Table 3.1. A few expressions are based on temperature and salinity and other are based on temperature, salinity and pressure.

Garcia (2003) suggests calculating the viscosity of brine using the equation proposed by Kumagai and Yokoyama (1999), fitted to experimental data on viscosities of aqueous NaCl solutions containing CO_2 at pressures up to 30MP:

$$\begin{aligned} \mu_{br} = {} & \mu_0 + (a + bT)M_{\mathrm{NaCl}} + (c + dT)M_{\mathrm{NaCl}}^{1/2} \\ & + (e + fT)M_{\mathrm{CO_2}} + (g + hT)M_{\mathrm{CO_2}}^2 + i(p - 0.1). \end{aligned} \quad (3.2.29)$$

Table 3.1 Several studies of brine viscosities

Reference	Fluid	T (°C)	S (g l^{-1})	P (MPa)
Phillips et al. (1981)	NaCl solution	10–350	<420	–
McCain (1991)	Brine	30–200	<350	–
Batzle and Wang (1992)	NaCl solution	<250	<460	–
Palliser and McKibbin (1988)	NaCl solution	<800	<1000	0.1–300
Kumagai and Yokohama (1999)	Brine	<20	<50	<30

where μ_{br} and μ_0 are the viscosities of brine and pure water in mPa-s, respectively, T is temperature in K, p is pressure in MPa, M_{NaCl} and M_{CO_2} are the molarities of NaCl and CO_2 in mol kg-respectively. The adjusted parameters, a through i, are: $a = 3.85971$, $b = 1.32561 \times 10^{-2}$, $c = -5.37539$, $d = 1.90621 \times 10^{-2}$, $e = 8.79552$, $f = -3.17229 \times 10^{-2}$, $g = -7.22769$, $h = 2.64498 \times 10^{-2}$, $i = 1.69956 \times 10^{3}$.

C. Brine enthalpy

Brine enthalpy,h_{br}, depends on pressure, temperature, salinity and dissolved CO_2. Practically, salt and CO_2 have opposing effects on brine enthalpy. Dissolution of salt is an endothermic reaction, so that salinity reduces the brine enthalpy, while the CO_2 dissolution in water is an exothermic reaction that, in turn, increases the brine enthalpy.

The effect of dissolved salts on the specific enthalpy of brine is taken into account by the approach developed by Michaelides (1981):

$$h_{br} = (1 - x_{\mathrm{NaCl}}) \cdot h_w + x_{\mathrm{NaCl}} \cdot h_{\mathrm{NaCl}} + \Delta h_{L,\mathrm{NaCl}}, \tag{3.2.30}$$

where h_w is the specific enthalpy of pure water, x_{NaCl} is mass fraction of NaCl, h_{NaCl} is the specific enthalpy of NaCl and $\Delta h_{L,NaCl}$ is the heat of NaCl dissolution.

A useful expression that takes into account the dissolution/precipitation of salts as well as dissolution of CO_2 is presented by Bielinski (2006). The enthalpy of brine can be calculated from the weighted contributions of the three components pure water, salt, and CO_2 (h_w,h_{NaCl},h_{CO_2}):

$$h_{br} = (1 - x_{\mathrm{NaCl}} - x_{\mathrm{CO_2}}) \cdot h_w + x_{\mathrm{NaCl}} \cdot (h_{\mathrm{NaCl}} + \Delta h_{L,\mathrm{NaCl}}) + x_{\mathrm{CO_2}} \cdot (h_{\mathrm{CO_2}} + \Delta h_{L,\mathrm{CO_2}}), \tag{3.2.31}$$

where $\Delta h_{L,CO_2}$ is the dissolution enthalpy of CO_2 in water. The effect of dissolved CO_2 on brine enthalpy is calculated by using the following empirical relationship:

$$\Delta h_{L,\mathrm{CO_2}} = -73696 + 564050 \cdot \left(\frac{T}{100}\right) + 703630 \cdot \left(\frac{T}{100}\right)^2 - 278820 \cdot \left(\frac{T}{100}\right)^3 + 42579 \cdot \left(\frac{T}{100}\right)^4. \tag{3.2.32}$$

The most common approach to calculate these thermo-physical properties is based on using the fundamental dimensionless Helmholtz free energy equations. Employing the thermodynamic definition of enthalpy:

$$h = A - T(\partial A/\partial T)_{\upsilon} - \upsilon(\partial A/\partial \upsilon)_T, \tag{3.2.33}$$

where v is the molar volume and the derivative of the Helmholtz free energy equations can be found in IAPWS (2009).

3.2.4 Thermodynamic Properties of CO_2–H_2O Mixture

Wendland et al. (1999) describe the phase diagram of CO_2–H_2O mixture in the temperature range of 250–310 K and a pressure up to 8 MPa. Figure 3.5 shows two pure component phase transition line, 3 binary transition lines, critical point, triple point and quadruple point. At temperatures lower than critical point, only two and three phases may exist, depending on the value of the pressure.

According to Wendland et al. (1999), H_2O rich liquid and CO_2 rich gas exist below the phase transition boundary, while H_2O rich liquid and CO_2 rich liquid exist above this boundary, while at a temperature below $\sim$283 K, the involved phases may include also a hydrate phase. Thus due to the complexity of the CO_2–H_2O system, the attempts to predict the overall behavior of system CO_2-Brine by a single equation of state have been very limited.

Duan et al. (2003, 2006) have published a series of papers on phase equilibrium and volumetric models of water–CO_2. The paper by Li and Duan (2007) establishes

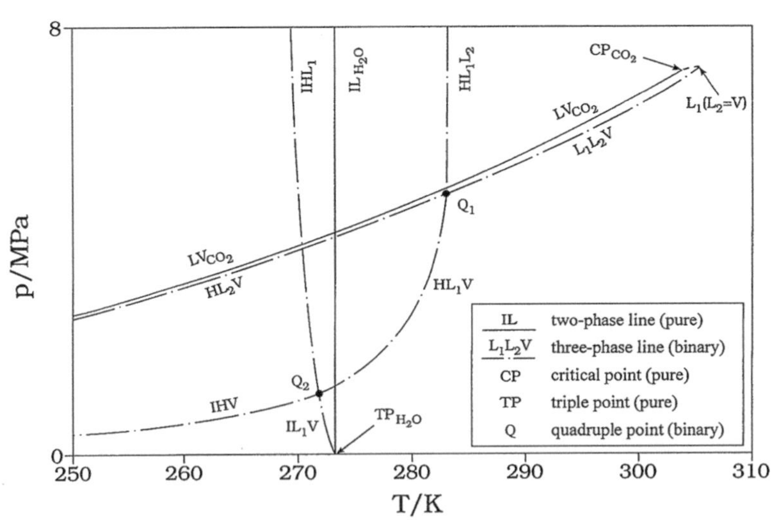

Fig. 3.5 Schematic diagram of the phase behavior of CO_2–H_2O system (Wendland et al. 1999)

a vapor-liquid phase equilibrium model for binary CO_2–H_2O–NaCl system in the range from 0 to 250 °C, from 0 to 1000 bar and from 0 to 5 mol kg^{-1} of NaCl. They review also volumetric models for CO_2–H_2O mixtures and adapt competitive models. Another prediction of vapor-liquid equilibrium and PVT properties, based on statistical fluid theory and Lennard-Jones potential equation of state, is presented by Sun and Dubessy (2012) and applied to water–CO_2 systems. Their model agrees well with data in the range of 323–623 K. Spycher and Pruess (2010) derive correlations for mutual solubility of CO_2 and chloride brine at relevant temperature (from 12 to 300 °C), pressure (from 1 to 600 bar), and salinity (between 0 to 6 mol kg^{-1} NaCl). They use experimental data for phase partitioning.

3.2.5 *Interfacial Properties*

Interfacial tension (IFT) is very sensitive to temperature, but also depends on pressure and salinity. A list of common IFT correlation for a CO_2–H_2O mixture at elevated pressure and temperature is given in Table 3.2.

Macleod (1923) suggested an empirical formula which is valid for a pure substance:

$$\sigma = k(\rho_l - \rho_v)^4, \qquad (3.2.34)$$

where ρ_l and ρ_v are the densities of the liquid and gas, respectively, σ is the surface tension and k is constant. Sugden (1930) expressed this constant as:

$$k = P/M, \qquad (3.2.35)$$

where M is the molecular weight of the substance and P is a characteristic of the substance called the parachor. This correlation was extended to IFT mixtures by Weinaug and Katz (1943) and it was found to be appropriate for low and moderate pressure. An empirical correlation by Chun and Wilkinson (1995) was found to be appropriate near and above the critical point.

$$\ln \sigma = a\ln p_{CO2} + b, \qquad (3.2.36)$$

Table 3.2 IFT of several studies

Reference	Temperature (°C)	Pressure (MPa)	Salinity (g l^{-1})
Hough et al. (1959)	–	Low and moderate	–
Chun and Wilkinson (1995)	5–71	0.1–18.6	–
Hebach et al. (2002)	5–62	0.1–20	–
Bachu and Bennion (2008)	20–125	2–27	<334

where a and b equal −0.2333 and 1.6487, respectively. This correlation is also useful for ternary systems like CO_2–H_2O-alcohol where a and b depends on the nature of the alcohol.

Hebach et al. (2002) developed apparatus for high accurate IFT in water and compressed CO_2 system. They claimed that their experimental error of measurements is less than 2 % and based on their results, they derived a regression function with high precision. Their IFT function for H_2O–CO_2 is:

$$\sigma = k_o\left(1 - \exp\left(k_1\sqrt{dd}\right) + k_2 \cdot dd + k_3 \cdot dd^2 + k_4 \cdot dd^3 + k_5 \exp(k_6(dd - 0.9958))\right),$$

$$\begin{aligned} \rho_{corr} &= \rho_{CO_2} + b_0(304\ (\mathrm{K}) - T)p^{b_1}, 25\ (\mathrm{kg\,m^{-3}}) < \rho_{CO_2} < 250\ (\mathrm{kg\,m^{-3}}), \\ \rho_{corr} &= \rho_{CO_2},\ \text{in all other cases}, \\ dd &= \left((\rho_{H_2O} - \rho_{corr})/1000\right)^2. \end{aligned} \tag{3.2.37}$$

Bachu and Bennion (2008) conducted 378 IFT measurements to derive a surface tension correlation depending on temperature, pressure and water salinity. They found that at a constant temperature and salinity, the IFT decrease significantly with pressure increasing up to the critical point and mildly decreasing to an asymptotic value above the critical pressure. They also found that, contrary to most gases, the IFT increases with salinity (although the dependence is somewhat erratic). IFT increases with temperature (except near the critical point). They propose:

$$\sigma = A \cdot P^{-B}, \tag{3.2.38}$$

where P is pressure, A and B were determined experimentally at different temperatures, and salinities. Although no attempt was made to find a function form of A and B, one can use their table to determine IFT under different conditions.

3.3 The Flow Model

Our objective in this section is to present the two phase flow model that describes the flow of the two fluid phases in the porous medium comprising the target formation of a GCS project. Like every model of transport of an extensive quantity in a porous medium domain, the core of such model is the macroscopic balance equation of the considered extensive quantity. Here, the core of the model consists of two mass balance equations, one for each of the two fluids.

3.3.1 The General Microscopic Balance Equation for an Extensive Quantity

We consider an extensive quantity, E, within a domain Ω bounded by a surface S. We shall use the symbol Ω to denote also the volume of a domain centered at a point x in the void space occupied by a fluid. Making use of the phonological approach, we write this balance in the form

$$\left\{\begin{array}{l}\text{Rate of}\\ \text{accumulation of}\\ E\text{ within }\Omega\end{array}\right\} = \left\{\begin{array}{l}\text{Net influx of}\\ E\text{ into }\Omega\\ \text{through }\mathcal{S}\end{array}\right\} + \left\{\begin{array}{l}\text{Net rate of}\\ \text{production of}\\ E\text{ within }\Omega\end{array}\right\} \tag{3.3.1}$$

which, using mathematical symbols, takes the form

$$\int_{\Omega(\mathbf{x},t)} \frac{\partial e}{\partial t} d\Omega = -\int_{\mathcal{S}(\mathbf{x},t)} e\mathbf{V}^E \cdot \mathbf{v}\, d\mathcal{S} + \int_{\Omega(\mathbf{x},t)} \rho\Gamma^E\, d\Omega, \tag{3.3.2}$$

where $\boldsymbol{e}$ is the quantity of E per unit volume of the phase, Γ^E denotes a source of E (= amount of E produced per unit mass of the phase), and V^E is the velocity of E.

Applying the Gauss theorem to the first term on the r.h.s. of the above equation, we obtain

$$\int_{\Omega(\mathbf{x},t)} \left(\frac{\partial e}{\partial t} + \nabla \cdot e\mathbf{V}^E - \rho\Gamma^E \right) d\Omega = 0. \tag{3.3.3}$$

By shrinking the volume $\Omega(\mathbf{x}, t)$ to zero around an arbitrary point, $\mathbf{x}$, we obtain (in vector notation)

$$\frac{\partial e}{\partial t} + \nabla \cdot e\mathbf{V}^E - \rho\Gamma^E = 0, \tag{3.3.4}$$

where all terms refer to the considered point at time t. Solving this PDE, within the framework of a well posed problem, will provide the distribution $e = e(\mathbf{x}, t)$ at all points within the considered phase occupying the void space or part of it.

At a point at the microscopic level, i.e., at a point in a phase continuum, the *flux*, $\mathbf{j}^E(\mathbf{x}, t)$, of E describes the amount of E passing through a unit area of the phase normal to the direction of $\mathbf{V}^E$, during a unit of time,

$$\mathbf{j}^E = e\mathbf{V}^E \tag{3.3.5}$$

in which $\mathbf{V}^E$ is the velocity of E.

The flux, $\mathbf{j}^E$ can be expressed as the sum of two fluxes:

$$\mathbf{j}^E(\equiv e\mathbf{V}^E) = e\mathbf{V} + e(\mathbf{V}^E - \mathbf{V}) = \mathbf{j}^E_{adv} + \mathbf{j}^E_{dif} \tag{3.3.6}$$

i.e., the sum of an *advective flux*, $\mathbf{j}^E_{adv}(= e\mathbf{V})$ and a *diffusive flux*, $\mathbf{j}^E_{dif}(= e(\mathbf{V}^E - \mathbf{V}))$.

The first expresses the flux of E as carried by the fluid moving at the fluid's velocity $\mathbf{V}$. The second flux, referred to as the *diffusive flux*, expresses the flux of E relative to this advective flux. The diffusive flux of a dissolved chemical species in a fluid phase will be further discussed in Sect. 3.4.

In view of (3.3.6), the balance Eq. (3.3.4) can be rewritten as

$$\frac{\partial e}{\partial t} = -\nabla \cdot (\mathbf{j}^E_{adv} + \mathbf{j}^E_{dif}) + \rho\Gamma^E. \tag{3.3.7}$$

The above equation is the microscopic differential balance equation of any extensive quantity, E, in a fluid continuum.

3.3.2 The Microscopic Mass Balance Equation

Here, $\Gamma^E = \Gamma^m \equiv 0$, as mass cannot be created, $\mathbf{j}^E = \mathbf{j}^m$, and $\mathbf{V}^E \equiv \mathbf{V}^m \equiv \mathbf{V}$, is the velocity of the phase. For this case, the diffusive mass of the phase vanishes, $\mathbf{j}^m_{dif} \equiv 0$, as there is no diffusive flux of the mass of the phase as a whole. Hence, the mass balance equation for the mass of a phase is

$$\frac{\partial \rho}{\partial t} = -\nabla \cdot \rho\mathbf{V}, \tag{3.3.8}$$

in which the left hand side is the rate of mass increase per unit volume of the phase in a small domain around the point, and the right hand side is the net influx through the bounding surface of the domain, per unit volume of the latter.

3.3.3 The General Macroscopic Balance Equation for an Extensive Quantity of a Fluid Phase

In the previous subsection, we used the phenomenological approach to develop E-balance equations at the microscopic level. The same approach will now be used to develop the corresponding equation at the macroscopic level, i.e., at a point in the porous medium considered as a *continuum*.

We shall generalize the discussion by considering an α-phase fluid that occupies the entire void space ($\theta_\alpha = \Omega_v/\Omega = \phi$), or only part of it, at a volumetric fraction

$\theta_\alpha(=\Omega_\alpha/\Omega)$ of an REV centered at a point within a porous medium domain. Also, E_α will be such that it can cross any $\alpha - \beta$ interface. The macroscopic E-balance equation can then be written in the form

$$\left\{\begin{array}{l}\text{Quantity of } E_\alpha\\ \text{accumulating}\\ \text{in } \Omega_\alpha\\ \text{during } \Delta t\end{array}\right\} = \left\{\begin{array}{l}\text{Net quantity of}\\ E_\alpha \text{ entering } \Omega_\alpha\\ \text{through } \mathcal{S}_{\alpha\alpha}\\ \text{during } \Delta t\end{array}\right\} + \left\{\begin{array}{l}\text{Net quantity of}\\ E_\alpha \text{ entering } \Omega_\alpha\\ \text{through } \mathcal{S}_{\alpha\beta}\\ \text{during } \Delta t\end{array}\right\} + \left\{\begin{array}{l}\text{Net production}\\ \text{of } E_\alpha \text{ in } \Omega_\alpha\\ \text{during } \Delta t\end{array}\right\}. \tag{3.3.9}$$

The above balance equation is now written for a small volume in a porous medium domain around a point, and for a small time interval, and then the volume and the time interval are shrunk to zero. The obtained balance equation takes the form of a *partial differential equation* that expresses the balance of E. at any point within the porous medium domain:

$$\underbrace{\frac{\partial \theta_\alpha e_\alpha}{\partial t}}_{(a)} = -\underbrace{\nabla \cdot [\theta_\alpha \mathbf{J}^E_{\alpha,tot}]}_{(b)} + \underbrace{f^E_{\beta\to\alpha}}_{(c)} + \underbrace{\theta_\alpha \rho_\alpha \Gamma^E}_{(d)}, \tag{3.3.10}$$

where e_α denotes the quantity of E_α per unit volume of fluid α-phase, θ_α denotes the fraction of the α-phase in the cross-section (assumed a good approximation of the volumetric fraction), and $\mathbf{J}^E_{\alpha,tot}$ denotes the total macroscopic flux of E, with and in the moving α-phase, *per unit phase area*. In the above balance equation: (a) is the rate of accumulation of E in the α-phase, (b) is the net inflow of e_α per unit time, (c) is the rate at which E is transferred into the α-phase from all other β-phases, and (d) is the rate of production of E in the α-phase. All terms are per unit volume of porous medium. Note that at the macroscopic level, the macroscopic flux is still per unit area of the considered phase, but the latter occupies only part of the cross-section through the porous medium domain.

In (3.3.10), the total E-flux,$\mathbf{J}^E_{\alpha,tot}$, consists of the three E-fluxes: advective, dispersive and diffusive,

$$\mathbf{J}^E_{\alpha,tot} = \mathbf{J}^E_{\alpha,adv} + \mathbf{J}^E_{\alpha,dis} + \mathbf{J}^E_{\alpha,dif} \tag{3.3.11}$$

We recall that there is no diffusive flux of the total mass of a phase.

In (3.3.10), the symbol $f^E_{\alpha\to\beta}$ denotes the rate of transfer of E per unit volume of porous medium, from the α-phase to the β-phase, across their common microscopic interface, by advection and diffusion:

$$f^E_{\alpha\to\beta} = \frac{1}{\Omega_0}\int_{S_{\alpha\beta}} \left[e(V_{\alpha i} - u_i) + j^E_{\alpha i}\right] \cdot v_i \, dS,$$

in which u_i is the velocity of the (microscopic) $S_{\alpha\beta}$ interfaces.

Note that we have introduced and shall be using throughout this chapter two different symbols for the flux of E: $\mathbf{j}^E$ and $\mathbf{J}^E$. The first for the microscopic flux and the second for the macroscopic one. However, for the sake of simplicity, *except for the flux*, we shall not use different symbols for the two levels; the correct level should be understood from the content of the equation itself recalling that each equation belongs only to *one* of the two levels, and so do the terms appearing in it. For example, if an equation includes a volumetric fraction, θ_α, or porosity, ϕ, it is obviously at the macroscopic level.

3.3.4 The Single Phase Flow Model

The discussion in this subsection is at the macroscopic level.

A. The mass balance equation

To obtain the for the case of a single fluid phase that occupies the entire void space, we insert $e_\alpha = \rho, \theta_\alpha = \phi$, $\mathbf{J}^m_{\alpha,dis} = \mathbf{J}^m_{\alpha,dif} \equiv 0$, $f^m_{\beta\to\alpha} \equiv 0$ and $\mathbf{J}^E_{\alpha,tot} = \mathbf{J}^m_{\alpha,adv} = \rho\mathbf{V}$ in (3.3.10), which leads to the mass balance equation

$$\frac{\partial\phi\rho}{\partial t} = -\nabla\cdot\phi\rho\mathbf{V} + \phi\rho\Gamma^m, \qquad (3.3.12)$$

in which the last term on the r.h.s. expresses sources of fluid mass per unit volume of porous medium. For point sources, we often express the source terms by $\Gamma^w = \sum_{(i)} Q^i(\mathbf{x}^i, t)\delta(\mathbf{x} - \mathbf{x}_i)$, where Q^i(dims. L^3/T) may represent a pumping well (with $Q^i < 0$), or recharge well (with $Q^i > 0$) located at point x^i, and $\delta(\mathbf{x} - \mathbf{x}^i)$ (dims. L^{-3}) is the *Dirac delta-function*. Another option, especially when there is only one or a small number of wells, is to remove the source term from the equation and to represent the flux through a well as a flux boundary condition on a cylindrical boundary, representing the well's screen.

It should be noticed that $\mathbf{V}$ in (3.3.12) is mass averaged velocity.

B. The flux equation–Darcy's law

The partial differential equation (3.3.12) involves the fluid's mass flux, $\mathbf{J}^m_{adv} = \rho\mathbf{q}$, where $\mathbf{q}(=\phi\mathbf{V})$ is referred to as the fluid's *specific discharge*. By writing the momentum balance equation for the fluid in the void space, and assuming that (1) the fluid is Newtonian, (2) inertial effects can be neglected, (3) the dispersive flux of momentum can be neglected, (4) the diffusive flux of momentum is expressed by the stress in the fluid, (5) only gravity force acts as a source of momentum, and (6) the fluid is Newtonian, such that the transfer of momentum from the fluid to the solid is proportional to the difference in velocity between the fluid and the solid (Bear and Fel 2012), we obtain the well-known Darcy's law:

$$\phi(\mathbf{V} - \mathbf{V}_s) = -\frac{\mathbf{k}}{\mu}(\nabla p + \rho g \nabla z), \tag{3.3.13}$$

which is a simplified version of the (macroscopic) momentum balance equation of the fluid occupying the entire void space. In this expression. $\mathbf{V}_s$ denotes the solid's velocity, with $\mathbf{V}_s = 0$ when the solid matrix is (or assumed to be approximately) stationary and non-deformable. The case of $\mathbf{V}_s \neq 0$ is addressed in Sect. 3.6. In the above expression,p denotes the fluid's pressure, ∇z denotes a unit vector, positive when directed upward, and μ denotes the fluid's *dynamic viscosity*.

The symbol $\mathbf{k}$ (dim L^2) appearing in (3.3.13) denotes the *permeability* of the porous medium. It is a property that depends on the configuration of the void space, say, on the pore size distribution, or on some mean pore size. For an *isotropic porous medium*, the permeability is a scalar, k. In the case of an *anisotropic porous medium*, $\mathbf{k}$ is a second rank symmetric tensor, with components k_{ij}.

By combining (3.3.12) and (3.3.13), we obtain the mass balance equation in the form

$$\frac{\partial \phi \rho}{\partial t} = \nabla \cdot \rho \frac{\mathbf{k}}{\mu}(\nabla p + \rho g \nabla z) + \phi \rho \Gamma^m, \tag{3.3.14}$$

in which the fluid's density is $\rho = \rho(p, T)$, and the porosity, ϕ, depends on the effective stress, which, in turn, depends on the pressure. Equation (3.3.14) is a single equation to be solved for the single variable p.

When the porous medium is deformable, the solid's velocity $\mathbf{V}_s$, is another variable and we have to introduce the mass balance of the solid matrix, solid matrix constitutive relations, etc. All this is discussed in detail in Sect. 3.6.

C. Initial and boundary conditions

Let $F \equiv F(x, y, z, t) = 0$ represent the equation of a (macroscopic) boundary surface between a considered porous medium domain and the exterior world. We shall use the symbol $\mathbf{v}$ (a vector) to denote the unit outward normal vector to this surface. A stationary boundary surface can be expressed in the form

$$F(x, y, z) = 0, \qquad \mathbf{v} = \frac{\nabla F}{|\nabla F|}. \tag{3.3.15}$$

To solve (3.3.14) as a single equation in the single variable, p, we have to present it as a *well posed problem*. This means that, in addition to the PDE (3.3.14), we have to provide

(1) *Initial conditions* for the variable $p = p(x, y, z, 0)$ at all points within the considered domain.
(2) *Boundary conditions*, i.e., conditions to be satisfied on the entire surface that serves as a boundary of the considered domain.

We assume that all boundaries are sharp surfaces between the considered porous medium domain and the external world.

The boundary may be divided into a number of segments, with a different type of condition on each of them.

Examples of boundary conditions are:

- Boundary of specified value of the variable

In terms of pressure at all points on the boundary surface:

$$p = p(x, y, x, t) \text{ on } f_1(x, y, z), \tag{3.3.16}$$

where $f_1 = f_1(x, y, z)$, a known function, describes the boundary's geometry. A boundary of this kind is called *boundary condition of the first type*, or *Dirichlet boundary condition.*

- Boundary of specified flux,

$$\mathbf{q}_r|_1 . \mathbf{v} = f_3(x, y, z, t), \tag{3.3.17}$$

where $\mathbf{q}_r$ is expressed in terms of the variable, e.g., the pressure, f_3 is a known function, and $\mathbf{v}$ denotes the external normal unit vector on the boundary surface. The relative specific flux,$\mathbf{q}_r$, has to be expressed by an appropriate motion equation, written in terms of p . In general, we assume $\mathbf{V}_s \approx 0$, so that $\mathbf{q}_r \to \mathbf{q}$.

For an impervious boundary, $f_3 = 0$ in (3.3.17).

Because the flux is expressed in terms of the gradient of the pressure, a boundary of this kind is called a Neumann boundary condition, or a boundary condition of the second kind.

3.3.5 Integro-Differential Balance Equation

In certain numerical codes for solving the kind of mathematical models considered here, e.g., the TOUGH2 (Pruess et al. 1991) and TOUGH + (Moridis et al. 2008) families of codes, the balance equation is presented in its *integro-differential form,* rather than in the form of a partial differential equation. Specifically, consider a porous medium domain, i.e., at the macroscopic level, of a finite volume Ω_n bounded by a closed surface of area $\partial\Omega_n$ with outward normal indicated by the unit vector $\mathbf{v}$. Let the void space in this domain be occupied partly by an α-fluid phase, at a volumetric fraction $\theta_\alpha (= \phi S_\alpha)$. An extensive quantity E (density e) is transported *in* and *with* the α-phase at the total flux $\mathbf{J}^E_{\alpha,tot}$. The same extensive quantity is present also in all other β-phases, including the solid s-phase, within Ω_n. Therefore, E may also be transferred from all phases to the considered α-phase, across their common microscopic interfaces. Within the α-phase, E is produced (e.g., by chemical reactions) at a rate $\theta_\alpha \rho_\alpha \Gamma^E_\alpha$ per unit volume of porous medium. We can write the balance of E within the α-phase contained in Ω_n, in the form

$$\underset{(a)}{\frac{d}{dt}\int_{\Omega_n} \theta_\alpha e_\alpha d\Omega_n} = \underset{(b)}{-\int_{\partial\Omega_n} \theta_\alpha \mathbf{J}^E_{\alpha,tot} \cdot \mathbf{v} d\partial\Omega_n} + \underset{(c)}{\int_{\partial\Omega_n} \hat{f}^E_{\beta\rightarrow\alpha} d\Omega_n} + \underset{(d)}{\int_{\Omega_n} \theta_\alpha \rho \Gamma^E d\Omega_n},$$

in which

(a) denotes the rate of added E in the fluid within $\partial\Omega_n$,
(b) denotes the net influx of E into Ω_n through $\partial\Omega_n$,
(c) denotes the rate at which E is transferred from all non-α phases into the α-phase in Ω_n, and
(d) denotes the rate at which E is produced within Ω_n.

The above equation is a balance equation for E in the α-phase, taken over the finite volume Ω_n bounded by the surface of area $\partial\Omega_n$.

3.3.6 The Two Phase Flow Model

The basic assumption underlying the modeling of a GCS project is that, following the injection of CO_2, the void space within the target formation is occupied by two fluids, the native brine and the injected CO_2, which, at the microscopic level, are separated by a sharp interface. These two fluids are *assumed immiscible,* in the sense that both remain as separate phases. As the injected CO_2 advances, a zone is created at every point in which the two fluids co-exist within the REV centered at the point. Nevertheless, it is assumed that dissolved chemical species do cross this interface, e.g., CO_2 dissolves in the brine, and H_2O "evaporates" into the CO_2. Accordingly, the core of the two-phase flow model is composed of the *two mass balance equations*—one for the indigenous brine, and one for the injected CO_2, regarded as immiscible fluids.

When two (practically) immiscible fluid phases occupy the void space in a porous medium domain, depending on the nature of the solid and of the fluids, one of the fluids remains always "closer" to the solid surface, attempting to spread out over it, while the other fluid tends to stay farther from the solid surface. The first is referred to as the *wetting fluid* (subscript w) while the other is called the *non-wetting fluid* (subscript n). For most rocks, the brine is the wetting fluid, while the CO_2 is the non-wetting one.

The term saturation, S_α, indicates the fraction of the void space occupied by the α-phase at a point, i.e., within an REV centered at the point. Thus,

$$S_w + S_n = 1. \tag{3.3.18}$$

A. Capillary pressure

A pressure exists at every (microscopic) point within the fluids that occupy the void space. A sharp interface is assumed to exist between the indigenous saline water and the invading CO_2, and a pressure jump exists across this interface as a consequence of the *surface tension* that exists within this (assumed sharp) interface. This pressure jump is referred to as *microscopic capillary pressure*, p'_c. At a point on the meniscus between the two fluids inside the void space, the value of p'_c is given by the *Young-Laplace formula*,

$$p'_c = p_n - p_w = \gamma_{wn}(\frac{1}{r'} + \frac{1}{r''}) = \frac{2}{r^*}\gamma_{wn}, \qquad (3.3.19)$$

in which γ_{wn} denotes the *surface tension* between the two fluids, and r', r'' are the two principal radii of curvature of this surface, with a radius considered positive when it lies within the n-fluid. The symbol r^* is a mean radius of curvature. When both radii are positive, the pressure is greater in the n-fluid, for which the surface is convex. We need to emphasize that (3.3.19) is strictly valid only when the fluids are stationary. Nevertheless, as a good approximation, it is common to assume that this relationship is valid also when the two fluids are in motion.

Following the phenomenological approach, we assume that (3.3.19) is valid also as a macroscopic expression. When the pressure in the two fluids within an REV centered at a point, are averaged over the REV, we obtain a (macroscopic) pressure difference, referred to as macroscopic capillary pressure,

$$p_c = p_c(S_w) = p_n(S_w) - p_w(S_w). \qquad (3.3.20)$$

This averaged, or macroscopic capillary pressures at a point (centroid of REV) in a porous medium domain expresses the difference between the averaged pressures in the two fluids: p_n and p_w. In analogy to the microscopic level expression, this difference is expressed by

$$p_c = p_n - p_w = \frac{2}{r^*}\gamma_{wn} cos\theta_{wn}, \qquad (3.3.21)$$

in which r^* is the mean radius of curvature, or a characteristic length of the void space, e.g., mean pore size or hydraulic radius of the void space, θ_{wn} is the *contact angle*, which expresses the relative affinity of the wetting fluid to the solid surface, relative to the non-wetting fluid, and γ_{wn} is the surface tension between the wetting and the non-wetting fluids. Note that we have used no special symbols to indicate average values as it is obvious that the equation is at the macroscopic level as the contact angle here is a macroscopic level parameter, and so is r^*.

Because both surface tension and contact angle depend on temperature and on the concentration of dissolved species in the fluids, p_c depends on the temperature and on the presence of solutes in the fluids,

$$p_c = p_n - p_w = p_c(S_w, \gamma_{wn}(T, c_\alpha^\gamma), \theta_{wn}(T, c_\alpha^\gamma)), \tag{3.3.22}$$

with $\gamma = 1, \ldots, NC$. Neglecting any effect of fluid composition, the structure of the pores, the effect of the void space configuration on the p_c-curve is the same, regardless of the nature of the two fluids. The effect of fluid properties may be expressed by

$$S_w(p_c|_{\text{fluids } n_1, w_1}) = S_w(\beta_{12} p_c|_{\text{fluids } n_2, w_2}),$$

in which

$$\beta_{12} = \frac{\gamma_{n_1 w_1} \cos\theta_{n_1 w_1}}{\gamma_{n_2 w_2} \cos\theta_{n_2 w_2}},$$

is a scaling factor, and $(n_1, w_1), (n_2, w_2)$ represent pairs of nonwetting and wetting fluids. When the contact angles remain unchanged, $\beta_{12} = \gamma_{n_1 w_1}/\gamma_{n_2 w_2}$.

It is obvious from the presence of r^* in (3.3.21) that the only way to determine the capillary pressure for any given porous medium is *experimentally*.

Figure 3.6 shows a water-air capillary pressure relationship $p_c = p_c(S_w)$. Draining a fully saturated ($S_w = 1$) sample, we follow the *boundary drainage curve*, until we reach a minimal saturation, called *irreducible wetting fluid saturation*, S_{wr}, below which S_w cannot be further reduced (except by heating and evaporation).

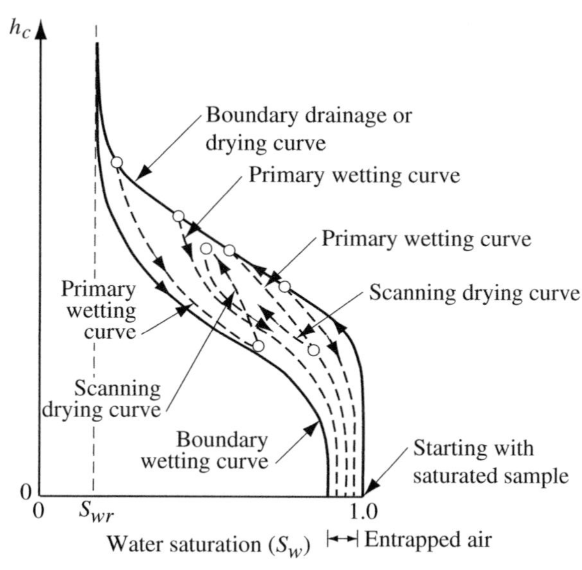

Fig. 3.6 Water-air capillary curve, showing hysteresis and entrapped air

At this saturation, the wetting phase is made up of isolated rings and the wetting phase permeability vanishes. We note the asymptotic shape of the curve as it approaches S_{wr}. As we re-wet the sample, we follow the *boundary wetting curve*. As we continue to wet the sample, we can never reach full saturation because some non-wetting fluid remains in the form of globules and ganglia at the *residual non-wetting saturation*, S_{nr}. At this saturation, the non-wetting fluid phase becomes discontinuous and its permeability vanishes.

For a given capillary pressure, a higher wetting fluid saturation is obtained during drainage than during imbibition.

At any point on the boundary drainage or wetting curves, in fact, also on a scanning curve, we can reverse the process and follow a scanning curve. The scanning curves are shown as dashed lines on the figure. In this way, the macroscopic relationship between capillary pressure and saturation, expressed by the capillary pressure curve, exhibits the phenomenon of *hysteresis*. This means that the capillary pressure at any given time depends also on the wetting-drying history of the particular sample under consideration. Hysteresis is a consequence of the pore-throat configuration of the void space (ink-bottle effect).

We have mentioned two points which are of special interest. Starting drainage from a fully saturated sample, stopping and reversing the process, the sample can never by fully re-wetted. *Entrapped non-wetting phase* will always remain. Similarly, a sample can never be fully drained (of the wetting fluid). The wetting fluid will always approach asymptotically *irreducible wetting fluid saturation*.

Various researchers have suggested analytical expressions for the general shape of fluid capillary pressure curves. These expressions involve coefficients that must be determined by fitting the analytical expression to experimental data. These are typically written in terms of the *effective*, or *reduced wetting fluid saturation*, defined as

$$S_e = \frac{S_w - S_{wr}}{1 - S_{nr}},$$

where S_{wr} is the irreducible wetting fluid saturation, usually fitted from experimental data. Various authors use slight variations on the denominator, which may be $(S_{ws} - S_{wr})$, or $(1 - S_{nr} - S_{wr})$ where S_{ws} is the is the saturated wetting fluid saturation and S_{nr} is the residual non-wetting fluid saturation. The two most widely used models are the Brooks and Corey (1964) and Van Genuchten (1980) models. The former is given by:

$$p_c = p_c^0 S_e^{-1/\lambda}, \tag{3.3.23}$$

where p_c^0 is the entry pressure parameter and λ is the pore size distribution index, which reflects the bundle of tubes which served as the origin of this equation; it can be obtained from the pore size distribution. However, λ and p_c^0 are most often determined experimentally.

The Van Genuchten (1980) model is given by:

$$p_c = p_c^0\left(S_e^{-1/m} - 1\right)^{1/n}, \tag{3.3.24}$$

in which *m, n* reflect also the pore size distribution, but are usually treated as empirical constants determined experimentally.

A useful way to express the capillary pressure curve as related to the saturation only, is by making use of the Leverett function (Leverett 1941), $J(S_w)$, or $J(S_w, \theta_{wn})$, i.e., with or without the effect of the contact angle. From (3.3.21), we obtain

$$p_c = p_c(S_{br}) = \frac{2\gamma_{br,CO_2}}{r^*}\cos\theta \Rightarrow J(S_w) = \frac{p_c}{2\gamma_{br,CO_2}\sqrt{\phi/k}\cos\theta},$$

in which $\sqrt{k/\phi}$ is a length characterizing the void space. This function now depends only on the geometry of the void space.

B. Mass balance equations

Making use of (3.3.10), which is the general macroscopic balance equation for an extensive quantity, *E*, of a phase in a multiphase system, we can now consider the model describing two phase flow—e.g., the indigenous brine and the injected CO_2—in the porous medium domain comprising the target geological formation. In this case the considered extensive quantity is the mass, m_α, of each of these α-phases.

With $\phi = \phi(x, y, z, t)$, denoting the porosity of the porous medium, $\theta_\alpha = \theta_\alpha(x, y, z, t)$ denoting the volumetric fraction of the α -fluid, and $S_\alpha = S_\alpha(x, y, z, t), \quad \alpha = w, n$, denoting the *saturations* of the fluids, we have

$$S_w + S_n = 1, \quad \theta_w + \theta_n = \phi. \tag{3.3.25}$$

From (3.3.10), it follows that the (macroscopic) mass balance equation for a fluid α-phase is:

$$\begin{aligned} &\frac{\partial}{\partial t}(S_\alpha\phi\rho_\alpha) = -\nabla.(\rho_\alpha\mathbf{q}_\alpha) - f_{\alpha\to\beta}^{m_\alpha} + S_\alpha\phi\rho_\alpha\Gamma'^\alpha, \\ &\alpha, \beta \equiv w, n = br, CO_2, \end{aligned} \tag{3.3.26}$$

in which ρ_α denotes α-fluid density, $\mathbf{q}_\alpha(\equiv\phi S_\alpha\mathbf{V}_\alpha)$ denotes the *specific discharge* of the α-phase, $\mathbf{V}_\alpha$ is the velocity of the α-phase, $f_{\alpha\to\beta}^{m_\alpha}$ is the rate of mass transfer from the α-phase to the β-phase, per unit volume of formation, and $\rho_\alpha\Gamma'^\alpha$ is the rate of production of mass of the α-phase, per unit volume of formation. Actually, in the models considered here, the only source of fluid mass is that due to injection at one, or at a small number of injection wells, but these wells will be introduced as boundary conditions.

Thus, there will be no sources of fluid mass *within* the considered domain, and the mass balance equation takes the form:

$$\begin{aligned}&\frac{\partial}{\partial t}(S_\alpha \phi \rho_\alpha) = -\nabla.(\rho_\alpha \phi S_\alpha \mathbf{V}_\alpha) - f^m_{\alpha\to\beta},\\ &\alpha, \beta \equiv w, n = br, CO_2.\end{aligned} \tag{3.3.27}$$

The mass of the α-phase that crosses into the β-phase is equal to the sum of masses of the individual γ-species present in the α-phase that cross into the β-phase, i.e., $f_{\alpha\to\beta} = \sum_{(\gamma)} f^\gamma_{\alpha\to\beta}$.

Note that so far we have the following variables:

(a) $S_\alpha(\mathbf{x}, t), \;\; \alpha = br, CO_2$ denotes two variables, constrained by

$$S_{br} + S_{CO_2} = 1. \tag{3.3.28}$$

(b) The densities $\rho_\alpha(\mathbf{x}, t)$ constitute two variables, which vary continuously as the pressure, temperature and concentrations of the various species comprising the fluid phases vary.

C. Fluids' advective mass fluxes

We recall that the term *advective mass flux* is used to denote the mass of a fluid phase that passes through a unit area of the cross section of a porous medium domain, per unit time. There is no diffusion of the mass of the fluid phase, and we neglect the dispersive flux of the total mass of a phase. In Sect. 3.4 we shall discuss the diffusive and dispersive fluxes of the mass of a chemical species dissolved in a fluid phase.

The advective flux of a single fluid that occupies the entire void space is discussed in Sect. 3.4B, leading to Darcy's law.

In the case of flow of two fluid phases that together occupy the entire void space, we need *two* flux equations, one for each phase. Furthermore, we have to take into account the (average of the) rate of momentum transfer across the microscopic interfaces separating the two fluid phases. This will lead to the following flux equations in two phase flow:

$$\begin{aligned}\mathbf{q}_{rw} &= -\frac{\mathbf{k}^w_w(S_w)}{\mu_w}\cdot(\nabla p_w + \rho_w g \nabla z) - \frac{\mathbf{k}^n_w(S_n)}{\mu_n}\cdot(\nabla p_n + \rho_n g \nabla z),\\ \mathbf{q}_{rn} &= -\frac{\mathbf{k}^w_n(S_w)}{\mu_w}\cdot(\nabla p_w + \rho_w g \nabla z) - \frac{\mathbf{k}^n_n(S_n)}{\mu_n}\cdot(\nabla p_n + \rho_n g \nabla z).\end{aligned} \tag{3.3.29}$$

However, it is usually assumed that we may neglect the exchange of momentum across the microscopic interface boundaries between the two fluid phases. Thus, the simplified form of the momentum balance equation for a fluid phase in two phase flow takes the form of Darcy's law (3.3.14), except that now we have two flux equations, one for each fluid:

$$\mathbf{q}_{rw} = -\frac{\mathbf{k}_w(S_w)}{\mu_w} \cdot (\nabla p_w + \rho_w g \nabla z),$$
$$\mathbf{q}_{rn} = -\frac{\mathbf{k}_n(S_n)}{\mu_n} \cdot (\nabla p_n + \rho_n g \nabla z). \tag{3.3.30}$$

For each fluid, the permeability at a point, now referred to as *effective permeability*, $\mathbf{k}_w(S_w), \mathbf{k}_n(S_n)$, is a coefficient that depends on the saturation of the respective fluid within the void space. In an inhomogeneous fluid, the dependence on saturation will vary also with location.

These two flux expressions are now employed to describe the flow of the two considered fluid phases in the mass balance equations (3.2.27).

Altogether, we have the following expressions for the advective mass fluxes for the brine and the CO_2:

$$\rho_\alpha \mathbf{q}_{\alpha r} \equiv \phi S_\alpha (\mathbf{V}_\alpha - \mathbf{V}_s) = -\rho_\alpha \frac{\mathbf{k}_\alpha(S_\alpha)}{\mu_\alpha} .(\nabla p_\alpha + \rho_\alpha g \nabla z),$$
$$\alpha = br, CO_2. \tag{3.3.31}$$

in which p_α is the pressure in the α-phase, $\mathbf{k}_\alpha(S_\alpha)$ denotes the *effective permeability tensor* of the α-phase in an anisotropic porous medium, $\mathbf{V}_\alpha$ denotes the α-phase velocity, and $z(\equiv x_3)$ denotes the vertical coordinate, positive upward. Note that as long as we assume that the porous medium comprising the target formation is anisotropic, *we may not use the concept of relative permeability*, as the relationship between effective permeability and permeability in single phase flow, may vary with direction.

We now assume that the velocity of the solid phase, $\mathbf{V}_s$, (due to matrix deformation) is much smaller than that of the fluid, and can, therefore, be neglected. Hence, the two flux, or specific discharge expressions used here are

$$\mathbf{q}_\alpha \equiv \phi S_\alpha \mathbf{V}_\alpha = -\frac{\mathbf{k}_\alpha(S_\alpha)}{\mu_\alpha} \cdot (\nabla p_\alpha + \rho_\alpha g \nabla z),$$
$$\alpha = br, CO_2. \tag{3.3.32}$$

D. Effective permeability

The (tensorial) coefficients $\mathbf{k}_w$ and $\mathbf{k}_n$, called effective permeabilities, are related to (microscopic) properties of the geometrical configuration of the portion of void space occupied by each fluid phase. For an isotropic porous media, the effective permeability reduces to a scalar.

In the case of an anisotropic porous medium, each of the components of the effective permeability varies differently in response to changes in saturation. In indicial notation, we write

$$k_{w,ij} = k_{w,ij}(S_w) \quad \text{and} \quad k_{n,ij} = k_{n,ij}(S_n).$$

For an isotropic porous medium, and only for such a medium, the effective permeabilities are scalars. In such case, the *relative permeability* is defined as the ratio of the effective permeability to permeability at saturation of the considered fluid, i.e.

$$k_{rw}(S_w) = \frac{k_w(S_w)}{k_{sat}} \quad \text{and} \quad k_{rn}(S_w) = \frac{k_n(S_w)}{k_{sat}},$$

where k_{rw} and k_{rn} are the wetting and non-wetting, respectively, relative permeabilities. Note that $0 \leq k_{rw} \leq 1$, and $0 \leq k_{rn} \leq 1$.

Figure 3.7 shows typical relative permeability curves for a wetting phase and for a non-wetting one.

Starting the drainage of a porous medium sample from full wetting fluid saturation, we note a rapid decline in k_{rw} as the larger pores are drained first, and the flow of the w-fluid takes place through the smaller pores. When the w-fluid saturation is below the *irreducible w-fluid saturation*, S_{wr}, the fluid remaining in the porous medium is in the form of isolated *pendular rings* and very thin films that cover the solid surface in the larger pores from which the w-fluid has already been drained. In this form, the fluid's permeability vanishes and the w-fluid constitutes a discontinuous, immobile phase that cannot transmit pressure.

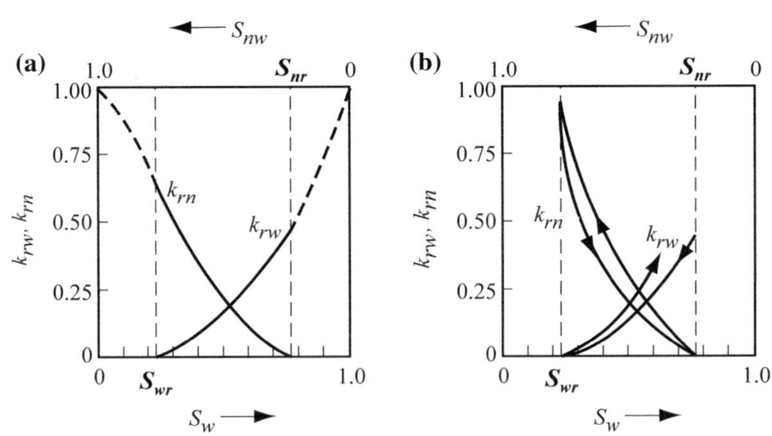

Fig. 3.7 Typical relative permeability curves for an isotropic porous medium: **a** without hysteresis, **b** with hysteresis

Thus, $k_{rw} = 0$ for $S_w \leq S_{wr}$. In reality, given enough time, the wetting fluid will continue to drain by gravity in the form of films, reducing the saturation to below the irreducible saturation. For $S_w = 1$, temporarily overlooking the meaning of the dashed portion of the k_{rw} curve, we have $k_{rw} = 1$, i.e., $k_{rw}(S_w)|_{S_w=1} \equiv k_{sat}$.

When a nonwetting fluid is being displaced by a wetting one to below a critical saturation value, S_{nr} referred to as *the residual nonwetting fluid saturation*, the latter breaks down into isolated blobs, or globules. Usually, these remain immobile under the pressure gradient that drives the wetting fluid. The value of S_{nr} is determined by properties of the nonwetting fluid and of the solid matrix.

The relative permeability curves have to be determined experimentally for each particular porous medium. However various investigators have suggested analytical expressions for the relationship between relative (or permeability and saturation). These expressions were usually obtained by analyzing simplified models of porous media, such as a bundle of parallel capillary tubes, or a network of such tubes. The results, while highlighting the main features of the sought relationship, always contain numerical coefficients that characterize the considered model. For a particular porous medium, the numerical values of the coefficients have to be determined by fitting the analytical expression to experimental curves. Analytical expressions (as compared with tables of experimental results) have the advantages that they can be used in analytical or semi-analytical solutions, and can more easily be used as input to numerical models.

Following are two examples, both for isotropic porous media:

- Gardner (1958) suggested the expression

$$K_w = \frac{a}{b + |\psi|^m},$$

where a, b and m are constants, with $m \approx 2$ for heavy clay soil, and $m \approx 4$ for sand.

- Childs and Collis-George (1950), for the flow of water in the unsaturated zone, suggested

$$K_w = B\frac{\theta_w^3}{\Sigma_{vs}},$$

where Σ_{vs} is the specific surface area of the soil and B is a coefficient.

- Van Genuchten (1980) and Mualem (1976) suggested

$$K_w = K_{sat}\, \tilde{S}_e^{1/2}\left[1 - \left(1 - \tilde{S}_e^{1/m}\right)^m\right]^2,$$

where all symbols have the same meaning as those in (3.3.24).

E. Initial and boundary conditions

As in the case of single phase flow (Sect. 3.4), a well-posed mathematical model requires information on initial and boundary conditions stated for a well specified domain. In the case of two-phase flow considered here, we have to solve two partial differential equations for two (independent) variables, either p_w, S_n or p_n, S_w.

Because this chapter is devoted to the mathematical modeling of a GCS project, no effort will be made to cover all possible boundary conditions for two-phase flow. The following discussion on initial and boundary conditions will be limited to those that are encountered in such projects.

- Initial conditions

Initial conditions are the known values of the state variables at $t = 0$. These should be assigned to all points within the considered (bounded) domain.

In the case of GCS, where we have to solve for two variables, we have to specify either p_{br}, S_{CO_2} or p_{CO_2}, S_{br} *at all points* of the considered domain. In general, initially, we do not have CO_2 in the target formation, so that $S_{CO_2} = 0.$ and we know the initial spatial distribution of the pressure in the brine $p_{br}|_{t=0}$.

- Boundary conditions

Boundary conditions have to be specified on *the entire* boundary that encloses the considered domain within the target formation—one condition for each of the two variables/fluids on every segment of the boundary. Often, the considered domain is the entire target formation. The bottom of the formation is usually considered impervious, and so is the ceiling.

We seldom encounter cases with naturally occurring, well specified, usually assumed vertical, boundaries at a reasonable distance upstream and downstream of the injection well (or wells), as well as laterally. When such boundaries occur, they are selected as boundaries of the modeled domain. However, in many cases, there is no naturally defined lateral boundary within a reasonable distance from the injection well(s). An artificially located boundary is then assumed, and a sensitivity analysis is conducted in order to determine the location of such boundary so as not to (significantly) affect the results of the investigations.

The injection well is also a boundary, composed of two parts: a screened part through which the CO_2 enters the formation, and an impervious part.

The total boundary, $\partial\Omega$, surrounding the domain of interest in the target formation, Ω, is composed of m segments, $\partial\Omega = \sum_m \partial\Omega_k$, on each of which we have to define a boundary condition for each of the two fluids. The commonly encountered boundaries in GCS projects are:

- Specified flow condition

Let the known specific discharge of an α-phase entering a portion of the boundary of a target formation, $\partial\Omega_k$, be a known $q_{\alpha 0}$ for $\alpha = CO_2$ or br. The condition on that portion of the boundary is then

$$\begin{aligned} q_{\alpha 0} &= -\left\{ \frac{\mathbf{k}_\alpha(S_\alpha)}{\mu_\alpha}.(\nabla p_\alpha + \rho_\alpha g \nabla z) \right\}.\mathbf{v}, \\ \alpha &= br, CO_2, \text{ on } \partial\Omega_k. \end{aligned} \tag{3.3.33}$$

Note that the above equation is written under the assumption of an anisotropic domain, with effective permeability, $k_{\alpha ij} = k_{\alpha ij}(S_\alpha)$.

On an impervious boundary segment, say, formation bottom and ceiling, and often at some sufficiently large lateral distance, $q_{\alpha 0} = 0$

Practically, a specified, non-zero, CO_2 flux is known only on the screened portion of an injection well. The remaining portion of the casing is regarded as an impervious boundary.

- Condition at an injection or pumping well

When we consider the screened portion of an injection or pumping well, as part of the formation's boundary, (1) the normal to the boundary is in the radial direction, and (2) the specific discharge of the CO_2 or the brine in the radial direction is assumed known, say, a uniformly distributed $q_{\alpha r0}$, by assuming that the total injected or pumped fluid is uniformly distributed on the screened portion of the well.

Thus, the condition is

$$\rho_\alpha q_{\alpha r0} = -\rho_\alpha \frac{k_\alpha(S_\alpha)}{\mu_\alpha} \frac{\partial p_\alpha}{\partial r}, \quad \alpha = CO_2, br. \tag{3.3.34}$$

Note that we assumed here that in the vicinity of the well, the formation is isotropic with respect to its permeability. Also, we are assuming the well to be vertical, so that the gravity term in Darcy's Law has been neglected. Otherwise, the radial coordinates version of (3.3.33) needs to be adopted.

- Boundary of specified pressure

In a GCS project, this kind of boundary occurs when the indigenous brine is not stationary within the formation. Then, upstream, there is an inflow surface on which the brine pressure, $p_{br,in}$, is known, and an outflow surface, downstream, on which the known pressure is $p_{br,out}$. The latter surface may be the natural outlet of the formation,. When this outlet is at a very large distance from the injection well, it is replaced by an artificial one specified at a "reasonable" distance from the injection area. The location of such boundary is usually specified such that the results of the simulated case are not (significantly) affected by the distance of this outflow boundary from the injection well(s).

F. Complete flow model

Assuming that the solid matrix is stationary and rigid, i.e., $V_{s=0} = 0, \quad \partial\rho_s/\partial t = 0$, the complete flow model can be stated as:

Determine the values of the 12 variables:

$$q_{br,i},\ q_{CO_2,i},\ S_{br},\ S_{CO_2},\ \rho_{br},\ \rho_{CO_2},\ p_{br},\ p_{CO_2}.$$

subject to the specified initial and boundary conditions.

We solve the problem by solving the following equations:

- Two (partial differential) mass balance Eq. (3.3.27), making use of:
- Six flux Eq. (3.3.32),
- One equation expressing the sum of saturations (3.3.28),
- One capillary pressure relationship, e.g., (3.3.21), and
- Two constitutive relationships for fluid density, e.g., (3.2.2) for CO_2 and (3.2.28) for saline water.

Thus, the *two* PDE have to be solved for selected *two* independent variables. Note that, in addition to the above equations, which relate the appearing variables explicitly in (3.3.27), we need to specify the domain geometry and B.C's, as well as the following information,

- Effective permeability: $k_\alpha = k_\alpha(S_\alpha), \quad \alpha = br, CO_2$, possibly for a heterogeneous medium, $k_\alpha = k_\alpha(\mathbf{x}, S_\alpha(\mathbf{x}, t)), \quad \alpha = br, CO_2$.
- Viscosity: $\mu_\alpha = \mu_\alpha(p_\alpha, T_\alpha, c_\alpha^\gamma), \alpha = br, CO_2$.

A discussion on degrees of freedom is presented in the Appendix. In the case considered here, we have *two* degrees of freedom to be solved for by the two partial differential equations (3.3.27). The selected two variables have to be *independent* of each other. For example, we may choose p_{br}, S_{CO_2} or p_{CO_2}, S_{br} as the two independent variables of the problem.

The solution of the two partial differential mass balance equations is subject to an appropriate set of initial and boundary conditions as described above.

In describing the solution procedure, we have not addressed how to compute the rate of transfer, $f_{\alpha\to\beta}, \quad \alpha, \beta = br, CO_2$. These mass transfers (dissolution of CO_2 in the saline water and evaporation of water into the CO_2 phase) depend on the composition of each phase. Therefore, it cannot be computed solely from phase mass balances, it requires compositional mass balances, which is the subject of next section.

3.4 The Reactive Transport Model

The previous section presented the flow equation, which expresses the conservation of mass of a fluid phase. If, instead of the whole phase, one is interested in specific chemical species dissolved in a phase that occupies the void space or part of it, the corresponding mass conservation equation is called *transport equation.* The transport equation governs the spatial distribution of chemical species and how this distribution evolves in time. When several species are transported simultaneously, they may react. The resulting phenomenon is called *reactive transport* and it entails several transport equations, one for each reacting species, and the equations that describe the chemical reactions among them and with immobile phases.

Mathematical modeling of solute transport follows the same approach as that of fluid flow in that solute mass conservation is expressed using equations like (3.3.2). However, two additional features need to be acknowledged:

- Chemical species are transported within a moving phase by advection, dispersion and molecular diffusion.
- Chemical species may react with each other, both within a phase (homogeneous reaction) and across phase interfaces (heterogeneous reactions). The latter include inter-phase mass transfer from one fluid phase to another (e.g., dissolution of supercritical CO_2 into the aqueous phase), or dissolution-precipitation reactions.

The natural state variable to express mass conservation of a species is *concentration*. There are a number of ways to express concentration, depending on whether mass is expressed in kg or in moles and whether this mass is expressed per unit volume, or per unit mass (of solution or solvent). Concentration may also be expressed as molar or mass fraction. Each definition has some advantages and disadvantages. Here we shall use *molar concentration*, or *molarity*, c_α^γ, which represents the number of moles of a γ species per unit volume of the α-phase. Actually, this is rather inconvenient for chemical reactions because in chemical data bases, concentrations are expressed as molalities, m^γ, moles of γ per kg of water.

3.4.1 Fluxes of Chemical Species

As mentioned earlier, chemical species are transported by advection, dispersion and molecular diffusion. Following is a brief discussion of these three processes.

A. Advection

Advection of a γ-species in an α-fluid phase refers to the dragging of the species by the flowing fluid phase, that is, the displacement of the species with the fluid's motion. The advective mass flux quantifies the mass of that species passing through a unit planar area of a porous medium, per unit time. Therefore, the advective flux of the γ-mass is expressed by;

$$\mathbf{q}^{\gamma}_{\alpha,adv} \equiv \theta_\alpha \mathbf{J}^{\gamma}_{\alpha,adv} = \mathbf{q}_\alpha c^{\gamma}_{\alpha}, \tag{3.4.1}$$

in which $\mathbf{q}_\alpha (= \theta_\alpha \mathbf{V}_\alpha)$ is the *specific discharge* of the α-phase, typically obtained from Darcy's law (3.3.13).

B. Molecular diffusion

Molecular diffusion (or simply diffusion) quantifies the net flux of particles of a specific species caused by their Brownian (random) motion within the host fluid. Therefore, it can be formally quantified by averaging the Brownian displacements. Since the movement is random, the number of particles being displaced within a fluid domain is proportional to the gradient of the particles' concentration. This is expressed by *Fick's law* of molecular diffusion *within* a phase:

$$\mathbf{j}^{\gamma}_{\alpha,diff} = -\mathcal{D}^{\gamma}_{\alpha} \nabla c^{\gamma}_{\alpha}, \quad \sum_{(\gamma)} \mathbf{j}^{\gamma}_{\alpha,diff} = \mathbf{0}, \tag{3.4.2}$$

where the scalar $\mathcal{D}^{\gamma}_{\alpha}$ is the *coefficient of molecular diffusion* (dims. L^2/T) of the γ-species within the α phase.

In general, $\mathcal{D}^{\gamma}_{\alpha}$ is a function of pressure and temperature. The value of this coefficient is very sensitive to the nature of the phase (Crank 1956). In liquid phases, $\mathcal{D}^{\gamma}_{\alpha}$ is not very sensitive to pressure, but is sensitive to the viscosity of the liquid. It ranges around 10^{-9} m^2 s^{-1} in aqueous phases, where it is not very sensitive to the nature of the species. Coefficients of diffusion in gases grow approximately as $T^{3/2}$ and are inversely proportional to the pressure. They decrease with the molecular weight of the diffusing species, and typically range around 1–9×10^{-5} m^2 s^{-1}. Diffusion also occurs in solids, where it is controlled by the nature of crystals and their defects.

Diffusion coefficients in supercritical CO_2 have been determined in the food industry. Therefore, they are available for species like benzene, caffeine, and the like. The measured diffusion coefficients are of the order of 10^{-8} m^2 s^{-1}. They decrease with pressure, and increase with temperature (Swaid and Schneider 1979; Feist and Schneider 1982).

In a porous medium, diffusion in the fluid phase is highly hindered by the solid phase, which reduces the actual area available for diffusion, and forces diffusion pathways to be tortuous and constrained at pore necks (e.g., Bear and Bachmat 1990). Thus, in a porous medium, we have

$$\mathbf{q}^{\gamma}_{\alpha,diff} \equiv \theta_\alpha \mathbf{J}^{\gamma}_{\alpha,diff} = -\theta_\alpha \mathcal{D}^{\gamma}_{\alpha} \mathbf{T}^*(\theta_\alpha) \cdot \nabla c^{\gamma}_{\alpha} = -\theta_\alpha \mathcal{D}^{*\gamma}_{\alpha}(\theta_\alpha) \cdot \nabla c^{\gamma}_{\alpha}, \tag{3.4.3}$$

where θ_α denotes the volumetric fraction of the considered α-phase, $\mathbf{T}^*(\theta_\alpha)$, a second rank symmetric tensor, represents the *tortuosity* of the porous medium and

$\mathcal{D}_\alpha^{*\gamma}(\theta_\alpha)$, a second rank symmetric tensor, is the coefficient of molecular diffusion within a phase *in a porous medium*.

For saturated flow, we replace θ_α by the porosity ϕ.

In an isotropic porous medium, the components of the tortuosity tensor, T^*_{ij}, may be represented as $T^*\delta_{ij}$, in which $T^*(\leq 1)$ is a scalar tortuosity, and δ_{ij} is the Kronecker delta (=unit tensor), with $\delta_{ij} = 1$ for $i = j$, and $\delta_{ij} = 0$ for $i \neq j$.

The tortuosity of a phase is a macroscopic geometrical coefficient that expresses the effects of the phase distribution within the REV. Hence, in systems with multiple phases, each of the tortuosity components is a function of the phase saturation. Millington (1959) relates the tortuosity in an isotropic porous medium to the volumetric fraction of the phase,

$$T^*(\theta_\alpha) = \frac{\theta_\alpha^{7/3}}{\phi^2}. \tag{3.4.4}$$

C. Dispersive flux, $\mathbf{J}^\gamma_{\alpha,disp}$

The *dispersive mass flux* (or flux due to mechanical dispersion) of a chemical species in a fluid phase is a *macroscopic* flux that reflects two facts: at the microscopic level, the concentration of a solute is not uniform within the fluid phase occupying the void space in an REV, and (2) the microscopic velocity also varies from point to point within that fluid. With $\overline{(.)}$ denoting the average of (.) over an REV, we can express the average macroscopic advective mass flux of a chemical species in a fluid phase, in the form $\overline{c^\gamma_\alpha \mathbf{V}_\alpha}$. Then, with $\hat{a}$ denoting the deviation (=perturbation) of a from the average, $\overline{a}$, we can write:

$$\overline{c^\gamma_\alpha \mathbf{V}_\alpha} = \overline{c^\gamma_\alpha}\,\overline{\mathbf{V}_\alpha} + \overline{\hat{c}^\gamma_\alpha \hat{\mathbf{V}}_\alpha}, \quad c^\gamma_\alpha = \overline{c^\gamma_\alpha} + \hat{c}^\gamma_\alpha, \quad \mathbf{V}_\alpha = \overline{\mathbf{V}_\alpha} + \hat{\mathbf{V}}_\alpha, \tag{3.4.5}$$

i.e., the average of the total (local) advective flux has been decomposed into an advective flux at the average velocity and a dispersive flux. This is a consequence of the averaging process that we have chosen to employ. Note that average values are assigned to the centroids of an REV, while the deviations from the average are at points within the considered phase inside the REV.

It is reasonable to imagine, that deviations from the mean concentration and velocity will correlate positively in the presence of a macroscopic (mean) concentration gradient. In such case, within the REV at a given time, points with high velocity (positive deviation from the mean) will contain water that has been brought from farther upstream (thus with higher concentration) than those with low velocity. The magnitude of the deviation in concentration should be proportional to the (mean) concentration gradient and the deviation in velocity should be proportional to the (mean) velocity. Therefore, it should be expected that the dispersive flux of a chemical species is driven by (1) the velocity of the fluid and (2) by the gradient in the species concentration gradient. Investigations over a period of about three decades, starting around the mid-50s (e.g., De Josselin de Jong 1958; Saffman 1959; Bear 1961a; Scheidegger 1961; Bear 1972; and Bear and Bachmat 1990), have led to the conclusion that the dispersive flux of a component (per unit area of fluid) in a porous medium can be expressed as a *Fickian-type* law (i.e., a law that resembles Fick's law of molecular diffusion in that the flux is driven by the concentration gradient) in the form:

$$\mathbf{q}^{\gamma}_{\alpha,disp} \equiv \theta_{\alpha}\mathbf{J}^{\gamma}_{\alpha,disp} = \theta_{\alpha}\overline{\hat{c}^{\gamma}_{\alpha}\hat{\mathbf{V}}_{\alpha}} = -\theta_{\alpha}\mathbf{D}\nabla c^{\gamma}_{\alpha}, \quad q^{\gamma}_{\alpha,disp,i} \equiv \theta_{\alpha}\overline{\hat{c}^{\gamma}_{\alpha}\hat{V}_{\alpha i}} = -D_{ij}\frac{\partial c^{\gamma}_{\alpha}}{\partial x_j}, \tag{3.4.6}$$

where the D_{ij} (dims. L^2/T) are components of a second rank symmetric tensor,**D**, called the *coefficient of mechanical (or advective) dispersion*. Equation (3.4.6) is valid for the general case of an anisotropic porous medium. The dispersion tensor is characterized by:

- It is *positive definite*. This is a consequence of thermodynamics: the rate of entropy production, $\dot{\mathcal{S}}$, is related to the thermodynamic driving force, $\boldsymbol{X}$, and the thermodynamic flux, $\boldsymbol{Y}$, (referred to by De Groot and Mazur (1962) as conjugated flux and force, respectively) by $\dot{\mathcal{S}} = Y_i X_i$. Here, the driving force is proportional to the negative concentration gradient, and, therefore,

$$\dot{S} = \chi\left(-D_{ij}\frac{\partial \overline{c}}{\partial x_j}\right) \times \chi\left(-\frac{\partial \overline{c}}{\partial x_j}\right) \geq 0, \quad \text{or} \quad \chi^2 D_{ij}\frac{\partial \overline{c}}{\partial x_j}\frac{\partial \overline{c}}{\partial x_i} \geq 0, \tag{3.4.7}$$

  in which χ is a parameter that depends on the extensive quantity considered; for each such quantity, it transforms the flux and the driving force, in the form of a gradient of an appropriate scalar considered into conjugated thermodynamic flux and force (De Groot and Mazur 1962). It is not only important to point out that the dispersion tensor must be positive definite (otherwise energy might be created), but also that dispersion is a dissipative processes. When dealing with reactive transport, we shall show that this is the magnitude that drives chemical reactions.
- It is *symmetric,* i.e. $D_{ij} = D_{ji}$. This is a consequence of the conjugated force and flux relation (De Groot and Mazur 1962).

Several authors (e.g., Nikolaevski 1959; Bear 1961a; Scheidegger 1961; Bear and Bachmat 1967, 1990) have derived an expression for the components D_{ij}, which we write in terms of fluxes in the form

$$D_{ij} = a_{ijk\ell}\frac{q_k q_\ell}{q}f(\text{Pe}, r), \quad q \equiv |\mathbf{q}|, \tag{3.4.8}$$

where the coefficient $a_{ijk\ell,\alpha} = a_{ijk\ell}(\theta_\alpha)$ is called *dispersivity* , which we are generalizing here for multiphase conditions, and r represents the ratio between characteristic lengths, in the direction of the flow and normal to it, within a pore, and **Pe** is a *Peclet number*, defined by: $\text{Pe} = V\Delta_f/\mathcal{D}_{diff}$; it expresses the ratio between the rates of transport of the considered chemical species, respectively, by advection and by diffusion. The characteristic length, Δ_f, is the *hydraulic radius* of the fluid-occupied portion of the void space, serving as a characteristic length of the void space, and $\mathcal{D}_{diff}$ denotes the coefficient of molecular diffusion in the fluid phase. However, as is common in practice for large scale problems, we shall assume $f(\text{Pe}, r) = 1$, so that the coefficient of dispersion is expressed in the form

$$D_{ij,\varepsilon} = a_{ijk\ell,\varepsilon}\frac{q_{k,\varepsilon}q_{\ell,\varepsilon}}{q_\varepsilon}, \quad q_\varepsilon \equiv |\mathbf{q}_\varepsilon|. \tag{3.4.9}$$

Considering the rate of entropy production, and following the discussion leading to (4.4.15a), we obtain

- $a_{ijk\ell}$ is positive definite. This means that all principal minors of $a_{ijk\ell}$ are positive.
- $a_{ijk\ell} = a_{ij\ell k}, \quad a_{ijk\ell} = a_{jik\ell}$.

Recently, Bear and Fel (2012) suggested that (3.4.9), in which D_{ij} is proportional to q, should be replaced by a similar expression but in which the proportionality is to q^2.

The coefficient $a_{ijk\ell}$ is a fourth rank tensor called *dispersivity*. It depends only on the geometry of the considered fluid phase inside the void space, and thus it is a function of phase volumetric fraction. In a three-dimensional space, this fourth rank tensor involves 81 components. However, because of symmetry conditions, only 36 of the 81 components are independent of each other. As the porous medium has more symmetry properties, the number of independent coefficients decreases.

For an isotropic porous medium, and a single fluid that occupies the entire void space, the 36 independent components of $a_{ijk\ell}$ reduce to *two* (e.g., Fel and Bear 2010). The two coefficients are designated as α_L and α_T, and are called the *longitudinal* and *transverse dispersivities*, respectively. They allow writing the dispersion tensor as:

$$D_{ij} = a_T q\delta_{ij} + (a_L - a_T)\frac{q_i q_j}{q}, \quad q = |\mathbf{q}|. \tag{3.4.10}$$

where δ_{ij} is the *Kronecker delta*.

Bear and Fel (2012) show that for an axially symmetric porous medium, the number of dispersivity coefficients reduce from 81 to 6.

We use the term *hydrodynamic dispersion* to denote the spreading at the macroscopic level that results from *both* mechanical dispersion and molecular diffusion. Actually, the two processes are inseparable. In the absence of fluid motion, only molecular diffusion takes place. Because molecular diffusion is a relatively slow process, its overall effect on dispersion is more significant at low velocities. Actually, it is molecular diffusion that makes dispersion an irreversible process.

3.4.2 *The Mass Balance Equations of Chemical Species*

The mass balance for a considered γ-species in an α-phase, can be expressed, with minor modifications from (3.3.10). For the mass of a chemical species, dissolved in an α-phase that occupies part of the void space, i.e., when $e = c_\alpha^\gamma$, this mass balance takes the form:

$$\frac{\partial(\theta_\alpha c_\alpha^\gamma)}{\partial t} = -\nabla \cdot \theta_\alpha \left(\mathbf{J}_{\alpha,adv}^\gamma + \mathbf{J}_{\alpha,diff}^\gamma + \mathbf{J}_{\alpha,disp}^\gamma \right) + f_{\beta\to\alpha}^\gamma + \rho_\alpha \phi \Gamma_\alpha^\gamma, \quad \alpha = CO_2, br, \tag{3.4.11}$$

where $\mathbf{J}_\alpha^\gamma$ denotes the mass of γ carried by the fluid α-phase, passing through a unit area of fluid in the cross-section, and $\theta_\alpha (=\phi S_\alpha)$ is the volumetric fraction of the α-phase, $f_{\beta\to\alpha}^\gamma$ denotes the rate of transfer of γ from all non-α phases to the α-phase, e.g., by dissolution, adsorption, etc., and $\rho_\alpha \phi \Gamma_\alpha^\gamma$ denotes the source term, i.e., production of γ in α (per unit volume of porous medium) by chemical reactions, decay, etc., possibly including chemical reactions.

Making use of the expressions for the mass fluxes discussed in the previous section, neglecting diffusion, the mass balance equation for a γ-chemical species in an α-fluid phase takes the form

$$\frac{\partial}{\partial t}(\phi S_\alpha c_\alpha^\gamma) = -\nabla \cdot \left[\mathbf{q}_\alpha c_\alpha^\gamma - \mathbf{D}_\alpha \cdot \nabla c_\alpha^\gamma \right] + f_{\beta\to\alpha}^\gamma + \rho_\alpha \phi \Gamma_\alpha^\gamma. \tag{3.4.12}$$

We shall refer to this equation also as the *transport equation*. This equation can be simplified further by subtracting the flow equation multiplied by c_α^γ. In the case of single phase flow ($S_\alpha = 1$), this leads to:

$$\phi \frac{\partial c_i}{\partial t} = -\mathbf{q} \cdot \nabla c_i + \nabla \cdot (\mathbf{D} \cdot \nabla c_i) + f_{ci}, \tag{3.4.13}$$

where several symbol changes have been made for the sake of simplicity to facilitate reactive transport equations below. First, the subscript α has been removed to highlight that (3.4.13) represents transport in a single phase. Second, all source and interphase transfer terms of the ith species have been lumped in a single term f_{ci} because interphase mass transfers (e.g., mineral or CO_2 dissolution) are treated as chemical reactions.

The behavior of this equation is illustrated in Fig. 3.8. Advection causes the plume (black line in Fig. 3.8 at t = 1) to advance, thus evolving to the blue line.

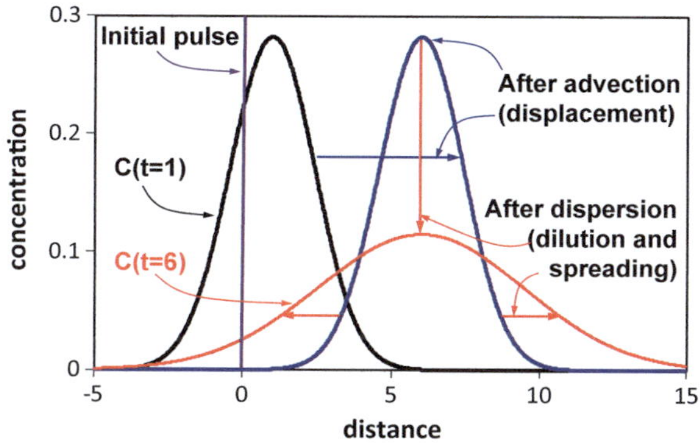

Fig. 3.8 One-dimensional conservative transport of a unit mass pulse. Conservative transport basically consists of advection, dispersion and molecular diffusion. Advection causes the displacement of the solute, as dragged by the mean fluid velocity. Dispersion and diffusion are driven by concentration gradient. They cause the plume to spread (increase the polluted volume) and, thus, dilute

Advection does not change the concentrations, but only the location of the plume. Dispersion and molecular diffusion cause the evolution of the blue to the red line, that is, spreading and dilution of the plume.

3.4.3 Coupling Chemistry to Transport: Simple Examples

The formulation and solution of a reactive transport problem is somewhat complicated from the mathematical point of view. Therefore, we start here with three simple examples that illustrate the most relevant features of reactive transport. We start by discussing linear sorption and retardation. We then discuss kinetic reactions and the Damkholer number. We conclude this section with the case of transport of two species in equilibrium with a mineral phase. These will allows us to introduce the key concept of mixing. These cases will then be generalized in Sects. 3.4.4 and 3.5.

Example 1 (Simple K_d model. The effect of sorption.) The simplest reactive transport model is the one corresponding to a solute A that sorbs onto the solid surface inside the void space:

$$A^{+} + (X-)^{-} \rightleftharpoons X - A \tag{3.4.14}$$

where A denotes the solute species and X^{-} denotes a sorption site. The sorbed solute concentration, of S_A (moles of A sorbed per unit mass of solid), equilibrates with the aqueous concentration according to the relationship:

$$S_A = K_d c_A \tag{3.4.15}$$

where K_d is the *distribution coefficient*, which encompasses the effects of non-ideal solution, the availability of sorption sites and the effect of expressing S_A per unit mass of solid. With these definitions, K_d varies with salinity and temperature. Moreover K_d may also vary with the concentrations of ions that form complexes with A.

Including sorption in the transport Eq. (3.4.13) simply implies realizing that every mol of A that sorbs onto the solid disappears from solution. Therefore, in this case,

$$f_c = -\rho_d \frac{\partial S_A}{\partial t} \tag{3.4.16}$$

where ρ_d is the bulk density, i.e., mass of solid per unit volume of formation. Substituting (3.4.16) into the solute transport Eq. (3.4.13), while using (3.4.15) for S_A leads to

$$(\phi + \rho_d K_d)\frac{\partial c_A}{\partial t} = -\mathbf{q} \cdot \nabla c_A + \nabla \cdot (\mathbf{D} \nabla c_A) \tag{3.4.17}$$

It is convenient to factorize ϕ on the left hand side and introduce $R = 1 + \rho_d K_d / \phi$ as *retardation factor*, so that the transport equation becomes

$$\phi R \frac{\partial c_A}{\partial t} = -\mathbf{q} \cdot \nabla c_A + \nabla \cdot (\mathbf{D} \nabla c_A) \tag{3.4.18}$$

Notice that this equation is identical to the one for a conservative solute (3.4.13), except that the left hand side is multiplied by R. Instantaneous sorption simply implies a change in the time scale. This means that the role of R is to slow down transport (hence its name!).

Example 2 (Equilibrium reactions. Mass Action Law. Activity and Fugacity. Mixing.) We consider a dissolution/precipitation reaction at equilibrium, where the immobile solid mineral S_{3s}, comprising the solid matrix, dissolves reversibly to yield ions B_1 and B_2 in a saturated porous medium. This system was considered by De Simoni et al. (2005) for obtaining an analytical solution to a reactive transport problem, which is similar to the one considered here. The basic reaction is:

$$S_{3s} \rightleftarrows B_1 + B_2. \tag{3.4.19}$$

Equilibrium is governed by the *Mass Action Law*, which states that, under equilibrium conditions, the products of activities raised to the power of the *stoichiometric coefficients* (i.e., the coefficients of the species in the expression of the reaction, are negative when on the left hand side). That is,

$$\frac{a_1 a_2}{a_3} = K \quad \text{or} \quad \log a_1 + \log a_2 - \log a_3 = \log K. \tag{3.4.20}$$

Let us elaborate on some of the thermodynamic concepts that have been introduced above:

The mass action law (MAL) describes the conditions for equilibrium, but it says nothing about the rate at which equilibrium will be reached. The MAL applies, often under different names, to all kinds of chemical reactions. Equation (3.4.20) is a form of MAL.

Activity: a_1, a_2, a_3 are the activities of species B_1, B_2, and S_{3s}, respectively. The activity can be viewed as an "effective" concentration. For aqueous species, the activity should, ideally, be equal to the molar fraction of the species. However, chemical data bases have been introduced for molalities, so that, ideally (for very dilute solutions), the activity is equal to the molality (i.e., moles of species per kg of water). As salinity increases, the "effective" concentration available for reaction decreases, so that the activity becomes $a_i = \gamma_i m_i$, where γ_i is the "activity coefficient", and m_i is the molality. For moderate salinities, the molality is very close to the molarity. Therefore, we shall write $a_i = \gamma_i c_i$. In general, however, one would need to write $a_i = \gamma_i c_i \rho/\omega_w$, where ρ and ω_w are fluid density and water mass fraction, respectively. Activity of minerals is much more complex to derive. For now, let us suggest that the activity of pure minerals is 1.0. The concept of activity also applies to gases, where it is called **fugacity**. The difference lies in that, instead of concentrations, chemical data bases express fugacity in terms of pressure (usually in bars), so that under ideal conditions, the fugacity of a gas is equal to its *partial pressure*. For high pressures, the behavior of gases, certainly CO_2, is not ideal and their partial pressure needs to be multiplied by a fugacity coefficient, analogous in concept to the activity coefficient.

The equilibrium constant, K, is a characteristic of every chemical reaction. It can be obtained from experiments (by means of (3.4.20)), or by minimizing the Gibbs free energy of the solution. In practice, it is obtained from chemical data bases. The selection of the database is a topic in itself. The most widely used databases for geochemical modeling are EQ3/EQ6 (Wolery 1979) and PHREEQE (Parkurst et al. 1980). But both are somewhat outdated and insufficient for CO_2 calculations. Therefore, many researchers have modified them to include recent data on CO_2. A review is provided by Krupka et al. (2010). It is important to realize that the equilibrium constant will depend (a little) on pressure and (possibly a lot) on temperature.

With these definitions, assuming that mineral S_3 is pure (i.e., $a_3 = 1$), we can re-write equilibrium as:

$$\log c_1 + \log c_2 = \log K/\gamma_1\gamma_2 = \log K^*, \tag{3.4.21}$$

where we use an apparent equilibrium constant K^* to indicate that we are using concentrations instead of activities.

The mass balance equations for the three species are

$$\phi\frac{\partial c_i}{\partial t} - \nabla\cdot(\mathbf{D}\cdot\nabla c_i) - \mathbf{q}\cdot\nabla c_i = -r, \quad i = 1, 2, \tag{3.4.22a, b}$$

$$\frac{\partial(S_3)}{\partial t} = r \tag{3.4.22c}$$

where r denotes the *reaction rate*, that is, moles of B_1 (and B_2) that precipitate in order to maintain equilibrium conditions throughout the system.

Note that the problem is now described by four equations: (3.4.21) and (3.4.22a, b and 3.4.22c) and four unknowns: r, c_1, c_2, and S_3. Moreover, these equations are intricately coupled. There is no expression for the reaction rate, r, as it results from the solution. The problem, formulated like this, is highly non-linear and coupled. Solving it requires eliminating some variables, and this is done by the use of components, and decoupling transport from chemistry. The procedure can be formalized as consisting of five steps.

Step 1: Definition of components. These are defined here as linear combinations of species that remain invariant with respect to equilibrium reactions. In our example, $c_1 - c_2$ is a component. Subtracting Eq. (3.4.22a, b) for i = 1 from the same equation for i = 2 yields

$$\phi\frac{\partial u}{\partial t} - \nabla\cdot(\mathbf{D}\cdot\nabla u) + \mathbf{q}\cdot\nabla u = 0, \tag{3.4.23}$$

where $u = c_1 - c_2$. Equation (3.4.23) implies that u is conservative. i.e., it is not affected by the reaction. Note that dissolution or precipitation of the mineral S_3 *equally* affects c_1 and c_2. Therefore, the difference $(c_1 - c_2)$ is not altered by dissolution or precipitation. We shall call u a *component*.

Step 2: Solving for components. In our example, this simply requires solving (3.4.23) for u. In general, when kinetic reactions also occur or when minerals are not present throughout the domain, this step may require iterations. For now let us assume that u can be computed.

Step 3: Speciation calculations. Once u has been obtained, the mass action law is used for obtaining the concentrations of the aqueous species. Speciation consists of solving the non-linear system of equations given by the mass action law and the definition of components. In our case:

$$c_1 - c_2 = u, \tag{3.4.24}$$

$$c_1 c_2 = K^*. \tag{3.4.25}$$

Assuming that K is independent of c_1 and c_2, the solution of (3.4.24) and (3.4.25) is

$$c_1 = \frac{u + \sqrt{u^2 + 4K}}{2}, \quad c_2 = \frac{-u + \sqrt{u^2 + 4K}}{2} \tag{3.4.26}$$

Again, this step can be quite complicated. Specific codes are designed for this step. PHREEQC (Parkhurst and Appelo, 1999) is the best known one.

Step 4: Evaluation of the reaction rate. Mixing. Substitution of the concentration of the secondary species, c_2, into its transport Eq. (3.4.22a, b) leads to

$$\frac{r}{\phi} = \frac{\partial^2 c_2}{\partial u^2}(\nabla^T u \cdot \mathbf{D} \cdot \nabla u) = f_{chem} f_{mix} \tag{3.4.27}$$

This implies that the reaction rate, r, can be written as the product of two factors: f_{chem} and f_{mix}, chemical and *mixing* factor, respectively. The *mixing factor* is defined as $f_{mix} = (\nabla^T u \cdot \mathbf{D} \cdot \nabla u)$. The *chemical factor* is defined by

$$f_{chem} = \frac{\partial^2 c_2}{\partial u^2} = \frac{2K}{\left(u^2 + 4K\right)^{3/2}} \tag{3.4.28}$$

Equation (3.4.27) deserves further discussion. First, it yields directly the rate of dissolution/precipitation reactions (recall that it is restricted to cases where chemical equilibrium holds locally). Furthermore, the reaction rate is always positive (i.e., precipitation occurs) (Rubin 1983). This is evident from Fig. 3.9 which displays another interesting feature. The equilibrium point can be obtained by drawing a line from the conservative mixing point towards the equilibrium line. The slope of this line is equal to the ratio of the stoichiometric coefficients and is equal to 1 in our example.

The rate of reaction depends on the chemistry, which controls f_{chem}, but also on transport processes that control the gradient of u. One of the most paradoxical

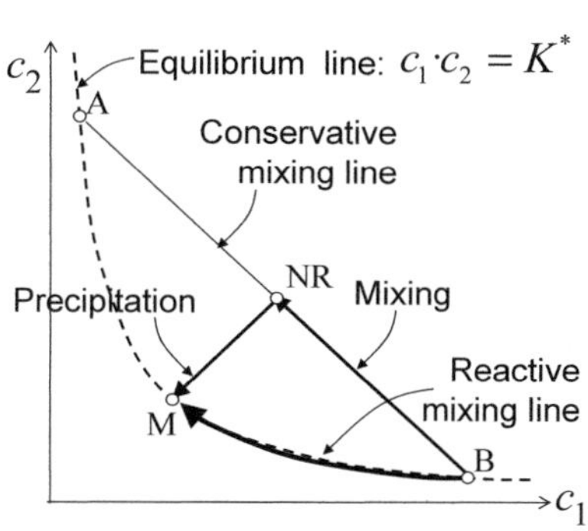

Fig. 3.9 Ideal mixing of two end members (A, B) that are in equilibrium with an ideal mineral. If mixing is non-reactive, the resulting mixture will be oversaturated (point NR). Returning to equilibrium (point M) requires precipitation. Actually, during transport, mixing is continuous (and slow) so that solution B will evolve to M along the reactive mixing line

features of (3.4.27) is that a reaction does not necessarily take place where concentrations attain their maximum values. In fact, the reaction rate equals zero when u is maximum or minimum.

Figure 3.9 emphasizes that mixing drives equilibrium reactions. If two end-member waters in equilibrium with a mineral (e.g., A and B) mix conservatively (i.e. without reactions) the resulting mixture will not be in equilibrium (e.g., point NR in Fig. 3.9). Therefore, mineral precipitation must occur to bring NR to the equilibrium line (point M). It is important to realize that the reaction would not have occurred if there had been no mixing. Thus mixing is defined, much like in the ordinary language, as the blending of resident and invading waters (in Chap. 4, we will see that the global rate of mixing is equal to the time derivative of the concentration variance, that is, mixing implies dilution of concentration extremes). This is different from dispersion that, also like in ordinary language, is defined as the spreading of the invading water (in Chap. 4, we will see that the global rate of dispersion is equal to the time derivative of the size of a plume, that is, dispersion implies spatial spreading of concentration). These two concepts (mixing and spreading) are equated in the Advection-Dispersion-Equation (3.4.13), and they are closely related (spreading favors dilution) but they do not need to be identical.

It is interesting to note that, in the absence of the first contribution on the right hand side of (3.4.27), all terms are proportional to the dispersion tensor, $\mathbf{D}$, thus strengthening the notion that mixing controls the rate of equilibrium reactions. In particular, the term $\nabla^T u \mathbf{D} \nabla u$, which is a scalar, can be used as a measure of the mixing rate. The latter is consistent with the concept of *dilution index*, as defined by *Kitanidis* [1994] on the basis of entropy arguments. This result also suggests that evaluating mixing ratios may help to properly identify not only the sources of water (Carrera et al. 2005), but also the geochemical processes occurring in the system.

Step 5: Computation of constant activity species. Ideal minerals are assumed to display a constant activity. Frequently, this is also assumed for gases or supercritical CO_2, whose activity (*fugacity*) depends on pressure or temperature, but not on the concentration of aqueous species. Therefore, they are termed *constant activity species* and their concentration cannot be derived from the mass action law, even if they are assumed to be in equilibrium with aqueous species. Instead, the concentrations of constant activity species must be derived from mass balance considerations. In our case, once r has been obtained, the concentration S_3 of the mineral is obtained by inserting the value of r in (3.4.22c).

To further illustrate these five steps, we employ them in a case where the transport equation can be solved analytically. Let us consider a three-dimensional homogeneous formation, with a constant porosity, ϕ, under uniform saturated flow conditions. The system is affected by an instantaneous point-like injection of water containing the same constituents as the initial indigenous water. Velocity is aligned with the x-coordinate, and the dispersion tensor is diagonal, with D_L and D_T, respectively, denoting the longitudinal and transverse dispersivities. The reactive transport system is governed by (3.4.21) and (3.4.22a, b and 3.4.22c).

Initially, we displace resident water in the formation by injecting a volume V_e of solution at the concentration $c_{iext} = c_{i0} + c_{i.e.}$. In order to find an analytical solution for small V_e, it is mathematically convenient to write the initial concentration condition, after equilibrium is reached at the injection point, as

$$c_i(\mathbf{x}, t = 0) = c_{i.e.} V_e \delta(\mathbf{x})/\phi + c_{i0}, \quad i = 1; 2, \tag{3.4.29}$$

where equilibrium must be satisfied at all points in the formation. This implies that

$$(c_{10} + c_{1e})(c_{20} + c_{2e}) = K, \quad \text{and} \quad c_{10}c_{20} = K. \tag{3.4.30}$$

This allows calculating $c_{\text{i.e.}}$ and u_e, which is the excess of injected component u that remains in the aquifer immediately after injection. The solution of this conservative transport problem is given by De Simoni et al. (2005)

$$u(\mathbf{x}, t) = u_0 + \frac{u_e V_e}{(2\pi)^{3/2} \sigma_L \sigma_T^2} exp\left[-\frac{1}{2} \left(\left(\frac{x - Vt}{\sigma_L} \right)^2 + \frac{y^2 + z^2}{\sigma_T^2} \right) \right]. \tag{3.4.31}$$

The expression of the local mineral mass precipitation rate, r, can be obtained by computing the gradient of (3.4.31) and substituting the result in (3.4.27) (Fig. 3.10).

The features of this solution are illustrated using the following data: we start with a resident water characterized by $c_{10} = 0.25$ (All concentrations are divided by $\sqrt{K}$ to express them in dimensionless form) and $c_{20} = 4.0$ ($u_0 = -3.75$); we then inject water from an external source, characterized by $c_{1ext} = 0.184$, $c_{2ext} = 5.434$ ($u_{ext} = -5.25$). Figure 3.10a depicts the dependence of concentrations on the normalized distance from the center of the (moving) plume, $(x - Vt)/\sqrt{2tD_L}$ (while $z = y = 0$). Concentrations are also shown for the non-reactive case. The spatial distribution of the (local) dimensionless reaction rate, $r/\phi\sqrt{K}$, is depicted in

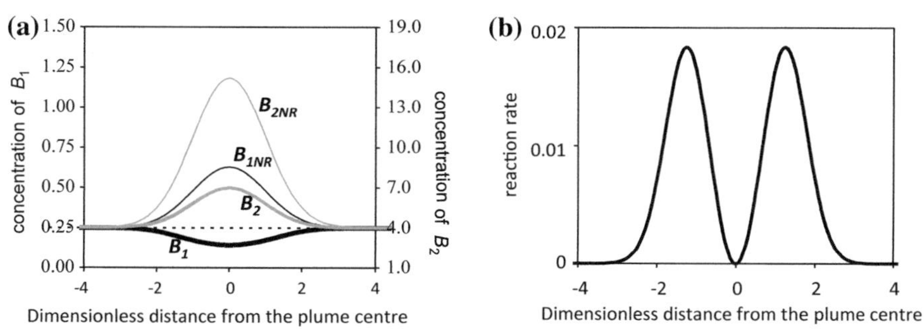

Fig. 3.10 Dimensionless concentrations (**a**) and reaction rate (**b**) in response to a pulse injection in a binary system. Precipitation causes a more significant concentrations decrease in the reactive case than in the non-reactive (NR) case. In fact c1, falls below initial and input values. Note that precipitation occurs on the plume edges, but not in the center (modified from De Simoni et al. 2005)

Fig. 3.10b. It is apparent that the system is chemically active (i.e., the reaction rate is significant) where concentration gradients are relevant, whereas no reactions occur when concentration gradients vanish. As a consequence, no reaction occurs at the (moving) center of the plume, that is, at the points of highest (or lowest) concentration values.

The sensitivity of the reaction rate to u_e for a given u_0 is presented in Fig. 3.11, with reference to a resident water with $u_0 = -20$ ($c_{10} = 0.05$; $c_{20} = 20.05$). The case with a negative $u_e(u_e/\sqrt{K} = -20)$ is considered first. From the plot it is clear that, at any given time, precipitation concentrates in a (three-dimensional) aureole around the moving center of the plume. The actual location and shape of this aureole depends on D_L, D_T and u_e. One should note that Fig. 3.11a displays an artificial symmetry, as coordinates are normalized by $\sigma = \sqrt{2Dt}$. Figure 3.11b depicts radial profiles of the reaction rate, r, for varying u_e. As expected the reaction rate is largest when u_e and u_0 have opposite signs, because this is the case with largest differences between resident and inflowing water.

Example 3 (Kinetic reactions. The Damkholer number) Contrary to equilibrium reactions, discussed above, the rate of kinetic reactions is solely controlled by the concentrations at the point where the reaction is occurring. That is, one can write the reaction rate as:

$$f_c = r(\mathbf{c}), \tag{3.4.32}$$

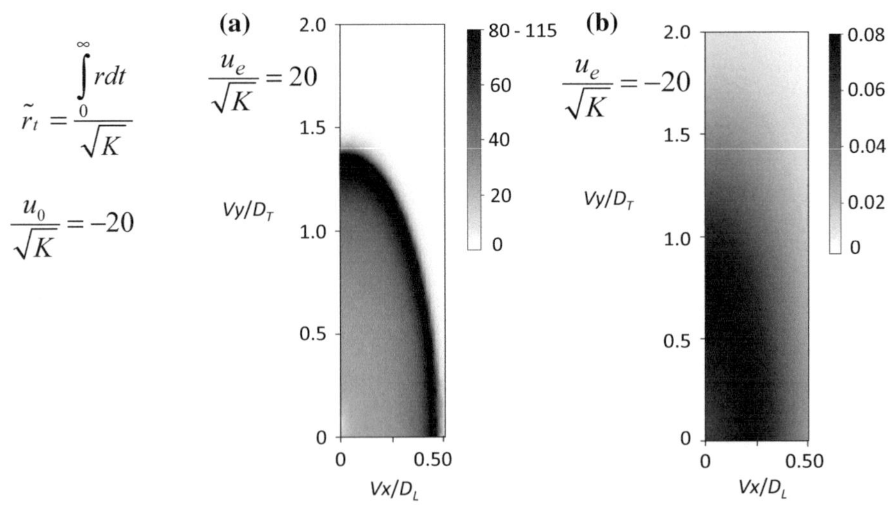

Fig. 3.11 Spatial distribution of cumulative dimensionless precipitation for two values of the modification of input with respect to resident concentrations. Note the variability in the appearance of the solution (modified from De Simoni et al. 2005)

where r is the number of moles of reactants that evolve into products per unit volume of aquifer and per unit time, and $\mathbf{c}$ is the vector of concentrations of the participating chemical species. There are numerous types of rate laws. A typical expression for mineral dissolution is (Steefel and Lasaga 1994):

$$r = \sigma\zeta e^{-\frac{E_a}{RT}} \sum_{j=1}^{N} k_j \prod_{i=1}^{N_j} a_i^{p_{ji}} \left(\Omega^{\theta} - 1\right)^{\eta_j} \tag{3.4.33}$$

where r is the mineral dissolution rate (moles of mineral per unit volume and per unit time), σ is the reactive surface area in contact with solution ($m^2\ m^{-3}$) E_a is the apparent *activation energy* of the overall reaction process; N is the number of terms (often reaction paths) in the experimental expression, and k_j an experimental constant characteristic of each mineral; $a_i^{p_{ji}}$ accounts for the catalytic effect of the N_j species (frequently H+) affecting the jth term, where the value of p_{ij} is determined experimentally; Ω is saturation (ratio of the activity product, left hand side in the MAL, to the equilibrium constant, $\Omega = a_1 a_2 / K a_3$ for the reaction of (3.4.21)); the parameters θ and η must be determined from experiments; usually, but not always, they are commonly assumed equal to 1. The term inside the parenthesis, called the "far-from-equilibrium function", reduces the reaction rate in a non-linear way, as the solution approaches equilibrium. The factor Σ takes on values of +1 or -1 depending on whether Ω is larger or smaller than 1 (precipitation or dissolution), respectively. At equilibrium, $\Omega = 1$, and, therefore, no reaction occurs. σ is, here, the reactive surface of the mineral per unit of volume of porous medium.

The above expressions can be written as a first order kinetic reaction when the concentration of the reacting species is the rate limiting factor. For the purpose of discussion, let us consider the nth order reaction whose rate is given by

$$r = kc^n. \tag{3.4.34}$$

In the absence of transport, the evolution of concentration is given by

$$\begin{aligned} c &= \left(\frac{1}{c_0^{n-1} - (n-1)kt}\right)^{1/n-1}, \quad \text{if } n \neq 1 \\ &\Rightarrow c = c_0 e^{-kt}. n = 1, \end{aligned} \tag{3.4.35}$$

The characteristic time for these reactions is

$$t_r = \frac{1}{kc_0^{n-1}}. \tag{3.4.36}$$

Therefore, in order to find out the relevance of a kinetic reaction, one should compare the reaction rate with the transport time. This comparison is performed with the Damkholer number, defined as:

$$Da = \frac{\tau}{t_r} = kc_0^{n-1}\tau, \tag{3.4.37}$$

where τ is a characteristic transport time. It should be noticed that the Damkholer number can also be defined as the reaction rate divided by a transport rate.

As characteristic transport time τ, it is common to use the advection time ($\tau = qL/\phi$, where L is the transport distance). However, when we are interested in pore scale reactions, it may be more appropriate to adopt the diffusion time ($\tau = L^2/D$). The resulting number would be the product the Da (defined with the advective time) and the Peclet number, ($Pe = qL/D\phi$). Thus, it is often termed $Da \times Pe$.

The Damkholer number is used for assessing whether a given kinetic reaction needs to be taken into account. When Da is small (much smaller than 1, say $Da < 0.1$), then one may assume that the reaction has barely modified the concentration. That is, transport is so fast that the reaction can be neglected. Reversely, if Da is large (e.g., much larger than 1, say $Da > 10$), then one may assume that the solute has been exhausted and that a new equilibrium has been reached, so that the reaction may be assumed to be an equilibrium reaction.

It is easy to confirm the above analysis for first order reactions, because the analytical solutions for solute transport used in Example 2, above, can be easily applied. Specifically, the solution for a pulse injection subject to first order kinetics is still (3.4.31), but multiplied by $\exp(-kt)$. The solution is displayed in Fig. 3.12 for several values of the Damkholer number.

The analysis is more complex, and the usefulness of the Damkholer number more limited for non-linear kinetics. The time evolution of solutes undergoing kinetic reactions of order 1, 2 and 3 are also displayed in Fig. 3.12. It is clear that Da close to 1 implies a sizeable reaction in all cases. However, it is not true that a large Da implies that the solution is close to equilibrium. Therefore, the Damkholer number is a useful indicator of the relative importance of kinetic reactions, but cannot be used blindly.

3.4.4 The Formulation of Reactions

This section contains the basic concepts needed for the mathematical formulation of chemical reactions and calculation of species concentrations within a fluid or fluids that occupy the void space of a porous medium domain. Vector and matrix notation are adopted here because it facilitates concise writing, which will be critical in subsequent sections. We hope that the section can be read and understood also by

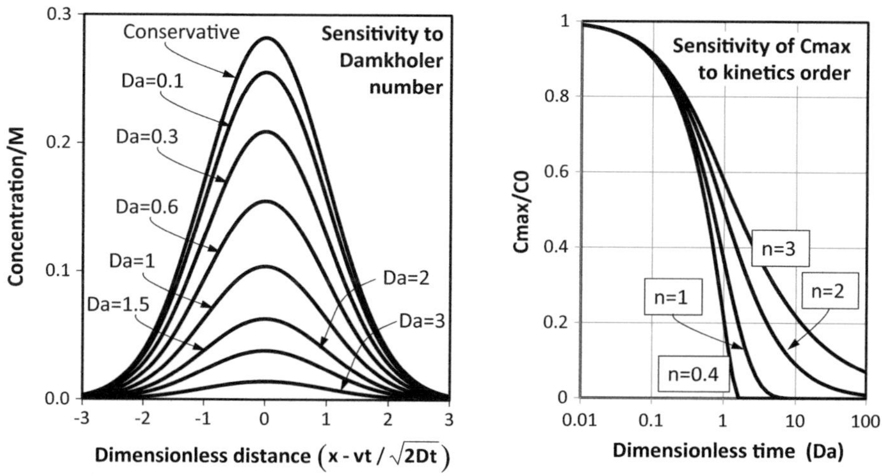

Fig. 3.12 Left, concentration versus (x − vt/s) for a pulse injection after t = D/v^2, but suffering first order kinetic decay. It is clear that kinetic reactions barely affect the solute if Da < 1, but make it disappear for Da > 3. This is not necessarily true for non-linear kinetics (*right*). Da = 1 implies a sizeable reaction, but note that significant amounts of solute remain for Da = 100 if n = 3. Note also the abrupt approach to zero for n < 1, which causes convergence problems to numerical solvers

those who are less familiar with matrix algebra. Readers may consult any textbook on matrix algebra (e.g., Meyer, 2000).

A. The stoichiometric matrix. Chemical equilibrium

A chemical reaction is represented by a "mass-balance" equation.

$$\sum_{j=1}^{j=n_s} S_{ij} q_j = 0, \quad i = 1, \ldots, n_r \tag{3.4.38}$$

where q_j is the chemical formula of the *j*th species,S_{ij} is its stoichiometric coefficient in reaction *i*, and n_s, n_r are the number of species and reactions, respectively. This equation can be written in a matrix-vector notation:

$$\mathbf{Sq} = \mathbf{0}, \tag{3.4.39}$$

where **S** is the $n_r \times n_s$ *stoichiometric matrix*, whose rows contain the stoichiometric coefficients in each reaction, and **q** a vector of size n_s, containing the chemical formula of all species. These concepts can be illustrated by means of two examples that will be used later.

Example 1 Consider the dissolution-precipitation of anhydrite assuming that other species can be ignored. Then, the chemical system consists of three species: SO_4^{2-},Ca^{2+} and $CaSO_4(s)$, and one reaction:

$$(R1)\quad CaSO_4(s) \Leftrightarrow Ca^{2+} + SO_4^{2-}.$$

The vector $\mathbf{q}$ is:

$$\mathbf{q} = \begin{pmatrix} CaSO_4(s) \\ SO_4^{2-} \\ Ca^{2+} \end{pmatrix},$$

and the stoichiometric matrix $\mathbf{S}$ is:

$$\mathbf{S} = \begin{pmatrix} & CaSO_4(s) & SO_4^{2-} & Ca^{2+} \\ R1 & -1 & 1 & 1 \end{pmatrix},$$

where it should be evident that, with these definitions, the reaction can be written either in the form $(R1)$ or in the format (3.4.39). Notice that this example is analogous to Example 2 in Sect. 3.4.3.

Example 2 This example is more relevant for carbon storage. Consider the H_2O–CO_2–Na–Cl–$CaCO_3$ system described by Duan and Li (2008), which may represent the saline water in a geological formation, in equilibrium with calcite. Injection of CO_2 will tend to dissolve calcite and increase salinity according to the following chemical reactions and interphase mass transfers:

R1: Calcite dissolution	$CaCO_3(s) + H^+ = Ca^{2+} + HCO_3^-$,
R2: CO_2 dissolution	$CO_2(aq) = CO_2(g)$,
R3: Carbonic equilibrium	$H^+ + HCO_3^- = H_2O + CO_2(aq)$
R4: Carbonate equilibrium	$HCO_3^- = H^+ +$
R5: Water dissociation	$H_2O = H^+ + OH^-$

The system consists of nine aqueous species (H_2O, Cl^-, CO_3^{2-}, H^+, OH^- HCO_3^- and $CO_{2(aq)}$), plus calcite ($CaCO_{3(s)}$), and $CO_{2(sc)}$. Therefore:

$$\mathbf{q}^t = \left(H_2O, CO_{2(sc)}, CaCO_{3(s)}, H^+, Na^+, Cl^-, Ca^{2+}, HCO_3^-, CO_2(aq), CO_3^{2-}, OH^-\right)$$

The stoichiometric coefficients of this chemical system can be described by the following stoichiometric matrix $\mathbf{S}$:

		H_2O	$CO_2(g)$	$CaCO_3(s)$	H^+	Na^+	Cl^-	Ca^{2+}	HCO_3^-	$CO_2(aq)$	CO_3^{2-}	OH^-
	R1	0	0	−1	−1	0	0	1	1	0	0	0
	R2	0	1	0	0	0	0	0	0	−1	0	0
S =	R3	1	0	0	−1	0	0	0	−1	1	0	0
	R4	0	0	0	1	0	0	0	−1	0	1	0
	R5	−1	0	0	1	0	0	0	0	0	0	1

(3.4.40)

Note that Cl^- does not participate in any reaction (i.e., it is conservative), and, therefore, the Cl^- column contains only zeros.

From these examples, it is clear that all the relevant information about the chemical system is contained in the vector of participating species, **q**, and the stoichiometric matrix, **S**, so that the set of chemical reactions is always described by (3.4.38), regardless of the actual complexity of the problem. This facilitates the encapsulation of all reactions in a simple expression. In the following sections, we shall make extensive use of these concepts.

The mass action law controlling thermodynamic equilibrium can be written as:

$$\prod_{j=1}^{j=n_s} a_j^{S_{ij}} = K_i \Rightarrow \sum_{j=1}^{j=n_s} S_{ij} \log a_j = \log K_i \quad i = 1, \ldots, n_e, \tag{3.4.41}$$

where a_j is the activity of the jth species, K_i is the equilibrium constant of the ith reaction, and n_e is the number of equilibrium reactions. Again, this equation can be written more concisely using matrix-vector notation:

$$\mathbf{S}_e \log \mathbf{a} = \log \mathrm{K}, \tag{3.4.42}$$

where subscript e indicates equilibrium. The mass action law (3.4.42) represents a set of mathematical equations that can be used for geochemical calculations. They consist of n_e equations (one for each equilibrium reaction, noting that we are now using the subscript e instead of r), and n_s unknowns (i.e., the activities of the vector **a**). Except for some rare cases, the number of reactions (n_e) is lower than the number of species (n_s). Therefore, we cannot calculate the activities of all species from just the mass action laws alone. However, if we knew the activities of ($n_s - n_e$) species (e.g., from mass balance calculations), we could calculate those of the other (n_e) species. We shall refer to primary species as the ($n_s - n_e$) known species and the (n_e) remaining species as secondary ones. Thus, by definition, the activities of the secondary species can be calculated from the activities of the primary ones through the mass action law. We can write the mass action laws in terms of primary and secondary species by splitting the matrix **S** and the vector **a** into two parts:

$$\mathbf{S}_e = (\mathbf{S}_{e1} | \mathbf{S}_{e2}), \tag{3.4.43}$$

$$\mathbf{a} = \begin{pmatrix} \mathbf{a}_1 \\ \mathbf{a}_2 \end{pmatrix}, \tag{3.4.44}$$

where $\mathbf{a}_1$ contains the activities of the ($n_s - n_e$) primary and $\mathbf{a}_2$ the activities of the (n_e) secondary species and where $\mathbf{S}_{e1}$ contains the stoichiometric coefficients for the primary and $\mathbf{S}_{e2}$ those for the secondary species. Substituting (3.4.43) and (3.4.44) into (3.4.42) gives:

$$\mathbf{S}_{e1} \log \mathbf{a}_1 + \mathbf{S}_{e2} \log \mathbf{a}_2 = \log \mathbf{K}. \tag{3.4.45}$$

Multiplying (3.4.43) by the inverse of matrix $\mathbf{S}_{e2}$, denoted by $\mathbf{S}_{e2}^{-1}$, yields:

$$\log \mathbf{a}_2 = \mathbf{S}_{e1}^{*} \log \mathbf{a}_1 + \log \mathbf{K}^{*}, \tag{3.4.46}$$

with

$$\mathbf{S}_{e1}^{*} = -(\mathbf{S}_{e2})^{-1}\mathbf{S}_{e1}, \tag{3.4.47}$$

$$\log \mathbf{K}^{*} = (\mathbf{S}_{e2})^{-1} \log \mathbf{K}. \tag{3.4.48}$$

In this way we obtain a new description of the same chemical system, which (1) is defined by equilibrium constants $\mathbf{K}^{*}$ instead of $\mathbf{K}$ and (2) involves a stoichiometric matrix for the primary species $\mathbf{S}_{e1}^{*}$ instead of $\mathbf{S}_{e1}$. In fact, we could also define a stoichiometric matrix for the secondary species ($\mathbf{S}_{e2}^{*}$ instead of $\mathbf{S}_{e2}$). However, by definition, $\mathbf{S}_{e2}^{*}$ equals the identity matrix and, hence, can be left out of Eq. (3.4.46).

To illustrate these concepts, let us consider Example 2. It is easy to verify that the inverse of $\mathbf{S}_{e2}$(last five columns in (3.4.40) is:

$$\mathbf{S}_{e2}^{-1} = \begin{pmatrix} 1 & 1 & 1 & 0 & 0 \\ 0 & -1 & -1 & 0 & 0 \\ 0 & -1 & 0 & 0 & 0 \\ 0 & -1 & -1 & 1 & 0 \\ 0 & 0 & 0 & 0 & 1 \end{pmatrix}.$$

After multiplying (3.4.40) by $-\mathbf{S}_{e2}^{-1}$, the stoichiometric matrix becomes:

$$\mathbf{S}_{e}^{*} = \left(\mathbf{S}_{e1}^{*} \mid -\mathbf{I}\right) =$$

	H_2O	$CO_2(g)$	$CaCO_3(s)$	H^+	Na^+	Cl^-	Ca^{2+}	HCO_3^-	$CO_2(aq)$	CO_3^{2-}	OH^-
$R1^*$	1	1	-1	-2	0	0	-1	0	0	0	0
$R2^*$	-1	-1	0	1	0	0	0	-1	0	0	0
$R3^*$	0	-1	0	0	0	0	0	0	-1	0	0
$R4^*$	-1	-1	0	2	0	0	0	0	0	-1	0
$R5^*$	-1	0	0	1	0	0	0	0	0	0	-1

where $R1^*$ is $R1 + R2 + R3$, but multiplied by -1, $R3^*$ is identical to the original $R2$, but multiplied by -1, etc. Whether these transformations make sense is arguable. The point is that we have made basic matrix operations, which is equivalent to arranging the equilibrium reactions in a way that ensures that the secondary species can be easily solved for in terms of primary species.

B. Components

The above equations facilitate the calculation of the activities of secondary species from the activities of the primary. However, if we do not know the latter, we have more unknowns (the concentrations of n_s species) than equations (n_e mass action law equations). To resolve this problem, we need to add mass balance equations, so that the number of unknowns will be equal to the number of equations. The problem is that reactions affect the mass of reacting species (and therefore the mass balance equations), which causes the resulting equations to be rather complicated. As discussed in Example 2 of Sect. 3.4.3, we overcome this difficulty by introducing *components*, defined here as chemical entities that are not affected by equilibrium reactions. This condition facilitates the expression of mass balances. Since equilibrium reactions establish relations among species, our definition of components is consistent with the definition in the Gibbs phase rule: the number of components is the number of chemically independent constituents of the system.

The meaning of components can be illustrated by considering the mass balances of all species undergoing chemical reactions in a closed system (that is, without mass entering or leaving the system):

$$\frac{\partial \mathbf{c}}{\partial t} = \mathbf{S}^T \mathbf{r}, \qquad (3.4.49)$$

where $\mathbf{c}$ is a vector of concentrations of all species is, $\mathbf{r}$ is a vector of the rates of all chemical reactions and $\mathbf{S}^T$ is the transpose of the stoichiometric matrix $\mathbf{S}$ (that is, columns and rows are exchanged). Equation (3.4.49) just expresses the change in concentration due to the chemical reactions. As vector notation is used, it is actually a set of equations (one equation for each chemical species). Note that the unknowns in (3.4.49) are the n_s concentrations ($\mathbf{c}$) and the n_e reaction rates ($\mathbf{r}$). Thus, in theory, we could calculate concentrations and reaction rates from the n_s mass balances (3.4.49) together with the n_e mass action laws (3.4.42) or (3.4.41), with the number of unknowns equal to the number of equations. Unfortunately, this system contains many unknowns and it is highly non-linear. Things would be much easier if reaction rates ($\mathbf{r}$) were eliminated from (3.4.49), just as we did in Example 2 of Sect. 3.4.3. This can be done by multiplying (3.4.49) by a full ranked matrix orthogonal to the matrix $\mathbf{S}^T$, that is:

$$\mathbf{U}\mathbf{S}^T = \mathbf{0}, \qquad (3.4.50)$$

where $\mathbf{U}$ is the $(n_s - n_e) \times n_s$ kernel matrix (or null space) of $\mathbf{S}^T$. As this matrix defines the components, we call it the *component matrix*. Multiplying (3.4.49) by $\mathbf{U}$ gives:

$$\frac{\partial \mathbf{Uc}}{\partial t} = \mathbf{0}. \tag{3.4.51}$$

We define a vector $\mathbf{u} = \mathbf{Uc}$, then $\partial \mathbf{u}/\partial t = 0$ and $\mathbf{u}$ is constant in time in case of a closed system. The vector $\mathbf{u}$ expresses the concentration of the components. This illustrates an important feature of the components: their concentrations do not depend on reaction rates ($\mathbf{r}$). This also gives a more physical meaning to the concept of components: Chemical reactions do not destroy or create components, but only change the concentrations of the species that define these components.

The matrix $\mathbf{U}$ is not unique. There are various methods to calculate it.

Method 1 (Gauss-Jordan elimination): The most widely used method, based on *Gauss-Jordan elimination* (Steefel and MacQuarrie 1996), consists of defining the component matrix as

$$\mathbf{U} = \left(\mathbf{I}\middle| -\mathbf{S}_1^T\left(\mathbf{S}_2^T\right)^{-1}\right) = \left(\mathbf{I}\middle|\left(\mathbf{S}_2^*\right)^T\right) \tag{3.4.52}$$

where $\mathbf{I}$ is the $(n_s - n_e) \times (n_s - n_e)$ identity matrix. It is easy to verify that this definition satisfies (3.4.50). The vector of concentrations of the components, $\mathbf{u}$, can be written as:

$$\mathbf{u} = \mathbf{Uc} = \left(\mathbf{I}\middle|\left(\mathbf{S}_1^*\right)^T\right)\begin{pmatrix}\mathbf{c}_1 \\ \mathbf{c}_2\end{pmatrix} = \mathbf{c}_1 + \left(\mathbf{S}_1^*\right)^T \mathbf{c}_2 \tag{3.4.53}$$

This equation allows us to calculate the component matrix, $\mathbf{U}$, of Example 2 presented earlier.

U=

	H_2O	$CO_2(g)$	$CaCO_3(s)$	H^+	Na^+	Cl^-	Ca^{2+}	HCO_3^-	$CO_2(aq)$	CO_3^{2-}	OH^-
H_2O	1	0	0	0	0	0	-1	1	0	1	1
TIC	0	1	0	0	0	0	-1	1	1	1	0
TCa	0	0	1	0	0	0	1	0	0	0	0
TH^+	0	0	0	1	0	0	2	-1	0	-2	-1
TNa	0	0	0	0	1	0	0	0	0	0	0
TCl	0	0	0	0	0	1	0	0	0	0	0

(3.4.54)

The diagonal matrix in the front portion of the above component matrix (i.e., the one coinciding with the primary species) ensures that these components are somehow associated with the primary species. This association is sometimes weak, but examining the components often provides some insight into the chemical system. For example, note that the second row of $\mathbf{U}$ (second component) coincides with the Total Inorganic Carbon (TIC), which is a typical measurement of natural waters. It is clear that the reactions do not affect the Total Inorganic Carbon. A decrease in HCO_3^- concentration, due to reaction $R2$, will be compensated by an increase in

CO_3^{2-}-concentration. For the particular selection of primary species in this example, the value of the fourth component (fourth row in (3.4.54)) represents the total acidity of the solution (its capacity to buffer external addition of alkalinity), and may be positive or negative. The sixth component (sixth row in (3.4.54)) coincides with the chloride concentration, which is the total chloride in the system. The meaning of the first component is more puzzling. We could imagine that, in a system initially consisting of only the primary species (i.e., zero concentrations for the secondary species), part of the water could be used up by forming, together with the other primary species, OH^- (by dissociation, $R1$), HCO_3^- (by means of $R2$), or $CO_3^=$ (by means of $R3$). This indicates that one does not need to seek any special meaning to components. Still, if primary species are chosen to be relevant elements of the system, the component concentrations will represent the total concentration of those species. That is, the sum of the concentration of the primary species plus all the complexes that can be formed with it (hence the T symbol to represent components).

Method 2 (Eliminate constant activity species): Method 1 is very convenient in the sense that it yields components that often have a well-defined meaning and, in reactive transport problems, are the same throughout the transport domain. This is very useful because the solute transport equation just expresses the mass balance of solute. Therefore, components defined by Method 1 can be transported conservatively. However, it is not convenient in cases, such as Example 1, where non-aqueous species participate in the reactions. In such cases, several options exist. The most common is to include those species in the chemical system. This is inconvenient, however, when the activity of those species is fixed (e.g., H_2O in Example 2, or a pure phase mineral in Example 1) because one is including as unknown a species that is not really unknown (its activity is fixed!). There are two options to overcome this difficulty:

(1) Reduce the set of primary species (Saaltink et al. 1998). The basic idea is to proceed as in Method 1, but then multiply the components' matrix by an *elimination matrix*
(2) Eliminate constant activity species from the outset (De Simoni et al. 2005). In essence, this method does not include constant activity species in the vector $\boldsymbol{q}$ of species, recall (3.4.38), and then proceeds as in Method 1.

We illustrate these approaches with Example 1.

Eliminating $CaSO_4(s)$ as a species yields,

$$\mathbf{q} = \begin{pmatrix} Ca^{2+} \\ SO_4^{2-} \end{pmatrix}, \quad \mathbf{S} = \begin{pmatrix} & Ca^{2+} & SO_4^{2-} \\ R1 & 1 & 1 \end{pmatrix}, \quad \mathbf{U} = \begin{pmatrix} & Ca^{2+} & SO^{2-} \\ U_{anh} & 1 & -1 \end{pmatrix}, \tag{3.4.55}$$

where we have denoted $U_{anh} = Ca^{2+} - SO_4^{2-}$. The meaning of this component is somewhat puzzling because Ca^{2+} and SO_4^{2-} appear to be two independent species. Actually, they are not, as equilibrium with anhydrite provides a strong link between them. In fact, dissolution of anhydrite will increase by the same amount the number of moles of Ca^{2+} and SO_4^{2-}. Therefore, U_{anh} will not be altered by dissolution (or precipitation) of anhydrite, which was indeed the way we selected components in the first place in Example 2 of Sect. 3.4.3.

Other methods. Molins et al. (2004) and Kräutle and Knabner (2005, 2007) describe other methods for computing the component matrix. Broadly speaking, the goal is to define components so that kinetic species do not affect all components. This facilitates reactive transport computations.

C. Speciation calculations

We have discussed extensively how to express reactions, the mass action law, and components. It should be noticed, however, that equilibrium was expressed in terms of activities (recall (3.4.41)), whereas components, and mass balances, in general, are expressed in terms of concentrations. Moreover, chemical analyses are presented in terms of concentrations. The objective of speciation calculations is to compute the concentrations of all species from information about the chemical system.

There are two types of iterative methods commonly used for solving non-linear systems: the *Picard* and *Newton-Raphson* methods. Both start from an initial guess of the solution, and method seeks to improve the previous approximate solution. The procedure is repeated until a satisfactory solution is obtained.

The **Picard method** is also called the "**sequential iteration**" or "**fixed point**" method. It requires writing the system of equations in the form $\mathbf{x} = \mathbf{g}(\mathbf{x})$, where $\mathbf{x}$ is the vector of unknowns and $\mathbf{g}$ is the algorithm function used for computing $\mathbf{x}$. The Picard method proceeds by computing $\mathbf{x}^{k+1} = g(\mathbf{x}^k)$ until a satisfactory solution is found. The method is typically used for computing $\mathbf{c}_2$, given $\mathbf{c}_1$. This calculation is required for reactive transport computations that are based on the "direct substitution approach" that will be discussed in Sect. 3.4.5. Basically, the calculation uses (3.4.46), where the activities are written in terms of concentrations as $\mathbf{a} = \gamma\mathbf{c}$:

$$\log \mathbf{c}_2 = \mathbf{S}_1^* \log \gamma_1 \mathbf{c}_1 + \log K^* - \log \gamma_2 \tag{3.4.56}$$

where γ_1 and γ_2 are vectors of activity coefficients of primary and secondary species, respectively. These are functions of all concentrations, e.g., according to the approach of Debye–Hückel or Pitzer. This equation would *be explicit* if the activity coefficients were known. As the dependence of activity coefficients on species concentrations is weak, the system is *almost explicit* and the Picard method should converge quickly. The algorithm becomes:

Step 0: Initialization. Set $\mathbf{c}_2^o$, $k = 0$

Step 1: Given $\mathbf{c}_2^k$ and $\mathbf{c}_1$, compute γ_1^k and γ_2^k

Step 2: Use γ_1^k, γ_2^k and $\mathbf{c}_1$ in (3.4.56) to compute $\mathbf{c}_2^{k+1}$

Step 3: Convergence check. If $\mathbf{c}_2^{k+1}$ is close to $\mathbf{c}_2^k$ (e.g., if $\left|c_{2i}^k - c_{2i}^{k+1}\right| < \varepsilon_i$, for all species i, where ε_i is a species dependent tolerance), then convergence has been reached and the algorithm stops. Otherwise, set $k = k+1$ and return to step 1.

This algorithm is quite straight forwards and usually converges easily. The only open issue is how to initialize the algorithm. Unit activity coefficients could be a good starting point. However, the algorithm should not be very sensitive to the initialization method and a good initialization will rarely do more than save a few iterations. The *convergence criterion*, ε_i is usually defined so as to ensure that the relative error will be small. For example ε_i may be chosen as $\varepsilon_i = \xi\left(c_{2i}^k + c_{2i}^{k+1}\right)/2$ where ξ is a relative tolerance (e.g., $\xi = 10^{-6}$).

The **Newton-Raphson** speciation method is based on linearizing the system of equations and solving the linear approximation. It is often used for computing **c** given **u**. This computation is required for reactive transport simulations based on sequential iteration methods, discussed in Sect. 3.4.5.

The system of equations to be solved includes the components' definition and mass action law, i.e.,

$$f_c(\mathbf{c}_1) = \mathbf{U}_1\mathbf{c}_1 + \mathbf{U}_2\mathbf{c}_2 - \mathbf{u} = 0. \tag{3.4.57}$$

$$\log \mathbf{c}_2 = \mathbf{S}_1^* \log \gamma_1 \mathbf{c}_1 + \log \mathbf{K}^* - \log \gamma_2. \tag{3.4.58}$$

We may see these equations as a non-linear system of n_s unknowns ($n_1 = n_s - n_e$ primary species concentrations and n_e secondary species) and n_s equations ($n_s - n_e$ component definitions and n_e mass action law equations). However, given that $\mathbf{c}_2$ is almost explicit, this system is usually solved in terms of $\mathbf{c}_1$. That is, we use Eq. (3.4.58), formally, to solve $\mathbf{c}_2$ as a function of $\mathbf{c}_1$ and then solve Eq. (3.4.57) for $\mathbf{c}_1$. Proceeding iteratively, the latter can be expanded as:

$$\mathbf{f}_C(\mathbf{c}_1^{k+1}) \approx \mathbf{f}_c(\mathbf{c}_1^k) + \frac{\partial \mathbf{f}_c}{\partial \mathbf{c}_1}\left(\mathbf{c}^{k+1} - \mathbf{c}_1^k\right) = 0. \tag{3.4.59}$$

Then, the algorithm proceeds as follows:

Step 0: Initialization. Set $k = 0$ and set initial guess for $\mathbf{c}_1^k$.

Step 1: Given $\mathbf{c}_1^k$, compute $\mathbf{c}_2^k$ by means of (3.4.58). We also need to compute $\partial\mathbf{c}_2/\partial\mathbf{c}_1$. This requires taking derivatives of (3.4.58) with respect to $\mathbf{c}_1$, which leads to the following system:

$$\left(\mathbf{I} + \frac{\partial \ln \gamma_2}{\partial \ln \mathbf{c}_2} - \mathbf{S}_{e1}^* \frac{\partial \ln \gamma_2}{\partial \ln \mathbf{c}_2}\right)\frac{\partial \ln \mathbf{c}_2}{\partial \ln \mathbf{c}_1} = \mathbf{S}_{e1}^*\left(\frac{\partial \ln \gamma_1}{\partial \ln \mathbf{c}_1} + \mathbf{I}\right) - \frac{\partial \ln \gamma_2}{\partial \ln \mathbf{c}_1}. \tag{3.4.60}$$

The equation is in logarithmic form, but can easily be transformed to non-logarithmic form through $dc_{2f}/dc_{1j} = (c_{2i}d\log c_{2i})/(c_{1j}d\log c_{1j})$.

Step 2: Given the vectors $\mathbf{c}_1^k$ and $\mathbf{c}_2^k$, compute the vector $\mathbf{f}_c^k$ by means of (3.4.57). Since $\mathbf{f}_c^k$ is the error in Eq. (3.4.57), it is often referred to as residual.

Step 3: Given $\partial\mathbf{c}_2/\partial\mathbf{c}_1$, compute $\Delta\mathbf{c}_1 = (\mathbf{c}_1^{k+1} - \mathbf{c}_1^k)$ by solving the system, equivalent to Eq. (3.4.59):

$$\left(\mathbf{U}_1 + \mathbf{U}_2\frac{\partial c_2}{\partial c_1}\right)\Delta\mathbf{c}_1 = -f_c^k. \tag{3.4.61}$$

Step 4: Compute $\mathbf{c}_1^{k+1} = \mathbf{c}_1^k + \Delta\mathbf{c}_1$. This computation may require some extra checks to ensure a robust solution. For example, if $c_{1i}^{k+1} \leq 0$, then one may choose $c_{1i}^{k+1} = \alpha_i c_{1i}^k$, where α_i is a *reduction factor* (for instance $\alpha_i = 0.1$). In fact, it is often safe to restrict the maximum change per iteration so that if $c_{1i}^{k+1} \leq \alpha_i c_{1i}^k$, then $c_{1i}^{k+1} = \alpha_i c_{1i}^k$ and if $c_{1i}^{k+1} \geq c_{1i}^k/\alpha_i$, then $c_{1i}^{k+1} = c_{1i}^k/\alpha_i$

Step 5: Convergence check. If $\mathbf{c}_1^{k+1}$ is close to $\mathbf{c}_1^k$ and/or $f_c^{k+1} \approx 0$, then stop. Otherwise, set $k = k + 1$ and return to step 1.

3.4.5 Coupled Equations

Reactive transport is the phenomenon resulting from the interaction and coupling between solute transport, outlined in Sect. 3.4.2, and chemical reactions, outlined in Sects. 3.4.3 and 3.4.4. Thus, equations for reactive transport can be viewed as the simultaneous mass balance of all chemical species while acknowledging that the mass may change due to reactions. Thus, the equations can be obtained starting from either the mass balance of reactive species in a closed system, without transport (Eq. (3.4.49) in Sect. 3.4.4), or from the transport equation of a conservative species (Eq. (3.4.13) in Sect. 3.4.2). The first approach requires adding terms for transport processes. The second one requires adding source/sink terms due to chemical reactions. Regardless of the starting point, it should be clear that reactive transport equations simply represent the mass balance of chemical species, subject to both transport processes and chemical reactions.

To write these mass balances, let us consider N_s chemical species that can react with each other through N_e equilibrium reactions and N_k kinetic reactions. The mass balances of every species i can be written as:

$$\phi \frac{\partial c_i}{\partial t} = L_i(c_i) + \sum_{j=1}^{j=N_e} S_{eij} r_{ej} + \sum_{j=1}^{j=N_k} S_{kij} r_{kj} \quad i = 1, \cdots, N_s\,, \tag{3.4.62}$$

where $L_i(c_i)$ is the linear operator for the transport terms and non-chemical sources/sinks of species i (it is zero for immobile species, such as minerals):

$$L_i(c_i) = \left\{ \begin{matrix} -\mathbf{q} \cdot \nabla c_i + \nabla \cdot (\mathbf{D} \cdot \nabla c_i) & \text{if } i \text{ is mobile,} \\ 0 & \text{if } i \text{ is immobile.} \end{matrix} \right\} \tag{3.4.63}$$

Note that, if one assumes the solid phases to be immobile, then $L_s = 0$.

It is more convenient to rewrite (3.4.62) using a matrix-vector notation, similar to Sect. 4.1:

$$\frac{\partial}{\partial t}(\mathbf{m}) = \mathbf{L} + \mathbf{S}_e^T \mathbf{r}_e + \mathbf{S}_k^T \mathbf{r}_k, \tag{3.4.64}$$

where $\mathbf{m}$ is a vector of concentrations expressed in moles per volume of porous media ($m_i = \phi c_i$). Since transport processes are phase dependent, it is convenient to subdivide vectors $\mathbf{m}$ and $\mathbf{L}$ according to the type of phase:

$$\mathbf{m} = \begin{pmatrix} \phi_l \mathbf{c}_l \\ \phi_s \mathbf{c}_s \end{pmatrix} \quad \mathbf{L} = \begin{pmatrix} L_l(\mathbf{c}_l) \\ \mathbf{0} \end{pmatrix} \tag{3.4.65}$$

where we assume that solid species are immobile (i.e., $L_s = 0$).

The kinetic reaction rates ($\mathbf{r}_k$) can be written explicitly as functions of concentration through the kinetic rate laws that are substituted into (3.4.64), equilibrium reactions cannot. Therefore, equilibrium reaction rates must be considered as unknowns, whose value can only be obtained by solving the whole problem. Therefore, the basic formulation of reactive transport contains $N_s + N_e$ unknowns. These are matched by the N_s mass balance equations of all species, expressed by (3.4.64), plus the N_e mass action laws (3.4.42), which relate reactant activities to reaction products.

The large number of coupled non-linear equations hinders direct use of the formulation of (3.4.64). The number can be reduced by making linear combinations of these equations. Examples are the formulations of Friedly and Rubin (1992), Saaltink et al. (1998), Fang et al. (2003), Molins et al. (2004), Kräutle and Knabner (2005, 2007) and De Simoni et al. (2005, 2007). All these formulations share the concept of using components to eliminate the equilibrium reaction rates by multiplying Eq. (3.4.64) by a component matrix. The concept was explained in Sect. 3.4.4 for the case without transport. For the case with transport, multiplying (3.4.64) by a components matrix $\mathbf{U}$ leads to:

$$\frac{\partial}{\partial t}(\mathbf{Um}) = \mathbf{UL} + \mathbf{US}_k^T \mathbf{r}_k. \tag{3.4.66}$$

If we wish to maintain a division between the various phases, we must split matrix $\mathbf{U}$ into submatrices for each fluid:

$$\mathbf{U} = (\mathbf{U}_l \quad \mathbf{U}_s),$$

which allows us to define the total concentration of components in every fluid:

$$\mathbf{u}_l = \mathbf{U}_l \mathbf{c}_l \qquad \mathbf{u}_s = \mathbf{U}_s \mathbf{c}_s\,,$$

so that:

$$\mathbf{Um} = \phi_l \mathbf{U}_l \mathbf{c}_l + \phi_s \mathbf{U}_s \mathbf{c}_s.$$

With these definitions we can rewrite (3.4.66) in the form:

$$\phi_l \frac{\partial \mathbf{u}_l}{\partial t} + \phi_s \frac{\partial \mathbf{u}_s}{\partial t} = L(\mathbf{u}_l) + \mathbf{U}\mathbf{S}_k^T \mathbf{r}_k. \tag{3.4.67}$$

Note that we have reduced the number of transport equations from N_s in, (3.4.62) to $(N_s - N_e)$ in (3.4.67) as matrix $\mathbf{U}$ has $(N_s - N_e)$ rows. Moreover, there are N_e mass action laws. Thus, the total number of equation is N_s which equals the number of variables (the concentrations of N_s species). As explained in Sect. 3.4.4, the mass action laws can be written in such a way that the concentrations of N_e secondary species are a function of the concentrations of $(N_s - N_e)$ primary species. These mass action laws can be substituted into (3.4.67), leading to $(N_s - N_e)$ equations and the same number of variables (the concentrations of $(N_s - N_e)$ primary species). However, this substitution has to be done with care because the *activities* of secondary species are explicit functions of the *activities* of the primary species (3.4.46), but this is not strictly true when written in terms of *concentrations* (3.4.56), because the activity coefficients (γ) depend on all concentrations, including those of primary species.

A special case of (3.4.67) is when all species of the chemical system pertain to the same fluid and all reactions are in equilibrium. In such case, (3.4.67) reduces to:

$$\phi_l \frac{\partial \mathbf{u}_l}{\partial t} = L(\mathbf{u}_l). \tag{3.4.68}$$

This equation, written in terms of concentrations of components in liquid ($\mathbf{u}_l$), has the same appearance as the transport equation for conservative solutes (3.4.13), that is, chemical reactions do not affect $\mathbf{u}_l$ and one could calculate $\mathbf{u}_l$ by means of algorithms and models for conservative transport. Moreover, transport of one component is assumed not to affect transport of the others. The concentrations of the individual species ($\mathbf{c}_l$) could be calculated from $\mathbf{u}_l$ by means of the equations explained in the previous section.

Unfortunately, most reactive transport problems in carbon storage are associated with two phase flow and heterogeneous chemical reactions. This makes the transport Eq. (3.4.67) of the components dependent on each other. Moreover, they are highly non-linear due to the combination of linear partial differential equations (i.e., transport equations) and non-linear algebraic equations (i.e., mass action laws).

To illustrate the components and component matrices with more than one fluid, the chemical system presented in Sect. 3.4.4 is extended here with gaseous and solid species and corresponding heterogeneous reactions, which are rearranged as:

$$
\begin{aligned}
&(R1) \quad H_2O = H^+ + OH^- \\
&(R2) \quad HCO_3^- = H^+ + CO_3^{2-} \\
&(R3) \quad CO_2(aq) + H_2O = H^+ + HCO_3^- \\
&(R4) \quad CO_2(g) = CO_2(aq) \\
&(R5) \quad CaCO_3(s) = Ca^{2+} + CO_3^{2-}
\end{aligned}
$$

The corresponding stoichiometric matrix is:

$$
\mathbf{S} = \begin{pmatrix}
 & H_2O & Cl^- & CO_2(aq) & H^+ & Ca^{2+} & OH^- & HCO_3^- & CO_3^{2-} & CO_2(g) & CaCO_3(s) \\
R1 & -1 & 0 & 0 & 1 & 0 & 1 & 0 & 0 & 0 & 0 \\
R2 & 0 & 0 & 0 & 1 & 0 & 0 & -1 & 1 & 0 & 0 \\
R3 & -1 & 0 & -1 & 1 & 0 & 0 & 1 & 0 & 0 & 0 \\
R4 & 0 & 0 & 1 & 0 & 0 & 0 & 0 & 0 & -1 & 0 \\
R5 & 0 & 0 & 0 & 0 & 1 & 0 & 0 & 1 & 0 & -1
\end{pmatrix}
$$

By applying Gauss-Jordan elimination (3.4.42) and using H_2O, Cl^-, $CO_2(aq)$, H^+, and Ca^{2+} sas primary species, we obtain the following component matrix:

$$
\mathbf{U} = \left(\begin{array}{c|ccccc|ccccc}
 & H_2O & Cl^- & CO_2(aq) & H^+ & Ca^{2+} & OH^- & HCO_3^- & CO_3^{2-} & CO_2(g) & CaCO_3(s) \\
\sum H_2O & 1 & 0 & 0 & 0 & 0 & 1 & 1 & 1 & 0 & 1 \\
\sum Cl & 0 & 1 & 0 & 0 & 0 & 0 & 0 & 0 & 0 & 0 \\
\sum C & 0 & 0 & 1 & 0 & 0 & 0 & 1 & 1 & 1 & 1 \\
\sum H^+ & 0 & 0 & 0 & 1 & 0 & -1 & -1 & -2 & 0 & -1 \\
\sum Ca & 0 & 0 & 0 & 0 & 1 & 0 & 0 & 0 & 0 & 1
\end{array}\right)
$$

$$
\underbrace{\qquad\qquad\qquad\qquad}_{\mathbf{U}_l} \quad \underbrace{\quad}_{\mathbf{U}_g} \quad \underbrace{\quad}_{\mathbf{U}_s}
$$

This gives the following total concentration of components in liquid, gas and solid.

$$
\mathbf{u}_l = \begin{pmatrix} c_{H_2O} + c_{OH^-} + c_{HCO_3^-} + c_{CO_3^{2-}} \\ c_{Cl} \\ c_{CO_2(aq)} + c_{HCO_3^-} + c_{CO_3^{2-}} \\ c_{H^+} - c_{OH^-} - c_{HCO_3^-} - 2c_{CO_3^{2-}} \\ c_{Ca^{2+}} \end{pmatrix} \quad
\mathbf{u}_g = \begin{pmatrix} 0 \\ 0 \\ c_{CO_2(g)} \\ 0 \\ 0 \end{pmatrix} \quad
\mathbf{u}_s = \begin{pmatrix} c_{CaCO_3(s)} \\ 0 \\ c_{CaCO_3(s)} \\ -c_{CaCO_3(s)} \\ c_{CaCO_3(s)} \end{pmatrix}
$$

In this example, water was included as a species and as a component because it is involved in several of the chemical reactions. Usually, as water is such a major constituent, the effect of reactions on the mass balance of water can be neglected. If so, it is more convenient to substitute this equation by the flow equation, that is, the mass balance equation of the aqueous phase.

Constant activity species deserve further discussion. Besides mass action laws, activities are also controlled by constraints for each phase. This constraint depends on the type of phase. A very dilute aqueous liquid can be seen as (almost) pure water and, therefore, the activity of water can be assumed unity. For moderately saline liquids, the activity of water equals its molar fraction. The activity of water in high salinity solutions needs to be calculated from the osmotic coefficient used in the model proposed by Pitzer (1973). For a gas phase, the sum of the partial pressures of all gaseous species must equal the total gas pressure. If the phase is pure (i.e., t consists only of one species), the partial pressure equals the total pressure. Major minerals are usually assumed to precipitate as a pure phase (see Example 2 in Sect. 3.4.3). Therefore, their activities can be assumed equal to unity.

If these constraints determine the activities (e.g., unit activity of water and/or minerals), the number of coupled transport equations can be reduced. Several methods have been proposed (e.g., Saaltink et al. 1998; Molins et al. 2004; De Simoni et al. 2007). The simplest and most illustrative one consists of calculating the *component matrix*, by applying Gauss-Jordan elimination as in the previous section, but using constant activity species as primary species. We divide the component matrix (3.4.52) into a part for the constant activity species (indicated by subscript 1) and a part for the rest (indicated by subscript 2):

$$\mathbf{U} = \left(\mathbf{I}\middle|(\mathbf{S}_1^*)^T\right) = \begin{pmatrix}\mathbf{U}_1\\ \mathbf{U}_2\end{pmatrix} = \begin{pmatrix}\mathbf{I} & \mathbf{0} & \left(\mathbf{S}_{1.1}^*\right)^T\\ \mathbf{0} & \mathbf{I} & \left(\mathbf{S}_{1.2}^*\right)^T\end{pmatrix}. \tag{3.4.69}$$

Likewise, we divide the concentration vectors:

$$\mathbf{m} = \begin{pmatrix}\mathbf{m}_1\\ \mathbf{m}_2\end{pmatrix} = \begin{pmatrix}\mathbf{m}_{1.1}\\ \mathbf{m}_{1.2}\\ \mathbf{m}_2\end{pmatrix} \quad \mathbf{c} = \begin{pmatrix}\mathbf{c}_1\\ \mathbf{c}_2\end{pmatrix} = \begin{pmatrix}\mathbf{c}_{1.1}\\ \mathbf{c}_{1.2}\\ \mathbf{c}_2\end{pmatrix}. \tag{3.4.70}$$

With these definitions we can rewrite Eq. (3.4.67) as:

$$\frac{\partial}{\partial t}\left(\mathbf{m}_{1.1} + \left(\mathbf{S}_{1.1}^*\right)^T\mathbf{m}_2\right) = \mathbf{L}\left(\mathbf{c}_{1.1} + \left(\mathbf{S}_{1.1}^*\right)^T\mathbf{c}_2\right) + \mathbf{U}\mathbf{S}_k^T\mathbf{r}_k, \tag{3.4.71}$$

$$\frac{\partial}{\partial t}\left(\mathbf{m}_{1.2} + \left(\mathbf{S}_{1.1}^*\right)^T\mathbf{m}_2\right) = \mathbf{L}\left(\mathbf{c}_{1.2} + \left(\mathbf{S}_{1.2}^*\right)^T\mathbf{c}_2\right) + \mathbf{U}\mathbf{S}_k^T\mathbf{r}_k. \tag{3.4.72}$$

Assuming that rate laws do not depend on constant activity species, the transport equations of components having constant activity species as primary species, (3.4.71) do not affect the other transport equations (3.4.72). Since the solution of (3.4.72) is independent of $\mathbf{c}_{1.1}$, the solution of (3.4.71), results from two properties. First, being primary, these species do not form part of the other components. Second, they do not affect any secondary species through the mass action law, because their activities are fixed, independent of their concentration. Therefore, we can first solve Eq. (3.4.72), in which the mass action laws have been substituted. Then, we can solve (3.4.71).

Let us illustrate the above by means of the above CO_2 example, where we have assumed that H_2O and $CaCO_3(s)$ have fixed activities equal to unity. For primary species, we choose these two fixed activity species, plus Cl^-, $CO_2(aq)$ and H^+. As explained in Sect. 3.4.4, we can write the mass action law in a form such that the activities of the secondary species are an explicit function of primary ones:

$$\begin{pmatrix} \log a_{Ca^{2+}} \\ \log a_{OH^-} \\ \log a_{HCO_3^-} \\ \log a_{CO_3^{2-}} \\ \log a_{CO_2(g)} \end{pmatrix} = \begin{pmatrix} -1 & 1 & 0 & -1 & 2 \\ 1 & 0 & 0 & 0 & 1 \\ 1 & 0 & 0 & 1 & -1 \\ 1 & 0 & 0 & 1 & -2 \\ 0 & 0 & 0 & 1 & 0 \end{pmatrix} \begin{pmatrix} \log a_{H_2O} \\ \log a_{CaCO_3(s)} \\ \log a_{Cl^-} \\ \log a_{CO_2(aq)} \\ \log a_{H^+} \end{pmatrix} + \log \mathbf{K}^*,$$

where $\mathbf{K}^*$ is the vector of equilibrium constants.

As the activities of H_2O and $CaCO_3(s)$ equal unity, we can also write the above, removing these species:

$$\begin{pmatrix} \log a_{Ca^{2+}} \\ \log a_{OH^-} \\ \log a_{HCO_3^-} \\ \log a_{CO_3^{2-}} \\ \log a_{CO_2(g)} \end{pmatrix} = \begin{pmatrix} 0 & -1 & 2 \\ 0 & 0 & 1 \\ 0 & 1 & -1 \\ 0 & 1 & -2 \\ 0 & 1 & 0 \end{pmatrix} \begin{pmatrix} \log a_{Cl^-} \\ \log a_{CO_2(aq)} \\ \log a_{H^+} \end{pmatrix} + \log \mathbf{K}^*.$$

This results in the following component matrix:

$$\mathbf{U} = \left(\begin{array}{l|cc|ccc|ccccc} & H_2O & CaCO_3(s) & Cl^- & CO_2(aq) & H^+ & Ca^{2+} & OH^- & HCO_3^- & CO_3^{2-} & CO_2(g) \\ \sum H_2O & 1 & 0 & 0 & 0 & 0 & -1 & 1 & 1 & 1 & 0 \\ \sum CaCO_3 & 0 & 1 & 0 & 0 & 0 & 1 & 0 & 0 & 0 & 0 \\ \hline \sum Cl & 0 & 0 & 1 & 0 & 0 & 0 & 0 & 0 & 0 & 0 \\ \sum C-Ca & 0 & 0 & 0 & 1 & 0 & -1 & 0 & 1 & 1 & 1 \\ \sum H^+ & 0 & 0 & 0 & 0 & 1 & 2 & 1 & -1 & -2 & 0 \end{array} \right)$$

Due to the structure of the component matrix (the left-hand part is the identity matrix), the last three components (labeled $\sum Cl$, $\sum C - Ca$, and $\sum H^+$) do not contain the constant activity species H_2O and $CaCO_3(s)$. This, together with the fact that the mass action laws can be written independent of the constant activity species as well, makes the transport equations of these last three components independent of the first two. Therefore, we can first calculate the concentrations of Cl^-, $CO_2(aq)$, H^+, and of all secondary species, from the transport equations of the last three components and the mass action laws. Then, we can calculate the concentrations of H_2O and $CaCO_3(s)$ from the transport equations of the first two components.

A situation may occur where some phases are present in only part of the domain. This is especially relevant for CO_2 storage. It is also important for minerals, which may be present only in some small fraction of the domain. They may disappear as a consequence of reactions. This means that the component matrix and/or the set of primary and secondary species may depend on space and time. This represents an added complexity, but not a conceptual difficulty.

3.5 The Energy Transport Model

Here, we are considering (1) two-phase flow, (2) reactive transport, and (3) deformations in a deep formation, which is at a temperature that follows from the earth geothermal gradient. The formation temperature as function of depth can be expressed as $T_{wb}(z) = T_{wh0} + \varpi z$, in which ϖ denotes the geothermal gradient, and T_{wh0} is the temperature at the wellhead (e.g., T_{wh0} = 2.5 K, $\varpi = -3\,\text{K}/100$ m).

Into this formation, we inject supercritical CO_2 which has its own temperature at ground surface (injection well-head). However, along its downward flow through the injection well, the temperature of the injected fluid may vary, as it may gain heat from the surrounding soil, or lose heat to the surrounding spoil, as the temperature there may be different than that of the fluid in the pipe. Altogether, the temperature of the CO_2 reaching the target formation may be different than that in the latter, so that we must consider non-isothermal flow conditions.

Furthermore, some of the chemical reactions may be exogenic or endogenic, thus affecting the temperature of the fluids and the solid matrix. Altogether, the above conditions calls for modeling under non-isothrermal conditions, i.e., with temperature as an additional variable.

In this chapter, we present the energy and heat sub-models, assuming that the two fluids are in thermodynamic equilibrium at every point in the solid matrix. This means that the temperature, T, is the same for both fluid phases and for the solid, so that only a single energy balance equation is required. This simplification may not be valid in fractured rock models (i.e., where fractures are not modeled explicitly), when solid matrix temperature is different from that of fracture fluids. This situation is not considered here, but can be addressed by non-local models of the kind discussed in Chap. 5 (see Sect. 5.3.2.1).

3.5.1 The Energy Balance Equation

The energy balance equation for the three-phase system considered here—the two fluids and the solid matrix is:

$$\frac{\partial}{\partial t}\left(\sum_{\alpha=brine,\mathrm{CO_2}} \phi S_\alpha \rho_\alpha h_\alpha + (1-\phi)\rho_s h_s\right) = -\frac{\partial J_i^H}{\partial x_i} + \sum_\alpha \rho_\alpha S_\alpha \phi \Gamma_\alpha^{\gamma\mathcal{E}}. \qquad (3.5.1)$$

in which $h_\alpha(= u_\alpha + p_\alpha v_\alpha)$ denotes the enthalpy of the α-phase, u_α denotes specific internal energy, h_s denotes the enthalpy of the solid phase, J_i^H denotes the total energy flux (per unit area of porous medium) in all phases.

In the 2nd term on the r.h.s. of (3.5.1), we include heat of chemical (exogenic/endogenic) reactions.

3.5.2 Energy Fluxes

At the microscopic level, the total energy flux is

$$j^{\mathcal{E}}_{total,i} = \rho\left(u + \frac{1}{2}V^2 + \varphi_{pot}\right)V_i - \lambda\frac{\partial T}{\partial x_i} - \sigma_{ij}V_j, \tag{3.5.2}$$

where, φ_{pot} denotes specific potential energy, σ_{ij} denotes the stress, and the last term denotes the diffusive flux of kinetic energy, leaving only the advective flux of internal energy, and the diffusive flux of internal energy (=conduction)

However, kinetic and potential energy are often neglected. Then, as a good approximation, we include in this total flux only the advective flux of internal energy, the diffusive (=conductive) flux energy and the dispersive flux of energy, i.e.,

$$J^H_i = -\lambda^{*H}_{ij}\frac{\partial T}{\partial x_j} + \phi\sum_{(\alpha)} S_\alpha h_\alpha \rho_\alpha V_{\alpha i}. \tag{3.5.3}$$

The first term on the r.h.s. of (3.5.3) denotes the sum of two fluxes: by conduction with $\lambda^*_{ij} = \lambda^*_{ij}(\phi, S_{w,}\lambda_w, \lambda_{nw}, \lambda_s)$, and by dispersion. The second term on the r.h.s. of (3.5.3) denotes advective flux of internal energy (expressed by the enthalpy) in the two fluid phases.

The flux expressed by (3.5.3) should be inserted in the energy mass balance Eq. (3.5.1).

3.5.3 Initial and Boundary Conditions

The boundaries of the domain of interest in the target formation were discussed in Sect. 3.3 in connection with the flow of the two fluid phases.

Here, we have one variable (i.e., one in addition to those of the flow sub-model) —the temperature, for which we have to solve *one* PDE, namely, the energy balance equation (3.5.1).

A. Initial conditions

These include the statement of the temperature at some initial time at all points within the considered domain.

B. Boundary conditions

The top and bottom of the target formation may permit heat conduction through them. One option is to consider the two impervious layers that bound the formation from above and below and assume that heat may be conducted through them from the formation to the surfaces that bound these layers, assuming that fixed (in time) temperatures are maintained on them. These temperatures may, for example correspond to those determined by the geothermal gradient.

Laterally, similar to the case of pressure discussed in Sect. 3.3, at some sufficiently large distance, we may assume that the initial temperature remains unchanged. The distance from the injection well to such boundary may be increased as the modeling period increases.

At the screened portion of the injection well, we assume that we know the rate at which CO_2 is injected through the screen. This fluid is injected at a known temperature and pressure, so that we know its enthalpy. The boundary condition expresses the continuity of energy flux across the screen (normal to the screen),

$$q_{CO_2,r}\rho_{CO_2,r=0}h_{CO_2,r=0} = -\lambda^{*H}\frac{\partial T}{\partial r} + \phi\sum_{(\alpha)} S_{CO_2}h_{CO_2}\rho_{CO_2}V_{CO_2,r} \tag{3.5.4}$$

At a point on the unscreened portion of the casing of the injection well, we have to equalize the energy flux by conduction form the fluid in the well to that through the casing to the formation. Since the casing is impervious to flow, the energy (here heat) flux is by conduction only. The condition can be expressed in the form

$$\overline{\lambda}_{pipe}\frac{T_{inside} - T}{d} = -\lambda\frac{\partial T}{\partial r}, \tag{3.5.5}$$

in which $\overline{\lambda}_{pipe}$ denotes some equivalent conductivity of the CO_2 in the pipe and the pipe/cement, d is some thickness, and λ denotes the equivalent thermal conductivity in the formation (i.e., depending on the conductivities of the two fluids and the saturations).

3.6 The Solid Matrix Deformation Model

The objective of this section is to present and discus the model that describes deformation and possibly failure in the geological formation as a consequence of CO_2 injection.

Injection of any fluid into a geologic formation causes an increase in fluid pressure in the vicinity of the injection well. The rise in pressure then propagates in the formation. Since the solid matrix comprising the formation is *deformable*, it deforms under the increased *stress* produced by the rise in fluid pressure. This is true whether the injected fluid is the same as the native fluid in the formation or not. The subject is relevant to the injection of any fluid, and especially for the injection of CO_2, because:

- The deformation of the solid matrix affects the storage of fluid(s) in the void space. The rise in pressure is associated with the increase in the amount of fluid or fluids in the void space.

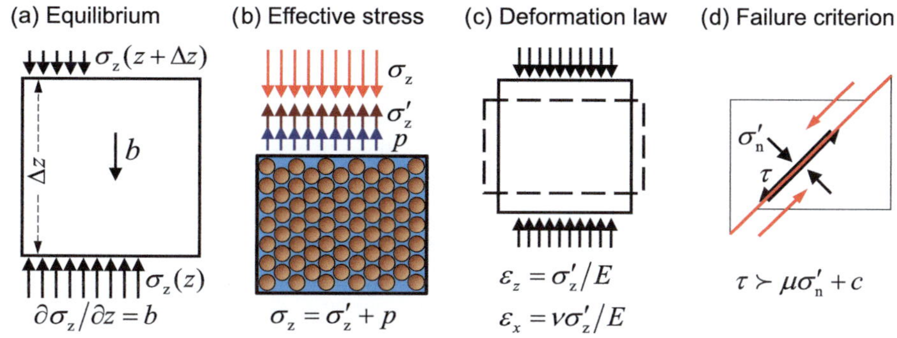

Fig. 3.13 The four basic building blocks of hydromechanical modeling

- The rise in stress and the associated deformation of the solid matrix may produce fractures in the solid matrix of both the storage formation and the caprock. This may compromise the storage.
- The deformation may manifest itself as land subsidence or upheaval of ground surface.

Mechanical deformation processes are based on four fundamental concepts (Fig. 3.13):

(1) The principle of *momentum conservation*. In practice, inertial effects are neglected, so that momentum conservation is expressed as *equilibrium of forces* (the sum of all forces in any direction must equal zero).
(2) Stresses are associated with deformations. The simplest law that expresses their relationship is *Hooke's law* that describes elastic behavior. However, other stress-strain relationships are possible, depending on the nature of the considered formation. Deformations must be geometrically *compatible* for the solid to remain continuous.
(3) *Effective stress*, expressed by Terzaghi's or Biot's laws, which controls deformation and failure.
(4) A *failure law* is required to express how the solid breaks when effective stresses exceed the rock's strength.

An important feature of the solid deformation model, compared to those based on mass or energy conservation, is that state variables–stress and strains–are not scalars, or vectors, but tensors. Therefore, let us start by introducing them in detail.

3.6.1 *Stress, Strain and Effective Stress*

We shall start by presenting a brief summary of the concept of stress in a porous medium. We then apply this concept to the case of GCS project in a geological formation.

Stress is the force exerted per unit surface area within a medium. As such, it is similar to pressure. However, we shall use pressure only within a fluid. Furthermore, fluid pressure, which expresses force perpendicular to the surface per unit area of the latter, is isotropic; it is independent of the orientation of the surface over which it is applied. Solids resist shear, so that stresses need not be isotropic. In general, they will not. In fact, they do not need to be perpendicular to the surface they are applied to. For convenience, the stress applied on a surface is decomposed into component perpendicular to the surface (normal stress) and parallel to it (shear stress).

Therefore, to define a stress state in a solid, we need to specify the normal and shear stress in each plane perpendicular to the coordinate axes (Fig. 3.14a). These define the stress tensor:

$$\sigma = \begin{bmatrix} \mathbf{T}_1 \\ \mathbf{T}_2 \\ \mathbf{T}_3 \end{bmatrix} = \begin{bmatrix} \sigma_x & \tau_{xy} & \tau_{xz} \\ \tau_{xy} & \sigma_y & \tau_{yz} \\ \tau_{zx} & \tau_{zy} & \sigma_z \end{bmatrix} = \begin{bmatrix} \sigma_{xx} & \sigma_{xy} & \sigma_{xz} \\ \sigma_{xy} & \sigma_{yy} & \sigma_{yz} \\ \sigma_{zx} & \sigma_{zy} & \sigma_{zz} \end{bmatrix} = \begin{bmatrix} \sigma_{11} & \sigma_{12} & \tau_{13} \\ \sigma_{21} & \sigma_{22} & \tau_{23} \\ \sigma_{31} & \sigma_{32} & \sigma_{33} \end{bmatrix} \tag{3.6.1}$$

Here, $\mathbf{T}_i$ is the vector representing the stress on the plane perpendicular to the *ith* coordinate axis (i = 1 for x, 2 for y and 3 for z). Therefore, we can write $\mathbf{T}_i = [\sigma_{i1} \quad \sigma_{i2} \quad \sigma_{i3}]$. Note the use of bold face symbols for vectors and tensors, and ordinary symbols with subscript for their (scalar) components. Here, we adopt the geomechanics sign convention, where compressive stresses are positive and shear stresses are positive when pointing in the same direction as a coordinate axis (in Fig. 3.14a, τ_{xy} is positive because it points in the direction of the +y axis). With this sign convention, the stress tensor is symmetric because $\tau_{xy} = \tau_{yx}$ (otherwise, the solid square in Fig. 3.14a would tend to rotate).

The stress tensor can be represented in 2-D (also in 3D, but it is less convenient) using the Mohr's Circle (Fig. 3.14b). We describe it here to gain some insight into the stress tensor and to operate with it. In order to plot a 2-D Mohr's Circle: (1) draw axes σ (horizontal) $-$ τ (vertical); (2) plot the points A (σ_x, τ_{xy}) and B (σ_{yy}, τ_{yx}) (blue points in Fig. 3.14b); (3) the intersection with the σ axis is the center of the circle (O) and the points A and B are on the circumference, so that OA is the radius.

The Mohr's circle can be used to conduct operations with the stress tensor. In particular, it can be used to obtain the normal and shear stresses (Fig. 3.14c) on any plane by imposing equilibrium with σ_x, σ_y and τ_{xy}. Such stresses can also be obtained by rotating point A twice the angle that the plane π forms with the plane represented by A. The new point (red point in Fig. 3.14b) represents the normal and shear stresses acting on π.

The above points out that the components of the stress tensor depend on the orientation of the coordinate axes. Combinations of stresses that do not depend on the coordinate system are called *invariants*. Two important invariants are the *mean*

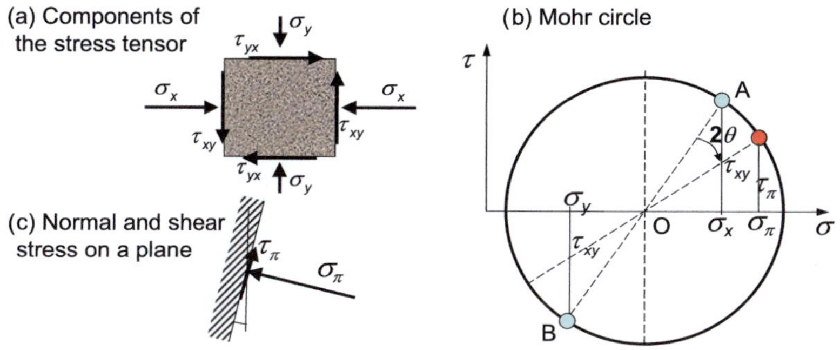

Fig. 3.14 The stress tensor: **a** components; **b** Mohr circle; **c** stresses acting on a plane other than the Cartesian coordinate axes

stress and the *deviatoric stress* (Fig. 3.15). The mean stress controls volume change. It is defined by:

$$\sigma_m = \frac{1}{3}\left(\sigma_x + \sigma_y + \sigma_z\right) = \frac{1}{3}tr(\boldsymbol{\sigma}) \tag{3.6.2}$$

where *tr*(.) denotes *trace* (sum of the diagonal terms of the tensor) of (.). The deviatoric stress tensor controls failure and plastic deformation and it is the portion of the stress tensor complementary to the mean stress, that is:

$$\boldsymbol{\sigma}_d = \boldsymbol{\sigma} - \sigma_m \mathbf{I} \tag{3.6.3a}$$

where **I** is the *identity matrix*. The equivalent deviatoric stress, σ_d, is defined from the second invariant of $\boldsymbol{\sigma}_d$ as:

$$J_2 = \frac{1}{2}tr\left(\boldsymbol{\sigma}_d^{\mathrm{t}} : \boldsymbol{\sigma}_d\right) = \frac{1}{3}\sigma_d^2 \tag{3.6.3b}$$

where superscript t denotes transpose. Despite its fancy definition, it can be checked that in 2D, the deviatoric equivalent stress is simply the radius of Mohr's circle (maximum shear stress) multiplied by $\sqrt{3}$ (Fig. 3.15).

An orientation of special interest is the one of *principal stresses*, which corresponds to the planes on which the shear stresses are zero. Obviously, they correspond to the points where Mohr's circle intersects the horizontal axis (see Fig. 3.15).

We have described stresses without reference to who transmits them. In reality, the total stress is transmitted by both the solid matrix and the fluid. This is best illustrated by Terzaghi's edometric experiment (Fig. 3.16). When a load is applied to a permeable medium, the load is initially resisted by the fluid and the solid

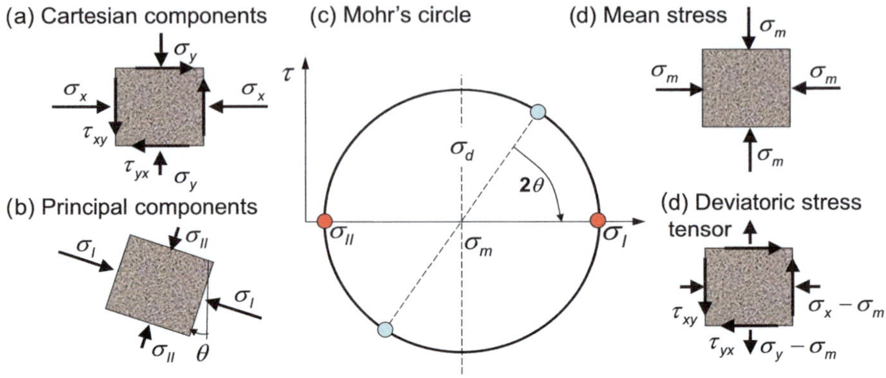

Fig. 3.15 Principal, mean and deviatoric stresses

matrix, proportional to their compressibility. If the fluid is water, its compressibility is low, so most of the load is taken by water, whose pressure increases instantaneously (Fig. 3.16b) this overpressure tends to dissipate, which causes the solid to take an increasing portion of the load and the deform (Fig. 3.16c, d).

In the above description, we have introduced the effective stress, which is the deformation producing stress. This experiment indicates that we have to distinguish between two stresses at a point in a porous medium domain:

- the *total stress* due to the load applied on the considered saturated porous medium domain, and
- the *effective stress*, which represent the stress effectively transmitted by the solid matrix. The latter is the one that controls deformation and failure.

They are related through *Terzaghi's* (1923) *law*, written here using Biot's (1941) notation:

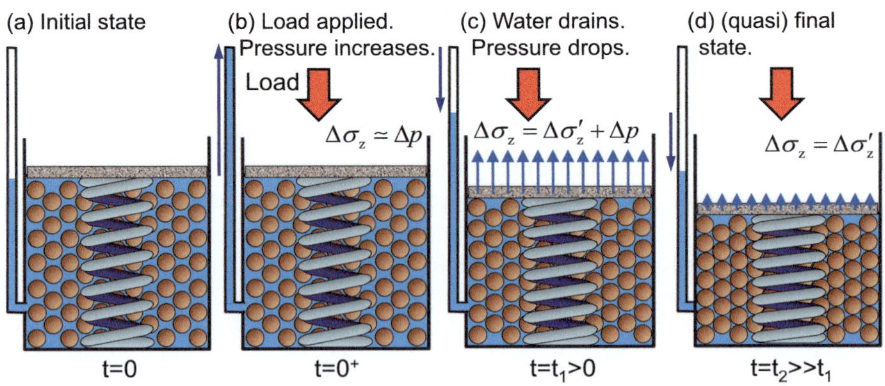

Fig. 3.16 Schematic description of the edometric test, illustrating effective and total stresses on a (soft) elastic porous medium

$$\boldsymbol{\sigma} = \boldsymbol{\sigma}' + \alpha p \mathbf{I} \tag{3.6.4}$$

where p is fluid pressure (or some kind of average pressure in the case of two fluids that together occupy the void space), and a factor α that has been the subject of much discussion. In fact, Terzaghi (1923) did not include it in his law (i.e. $\alpha = 1$) while Biot (1956) assumed it to be equal to the porosity, which would be inconsistent with the results of edometric experiments (Fig. 3.16). The usual assumption is to take $\alpha = 1 - K_{dr}/K_s$ (e.g., Jha and Juanes, 2014; Kim et al. 2013). Here, K_{dr} is the drained bulk modulus (inverse of compressibility) of the porous medium and K_s is that of the solid grains. For simplicity, we shall assume $\alpha = 1$.

When two fluids occupy the void space simultaneously, the total stress (tensor), $\boldsymbol{\sigma}$, at a point within a porous medium domain is the sum of the stresses in the two or three phases occupying the REV centered at that point. We then have

$$\boldsymbol{\sigma} = \boldsymbol{\sigma}_s + \boldsymbol{\sigma}_w + \boldsymbol{\sigma}_{nw} \tag{3.6.5}$$

where all terms express phase *averages*.

Deformation is the change in length of a body per unit length. If both the change and the original length are taken along the same direction, then we refer to as *normal strain*. If not, we denote it *shear strain*. These concepts are illustrated in Fig. 3.17. Like stress, strain is a second rank symmetric tensor, $\boldsymbol{\varepsilon}$. It is related to the *displacement vector*,$\mathbf{u}$, by

$$\boldsymbol{\varepsilon} = \frac{1}{2}\left[\nabla\mathbf{u} + (\nabla\mathbf{u})^T\right] \tag{3.6.6}$$

The components of $\mathbf{u}$ are the displacements in the three coordinate axes, $(..)^T$ stands for transpose of (..) and the tensor $\nabla\mathbf{u}$ is the gradient of $\mathbf{u}$.

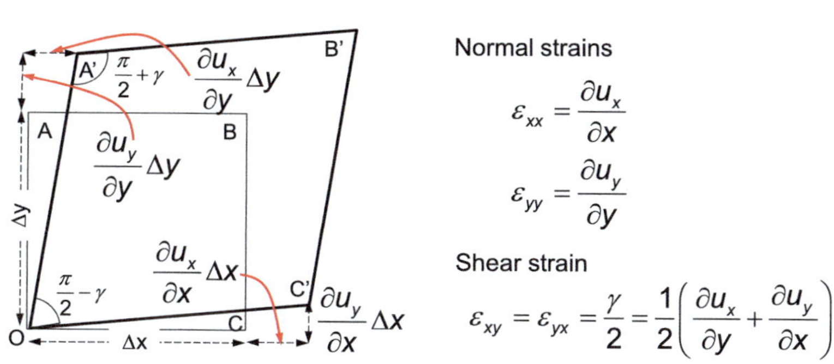

Fig. 3.17 Components of the strain tensor when the square OABC deforms to OA′B′C′. Normal strains represent the change in length of the solid per unit length. Shear strains represent the change in the angle of two lines initially perpendicular (actually, half the angle)

Change of volume can be characterized from the *strain tensor*. In fact, normal strain can be seen as the change in volume (per unit volume) along each axis. Shear strains do not produce changes in volume. Therefore,

$$\frac{\Delta V}{V} = \varepsilon_v = 3\varepsilon_m = \varepsilon_{xx} + \varepsilon_{yy} + \varepsilon_{zz} \tag{3.6.7}$$

where we have implicitly defined the *volumetric strain*,ε_v, and the *mean strain*, ε_m. These are specifically important for coupling between flow and deformation because most volume change is associated to reduction of porosity (i.e.,$\Delta V/V = \varepsilon_v \simeq \Delta\phi$). Thus, volumetric deformation implies a change in the amount of fluid stored. The remaining portion of the strain tensor is the *displacement vector*, similar to what we defined for the stress tensor, that is:

$$\boldsymbol{\varepsilon}_d = \boldsymbol{\varepsilon} - \varepsilon_m \mathbf{I} \tag{3.6.8}$$

Strains are produced by changes in stresses, which is the subject of the next section.

3.6.2 The Deformation Model

An essential element of the hydromechanical model is the fact that the solid deforms in response to changes in (effective) stresses. There are several models that relate strains and stresses, depending on whether:

(1) Deformations are recovered after stresses cease, i.e., elastic or plastic behavior.
(2) Deformations occur slowly in response to stress changes, i.e., viscous behavior.
(3) Deformations are a linear or non-linear function of stress.
(4) Failure occurs suddenly (fragile), or slowly (ductile).

In parallel, numerous other criteria can be used to define the (hydro) mechanical behavior of materials: soft, stiff, hard, resilient, etc. Combinations of these criteria are also used (viscoelastic, etc.). Some of these behaviors are described by the strain-stress curve shown in Fig. 3.18. The need for such a large number of adjectives points out the broad range of mechanical behaviors. Here, we describe in some detail the simplest such model, namely the *isotropic Hooke's linear elastic model.*

According to Hooke's model, strain and stress are proportional. In the case of 1D stress (zero stresses in the y and z directions):

$$\varepsilon_x = \sigma'_x / E \tag{3.6.9}$$

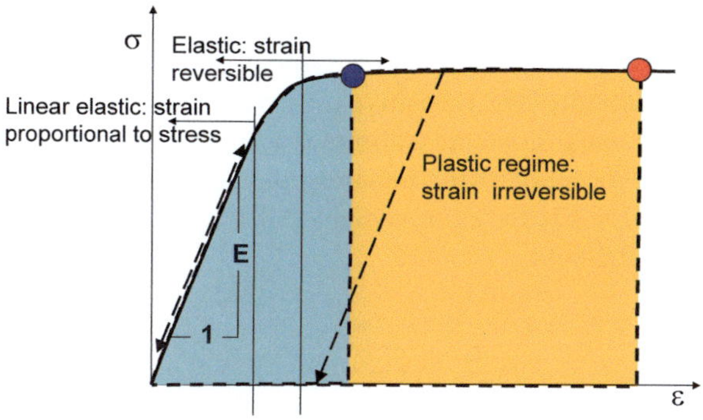

Fig. 3.18 Stress-strain relationship. Most of the deformation behavior of any solid is captured by the stress-strain curve. Most geomaterials behave elastically for small deformations and linearly for very small deformations. Beyond the elastic limit, part of the deformation is not recoverable (plastic behavior). The area under the curve is the energy (per unit volume) absorbed by the solid. If deformation at failure is small (*blue dot*), this energy is small and the solid is brittle (*fragile*). If this failure strain is large (*red dot*), the solid is ductile and can absorb a large energy prior to failure

where E is the *Young's modulus* and we have used effective stress, σ'_x (i.e., the one that causes deformation). As shown in Fig. 3.13a, 1D shortening in the x direction will produce an expansion in the y and z directions of magnitude:

$$\varepsilon_{yy} = \varepsilon_{zz} = -\nu\varepsilon_{xx} \tag{3.6.10}$$

where ν is the *Poisson ratio*. Shear strains are produced by shear stresses:

$$\varepsilon_{xy} = \frac{1}{2G}\tau_{xy} \tag{3.6.11}$$

where $G = E/[2(1+\nu)]$ is the *shear modulus*. Notice that the superscript (′) is not needed here because the effective shear stress is identical to the total shear stress. Superimposing the deformations caused by all components of the stress tensor, leads to

$$\varepsilon_x = \frac{1}{E}\left[\sigma'_x - \nu\left(\sigma'_y + \sigma'_z\right)\right] \tag{3.6.12}$$

With similar expressions for ε_y and ε_z. Adding these three expressions and recalling the definitions of mean stress (3.6.2) and volumetric strain (3.6.7), we can write:

$$\varepsilon_v = \frac{3(1-2\nu)}{E}\sigma'_m = \frac{\sigma'_m}{K} \tag{3.6.13}$$

where we have implicitly defined the bulk modulus $K = E/3(1-2\nu)$. It is worth noting that we have introduced three moduli (E, K and G) that relate stresses to strains. By definition, they have units of pressure (they are of the order of a few 10's of Gpa in rocks at depths around 1 km, typical of CO_2 storage). We have also introduced the Poisson ratio (v), a positive dimensionless parameter less than 0.5, at which the solid becomes incompressible (see (3.6.13)). Any two of these parameters suffice to define all of them for isotropic media. The choice is based on convenience. Actually, it is most convenient to write the equilibrium conditions by using the Lamé coefficient:

$$\lambda = \frac{Ev}{(1+v)(1-2v)} \tag{3.6.14}$$

A second Lamé coefficient, often denoted μ, is identical to the shear modulus, G.

Using the Lamé coefficient, Hooke's law can be written to express the stress tensor as a function of the strain tensor:

$$\boldsymbol{\sigma}' = K\varepsilon_v\mathbf{I} + 2G\left(\boldsymbol{\varepsilon} - \frac{\varepsilon_v}{3}\mathbf{I}\right) = 3K\varepsilon_m\mathbf{I} + 2G\boldsymbol{\varepsilon}_v = 2G\boldsymbol{\varepsilon} + \lambda\varepsilon_v\mathbf{I} \tag{3.6.15}$$

or, reversely, to write the strain tensor as a function stresses

$$\boldsymbol{\varepsilon} = \frac{1+v}{E}\boldsymbol{\sigma}' - \frac{3v}{E}\sigma_m\mathbf{I} = \frac{\sigma_m'}{3K}\mathbf{I} + \frac{1}{2G}\left(\boldsymbol{\sigma}' - \sigma_m'\mathbf{I}\right) \tag{3.6.16}$$

Other generalizations of these expressions are possible. The last one facilitates acknowledging that changes in temperature may also produce deformation. In this case, the simplest approach is to add thermal strains to (3.6.16), which leads to (Biot 1956):

$$\boldsymbol{\varepsilon} = \frac{1+v}{E}\boldsymbol{\sigma} - \frac{3v}{E}\sigma_m\mathbf{I} - \frac{1-2v}{E}p\mathbf{I} - \alpha_T\Delta T\mathbf{I} \tag{3.6.17}$$

where α_T is the linear thermal expansion coefficient and ΔT is the change in temperature. Notice that we have written (3.6.17) in terms of total stresses to emphasize that temperature is similar to pressure in producing strains (both add an isotropic strain). The main difference is that pressure driven strains are affected by elastic parameters, while thermal strains are not. This means that, for a given temperature change, thermal stresses are going to be proportional to the stiffness of the soil. This is best expressed by solving (3.6.17) for the effective stress tensor, which yields:

$$\boldsymbol{\sigma}' = 2G\boldsymbol{\varepsilon} + \lambda\varepsilon_v\mathbf{I} - (2G + 3\lambda)\alpha_T\Delta T\mathbf{I} \tag{3.6.18}$$

It is easy to check that $K = 2G + 3\lambda$, which makes thermal stresses proportional to the solid matrix compressibility, to the thermal expansion coefficient and to the temperature change.

A second relevant generalization of (3.6.15) and (3.6.16) is the case where the solid is not isotropic. The generalized Hooke's law reads:

$$\boldsymbol{\sigma}' = \mathbf{C} \circ \boldsymbol{\varepsilon} \tag{3.6.19}$$

where **C** is the *stiffness tensor* and the symbol ∘ denotes double inner product. In indicial notation:

$$\sigma_{ij} = \sum_k \sum_l c_{ijkl} \varepsilon_{kl} \tag{3.6.20}$$

Note that the coefficient **C** is a fourth order tensor. In three dimensions, it consists of 81 components. Because of symmetry, $c_{ijkl} = c_{ijlk} = c_{jikl} = c_{klij}$, so that only 21 of them are independent. This number is still pretty high, which makes (3.6.19) difficult to use. Therefore, the most frequent option is to work with the isotropic model discussed earlier or, at most, axially symmetric

3.6.3 *Failure*

Prior to analyzing more complex behaviors, it is necessary to recall failure conditions. *Failure* will occur along a given plane when σ and τ along that plain meet a failure criterion. The most typical is Coulomb's criterion (Fig. 3.19), according to which failure occurs if

$$\sigma_\pi > \mu \sigma'_\pi + c, \tag{3.6.21}$$

where μ is the *friction coefficient* and c is *cohesion*. Actually, the condition (3.6.21) can be checked graphically using Mohr's circle (Fig. 3.19), which also yields the orientations of the two conjugate planes on which sliding will take place.

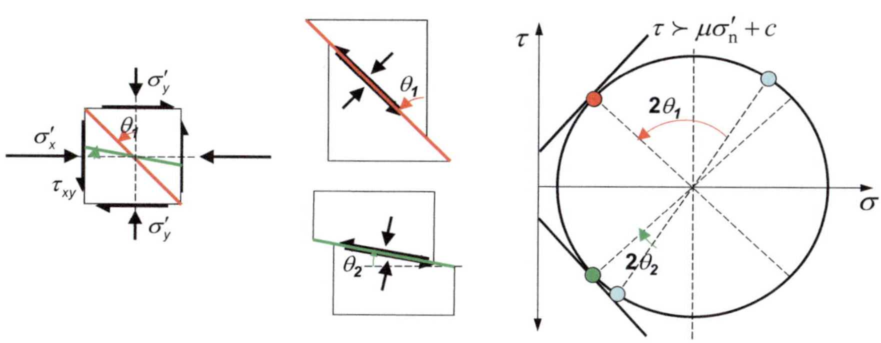

Fig. 3.19 Mohr–Coulomb failure criterion. Mohr's circle can be used to define the two orientations of the planes where the failure criterion is met (i.e. when the circle becomes tangent to $\tau = \mu\sigma'_n + c$, *green* and *red lines*)

Failure implies that, momentarily, shear resistance is deactivated, so that shear stress increases at the points adjacent to the failure one, which may also fail. This occurs when the material is fragile and produces a (micro) seism. If the material is ductile, it will behave plastically (recall Fig. 3.18).

Fluid injection may produce *seismicity* because it causes fluid pressures to increase. This is demonstrated as a shift of the Mohr circle to the left, closer to the failure criterion. Shear failure, leading to slip along the planes of a fracture, occurs when the Mohr circle becomes tangent to the failure envelope (Fig. 3.19). This may occur in a favorably oriented cohesionless pre-existing fracture or in intact rock if the *deviatoric stress* is sufficient to overcome cohesion. A second kind of failure occurs when the least principal stress equals the rock tensile strength, σ'_t, which creates a hydrofracture perpendicular to the *smallest principal stress*. If the least principal stress is horizontal, hydrofractures will be vertical, and vice versa (Klee et al. 2011). Generally, shear failure in pre-existing fractures occurs before failure of intact rock, even when they are not favorably oriented (Rutledge and Phillips 2003).

The mode of faulting is a consequence of the initial stress tensor. Depending on the relative magnitude of the vertical stress with respect to the two horizontal principal stresses, three cases can be distinguished: normal, strike slip and reverse faulting.

- Normal faulting occurs when the vertical stress is the maximum principal stress.
- Strike slip faulting occurs when the vertical stress is the intermediate principal
- Reverse faulting occurs when the vertical stress is the minimum principal stress. The latter may take place in compressional regimes where lateral deformation is constrained in the direction perpendicular to compression.

A compilation of the present-day stress field was carried out by the World Stress Map Project (Zoback 1992). However, local variations must be expected in response to local stiffness heterogeneity. Therefore, an assessment of the local initial stress tensor must be made in every case.

3.6.4 Equilibrium

As mentioned at the beginning of Sect. 3.6, if inertial forces are neglected, momentum conservation becomes equilibrium of forces acting on (any portion of) a body. Equilibrium expresses that the forces acting on that body must be balanced by the stresses within the body, which is usually written as:

$$\nabla \cdot \boldsymbol{\sigma} = \mathbf{b} \tag{3.6.22}$$

where **b** is the vector of body forces. This expression assumes that tensile stresses are positive. Adopting the sign convention of geomechanics (positive stresses are compressive), the sign of **b** must be changed. Typically, body forces are restricted to gravity forces:

$$b_x = b_y = 0,\ b_z = -g[\rho_s(1-\phi) + \rho_w \phi S_w + \rho_{nw} \phi S_{nw})] \tag{3.6.23}$$

The coupled hydromechanical equation can be obtained from (3.6.22), substituting $\boldsymbol{\sigma}$ by means of Terzaghi's law (3.6.4) and then using Hooke's law, as written in (3.6.18), to express σ' as a function of deformations. These are substituted by the compatibility constraints (3.6.6), which allows one to write strains as a function of displacements and leads to:

$$G\nabla^2\mathbf{u} + (G+\lambda)\nabla(\nabla\cdot\mathbf{u}) + \mathbf{b} = \nabla p + (2G+3\lambda)\alpha_T\nabla T \tag{3.6.24}$$

Several issues deserve discussion in this equation. First, notice that we have written it using Lamé parameters. We could do it using any other pair of elastic parameters, but it would have been somewhat more cumbersome. Second, Eq. (3.6.24) still represents an equilibrium equation. It is, still, a vector equation, that is, three equations in 3-D, each representing equilibrium of forces (per unit volume) in each direction. Third, pressure and temperature gradients (not p and T, their gradients!) can be viewed as body forces. Fourth, in the case of multiphase flow, which pressure should be used, that of CO_2 or that of water? In this context, it is important to recall that that p in (3.6.24) comes from Terzahi's Law. In the multiphase flow case, the total stress tensor at a point within a porous medium domain is the sum of the stresses in the three phases occupying the REV centered at that point, i.e.,

$$\boldsymbol{\sigma} = \boldsymbol{\sigma}' + ap_w\mathbf{I} + (1-a)p_n\mathbf{I} \tag{3.6.25}$$

where numerous expressions are available for a (e.g., Bishop and Blight 1963; but see also Gray and Schrefler 2007). In fact, (3.6.24) remains valid if we redefine $p = ap_w + (1-a)p_n$.

In examining (3.6.24), one may conclude that coupling between flow, energy transport and deformation is one-way. That is, it looks that it is necessary to solve, first for pressures using the (multiphase) flow equation, second for T, and finally, use these to solve for **u**. In reality, it is not quite like that. The three equations are tightly coupled. Pressure and temperature variations induce deformation of the porous media, changing its porosity. An expression for porosity variation can be obtained from the mass balance of solid present in the medium

$$\frac{\partial}{\partial t}(\rho_s(1-\phi)) + \nabla\cdot\mathbf{j}_s = 0 \tag{3.6.26}$$

where $\mathbf{j}_s$ is the flux of solid, which can be expressed as the velocity of the solid multiplied by the volumetric fraction occupied by the solid phase and its density

$$\mathbf{j}_s = \rho_s(1-\phi)\frac{d\mathbf{u}}{dt} \tag{3.6.27}$$

where **u** is the displacement vector. Combining Eqs. (3.6.26) and (3.6.27) using the material derivative for the solid, we obtain an expression for porosity variation is obtained

$$\frac{D_s \phi}{Dt} = \frac{(1-\phi)}{\rho_s} \frac{D_s \rho_s}{Dt} + (1-\phi) \nabla \cdot \frac{d\mathbf{u}}{dt} \tag{3.6.28}$$

This equation conveys that porosity varies due to both volumetric deformation (second term r.h.s.) and solid density variation (first term). Accounting for solid density variations due to temperature changes may be of particular importance in formations with a very low porosity, where the relative error of neglecting the first term can be significant. A typical dependence of solid density as a function of temperature and pressure is (Olivella et al. 1994):

$$\rho_s = \rho_{s_0} \exp\left(-\alpha_s (T - T_0) + \frac{1}{K_s} (p - p_0) \right) \tag{3.6.29}$$

where ρ_{s_0} is the solid density corresponding to the reference temperature and pressure, T_0 and p_0, respectively, α_s is the thermal expansion coefficient of the grains and $1/K_s$ is the grains compressibility, which is usually neglected because it is much smaller than the bulk compressibility of the porous medium. In summary, the point is that porosity variations in the flow problem are governed by mechanics.

In order to accommodate (3.6.28) and to incorporate more complex deformation behaviors, it may be convenient to formulate the mechanical problem using time derivatives, which can be obtained by taking derivatives with respect to time in (3.6.24).

$$G\nabla^2 \frac{\partial \mathbf{u}}{\partial t} + (G + \lambda) \nabla \left(\frac{\partial \varepsilon_v}{\partial t} \right) + \mathbf{b} - \nabla \left(\frac{\partial p}{\partial t} \right) + (2G + 3\lambda) \alpha_T \nabla \frac{\partial T}{\partial t} = \mathbf{0} \tag{3.6.30}$$

This equation needs to be solved together with the mass balance equation for the fluids involved (i.e., water and CO_2). If flow takes place under non-isothermal conditions, then we add T as a primary variable and we add the energy balance equation.

3.6.5 *Initial and Boundary Conditions*

Like any flow and transport model, the set of equations describing deformation have to be solved subject to specified boundary and initial conditions. Here we shall mention only the boundary conditions that relate to the stress-strain model, as those related to the flow have been mentioned earlier.

A. Initial conditions

In order to solve the mechanical problem, we need information on the initial total stress, or the initial effective stress field in the considered formation. Because one of the principal stresses is the vertical one, the total stress field can be defined by the three principal components of the total stress: the vertical total stress and the two horizontal total stresses, which are orthogonal to each other.

B. Boundary conditions

On the ceiling (aquifer-caprock contact) of the formation, we assume that there is no jump in the component of the total stress normal to the boundary. It is frequently assumed that the material on top of this boundary has low shear strength (otherwise, the geologic layers above this boundary need to be modeled in detail), the total stress on this boundary is the stress produced by the weight of the rock and fluids above this surface (=the vertical *lithostatic stress*). By making this assumption, we allow this surface to be displaced. Also, the effective stress along this surface, as everywhere, can be calculated by (3.6.8). In the cases in which the caprock integrity is of interest, the caprock has to be modeled. Then, the mechanical boundary condition of vertical lithostatic stress is applied on the top of the caprock, instead of on the top of the aquifer. This condition involves the matching of normal stress between the two layers.

The target formation considered is often assumed to be bounded from below by a stationary impervious surface. On this boundary, it is typical to assume no vertical displacement. A condition of zero displacement normal to the boundary can be assigned to lateral boundaries. The modeler needs to assess carefully the nature of these boundary conditions. In general, it is assumed that outer and bottom boundary conditions are not important in extensive aquifers because the boundary is far enough so that the solution is practically not affected by it. However, we need to bear in mind that (a) failure open occurs in the formation beneath the aquifer, which may be critically stressed (see discussion by Vilarrasa and Carrera 2015), and (b) contrary to the flow equations, the equilibrium equation is an elliptic equation, so that actions at one point are transmitted instantaneously to the rest of the domain, so that the "mechanical" boundary, may be much further away than the hydraulic boundary.

3.7 Concluding Remarks

Altogether, four models have been presented in this chapter:

- The flow model
- The solute transport model
- The energy transport model
- The deformation model.

We may envision them as partial models, or submodels, of a single comprehensive model. Is this approach justified? Are there conditions under which it will be more efficient to investigate some of these submodels separately? Because of the complexity of the models, it may certainly be of interest to investigate, at least at the preliminary stages of project planning, the use of one or two of these submodels prior to undertaking the solution of the comprehensive model. For example, as fluid pressure varies within the target formation, deformation of the solid matrix will occur. The main objective of the stress-strain analysis is to investigate induced seismicity or possible damage to the ceiling of the formation, enhancing leakage of CO_2 to upper layers, to a fresh water aquifer, or to ground surface. One may ask, however, to what extent do the chemical reactions affect stress changes within the formation? Perhaps a sufficiently good approximation is to conduct the stress-strain analysis under the assumption that chemical reactions may be neglected. Thus coupling two-phase flow with deformation, possibly under non-isothermal conditions may be justified.

Another question is the effect of temperature. Do we need to take temperature changes, which certainly occur, into account when considering chemical reactions?

Some of these options are summarized in Table 3.3.

Table 3.3 Possible model couplings addressed in this chapter

	H—two phase flow	T—non-isothermal	C—reactive transport	M-mechanical	
H	+	–	–	–	Flow, isothermal
HT	+	+	–	–	Flow, non-isothermal
HTC	+	+	+	–	Flow, reactive, non-isotherm
HTCM	+	+	+	+	Comprehensive
HC	+	–	+	–	Flow, reactive, isothermal
HM	+	–	–	+	Flow, mechanical, isothermal

Appendix: Primary Variables and Degrees of Freedom

Extension of Gibbs Phase Rule

As shown throughout this chapter, a rather large number of variables are required in order to describe the complete behavior of a problem of transport that involves one or two fluid phases, a large number of chemical species, non-isothermal conditions, and a deformable solid matrix.

However, if we assume that *locally*, i.e., at every (macroscopic) point in the considered porous medium domain, all phases and chemical species are in *thermodynamic equilibrium*, or when the rate of transformation of the system from one state to another is sufficiently slow so that it can be assumed to be continuously close to equilibrium, this number can be significantly reduced, thus simplifying the task of solving the mathematical model.

The number of *degrees of freedom* is the smallest number of independent variables needed to fully define a system's present and future behavior. We refer to these variables as *primary variables*. The values of the primary variables are obtained by solving *partial differential balance equation*. Values of all other system's state variables can be obtained from the primary ones through the use of constitutive relationships and definitions.

Gibbs phase rule states that (at the microscopic level of description) the state of a system composed of NP phases and NC non-reacting components, under conditions of equilibrium, is fully defined by NF state variables, with NF, which is the number of *degrees of freedom* of the problem, determined by the relationship

$$\mathrm{NF} = \mathrm{NC} - \mathrm{NP} + 2 \tag{3.8.1}$$

Bear and Nitao (1995), on the basis of balance considerations and thermodynamic relationships, showed that under (exact of approximate) conditions of thermodynamic equilibrium among all phases and components present in a deformable porous medium domain under non-isothermal conditions, the number of degrees of freedom, NF, in a problem of non-isothermal mass transport, involving NP fluid phases and NC chemical species, is given by the relationship

$$\mathrm{NF} = \mathrm{NC} + \mathrm{NP} + 4 \tag{3.8.2}$$

Under conditions of *non-equilibrium* between the phases, this rule becomes

$$\mathrm{NF} = \mathrm{NC} \times \mathrm{NP} + 2 \times \mathrm{NP} + \mathrm{NC} + 4 \tag{3.8.3}$$

In both cases, when Darcy's law is used to determine the velocities of the fluid phases, NF is reduced by NP. When the solid matrix is non-deformable, NF is reduced by 3, leading to the relationship

$$NF = NC + 1 \tag{3.8.4}$$

Under isothermal conditions, the rule reduces to

$$NF = NC \tag{3.8.5}$$

These rules are, thus, extensions of the *Gibbs phase rule* mentioned in Sect. 3.2 to phenomena of transport in porous media. The number of degrees of freedom for reactive transport problems was also discussed by Saaltink et al. (1998), and by Molins et al. (2004).

For a system with chemical reactions, let NS be the number of reacting species and NR_{eq} denote the number of equilibrium reactions. Then, by expressing the reactions in the form a canonical set of equations, and using of the *law of mass action*, there are $NC = NS - NR_{eq}$ independent species concentrations. Thus, the number of degree of freedom, NF, in the above equations for a non-reacting system, still applies to a reacting system, as long as we use

$$NC = NS - NR_{eq} \tag{3.8.6}$$

Once the number of degrees of freedom has been determined for a given problem, we select the most convenient variables to be declared as primary ones, and identify the (same number of) balance (partial differential) equations which have to be solved in order to determine the values of these variables. All other variables are, subsequently, determined by using the remaining equations—constitutive relations and definitions.

Degrees of Freedom for Phases in Motion

Essentially, in a porous medium, a system that undergoes changes in time due to motion of the phases can never be strictly in complete/exact thermodynamic equilibrium (Bear and Nitao 1995). Conditions of mechanical nonequilibrium prevail as a consequence of the transfer of momentum from the moving fluid to the solid by viscous forces. This gives rise to pressure gradients at the microscopic level within the REV. Temperature gradients may also occur because of viscous dissipation. If these pressure and temperature gradients are large, the system will be far from chemical and thermal equilibrium. In a multiphase REV, flow can cause some phases to be under nonequilibrium conditions, and some of the phases may be in nonequilibrium with each other. Table 3.4 summarized degrees of freedom for the non-equilibrium case.

As seen in the above table, the number of degrees of freedom depends on the type of problem that is being modeled and on the simplifications involved. e.g., whether or not the solid is assumed to be deformable, in an isothermal case and when Darcy's law is employed (Table 3.4).

Table 3.4 Degrees of freedom for non-equilibrium case

Deformable solid	Using Darcy's law	Isothermal	NF
No	No	No	$NC \times NP + NC + 2NP + 1$
No	No	Yes	$NC \times NP + NC + NP$
No	Yes	No	$NC \times NP + NC + NP + 1$
No	Yes	Yes	$NC \times NP + NC$
Yes	No	No	$NC \times NP + NC + 2NP + 4$
Yes	No	Yes	$NC \times NP + NC + NP + 3$
Yes	Yes	No	$NC \times NP + NC + NP + 4$
Yes	Yes	Yes	$NC \times NP + NC$ +3

When the phases are in thermal equilibrium with each other, the value of NF is reduced in all of the above cases by the amount NP. When the components absorbed on a solid are in chemical equilibrium with the fluid phases, the value of NF is reduced in all of the above cases by the amount of NC.

Degrees of Freedom Under Approximate Chemical and Thermal Equilibrium

Under the assumption that at every point within the domain and at every instant of time, the system is in *approximate thermodynamic equilibrium* between the averaged (over the REV) behavior of the phases and the components at that point, the number of degrees of freedom is as follow:

For a reactive system, the number of components, NC, is determined by reducing the number of the reactive species (NS) by the number of equilibrium equations (NE)—NC = NS − NE (Table 3.5).

Table 3.5 Degrees of freedom for equilibrium case

Deformable solid	Using Darcy's law	Isothermal	NF
No	No	No	$NC + NP + 1$
No	No	Yes	$NC + NP$
No	Yes	No	$NC + 1$
No	Yes	Yes	NC
Yes	No	No	$NC + NP + 4$
Yes	No	Yes	$NC + NP + 3$
Yes	Yes	No	$NC + 4$
Yes	Yes	Yes	$NC + 3$

References

Adams JJ, Bachu S (2002) Equation of state for basin geofluids: algorithm review and intercomparison for brines. Geofluids 2:257–271

Altunin VV (1975) Thermophysical properties of carbon dioxide. Publishing House of Standards (in Russian), Moscow

Altunin VV, Sakhabetdinov MA (1972) Vicosity of liquid and gaseous carbon dioxide at temperatures 220–1300 K and pressures up to 1200 bar. Teploenergetika 8:85–89

Bachu S, Bennion DB (2008) Interfacial tension between CO_2, freshwater, and brine in the range of pressure from (2 to 27) MPa, temperature from (20 to 125)°C, and water salinity from (0 to 334 000) mgL^{-1}. J Chem Eng Data 54(3):765–775

Batzle M, Wang Z (1992) Seismic properties of pore fluids. Geophysics 57:1396–1408

Bear J (1961) On the tensor form of dispersion. J Geophys Res 66(4):1185–1197

Bear J (1972) Dynamics of fluids in porous media. Elsevier, New York

Bear J, Bachmat Y (1967) A generalized theory on hydrodynamic dispersion. In: Proceedings of I. A.S.H. symposium on artificial recharge and management of aquifers, IASH Publ. No. 72, Mar 1967, pp 7–16

Bear J, Bachmat Y (1990) Introduction to modeling of transport phenomena in porous media. Kluwer Academic Publications, Dordrecht

Bear J, Fel L (2012) A phenomenological approach to modeling transport in porous media. Transp Porous Media 92(3):649–665

Bear J, Nitao JJ (1995) On equilibrium and primary variables in transport in porous media. Transp Porous Media 18(2):151–184

Bielinski A (2006) Numerical simulation of CO_2 sequestration in geological formations. Dissertation, Institut Fur Wasserbau, Univesitat Stuttgart

Biot MA (1941) General theory of three-dimensional consolidation. J Appl Phys 12(2):155–164

Biot MA (1956) Thermoelasticity and irreversible thermodynamics. J Appl Phys 27(3):240–253

Bishop AW, Blight GE (1963) Some aspects of effective stress in saturated and partly saturated soils. Geotechnique 13:177–197

Brooks RH, Corey AT (1964) Hydraulic properties of porous media. Colorado State University, Fort Collins

Carrera J, Alcolea A, Medina A, Hidalgo J, Slooten LJ (2005) Inverse problem in hydrogeology. Hydrogeol J 13(1):206–222

ChemicaLogic (1999) ChemicaLogic Corporation www.chemicalogic.com. http://www.chemicalogic.com/Pages/DownloadMollierCharts.aspx

Childs EC, Collis-George N (1950) The permeability of porous materials. Proc R Soc A 201:392–405

Chun BS, Wilkinson GT (1995) Interfacial tension in high-pressure carbon dioxide mixtures. Ind Eng Chem Res 34(12):4371–4377

Cooper JR (1982) Representation of the ideal has thermodynamic properties of water. Int J Thermophys 3:35–43

Crank J (1956) Mathematics of diffusion. Oxford University Press, Oxford

De Groot SR, Mazur P (1962) Non-equilibrium thermodynamics. North-Holland Pub. Co., Amsterdam

De Josselin de Jong G (1958) Longitudinal and transverse diffusion in granular deposits. Trans Am Geophys Union 39(1):67–74

De Simoni M, Carrera J, Sánchez-Vila X, Guadagnini A (2005). A procedure for the solution of multicomponent reactive transport problems. Water Resour Res 41:W11410. doi:10.1029/2005WR004056

De Simoni M, Sanchez-Vila X, Carrera J, Saaltink MW (2007) A mixing ratios-based formulation for multicomponent reactive transport. Water Resour Res 43:W07419. doi:10.1029/2006WR005256

Denbigh K (1981) Principles of chemical equilibrium: with applications in chemistry and chemical engineering, 4th edn. Cambridge University Press, Cambridge

Duan Z, Sun R (2003) An improved model calculating CO_2 solubility in pure water and aqueous NaCl solutions from 273 to 533 K and from 0 to 2000 bar. Chem Geol 193:257–271

Duan Z, Sun R, Zhu C, Chou IM (2006) An improved model for the calculation of CO_2 solubility in aqueous solutions containing Na+, K^+, Ca^{2+}, Mg^{2+}, Cl^- and SO_4^{2-}. Mar Chem 98:131–139

Duan Z, Li D, (2008) Coupled phase and aqueous species equilibrium of the H_2O–CO_2–NaCl–$CaCO_3$ system from 0 to 250 0C, 1 to 1000 bar with NaCl concentrations up to saturation of halite. Geochimica et Cosmochimica Acta 72:5128–5145. doi: 10.1016/j.gca.2008.07.025

Fang Y, Yeh GT, Burgos WD (2003) A general paradigm to model reaction-based biochemical processes in batch systems. Water Resour Res 39:1083. doi:10.1029/2002WR001694

Feist R, Schneider GM (1982) Determination of binary diffusion coefficients of benzene, phenol, naphthalene and caffeine in supercritical CO_2 between 308 and 933 K in the pressure range 90 to 160 gar with supercritical fluid chromatography (SFC). Sep Sci Technol 17(1):261–270

Fel L, Bear J (2010) Dispersion and dispersivity tensors in saturated porous media with uniaxial symmetry. Transp Porous Media 85(1):259–268

Friedly JC, Rubin J (1992) Solute transport with multiple equilibrium-controlled or kinetically controlled chemical reactions. Water Resour Res 28(6):1935–1953

Garcia JE (2003) Fluid dynamics of carbon dioxide disposal into saline aquifers. Report Number: LBNL-54280, 157 pp. Berkeley, CA, USA

Gardner WR (1958) Some steady state solutions of the unsaturated moisture flow equation, with application to evaporation from a water table. Soil Sci 85:228–232

Gray WG, Schrefler BA (2007) Analysis of the solid phase stress tensor in multiphase porous media. Int J Numer Anal Methods Geomech 31:541–581

Hebach A, Oberhof A, Dahmen N, Kögel A, Ederer H, Dinjus E (2002) Interfacial tension at elevated pressures measurements and correlations in the water + carbon dioxide system. J Chem Eng Data 47(6):1540–1546

Hough EW, Heuer GJ, Walker JW (1959) An improved pendant drop, interfacial tension apparatus and data for carbon dioxide and water. Trans Am Inst Min Metall Pet Eng 216:469–480

Jha B, Juanes R (2014) Coupled multiphase flow and poromechanics: a computational model of pore-pressure effects on fault slip and earthquake triggering. Water Resour Res 50(5):3776–3808. doi:10.1002/2013WR015175

Kim J, Tchelepi HA, Juanes R (2013) Rigorous coupling of geomechanics and multiphase flow with strong capillarity. SPE J 18(6):1123–1139. doi:10.2118/141268-PA

Kitanidis PK (1994) The concept of the Dilution Index. Water Resour Res 30(7):2011–2026

Klee G, Bunger A, Meyer G, Rummel F, Shen B (2011) In situ stresses in borehole blanche-1/south Australia derived from breakouts, core discing and hydraulic fracturing to 2 km depth. Rock Mech Rock Eng 44(5):531–540

Kräutle S, Knabner P (2005) A new numerical reduction scheme for fully coupled multicomponent transport-reaction problems in porous media. Water Resour Res 41:W09414. doi:10.1029/2004WR003624

Kräutle S, Knabner P (2007) A reduction scheme for coupled multicomponent transport-reaction problems in porous media: generalization to problems with heterogeneous equilibrium reactions. Water Resour Res 43:W03429. doi:10.1029/2005WR004465

Krupka KM, Cantrell KJ, McGrail BP (2010) Thermodynamic data for geochemical modeling of carbonate reactions associated with CO_2 sequestration. In: PNNL-19766, Pacific Northwest National Laboratory, Richland, Washington

Kumagai A, Yokoyama C (1999) Viscosities of aqueous NaCl solutions containing CO_2 at high pressures. J Chem Eng Data 44:227–229

Leverett MC (1941) Capillary behaviour in porous solids. AIME Trans 142:152–169

Li D, Duan ZH (2007) The speication equilibrium coupling with phase equilibrium in the H_2O–CO_2–NaCl system from 0 to 250°C and from 0 to 1000 bar. Chem Geol 244:730–757

Macleod DB (1923) On a relation between surface tension and density. Trans Faraday Soc 19 (July):38–41

McCain WD Jr (1991) Reservoir fluid property correlations-state of the art. SPE Reservoir Eng 6:266–272

Meyer CD (2000) Matrix analysis and applied linear algebra. SIAM. ISBN 0-89871-454-0. http://matrixanalysis.com/
Michaelides EE (1981) Thermodynamic properties of geothermal fluids. Geotherm Resour Council Trans 5:361–364
Millington RJ (1959) Gas diffusion in porous media. Science 130:100–102
Molins S, Carrera J, Ayora C, Saaltink MW (2004) A formulation for decoupling components in reactive transport problems. Water Resour Res W10301 doi: 10.1029/2003WR002970
Moridis GJ, Kowalsky M, Pruess K (2008) TOUGH + Hydrate v1. 0 user's manual. Report LBNL-0149E, Lawrence Berkeley National Laboratory, Berkeley, CA
Mualem Y (1976) A new model for predicting the hydraulic conductivity of unsaturated porous media. Water Resour Res 12(3):513–522. doi:10.1029/WR012i003p00513
Nikolaevski VN (1959) Convective diffusion in porous media. J Appl Math Mech 23:1042–1050
Olivella S, Carrera J, Gens A, Alonso EE (1994) Nonisothermal multiphase flow of brine and gas through saline media. Transp Porous Media 15(3):271–293
Palliser C, McKibbin R (1988) A model for deep geothermal brines III: thermodynamic properties-enthalpy and viscosity. Transp Porous Media 33:155–171
Parkhurst DL, Appelo CAJ (1999) User's guide to PHREEQC (Version 2): a computer program for speciation, batch-reaction, one-dimensional transport, and inverse geochemical calculations. USGS Water-Resources Investigations Report 99-4259
Parkurst DL, Thorstenson DC, Plummer LN (1980) PHREEQE, a computer program for geochemical calculations. US Geological Survey Water Resources Investigations Report, 80, 96
Peng DY, Robinson DB (1976) A new two-constant equation of state. Ind Eng Chem Fundam 15:59–64
Phillips SL, Igbene A, Fair JA, Ozbek H, Tavana MA (1981) Technical data book for geothermal energy Uuilization. Report: LBNL-12810. Berkeley, CA, USA
Pitzer S (1973) Thermodynamics of electrolytes. I. Theoretical basis and general equations. J Phys Chem 77(2):268–277
Potter RWII, Brown DL (1977) The Volumetric properties of sodium chloride solutions from 0 to 500°C at pressures up to 2000 bars based on regression of available fata in literature. US Geological Survey Bulletin 1421C
Prausnitz JM, Lichtenthaler RN, Gomes de Azvedo E (1986) Molecular thermodynamics of fluid-phase equilibria, 2nd edn. Prentice Hall, New Jersey
Pruess K (1991) TOUGH2—a general-purpose numerical simulator for multiphase fluid and heat flow. Larence Berkeley Laborary, University of California, LBL-29400, UC-251
Redlich O, Kwong JNS (1949) On the thermodynamics of solutions. Chem Rev 44:233–244
Rowe AM, Chou JCS (1970) Pressure-volume-temperature-concentration relation of aqueous NaCl solution. J Chem Eng Data 15:61–66
Rubin J (1983) Transport of reacting solutes in porous media: relation between mathematical nature of problem formulation and chemical nature of reactions. Water Resour Res 19 (5):1231–1252
Rutledge JT, Phillips WS (2003) Hydraulic stimulation of natural fractures as revealed by induced microearthquakes, Carthage Cotton Valley gas field, east Texas. Geophysics 68(2):441–452
Saaltink MW, Ayora C, Carrera J (1998) A mathematical formulation for reactive transport that eliminates mineral concentrations. Water Resour Res 34:1649–1656
Saffman PG (1959) A theory of dispersion in a porous medium. J Fluid Mech 6(3):321–349
Scheidegger AE (1961) General theory of dispersion in porous media. J Geophys Res 66:3273–3278
Span R, Wagner W (1996) A new equation of state for carbon dioxide covering the fluid region from the triple-point temperature to 1100 K at pressures up to 800 MPa. J Phys Chem Ref Data 25:1509–1596
Spycher N, Pruess K (2010) A phase partitioning model for CO_2-brine mixtures at elevated temperatures and pressures: application to CO_2-enhanced geothermal system. Transp Porous Media 82:173–196

Spycher N, Pruess K, Ennis-King J (2003) CO_2-H_2O mixtures in the geological sequestration of CO_2. I. Assesment and calculation of mutual solubilites from 12 to 100°C and up to 600 bar. Geochim Cosmochim Acta 67(16):3015–3031

Steefel CL, Lasaga AC (1994) A coupled model for transport of multiple chemical species and kinetic precipitation/dissolution reactions with application to reactive flow in single phase hydrothermal systems. Am J Sci 294:529–592

Steefel CI, MacQuarrie KTB (1996) Approaches to modeling reactive transport. Rev Mineral 34:83–129

Sugden S (1930) The parachor and valency, vol 2. G. Routledge, London

Sun R, Dubessy J (2012) Prediction of vapor-liquid equilibrium and PVTx properties of geological fluid system with SAFT-LJ-EOS including multi-polar contribution. Part II: application to H_2O–NaCl and CO_2-H_2O–NaCl system. Geochim Cosmochim Acta 88:130–145

Swaid I, Schneider GM (1979) Determination of binary diffusion coefficients of benzene and some alkylbenzenes in supercritical CO_2 between 308 and 328 K in the pressure range 80 to 160 bar with supercritical fluid chromatography (SFC). Ber Bunsenges Phys Chem 83(10):969–974

Terzaghi KV (1923) Die berechnung der durchlassigkeitsziffer des tones aus dem verlauf der hydrodynamischen spannungserscheinungen. Sitzungsberichte der Akademie der Wissenschaften in Wien, Mathematisch-Naturwissenschaftliche Klasse. Abteilung IIa 132:125–138

The International Association for the Properties of Water and Steam (2009) Revised release on the equation of state 2006 for H_2O Ice Ih. A minor revision to Revised Release on the IAPWS Industrial Formulation 1997 for the Thermodynamic Properties of Water and Steam, 2006

Van Genuchten MT (1980) A closed-form equation for predicting the hydraulic conductivity of unsaturated soils. Soil Sci Soc Am J 44(5):892–898

Vilarrasa V, Carrera J (2015) Geologic carbon storage is unlikely to trigger large earthquakes and reactivate faults through which CO_2 could leak. Proc Natl Acad Sci USA 112(19):5938–5943

Weinaug CF, Katz DL (1943) Surface tensions of methane-propane mixtures. Ind Eng Chem 35(2):239–246

Wendland M, Hasse H, Maurer G (1999) Experimental pressure-temperature data on three- and four- phase equilibria of fluid, hydrate, and ice phases in the system carbon dioxide-water. J Chem Eng Data 44:901–906

Wolery TJ (1979) Calculation of chemical equilibrium between aqueous solutions and minerals: the EQ3/EQ6 software package. Livermore, University of California, Lawrence Livermore Laboratory (UCRL 52658), 41 pp

Zarembo VI, Federov MK (1975) Density of sodium chloride solutions in the temperature range 25–350°C at Pressures up to 100 kg/cm^2. J Appl Chem 48:2021–2024

Zoback ML (1992) First-and second-order patterns of stress in the lithosphere: the World Stress Map Project. J Geophys Res Solid Earth 97(B8):11703–11728

Chapter 4
Mathematical Modeling: Approaches for Model Solution

Auli Niemi, Zhibing Yang, Jesus Carrera, Henry Power, Christopher Ian McDermott, Dorothee Rebscher, Jan Lennard Wolf, Franz May, Bruno Figueiredo and Victor Vilarrasa

Abstract The governing equations and mathematical models describing CO_2 spreading and trapping in saline aquifers and the related hydro-mechanical and chemical processes were described in Chapt. 3. In this chapter, the focus is on methods for solving the relevant equations. The chapter gives an overview of the different approaches, from high-fidelity full-physics numerical models to more simplified analytical and semi-analytical solutions. Specific issues such as modeling coupled thermo-hydro-mechanical-chemical processes and modeling of small-scale processes, such as convective mixing and viscous fingering, are also addressed.

A. Niemi (✉) · Z. Yang · B. Figueiredo
Department of Earth Sciences, Uppsala University, Villavägen 16, 75236 Uppsala, Sweden
e-mail: Auli.Niemi@geo.uu.se

Z. Yang
e-mail: zhibing.yang@hyd.uu.se

B. Figueiredo
e-mail: bruno.figueiredo@geo.uu.se

J. Carrera · V. Vilarrasa
Groundwater Hydrology Group (GHS UPC-CSIC), Institute of Environmental Assessment and Water Research (IDAEA), Spanish National Research Council (CSIC), Barcelona, Spain
e-mail: jesus.carrera.ramirez@gmail.com

V. Vilarrasa
e-mail: victor.vilarrasa@upc.edu

H. Power
Department of Mechanical Engineering, School of Mechanical, Materials, Manufacturing Engineering and Management, The University of Nottingham, University Park, Nottingham NG7 2RD, UK
e-mail: Henry.Power@nottingham.ac.uk

C.I. McDermott
School of Geoscience, Edinburgh Collaborative of Subsurface Science and Engineering (ECOSSE), University of Edinburgh, Edinburgh, UK
e-mail: cmcdermo@staffmail.ed.ac.uk

A. Niemi et al. (eds.), *Geological Storage of CO_2 in Deep Saline Formations*, Theory and Applications of Transport in Porous Media 29,
DOI 10.1007/978-94-024-0996-3_4

Finally, illustrative examples of modeling real systems, with different types of modeling approaches, are presented.

4.1 Different Approaches for Modeling CO_2 Geological Storage

Zhibing Yang, Auli Niemi and Jesus Carrera

The previous chapter has provided the mathematical description of the processes involved in geological storage of CO_2 in deep saline formations. These descriptions result in a set of governing partial differential equations which need to be solved together with appropriate initial and boundary conditions relevant to the systems to be modeled. This chapter discusses modeling approaches to solve these equations with emphasis on modeling large-scale systems, whose goals can range from theoretical questions, such as questions related to CO_2 dissolution or reservoir deformation, to practical issues like operational management, site screening, capacity estimation and risk assessment. This requires taking into account relevant physical processes and obtaining output from the model solutions for quantities and their uncertainties regarding issues like CO_2 mass inventory, CO_2 plume extent and pressure buildup.

As indicated by the mathematical representations in Chap. 3, CO_2 storage modeling is complicated by the presence of several multiphase flow regimes, coupled to other processes, and nonlinearity as well as parameter heterogeneity at several spatial scales, which generally hinders exact solutions. Therefore, approximate solutions often need to be sought for practical models. Generally, approximate solutions can be achieved by system simplification and/or numerical discretization (Nordbotten and Michael 2011) and may be categorized into analytical (and semi-analytical) and numerical solutions. The former are easy to use and provide insight into the nature of the problem and its solution. However, in many cases, numerical approximations are also needed because they are versatile, able to treat various boundary conditions and also to handle complex geomechanical and geochemical processes and their coupling.

D. Rebscher · J.L. Wolf · F. May
Bundesanstalt fuer Geowissenschaften und Rohstoffe (BGR), Hannover, Germany
e-mail: Dorothee.Rebscher@bgr.de

J.L. Wolf
e-mail: JanLennard.Wolf@bgr.de

F. May
e-mail: franz.may@bgr.de

In this section, we categorize the different modeling approaches for CO_2 storage into (1) high-fidelity hydrodynamic modeling, (2) reduced-physics modeling, (3) analytical modeling and (4) other modeling approaches. It can be noted that sometimes it is optimal if a combination of different approaches is used for modeling the complex system, to obtain an integrated understanding.

4.1.1 High-Fidelity Hydrodynamic Modeling

The high-fidelity hydrodynamic modeling approach refers here to numerical modeling in which three-dimensional balance equations describing multiphase flow and transport processes at a suitable scale are solved with sufficient accuracy and minimum degree of system simplification. In this kind of modeling, available geological information concerning the reservoir reservoir rock properties and boundaries as well as geometrical and structural features is taken into account to a degree that is the maximum computationally affordable. In other words, a minimum set of simplifications in terms of the hydrodynamic behavior and system characteristics is applied.

In order to solve the three-dimensional mass and energy balance equations, the spatial domain is discretized into a finite number of nodes/elements or cells and the continuous partial differential equations (PDEs) are converted to discrete equations typically using the finite differences method (FDM), the finite element method (FEM) or the finite volumes method. Eventually, the PDEs are represented by a set of (non-)linear algebraic equations at discrete time steps, which can be solved to yield numerical solutions at each node/element for a desired simulation period.

4.1.1.1 Simulation Codes for Hydrodynamic Modeling of CO_2 Storage

Historically, many simulation codes have been developed by petroleum engineers and groundwater hydrologists to obtain numerical solutions to the multiphase flow and transport problems. During the last decade, many of these codes have been extended to incorporate the ability to simulate the CO_2 storage system, since essentially the same basic governing equations are used to represent the system and only the fluid properties and associated phase behaviors need to be changed. Each code has different features that are included. The codes can be broadly divided into two categories: research codes [e.g. TOUGH2 (Pruess et al. 1999), CODE_BRIGHT (Olivella et al. 1996), PFLOTRAN (Hammond et al. 2007), OpenGeoSys (Kolditz et al. 2012a), GPRS (Cao 2002)] and commercial codes [e.g. ECLIPSE (Schlumberger 2012)]. Here we give an example for each category, while a complete list of codes is not pursued.

One of most widely used research codes is TOUGH2 (Pruess et al. 1999), developed by Lawrence Berkeley National Laboratory. It is a general-purpose simulation code for non-isothermal, multi-component, multi-phase fluid flow in porous and fractured media. It employs integral finite difference for spatial discretization, and uses fully implicit, first-order finite differences for temporal discretization. The non-linear equations for each time step are solved using a Newton–Raphson iterative method with adaptive time step size. It has been coupled to a reactive transport simulator [TOUGHREACT (Xu et al. 2014)], using the formulation of Saaltink et al. (1998), described in Sect. 3.4. ECO2N (Pruess 2005) is a fluid property module for the TOUGH2 simulator for applications to geologic sequestration of CO_2 in saline aquifers. It describes the thermodynamics and thermophysical properties of H_2O–NaCl–CO_2 mixtures, and accurately reproduces fluid properties for the temperature, pressure and salinity conditions of interest for geological sequestration (10 °C $\leq$ T $\leq$ 110 °C; P $\leq$ 600 bar; salinity up to full halite saturation) and ECO7CMA (Freifeld et al. 2013) have extended the properties to even greater depths and larger temperatures. Phase conditions considered include a single (aqueous or CO_2-rich) phase, as well as two-phase mixtures. Local equilibrium solubility is applied to treat phase partitioning between the aqueous and the CO_2-rich phase as well as to handle dissolution or precipitation of salt. An option for modeling associated changes of porosity and permeability is also included. In ECO2N, no distinction is made with regard to whether the CO_2-rich phase is liquid or gas. In ECO2M (Pruess 2011), an enhanced version of ECO2N, all possible phase conditions for brine–CO_2 mixtures, including transitions between super- and sub-critical conditions, and phase change between liquid and gaseous CO_2 are described, which allows for more accurate modeling of CO2 leakage to the shallow subsurface. Essentially, CODE_BRIGHT incorporates the same suite of processes but its coupling to mechanical deformation is direct. It is also coupled to reactive transport (Saaltink et al. 2004) and was modified as part of the MUSTANG project to incorporate CO_2 as a fluid phase.

One of the extensively used simulators in the petroleum industry is ECLIPSE (Schlumberger 2012). It is a fully implicit, three dimensional, general purpose simulator and has two modules: one for black oil simulation (E100) and the other for compositional simulation (E300). E100 assumes three components, water, oil and gas in a three-phase system of liquid, gas and gas in solution. When applied to the CO_2 storage system, E100 essentially uses a gas–oil system with the oil phase given the brine properties and the gas phase given the CO_2 properties (Singh et al. 2010). E300 is a compositional simulator with a cubic equation of state and features like pressure-dependent permeability values. E300 includes an option called CO2STORE which can handle mutual solubility of CO_2 and water and accurately calculate the fluid properties (density, viscosity, compressibility, etc.) of pure and impure CO_2 as a function of temperature and pressure. It also includes functionality to describe the dry-out and salt precipitation phenomena. Accurate calculations of fluid properties and mutual solubility are required for accurate hydrodynamic modeling of CO_2 storage system. However, it is worth noting that compositional simulations with a sophisticated equation of state, though very accurate in representation of the fluid properties, will correspond to much more expensive computational burden than that in black oil simulations (Hassanzadeh et al. 2008). An efficient thermodynamic model (Hassanzadeh et al. 2008) has been developed to reproduce the PVT data for the CO_2–brine mixture to be used in black-oil simulations for saving computational time.

4.1.1.2 An Illustrative Example

We here present an example of full-physics hydrodynamic modeling, using the TOUGH2 model, for the purpose of illustrating the flow regimes and the spatial distributions of quantities of practical interest such as gas saturation and pressure perturbation. A 'disk-shaped', vertically bounded formation is considered. Some of the main parameters for the simulation are summarized in Table 4.1. The parameters are chosen to represent typical scenarios of industrial scale CO_2 storage, except that formation water salinity is ignored in this example. The radial extent is set to a large value and is infinite acting for the example considered here. CO_2 storage is simulated as injection along a vertical well perforated through the whole formation thickness.

Figure 4.1 presents the simulated spatial profile of vertically averaged CO_2 saturation and pressure increase for the example CO_2 injection scenario, which can be considered representing an industrial scale CO_2 injection. As can be seen in Fig. 4.1, there exists a dry-out zone (free of aqueous phase) around the injection well. In this dry-out zone all water has been either displaced outwards or vaporized into the CO_2-rich (gas) phase. In this example, salinity is not considered. When salinity is considered, the salt that is originally dissolved in the brine would have precipitated in the dry-out zone and would thus reduce the porosity and permeability. The radius of dry-out zone is on the scale of ~ 100 m at the end of the injection period (say e.g. 50 years). Surrounding the dry-out zone is a region where the gas phase and the aqueous phase co-exist. The radius of this two-phase flow region is typically several kilometers at the end of the injection period. Outside of the two-phase region only brine exists with single phase brine flow.

Table 4.1 Modeling parameters for the illustrative example

Parameter	Value and unit	Parameter	Value and unit
Initial pressure	160 bars	Temperature	45 °C
Permeability	100 mD	Porosity	0.12
Thickness	50 m	Injection time	50 years
Res. brine sat.	0.3	van Genuchten param. m	0.41
Entry pressure	9983 Pa	Salt mass fraction	0.0
Radial extent	250 km	Injection rate	1.0 Mt/a

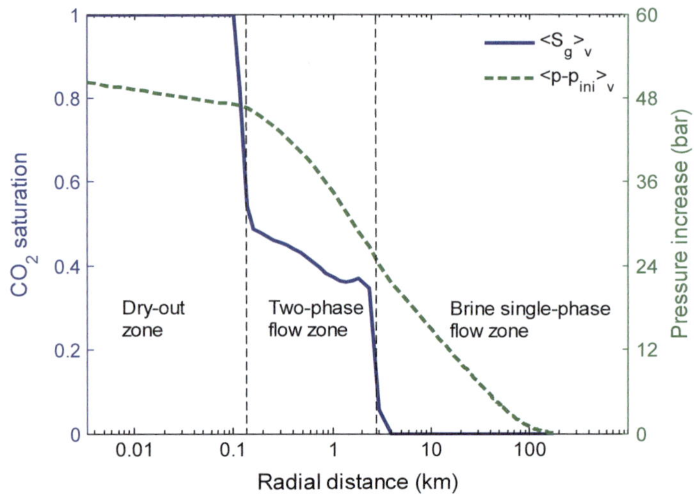

Fig. 4.1 Radial profile of CO_2 saturation and pressure increase (vertically averaged) at 50 years for an injection scenario with rate 1 Mt/a based on the high-fidelity modeling of an ideal disk-shaped infinite-acting aquifer using TOUGH2/ECO2N

4.1.1.3 Application Examples

Examples of using the TOUGH2 simulator and its extensions for obtaining numerical solutions to study varying aspects of geological storage of CO_2 are widely available in the literature, and several examples are presented throughout this book. For example, Zhou et al. (2010) performed a basin-scale simulations for multiple-site CO_2 injection in the Mount Simon aquifer in the Illinois Basin. They simulated CO_2 injection with a rate of 5 Mt/year/well and a total number of 20 wells, given the thick Mount Simon aquifer (300–730 m) confined by the low-permeability Eau Claire caprock. Both the plume-scale processes (i.e. hydrodynamic interactions between supercritical CO_2 and formation brine) and the basin-scale processes (i.e. large-scale pressure build-up and brine migration) were modeled. Numerical solutions were obtained with regard to CO_2 mass distribution, CO_2 plume sizes and fluid pressure changes for the aquifer as a whole, which can facilitate evaluation of the injection strategies and the large-scale hydrogeological impact. Yang et al. (2015) and Tian et al. (2016) presented a comprehensive simulation workflow, where TOUGH2/ECO2N is used together with simpler models (semi-analytical solutions and vertical equilibrium models), and estimated the CO_2 storage capacities of the Baltic Sea Basin and the South Scania site, given the constraints due to pressure build-up and long-term CO_2 containment.

The ECLIPSE reservoir simulator package has also been applied in many studies for numerical solutions of modeling CO_2 storage. For instance, Juanes et al. (2006) used the ECLIPSE black oil simulator to investigate the role of relative permeability hysteresis on CO_2 residual trapping for both a synthetic geological formation and a more realistic geological model (the PUNQ-S3 Case Studies). Shamshiri and Jafarpour (2012) employed the compositional simulator E300 for modeling CO_2 injection and migration with the aim of developing optimized injection rate allocation for improved residual and dissolution trapping as well as decreased risk of CO_2 plume approaching leakage pathways.

4.1.1.4 Remarks on High-Fidelity Hydrodynamic Modeling

High-fidelity modeling with sophisticated numerical simulators, can yield simulation results that are the most accurate achievable, provided that all the relevant processes are considered and space and time are properly discretized. It can also serve as a benchmark for comparison when we attempt to simplify the mathematical description for gaining insight into the interplay between parameters/processes. However, CO_2 storage systems are quite complex to model with physical processes spanning a wide range of spatial and temporal scales (Nordbotten and Michael 2011). Even though simulators are continuously becoming increasingly powerful in terms of considering a multitude of physical processes and handling larger grids, there are still limitations to integrate the effect of small-scale hydrodynamic processes into large scale models (this requires upscaling, which is the subject of Chap. 5). To this end, continued development and enhancement of simulation codes continues to be important.

High-fidelity, high-resolution numerical simulations at the reservoir or basin scale require solving the 3D non-linear partial differential equations with large grids, which points to the need for a significant amount of computational power and resources. Efficient modeling and performance optimization are often pursued. Parallel computing is becoming more and more important with the need for more accurate simulation results and better computing performances. Nowadays several codes offer parallel computing options (e.g. Lu and Lichtner 2007; Zhang et al. 2008). Zhou et al. (2010), Yamamoto et al. (2009), Yang et al. (2015), and Tian et al. (2016) are examples of full basin-scale integrated modeling of CO_2 storage using parallel simulation codes.

For full basin-scale 3D models it would be restrictive, if not impossible, to incorporate the multi-scale heterogeneity and scale-dependent processes. Even though computers are becoming more and more powerful, the computational resources needed for accurate solutions for practical basin-scale modeling of CO_2 storage are often beyond the given computational capabilities. Besides, there will be a lack of geological information for the parameters that are used in fine spatial resolution models. Oftentimes, simplification through parameter averaging or process upscaling is unavoidable.

Finally, it should be pointed out that different numerical simulators may show discrepancies because of factors such as the way fluid properties are calculated or differences in the numerical solution or gridding and time stepping. In practice, differences in the way different modelers interpret the setting of a given problem may be even more important. In order to understand the impact of both numerical and modeling differences it is important to perform code inter-comparison studies of model solutions for benchmark problems (Pruess et al. 2004; Class et al. 2009; Nordbotten et al. 2012).

4.1.2 Reduced-Physics Modeling

As pointed out in the previous section, high-fidelity full physics modeling may not always be viable or necessary. In such situations, system simplification can be made through reduction of the number of processes considered by keeping only those that are dominant and essential for the objectives of the study. System simplifications can also be done by replacing the complex domain geometries with simpler ones and reducing the dimension of the model (e.g. from 3D to 2D and even 1D). A set of seven different system simplifications for practical modeling of CO_2 storage is summarized and discussed in Celia and Nordbotten (2009). Invoking subsets of simplifications, we can modify the general governing equations to obtain a new and simpler set of equations which may render analytical and/or semi-analytical solutions or considerably easier numerical solutions. This can bring great computational benefits, especially if a Monte Carlo approach is required for uncertainty quantification (see Chap. 5).

In this section, we describe two simplified numerical modeling approaches for practical modeling of CO_2 storage. Analytical solutions are discussed in the next Sect. 4.1.3.

4.1.2.1 Vertical Equilibrium Approach

Given that for CO_2 storage projects, the lateral length scale of a storage formation is typically much larger than the thickness of the formation, it is convenient to make the so-called vertical equilibrium assumption (Lake 1989; Yortsos 1995), which is the CO_2 storage version of the Dupuit approximation. This approximation is based on the assumption that the timescale for gravity segregation is much shorter than that of lateral flow. This means that the vertical distribution of quantities, such as pressure and fluid saturation, can be defined from vertically averaged quantities (Gasda et al. 2011). To further simplify the system of analysis, it may also be convenient to assume that there exists a macroscopic sharp interface between the CO_2 and brine. This assumption may be reasonable when the capillary fringe between CO_2 and brine is very small compared to the thickness of the domain. Figure 4.2 presents a schematic showing the vertically averaged CO_2-brine system where B is the thickness of the formation, b is the thickness of the CO_2 plume, and d_B, d_M, and d_T are the distances between the datum z = 0 and the bottom surface, the macroscopic interface and the top surface, respectively.

Starting from the three-dimensional mass balance equation for phase α ($\alpha = a$ for the aqueous phase and $\alpha = g$ for the CO_2-rich, or gas, phase), which was introduced in Sect. 3.3 (Eq. 3.3.26), while neglecting dissolution, we have:

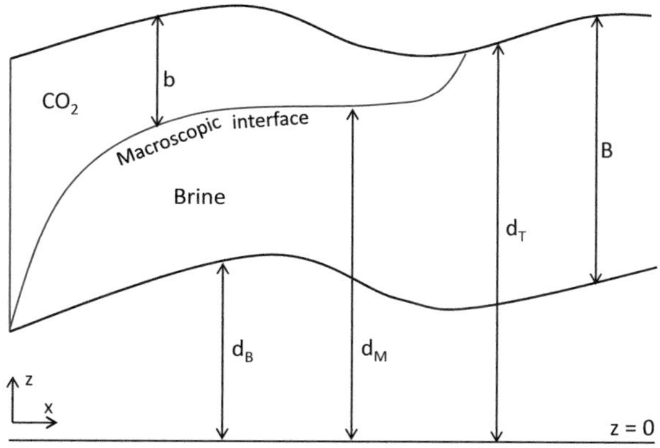

Fig. 4.2 Schematic for vertically averaged configuration of macroscopic regions of CO_2 and brine [modified after Gasda et al. (2009)]

$$\frac{\partial(\phi \rho_\alpha S_\alpha)}{\partial t} + \nabla \cdot (\rho_\alpha \mathbf{q}_\alpha) = \rho_\alpha G_\alpha. \tag{4.1.1}$$

In the above equation, ϕ is porosity, ρ_α is the fluid density of α-phase, S_α is the saturation of α-phase, $\mathbf{q}_\alpha$ is the flux calculated by Darcy's law and G_α is the source or sink term. Vertical integration of Eq. (4.1.1) for each phase over the formation thickness gives (Gasda et al. 2009):

$$\begin{aligned} &\phi \rho_a \bar{\beta}_a (B-b) \frac{\partial p_a}{\partial t} + \phi \rho_a (1 - S_{ar}) \frac{\partial}{\partial t}(B-b) + \nabla \cdot \rho_a \bar{\mathbf{q}}_a \\ &- \rho_a q_{a,v}\big|_{d_B} + \rho_a q_{a,v}\big|_{d_T} = \int_{d_B}^{d_M} \rho_a G_a dz \end{aligned} \tag{4.1.2a}$$

$$\begin{aligned} &\phi \rho_g \bar{\beta}_g b (1 - S_{ar}) \frac{\partial p_g}{\partial t} + \phi \rho_g (1 - S_{ar}) \frac{\partial b}{\partial t} + \nabla \cdot \rho_g \bar{\mathbf{q}}_g \\ &- \rho_g q_{g,v}\big|_{d_B} + \rho_g q_{g,v}\big|_{d_T} = \int_{d_B}^{d_M} \rho_g G_g dz \end{aligned} \tag{4.1.2b}$$

In Eq. (4.1.2), S_{ar} is the residual brine saturation, p_α is the fluid pressure of α-phase, $\bar{\beta}_\alpha$ is the vertically average bulk compressibility of phase α, $q_{\alpha,v}$ is the vertical component of the flux of phase α, and $\bar{\mathbf{q}}_\alpha$ is the vertically integrated horizontal fluxes for phase α which can be obtained from

$$\bar{\mathbf{q}}_a = -\frac{(B-b)\mathbf{k}}{\mu_a} \left[\nabla p_{top} + \rho_a g \nabla d_T - (\rho_a - \rho_g) g \nabla b \right] \tag{4.1.3a}$$

$$\bar{\mathbf{q}}_g = -\frac{bk^*_{r,g}\mathbf{k}}{\mu_g}\left[\nabla p_{top} + \rho_g g \nabla d_T\right] \tag{4.1.3b}$$

where $\mathbf{k}$ is the intrinsic permeability tensor, $k^*_{r,g}$ is the relative permeability of the CO_2-rich (gas) phase evaluated at gas saturation $(1 - S_{ar})$, μ_α is the viscosity of α-phase, g is acceleration due to gravity and p_{top} is the fluid pressure along the top of the formation. Under the vertical equilibrium assumption, the fluid pressure p_α in Eq. (4.1.2) is related to p_{top} according to vertical static distribution.

It must be emphasized that the above descriptions of VE assumption are limited to the scenario of primary drainage, corresponding to the CO_2 injection period. For the post-injection period, the fluid region occupied by CO_2 needs to be divided into a mobile CO_2 region and residual CO_2 region. The above Eqs. (4.1.2) and (4.1.3) also need to be reformulated to include residual trapping.

Equations (4.1.2) and (4.1.3) together describe the vertically integrated flow system in a formation and can be solved for pressure and for the thickness of the CO_2 plume with appropriate initial and boundary conditions using numerical methods such as FEM and FDM. Compared with the 3D full-physics modeling approach, the sharp-interface vertical equilibrium approach requires significantly less computational resources due to the fact that (1) the two-dimensional equations are much more efficient to solve and (2) the nonlinearities resulting from local relative permeability and capillary pressure in the 3D representation are greatly reduced (Lake 1989; Yortsos 1995; Celia and Nordbotten 2009; Gasda et al. 2009, 2011).

The effectiveness of the VE approach for exploring CO_2 injection and migration at large spatial and temporal scales has been demonstrated in a modeling study (Gasda et al. 2012) where the VE model was modified to take into account the effect of capillary fringe and convective dissolution. The study applied the modified VE modeling approach to the Johansen Formation and investigated the relative-importance of small scale processes, including residual trapping, capillary fringe and convective dissolution, on long-term CO_2 storage security. Yang et al. (2015) and Tian et al. (2016) in turn compared numerical solutions obtained by a VE model and a 3D TOUGH2 model, and demonstrated the usefulness of the VE approach. Further improvements to the VE approach are discussed by Nilsen et al. (2016), which introduces the MRST-CO2LAB family of codes that is designed to interact with oil industry codes, and Andersen et al. (2015), which explicitly acknowledges the high compressibility of CO_2 by simulating its density variations.

When the storage system is further simplified (e.g. assuming a horizontal, homogeneous formation with uniform thickness), the sharp-interface vertical equilibrium approach can also allow approximate analytical solutions for plume migration in the three-dimensional radially symmetrical or two-dimensional systems. These will be further discussed in Sect. 4.1.3.

4.1.2.2 Single-Phase Flow Modeling for Far-Field Pressure Buildup

Injection of a large volume of CO_2 induces large scale pore pressure buildup and associated change in the stress field, which may become a limiting factor for the CO_2 storage capacity. Far-field brine displacement into fresh-water aquifers or brine leakage through unplugged wells or fault zones may be induced by large-scale pressure buildup, even if the injected CO_2 is safely trapped. Here, far-field may be defined as regions outside of the CO_2 plume, i.e. the single-phase brine flow regions as indicated in Fig. 4.1.

The CO_2-brine two-phase complications are not important when the objective of the modeling study is to evaluate the far-field pressure impact and to determine the Area of Review (defined as the regulatory region surrounding the CO_2 storage where there may be detrimental effects on groundwater resources) at the regional/basin scale, given that the CO_2 plume size is relatively small compared to the spatial scale of interest. It has been shown (e.g. Chang et al. 2013; Yang et al. 2013a; Huang et al. 2014) that single-phase models may be sufficient for prediction of far-field pressure perturbation. The key point is calculating the volume of the CO_2 plume in order to know the volume of displaced brine and therefore determine the induced pressure buildup in the far-field. The volume of the CO_2 plume is not a straightforward calculation because the dependence is two-way: CO_2 density depends explicitly on fluid pressure, but fluid pressure also depends on density, because density controls the plume volume, and thus the fluid pressure through the volume of water that needs to be displaced. This nonlinear problem can be solved using an iterative scheme (Vilarrasa et al. 2010a). Single-phase flow equations are much easier to solve than multi-phase flow ones and offer significant computational advantage. For example, single-phase numerical modeling has been applied to evaluate the potential impact of CO_2 injection in the Texas Gulf Coast Basin on the shallow groundwater resources (Nicot 2008) and to estimate the pressure build-up at South Scania site in Sweden by Yang et al. (2013a). Yang et al. (2013a) also show the comparison and good agreement in predictions for far-field pressure build-up between the full two-phase simulation with TOUGH2 and the single phase with simple Theis solution (see Chap. 7). However, it should be pointed out that while the single-phase flow models are appropriate for predicting pressure buildup in the single-flow domain of the far-field, they are obviously not appropriate for pressure prediction near the CO_2 injector (which may be important for the analysis of mechanical failure) where processes and effects such as brine evaporation, viscosity contrast and two-phase flow are important.

4.1.3 Analytical Solutions

Development of analytical models for flow and transport problems in hydrogeology has been an important and continuous effort since decades ago. Analytical solutions can provide physical insight into the balance of the physical driving mechanisms or for verifying numerical models for special cases. They are also extensively used for the interpretation of characterization tests (see Chap. 7). Even though exact analytical solutions to the general two-phase flow equations are not achievable, derivation of approximate analytical solutions with simplifying assumptions to the two-phase flow equations under certain initial and boundary conditions is still desirable. In contrast to numerical models, analytical methods allow for rapid evaluation and can be employed to study parameter sensitivity for quick site screening and evaluation. Thus, analytical solutions can be useful to support decision making concerning the operation of CO_2 injection projects and provide guidance for scenario selection for more detailed full-physics modeling (see e.g. Yang et al. 2013a, 2015). In some cases, it is also beneficial to incorporate analytical solutions in a numerical framework to form a hybrid numerical-analytical approach. An example of such approach is given in Gasda et al. (2009), where analytical solutions of sub-scale flow through leaky wells are embedded into a vertical equilibrium numerical model.

For application to CO_2 storage problems, approximate analytical solutions have been used to model CO_2-brine interface dynamics during injection (e.g. Nordbotten et al. 2005; Nordbotten and Celia 2006; Dentz and Tartakovsky 2008; Houseworth 2012; Vilarrasa et al. 2013a), pressure buildup (e.g. Vilarrasa et al. 2010a, 2013a; Mathias et al. 2011; Nordbotten et al. 2005; Nordbotten and Celia 2006; Dentz and Tartakovsky 2008; Houseworth 2012; Yang et al. 2015; Tian et al. 2016), brine leakage through abandoned wells (Cihan et al. 2011) and post-injection CO_2 plume migration that incorporate capillary trapping (Hesse et al. 2008; MacMinn and Juanes 2009; Juanes et al. 2010) and/or CO_2 dissolution (MacMinn et al. 2011). Each of the analytical solutions employs a different set of assumptions. In the following, we focus on CO_2 plume shapes and pressure buildup, and introduce some of the (semi)-analytical models.

4.1.3.1 CO_2–Brine Interface Dynamics During Injection

For the derivation of the analytical solutions describing CO_2 plume migration, both during injection and after injection, the sharp interface assumption is typically made. The abrupt interface approximation considers that the two fluids, CO_2 and brine in this case, are immiscible and separated by a sharp interface and capillary effects are usually neglected. The fluid regions occupied by CO_2 or brine are assumed to have constant saturations, written below for simplicity as 0 for the aqueous phase saturation in the CO_2 occupied region, and vice versa.

The two standard solutions for the brine interface are those of Nordbotten et al. (2005) and Dentz and Tartakovsky (2008). The former derived their solution neglecting the gravity term in the flow equation, thus emphasizing viscous dissipation, and approximating the aquifer response by Cooper Jacobs approximation (see also Chap. 7). In addition, they impose (1) volume balance, (2) gravity override

(CO_2 plume travels preferentially along the top) and (3) they minimize energy at the well. The fluid pressure applies over the entire thickness of the aquifer and fluid properties are vertically averaged. The vertically averaged properties are defined as a linear weighting between the properties of the two phases. Nordbotten et al. (2005) write their solution as a function of the mobility, ratio of relative permeability to viscosity, of each phase. For the case of an abrupt interface where both sides of the interface are fully saturated with the corresponding phase, the relative permeability is 1 and the mobility becomes the inverse of the viscosity of each phase, assumed constant. Under this conditions, the thickness of the CO_2 plume is:

$$b = B\frac{\mu_c}{\mu_w - \mu_c}\left(\sqrt{\frac{\mu_w Q_0 t}{\mu_c \phi \pi B r^2}} - 1\right) \tag{4.1.4}$$

where B is the thickness of the formation, r is the radial distance, and Q_0 is the volumetric injection rate.

Dentz and Tartakovsky (2008) derived an analytical solution by adopting complementary assumptions. That is, they invoked of the Dupuit assumption and the quasi-steady approximation for vertical equilibrium, thus neglecting vertical viscous dissipation and emphasizing buoyancy forces. They find that the thickness of the CO_2 plume, b, for radially symmetrical three-dimensions is equal to

$$b = B(1 - \Lambda_{cw}\ln[r/r_b]) \tag{4.1.5}$$

where $r_b = \sqrt{2Q_0 t/\pi\varphi B\Lambda_{cw}(e^{2/\Lambda_{cw}} - 1)}$ is the radius of the CO_2 plume at the base of the aquifer, and $\Lambda_{cw} = Q_0(\mu_a - \mu_g)/2\pi k B^2 g(\rho_a - \rho_g)$ is a dimensionless parameter that measures the relative importance of viscous and gravity forces.

It should be noted that the solution given by Eq. (4.1.5) is best in the buoyancy-dominated flow regime, that is for low injection rate, high permeability, or far from the injection well, where buoyancy dominates over viscous forces. Such conditions can be found in storage sites like Sleipner, Norway, where the reservoir has a high permeability and even though 1 Mt of CO_2 is stored annually, gravity forces dominate (Cavanagh et al. 2015). Nevertheless, such high permeability reservoirs are scarce, and since the injection rate will have to be as large as geomechanically permissible in order to achieve better CO_2 storage efficiency, viscous forces are likely to dominate in most storage sites near the injection well. Still, Eq. (4.1.5) should be the formulation of choice for prediction of CO_2 plume evolution at the regional scale.

An inherent assumption in the analytical solutions given in Dentz and Tartakovsky (2008) and Nordbotten et al. (2005) is that volumetric injection rate and the CO_2 properties (density and viscosity) are kept constant. In reality, however, the properties of CO_2 can vary significantly as the pressure changes and thus the volumetric injection rate cannot be prescribed. The compressibility of CO_2 can be one or two orders of magnitude larger than that of brine for typical injection depth. Vilarrasa et al. (2010a) proposed an iterative method to account for CO_2

compressibility in these analytical solutions. Thus, the actual mean CO_2 density and CO_2 viscosity of the plume are used, leading to more accurate estimations of the CO_2 plume position and pressure buildup than when CO_2 compressibility is neglected (Vilarrasa et al. 2013a). Further extended the solution to cases where the CO_2 density can vary in space and time as the fluid pressure evolves. A further merit of this semi-analytical solution is that it allows for an uneven distribution of mass injection rate along the vertical well. It employs vertical discretization (layers) of the formation thickness to include buoyancy flow between layers and calculates the pressure build up and interface position in a step-wise manner in time. For more details, interested readers are referred to Vilarrasa et al. (2013a).

4.1.3.2 Pressure Buildup

Injecting a large volume of CO_2 in deep saline formations causes significant pore pressure buildup near the injection well and associated change in the stress field. This may induce tensile or shear failure of the caprock (Rutqvist et al. 2008; Vilarrasa et al. 2010b) and reactivation of existing fractures or faults (Rutqvist and Tsang 2002; Streit and Hillis 2004; Cappa and Rutqvist 2011). The storage capacity of a formation is thus limited by, among other factors, the maximum allowed pressure buildup.

Pressure buildup can be easily calculated by means of a semi-analytical solution, once the interface position has been obtained, by simply integrating Darcy's law, typically assuming radial flow (Hesse et al. 2008; Mathias et al. 2011; Cihan et al. 2011; Mathias et al. 2013). Vilarrasa et al. (2010a) show that pressure buildup for a constant injection into an infinite homogeneous formation equals

$$p - p_0 = \frac{Q_0 \mu_w}{2\pi B k} \begin{cases} \ln\left(\frac{R}{r_0}\right) + \sqrt{\frac{\mu_c \phi \pi B}{\mu_w V(t)}}(r_0 - r_b) + \frac{\mu_c}{\mu_w} \ln\left(\frac{r_b}{r}\right), & r < r_b \\ \ln\left(\frac{R}{r_0}\right) + \sqrt{\frac{\mu_c \phi \pi B}{\mu_w V(t)}}(r_0 - r), & r_b \le r < r_0 \\ \ln\left(\frac{R}{r}\right), & r \ge r_0 \end{cases} \tag{4.1.6}$$

where, p is the pressure, p_0 is the initial pressure, $V(t)$ is the volume injected up to time t, and r_0 and r_b are the radii of the top and bottom of the plume, which can be obtained from Eqs. (4.1.4) and (4.1.5), respectively, for the formulations of Nordbotten et al. (2005) and Dentz and Tartakovsky (2008), discussed above. Details on these approximations, including how to acknowledge that CO_2 density varies with depth and radial distance, are discussed by Vilarrasa et al. (2010a).

It can also be seen that Eq. (4.1.6) simply represents a generalization of Thiem's solution for transient problems because $R = \sqrt{2.25 k \rho_w g t / \mu_w S_s}$ (see Sect. 7.1.3). As such, it simplifies viscous dissipation effects near the injection well, such as vertical flow components at the plume edge or partial penetration (all analytical approximations, and most numerical solutions assume that flow is uniformly distributed along the vertical at the injection well, whereas it should concentrate at the top for low flow rates and at the bottom for high flow rates in partially penetrating wells). These effects are acknowledged by the semi-analytical solutions of Vilarrasa et al. (2013a).

4.1.4 *Other Modeling Approaches*

Apart from the analytical solutions and comprehensive high-fidelity numerical models discussed above, alternative methods have also been developed for modeling CO_2 migration. Most notable of these are perhaps invasion percolation (IP) models and streamline-based approaches. They are complementary to the methods discussed above and can give additional understanding to the system in special cases. A brief introduction to these methods is given below.

4.1.4.1 Invasion Percolation Approach

Invasion percolation (IP) was introduced in the 1980s (Wilkinson and Willemsen 1983) as a model for describing slow fluid displacement (i.e. negligible viscous forces) in porous media at the microscopic (pore) scale. Based on the classical percolation theory, IP takes into account the fluid transport process by considering a path of least capillary resistance. Application of the IP models requires an infinitesimal flow rate where the capillary forces dominate over the viscous forces. For typical CO_2 injection scenarios, such as the illustrative example given in Sect. 4.1.1, the pressure gradient is about 10^5 Pa/m close to the injection well but decreases to $\sim 10^3$ Pa/m in the two-phase flow zone. Upon comparison with the entry capillary pressure of $\sim 10^4$ Pa, this means that, at the sub-meter scale the capillary dominance condition will usually be satisfied. In a general case, one can calculate the dimensionless capillary number to check the relative importance of capillary forces over viscous forces (Lenormand et al. 1988). IP models are particularly well suited to study the effects of small scale heterogeneity. In fact, Kueper and McWhorter (1992) applied the IP model at the macroscopic scale to perform large-scale averaging of capillary pressure functions for unsaturated flow.

IP have been used for CO_2 storage simulations. A macroscopic IP model (Yang et al. 2013b) was used to upscale capillary pressure and relative permeability relationships for CO_2 migration in a multimodal heterogeneous field. An IP simulator has also been developed (Singh et al. 2010) and applied to the Sleipner Layer 9 benchmark problem.

4.1.4.2 Streamline-Based Approach

Streamline-based models have been also developed to simulate CO_2 storage (Jessen et al. 2005; Obi and Blunt 2006). In streamline-based approaches an operator splitting method is typically used. The basic idea is to decouple the advective from dispersive/reactive terms in the saturation transport equation, transforming a three-dimensional system into a series of one-dimensional equations. At each time step, the pressure equation and flow velocity fields are solved and streamlines are traced on the underlying grid. Then one-dimensional advective transport of saturation and concentration is numerically solved along the streamlines, possibly including exchange with adjacent streamlines explicitly in time. After accounting for the dissolution and reactions, the saturation and concentration can then be mapped to the underlying grid. These are then used to solve for the total flow (aqueous plus CO_2 phases). The two steps (flow computations and saturation transport) are repeated through discrete time steps. The streamline-based approach has been demonstrated, for example, by simulation of CO_2 storage in a highly heterogeneous million-grid-block model of a North Sea reservoir (Obi and Blunt 2006). The main advantage of the streamline-based approach lies in its computational efficiency. While the method is usually solved under the assumptions of incompressible flow and negligible capillary pressure, neither one is strictly needed and may be a significant source of error for CO_2 transport prediction in some cases.

4.2 Modeling of the Coupled Processes

Christopher Ian McDermott

4.2.1 Introduction

During CO_2 injection into a saline aquifer, Thermal, Hydraulic, Mechanical and Chemical (THMC) processes are operating on the system (see Chap. 2, 3). These processes also influence each other. We commonly use the term coupled processes when the parameters of one process depend on the solution of another one and with coupled THMC processes we mean the couplings between these four processes.

To attempt to model the behavior of such a system some estimation as to the importance of different processes and their interaction needs to be made. It is not possible to model in detail every single process and phenomena present. Therefore, decisions need to be taken which processes and phenomena are included and how the model is to represent the coupling between the processes. Key decision in the modeling process is not so much finding the most accurate mathematical representation of the system, but the decision which processes can be left out without losing key characteristics of the system. Below we will briefly discuss the importance of coupled processes, the basis behind their division and the numerical methods for solving these processes and their coupling.

4.2.2 *Brief Overview of Coupled Processes*

Several researchers have described a number of physical phenomena as a result of the coupled THMC processes interacting with one another, a key pioneering paper being (Tsang 1991). Since then, there have been several hundreds of publications on this topic. The interaction of coupled processes leading to physical phenomena is illustrated in Fig. 4.3 (see also Fig. 2.3 in Chap. 2 for the basic concept of coupled processes in CO_2 geological storage). The diagram in Fig. 4.3 demonstrates the fact that all the key processes are linked via material behavior and coupled in various ways. This material behavior (physically observable phenomena), leads to a measureable, observable change in the properties of the continuum as a response to the change of two or more of the field variables of temperature, pressure, stress or chemistry. This behavior is also influenced in the storage settings we are considering by the complex three dimensional heterogeneous geology which the saline aquifer forms part of. The geology defines the distribution of the different facies present and their associated material parameters. We can expect that all of the physical phenomena illustrated in the figure below, will be operating in a caprock during CO_2 geological storage.

As the material characteristics change the interaction of the processes may either amplify the effect of a certain phenomenon, or diminish it through a feedback

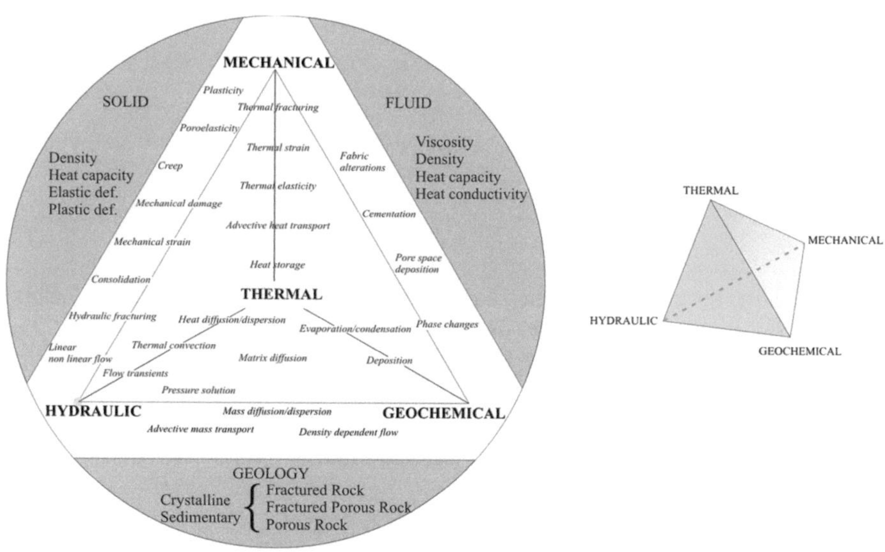

Fig. 4.3 Illustration of coupled processes effects

mechanism. For instance, temperature impacts fluid flow through viscosity, with increasing temperature leading to lower viscosity, which in turn leads to increased fluid flow and to an increased control of the fluid temperature on the system. To try to investigate a system by only considering one of the processes leads to an oversimplified representation.

Modeling systems such as these requires an understanding of the key processes operating, the scale at which the processes are operating, the magnitude of the fields gradients present (fluid pressure, temperature, stress, chemical concentration) and the different strata through which the processes are operating. For example, in flow dominated systems advective heat transport may be much more rapid than the convective transport of heat.

Any modeling approach also needs to take the different spatial and temporal scales into account. Complicated balance equations describing pore level processes may be solvable but must be up-scaled (see Chap. 5) to be useful for large scale modeling of a CO_2 storage site. Methods of solution include the application of multiple mesh approaches, adaptive mesh refinement, and hybrid analytical and numerical solutions in attempts to capture key behavior of coupled process systems using the computational power available.

A thermal, hydraulic, mechanical or chemical process is described by a balance equation based on conservation of mass or energy. The key basic balance equations for the THMC processes are described below. Generally there will be one thermal process (T), one, two or in some cases three fluid flow processes (H) depending on how multi-phase flow is dealt with (see also Fig. 2.3), one mechanical process (M), and (n) chemical (C) processes included depending on the number of chemical species considered.

4.2.3 Thermal Process

For heat transport (T) the balance equation is given by Eq. (4.2.1).

$$D_T \Delta T - c^w \rho^w \nabla \cdot (qT) - \rho Q_T = c^m \rho^m \frac{\partial T}{\partial t} \tag{4.2.1}$$

where c^m is the specific heat of the saturated porous rock, c^w is the specific heat of the fluid, D_T is the heat diffusion dispersion tensor for the porous medium, T is temperature, ρ^w is fluid density, ρ^m is the density of the saturated porous rock, Q_T is a heat source or sink and q is the advective fluid velocity. Here, after de Marsily (1986), the heat diffusion dispersion tensor contains a component for pure diffusion and a component for dispersion due to advection, i.e.

$$D_\alpha = \lambda_m + q_\alpha \beta_\alpha \tag{4.2.2}$$

where β is the heat diffusion dispersion coefficient in the α direction, λ_m is the isotropic heat conductivity of the porous medium, q_α is the advective flow velocity in the α direction, and D_α is the heat dispersion coefficient in the α direction. The value β_α is the product of the directional (longitudinal or transverse) dispersion coefficient, with $c^w \rho^w$.

Equation (4.2.1) represents the change in the amount of energy in a continuum at a discrete location in a unit volume as a consequence of heat entering, leaving or being stored in the discrete unit volume either through conduction, radiation diffusion or convection.

4.2.4 Hydraulic Process

For single phase fluid flow (H), the hydraulic process is similarly described as being the change in the mass of fluid in a continuum at a discrete location as a consequence of fluid entering, leaving or being stored in the discrete unit volume either through advection, or addition/removal through a source term.

$$S_s \frac{\partial p}{\partial t} - \nabla \cdot \left(\frac{k}{\mu} (\nabla p + \rho g \nabla z) \right) = Q \tag{4.2.3}$$

where S_s is the specific storage coefficient, k denotes the permeability tensor, p is the fluid pressure, and Q is a source/sink term. This equation is valid for a saturated, non-deforming porous medium with heterogeneous hydraulic conductivity.

For multiphase flow including some solid deformation in the pore space

$$S_\alpha \rho_\alpha \nabla \cdot \frac{\partial \mathbf{u}}{\partial t} + \frac{\partial (\phi S_\alpha \rho_\alpha)}{\partial t} + \nabla \cdot (\rho_\alpha \mathbf{q}_\alpha) - \rho_\alpha Q_\alpha = 0 \tag{4.2.4}$$

The fluid velocity $\mathbf{q}_\alpha$ is a non-linear function of the pressure gradient, after Darcy's law as given above

$$\mathbf{q}_\alpha = \phi S_\alpha \mathbf{v}_\alpha = \frac{Q_\alpha}{A} = -\frac{k_{r\alpha} \mathbf{k}}{\mu_\alpha} (\operatorname{grad} p_\alpha - \rho_\alpha \mathbf{g}) \tag{4.2.5}$$

where $k_{r\alpha} = f(S_\alpha)$ is discussed previously (Chap. 3). Here for the wetting phase

$$p_\alpha = p_w \tag{4.2.6}$$

and for the non-wetting phase, CO_2 in our example

$$p_\alpha = p_w + p_c \tag{4.2.7}$$

where p_c is expressed as a negative suction pressure.

4.2.5 Mechanical Processes

For mechanical processes (M), we again describe the deformation of a continuum as a function of the stress applied, which can be shown to be an energy balance equation. Again the basis for the analysis is that of the energy balance in a continuum at a discrete location as a consequence of stress and deformation changes within a discrete unit volume.

$$\nabla \cdot \sigma - \rho g = 0 \tag{4.2.8}$$

where σ is the stress, ρ is the medium density and g is the acceleration due to gravity.

4.2.6 Chemical Transport

Chemical transport (C) is described as being the change in the mass of a chemical species in a continuum at a discrete location as a consequence of fluid that species entering, leaving, being stored or being removed from the discrete unit volume either through advection, diffusion, sorption, or addition/ removal through reactivity, usually represented as a source term.

$$\frac{\partial C}{\partial t} = \nabla \cdot (D\nabla C) - q \cdot \nabla C + C_s \tag{4.2.9}$$

where C is the concentration, **D** is the dispersion tensor, q is the advective velocity, C_s is a concentration source/sink.

4.2.7 Solution of the Equation Systems

For each of the individual processes given above there are standard analytical solutions when considering the equation alone, e.g. Häfner et al. (1992), Bronstein and Semedjajew (1977). Coupling occurs when the solution of one process depends on the solution of another process. The link between the solutions of these processes is via the constitutive relationships, whereby the behavior of the material, fluid or energy is described mathematically as a function of the field variables of the processes involved in the solution. This makes the equations highly nonlinear.

Beyond a number of analytical solutions for idealized conditions, in heterogeneous systems the equations can either be solved in a staggered or non-staggered (monolithic) solution. A staggered solution involves the splitting of the system of equations into the individual processes. For each process the field variable is evaluated advanced with time, and the solution iterated internally between the split equations until a certain degree of accuracy in the solution is reached. Simplifications of the solution procedure using explicit results can be useful, e.g. during creation of the elasto-plastic matrix during deformation, but must be applied with caution. For a non-staggered solution all the equations are combined into a single mass or energy balance formulation for solution and the whole problem is advanced simultaneously in time. In both cases strict time control is necessary and a number of numerical stability criteria need to be observed related to the geometrical discretization and the amount of energy or mass being moved around per time interval.

It can be pointed out that caution must be applied in overstating the need for mathematical accuracy in a geological systems given the parametrical uncertainty at several levels of the equation systems. Computationally expensive solution procedures valid for the aeronautical industry where the constitutional behavior and the parameterization of the material properties is known to within very tight limits may not be a priority when the constitutional behavior is dependent on the geology of the system and the uncertainty of material properties bound several orders of magnitude.

The processes will also vary in terms of their spatial impact and the rate of change of the field variable. For instance during CO_2 injection, the fluid pressure change may move rapidly through the porous media and impact a wide area, while the thermal change will be localized around the borehole, and chemical reactions associated with the injection of the CO_2 will be limited to the contact of the CO_2 being injected and the in situ reservoir fluids.

In terms of chemical changes, equilibrium solutions are often assumed in the multi-species simulations, basically because the complexity of the chemical system is at the limits of our reasonable computational possibilities. Given the possible thousands of interacting chemical species, key reactions need to be identified. These can be rate-limited but again there is much discussion regarding reaction rates and surface areas available to reaction. Unfortunately it appears that most lab based parameterization of reaction rates suggest much more rapid reaction of minerals than derived from natural analogues. The location of the growth or dissolution of minerals within the porous media can have a significant impact on the constitutive relationships such as the intrinsic permeability of the system, or the mechanical stability of the system. If a mineral grows in a fracture, the physical size of mineral growth filling or dissolving from the fracture is linked to the permeability by a cube of the fracture aperture. Modeling such systems can easily lead to a complexity not supported by available data, and at best an understanding of the impact of key material phenomena can be gained.

Likewise phenomena vary in terms of the spatial distance over which it occurs, as well as the rate of change. For instance during rapid advective flow of CO_2 through a fracture, the CO_2 will cover a large distance in the direction of the fracture whilst diffusion normal to the fracture into the rock matrix will be operating at a time scale orders of magnitude slower than the flow.

As would be expected, seeking for efficient solutions of highly nonlinear partial differential equations has been of great scientific interest since the beginning of the last century. Numerous publications and journals are dedicated to this theme. There exist already a number of industrial and research codes developed and tested for the solution of the coupled processes. Prominent examples include the developments of the (1) TOUGH2 family of codes, including TOUGHREACT (Xu et al. 2014) for solving coupled hydro-chemical processes in multiphase CO_2 storage scenarios and TOUGH-FLAC (Rutqvist 2011) for hydro-mechanically coupled systems (http://esd.lbl.gov/research/projects/tough); (2) Open GeoSys (http://www.opengeosys.org) (Kolditz et al. 2012a) and Code Bright (http://www.etcg.upc.edu/recerca/webs/code_bright) (Olivella et al. 1996). These codes require expert users able to clearly define the systems they are attempting to model and with an understanding of which phenomena to include and exclude. There is no "off the shelf" solution for the selection of the codes which can be used. Several international collaboration projects have been looking at different numerical code possibilities and benchmarking their success to improve confidence of use. Examples include http://www.cgseurope.net/ and the IEAGHG modeling network.

In Sect. 4.4 we will present some example applications of coupled modeling with these models, addressing the coupled hydro-chemical (HC) system and the coupled hydro-mechanical (HM) system.

4.3 Modeling of the Small Scale Effects

Henry Power, Jesus Carrera and Christian Ian McDermott

In this section modeling of two important small-scale processes related to geological storage of CO_2 are discussed, namely the processes of convective dissolution and viscous fingering.

4.3.1 Convective Dissolution

Jesus Carrera

CO_2 injection in saline aquifers leads to a CO_2 plume that is less dense than the brine and floats on it while spreading horizontally. CO_2 solubility is quite high, so that a significant portion of CO_2 dissolves into the brine at the interface between the two fluids. CO_2 dissolution is favourable for several reasons. First, by reducing the volume and pressure of the free phase, dissolution reduces the risk of CO_2 phase leaking upwards. In fact, migration of dissolved CO_2 is hard both because the viscosity of brine is much larger than that of any CO_2 phase and because CO_2–rich brine is 1–2 % denser than resident brine, so that it will tend to sink. Second, dissolution facilitates that CO_2 transforms to more stable species such as bicarbonate or, if geological conditions are favourable, mineral carbonates.

CO_2 dissolution into the brine is initially controlled by diffusion from the CO_2-brine interface, which is a slow process with mass flux evolving proportional to $t^{-1/2}$, t being time. For this reason, early estimates of dissolution were quite pessimistic, predicting it would take some 1000 years for dissolution to be a relevant process (IPCC 2005). However, as the amount of dissolved CO_2 builds up, the resulting conditions (denser liquid on top) are unstable, so that the CO_2 rich brine will tend to sink, thus promoting the convective transport of dissolved CO_2 away from the CO_2 phase plume. In fact, Riaz et al. (2006) showed that this mechanism can reduce significantly the dissolution time.

The question is how long it takes for the instability to develop. Riaz et al. (2006) found that the onset of instability may take some 100 years. However, they made a number of simplifications that may be critical. The first one is the analogy to a heat transport problem. This is a common simplification, because of the tradition of studying convective transport of heat. In a heat transport model, fluid may be assumed to consist of a single component. As such, heat transport and fluid flow are basically linked through the buoyancy term. However, in mass transport problems the fluid must be viewed as consisting of at least two components (i.e. brine and CO_2). Since the flow equation expresses the mass balance of the whole fluid phase, additions of any of the two components must be accounted for in it. Specifically, while an impervious boundary (i.e. a boundary with zero brine flux) can be treated as a zero mass flux in thermal problems, it must allow for fluid mass flux (CO_2 component) in mass transport problems (Hidalgo et al. 2009b). The second issue refers to the simplifications assumed for flow and transport problems. For flow, fluid is considered incompressible and the Boussinesq approximation valid. However, acknowledging compressibility of the fluid helps in simulating the pressure rise caused by the influx of CO_2, which helps promoting CO_2 flux downwards. Boussinesq approximation may affect the transient solution (Johannsen 2003a), although it is valid for the typical range of values of the Rayleigh number (see Landman and Schotting 2007). For transport, dispersion is neglected. Hydrodynamic dispersion accounts for the effects of the deviations from the mean flow caused by small scale heterogeneity on solute transport. Heterogeneity is present in all natural systems; therefore, dispersion has to be included in any realistic transport formulation. Notice that in thermal analogies, which are the base of many CO_2 dissolution models, dispersion is often neglected. This is arguably done (See, e.g., Ferguson and Woodbury 2005; Hidalgo et al. 2009a) because of the relatively large value of thermal conductivity. However, dispersion cannot be disregarded in solute transport because it can be much larger than molecular diffusion. In fact, dispersion can be artificially increased if a fluctuating injection regime is adopted (Dentz and Carrera 2003).

In the following, we summarize the work of Hidalgo et al. (2009a) on how a more realistic representation of CO_2 dissolution affects the time for the onset of convection and the estimation of the CO_2 flux across the interface.

4.3.1.1 Governing Equations

The compressible density-dependent flow and advective–diffusive–dispersive transport equations that govern the CO_2 dissolution in a saline aquifer are written as (recall Chap. 3):

$$\rho S_p \frac{\partial p}{\partial t} + \rho\theta\beta_\omega \frac{\partial \omega}{\partial t} = -\nabla \cdot (\rho \mathbf{u}) \tag{4.3.1}$$

$$\rho\theta \frac{\partial \omega}{\partial t} = -\rho \mathbf{u} \cdot \nabla\omega + \nabla \cdot (\rho(\theta D_m + \mathbf{D})\nabla\omega) \tag{4.3.2}$$

where ρ is fluid density, S_p the specific pressure storativity ($S_p = S_s \rho_{wg}$, where S_s is the traditional specific storativity), p pressure, θ the volumetric fluid content, $\beta_\omega = (1/\rho)\partial\rho/\partial\omega$ expresses the brine density dependence on CO_2 mass fraction, considered constant, ω the CO_2 mass fraction, $\mathbf{u}$ the Darcy velocity and $\mathbf{D}$ the hydrodynamic dispersion tensor, as described in Chap. 3.

The dissolution interface between the CO_2 and the brine is located to be at the top of the domain, i.e. z = 0, and it is modeled as a prescribed concentration boundary. The domain is conceptually considered semi-infinite, i.e. there is no interaction with the lower and lateral boundaries. Therefore, boundary conditions are written as:

$$\begin{array}{ll} -\rho \mathbf{u}|_b \cdot \mathbf{n} = m_s & \text{if } z = 0, \\ \rho \mathbf{u}|_b \cdot \mathbf{n} = 0 & x, z \to \infty, \end{array} \tag{4.3.3}$$

for flow and

$$\begin{array}{rcll} \omega|_b & = & \omega_s & \text{if } z = 0, \\ (-\rho \mathbf{u}\omega + \rho(\theta D_m + \mathbf{D})\nabla\omega)|_b \cdot \mathbf{n} & = & 0 & x, z \to \infty, \end{array} \tag{4.3.4}$$

for transport. In these equations $\mathbf{n}$ is the unit vector normal to the boundary pointing outwards, $m_s [ML^{-2}T^{-1}]$ is the CO_2 mass flux across the top boundary, ω_s is CO_2 solubility in brine (highly sensitive to both pressure and salinity, see Sect. 3.2) and $|_b$ indicates evaluation at the boundary.

The conceptual model proposed by Eqs. (4.3.1)–(4.3.4) introduces significant modifications with respect to previous CO_2 dissolution models. First, governing equations acknowledge fluid compressibility and hydrodynamic dispersion. Second, boundary conditions include the CO_2 mass flux across the top boundary. This CO_2 flux is

$$m_s = (-\rho \mathbf{u}\omega + \rho(\theta D_m + \mathbf{D})\nabla\omega)|_b \cdot \mathbf{n}. \tag{4.3.5}$$

Subtracting (4.3.3) multiplied by ω from (4.3.5) yields

$$m_s = \frac{1}{1-\omega}\rho(\theta D_m + D_{zz})\frac{\partial \omega}{\partial z}, \tag{4.3.6}$$

where it has been imposed that the concentration gradient is vertical.

For analysis purposes, it is convenient to write governing equations in a dimensionless form. We adopt the characteristic space and time, respectively, as

$$L_s = \frac{\theta D_m + \alpha_L u_b}{u_b}, \qquad t_s = \frac{\theta(\theta D_m + \alpha_L u_b)}{u_b^2} \tag{4.3.7}$$

where $u_b = k\Delta\rho g/\mu$ is the modulus of velocity when CO_2 is dissolved at saturation ($\omega = \omega_s$), and $\Delta\rho$ is the density contrast between the brine and the CO_2-saturated brine (see Sect. 4.2 for values of both densities). This choice leads naturally to the following dimensionless variables (prime denotes dimensionless variable): $p' = pS_p/\theta$, $\rho' = \rho/\Delta\rho$, $\mathbf{u}' = \mathbf{u}/u_b$, $(x', z') = (x, z)/L_s$ and $t' = tu_b/\theta L_s$.

Using these scaling factors, the dimensionless form of flow and transport equations becomes

$$\frac{\partial p'}{\partial t'} + \frac{\partial \omega'}{\partial t'} = -\frac{1}{\rho'}\nabla' \cdot (\rho'\mathbf{u}') \tag{4.3.8}$$

$$\frac{\partial \omega'}{\partial t'} = -\mathbf{u}' \cdot \nabla'\omega' + \frac{1}{\rho'}\nabla' \cdot (\rho'((1-b_L)\mathbf{I} + b_L\mathbf{D}')\nabla'\omega'), \tag{4.3.9}$$

where b_L is defined as

$$b_L = \frac{\alpha_L}{L_s} \tag{4.3.10}$$

with the corresponding dimensionless boundary conditions. Notice that, because we are using a semi-infinite domain, we do not use a Raleigh number. Still, for comparison purposes, we define it as $R_a = u_b L_s/\theta D_m$, where D_m is the molecular diffusion coefficient.

4.3.1.2 Numerical Analysis

CO_2 dissolution was simulated using the code Transdens (Hidalgo et al. 2005). Transdens solves density-dependent flow and transport problems using a finite element discretization in space and weighted finite difference in time. Reverse time weighting by Saaltink et al. (2004) is used to minimize mass balance error during time integration. Darcy velocity is computed using the consistent velocity algorithm by Knabner and Frolkovic (1996). The code was validated using the usual density-dependent problems benchmarks and by comparison with other codes.

Instabilities in the system were not seeded explicitly, but by the propagation of numerical errors. Simulations were carried out on a squared domain of 10 × 10 m (dimensionless size between 17.86 and 2000 depending on b_L), which proved to be large enough to warrant that there was no interaction with the lower boundary. The domain was discretized with a mesh of rectangular finite elements of 101 × 201 nodes. For comparison purposes, parameters were chosen equal to those of Riaz et al. (2006) ($H = 10$ m, $k = 3.0581 \times 10^{-13}$ m^2, $\theta = 0.3$, $D = 10^{-9}$ m^2/s, $\mu = 5 \times 10^{-4}$ Pa s, $\Delta\rho$ kg/m^3), details are shown by Hidalgo and Carrera (2009).

Our findings are summarized in Figs. 4.4, 4.5 and 4.6. Figure 4.4 displays the growth of fingers. Figure 4.5 displays the time evolution of CO_2 mass flux. Dissolution is initially controlled by diffusion. Therefore, it decays as $t^{-1/2}$. However, when fingering develops, the dimensionless mass flux fluctuates around 35 ω_s, which is indeed a very simple expression. Notice that it is not sensitive to dispersion, but basically to permeability and buoyancy forces. Figure 4.6 displays the onset time of convective dissolution. Again a simple expression $t_{onset} = 5619–5731b_L$ fits data quite well.

The topic has been extensively studied since Riaz et al. (2006) and Hidalgo et al. (2009a). High resolution simulations were performed by Pau et al. (2010), the effect of geochemistry and reactive transport was studied by Ghesmat et al. (2011), Hidalgo et al. (2015), and Fu et al. (2015); the effect of the capillary transition zone was studied by Elenius et al. (2014) and Emami-Meybodi and Hassanzadeh (2015). Visualization experiments have been performed after Neufeld et al. (2010) and MacMinn et al. (2011). Rasmusson et al. (2015) studied the prerequisites for convective mixing for a wide range of field conditions.

The summary of all this work is that the phenomenon is complex. Szulczewski et al. (2013) identify seven dissolution regimes. But, for most practical purposes, three sequential regimes are dominant:

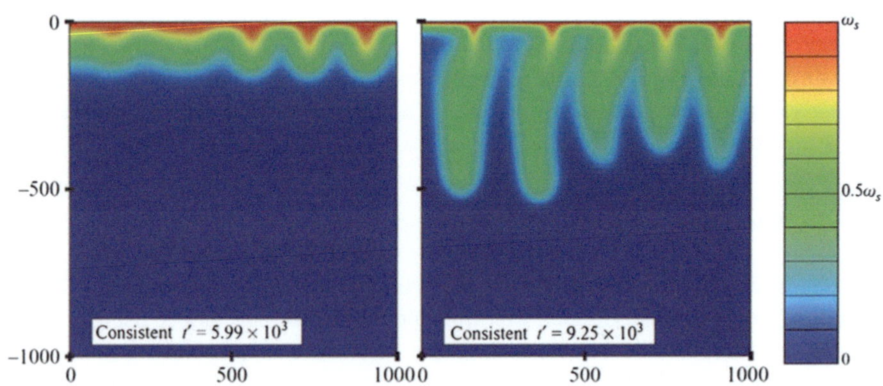

Fig. 4.4 Onset of fingering

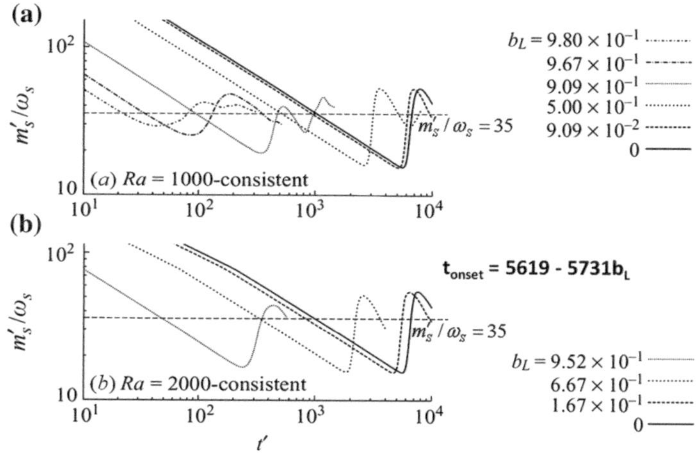

Fig. 4.5 Time evolution of the CO_2 mass flux (dimension rate) for Rayleigh number of **a** 1000 and **b** 2000. The CO_2 dissolution flux is initially controlled by diffusion (dispersion) but increases to values that fluctuate around $m'_s = 35\omega_s$ after the onset of fingering. While the onset time depends on dispersion (b_L), the mass flux does not

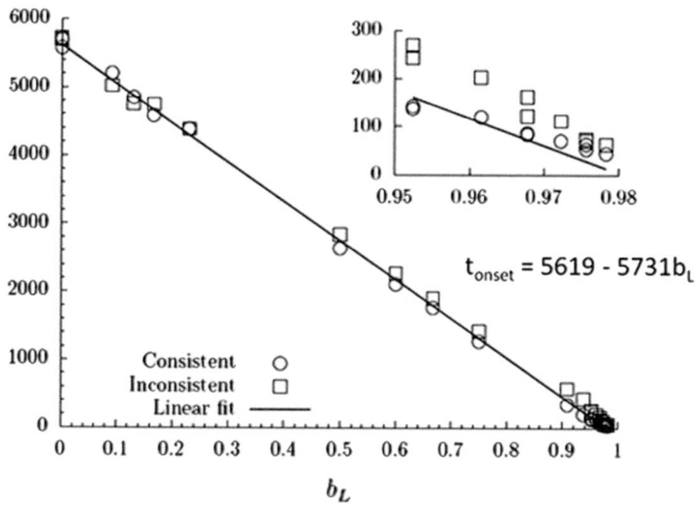

Fig. 4.6 Dimensionless time for the offset of convection versus dimensionless dispersion (Hidalgo and Carrera 2009). Notice that dispersion brings a dramatic decrease in the time for the onset of convection

1. *Diffusive Regime* Dissolution is controlled by diffusion away from the CO_2 plume until the onset of fingering. The time for this onset is greatly reduced by dispersion, which is important because dispersion can be increased by simply adopting a fluctuating injection regime. It is also reduced by capillarity. In fact,

visualization experiments suggest that it is virtually simultaneous with the development of the plume.

2. *Convective Dissolution* The mass flux after fingering develops is constant (except for capillarity effects) and virtually insensitive to dispersion. It solely depends on permeability and buoyancy and equals $m = \alpha\omega_s\rho u_b$, where α was evaluated as 0.0142 in the work of Hidalgo and Carrera (2009), but is usually taken as 0.0175, resulting from the high resolution simulations of Pau et al. (2010).
3. *Shutdown Regime* As concentration below the interface starts increasing, the density difference drops and so does u_b. This can be assumed to occur when the fingers reach the aquifer bottom. After this time, dissolution is approximately proportional to the difference between the average concentration in the brine and the saturation concentration, ω_s.

4.3.2 Viscous Fingering

Christopher Ian McDermott and Henry Power

In this section we will consider the interfacial instability between two immiscible fluids in a porous media, during the displacement of brine (wetting fluid) by an injected supercritical CO_2 (non-wetting fluid). Supercritical CO_2 has a lower density and viscosity than the surrounding brine. Therefore, at moderate capillary number, as the plume of the injected CO_2 evolves, the interface between the two fluids will develop viscous fingering patterns. Besides, the supercritical CO_2 in contact with the brine dissolves into the liquid phase, having a large contact surface area due to the fingering structure of CO_2–brine interface.

Viscous fingering instability of immiscible fluids displacement in porous media can be studied by numerical simulation of the multiphase flows formulation, in terms of the saturation index S_w where a flow region of overlapping between the fluids is considered for $0 \leq S_w \leq 1$ without a specific definition of the fluid interface. This type of approach has been reported in the literature by several authors, including Huang et al. (1984), Yortsos and Huang (1984) and more recently Garcia and Pruess (2003) and Riaz et al. (2006). By use of this type of formulation, however, direct implementation of interfacial processes is not possible. Several numerical techniques have been proposed in the literature to include the existence of an interface, with the objective of incorporating some of the interfacial processes in the formulation [for more details see McDermott et al. (2011)]. In the next section we will discuss different approaches of front tracking for multiphase flow and present an illustrative example of application of one of the methods.

4.3.2.1 Front Tracking Numerical Algorithms for Multiphase Flow Simulations of Viscous Fingering

Modeling two-phase flow using standard grid based numerical techniques presents a problem due to the sharpness of the front developed by the replacement of one fluid with another, balanced against the need to discretize the model into grid points and elements. One of the issues is that changes due to coupling to other processes, such as mechanical faulting and pressure release of the fluid, is likely to occur as the front passes. Depending on the resolution of the fluid flow grid, this time dependent signal may be missed or smeared with other signals loosing information on the integrity of the reservoir. In finite element approaches lower order based interpolation functions often fail to represent the sharp front and this can also lead to oscillations around the true solution. Finite volume methods can avoid these oscillations, but there are issues concerning the relative computational expense of these formulations and difficulties in representing smoothly varying heterogeneity fields while minimizing discretization overheads. Mixed finite element solutions whereby both the velocity field and the pressure solution are considered primary variables are finding some acceptance; however they are computationally more complicated to implement, especially with respect to solver capabilities (Younes et al. 2010).

Three main approaches have been adopted to address this problem. The most widely adopted approach is that of **grid refinement,** or adaptive mesh refinement in the vicinity of the front. The geometry of the grid is locally adapted to better represent the numerical processes operating at a local scale and represent steep gradients within the model. Recent examples for highly heterogeneous fields include Chen et al. (2003) and Durlofsky et al. (2007). Such adaptive grid methods allow the discretization scale to follow the front through the model, however the front location will always be approximated within the scale of the discretization used. Such methods are extremely useful but can be computationally expensive and make coupling to other processes such as thermal, mechanical, reactive chemical (TMC) more complicated due to the constant alteration in grid size and location.

Interface tracking methods have been developed that apply sophisticated reconstruction algorithms based on either the ratio of volumes of a fluid in an element (volume of fluid methods) or the advective velocity of the interface throughout the modeling regime (level set method). Meakin and Tartakovsky (2009), with references therein, review these approaches and conclude that there has actually been very little application of these techniques to multiphase fluid flow in fractured and porous media. Recent examples of their application include Huang and Meakin (2008), Huang et al. (2005) and Unverdi and Tryggvason (1992). Glimm et al. (1999) tracked the front throughout the computational domain using a grid based interface reconstruction based on information in the grid element and the information from surrounding elements. Interface tracking methods address the problem of the location of the front based on the volumetric fluid fluxes into and out of an element, the understanding that there must be continuity between elements and assumptions as to the shape of the front.

McDermott et al. (2011) in turn presented an alternative development of a front tracking based on a **local one-dimensional analytical approximation** of the fluid interface to add information to the model and to predict the location and shape of the front within the elements. The assumption removes the necessity for adaptive mesh refinement and the need for further sophisticated reconstruction of the front surface. The information on the geometry of the front surface under the conditions given in the element is being predicted by the locally one-dimensional representation of the front. The method increases the accuracy of the prediction of the front location, but is still bound by the overall accuracy of the numerical method applied to determine the primary variables which the analytical solution depends upon.

In the approach, the standard two-phase flow equations described in Chap. 3 are solved using the IMPES (implicit pressure explicit saturation) formulation (e.g. Helmig 1997). The saturation equation is formulated in terms of the volume of replacing fluid in the discretised elements. This volume may either be predicted using a standard first order approximation, or by using the analytical solution of the location of the front directly and integrating under this front. The latter approach works for homogenous conditions, and predicts radial flow better than the standard numerical methods using full up-winding schemes, however, still requires further development for heterogeneous conditions.

The unknowns of the wetting phase pressure and the non-wetting phase saturation are solved sequentially using different approaches. For the pressure formulation the Full upwind galerkin (FUG) finite element method is used, providing maximum mobility and for the solution of the saturation equation we introduce extra information in the model by including an analytical derivation of the shape function for the evaluation of the saturation front. The details of the approach and the numerical implementation are given in McDermott et al. (2011) and are not repeated here.

The advantage of using an analytical derivation for the location of the saturation front is that it removes the necessity to refine the mesh in the locality of the saturation front whilst still maintaining the sharpness of the front without numerical oscillations. The method, however, still has the requirement that the Courant time criteria apply for the advective flux of the front (Kolditz 2001), and, as is the case using a FUG scheme, numerical diffusion is introduced. In addition, the accuracy of the location of the front prediction is dependent on the accuracy of the numerical solution of the primary variables.

To demonstrate the use of this hybrid method for the solution of two phase flow and the prediction of the front, we assume simplistic conditions. First it is assumed that there is no pressure difference across the liquid–liquid phase boundary, i.e. capillary pressure effects are negligible. The second assumption is that the solid–liquid–liquid contact angles have no significant impact on the flow characteristics. This approximation means that the capillary pressure term P_c is neglected in the pressure formulation. It needs to be pointed out that at the pore scale it would be necessary to include these effects, as discussed in detail by Meakin and Tartakovsky (2009) and references therein. Niessner and Hassanizadeh (2008) in turn examine the role of fluid–fluid interfaces and the impact they can have, such as hysteresis. At the macroscopic scale, given the heterogeneity of geological medium and the assumption of generally continuous fluid phases, these simplifications can, however, be considered valid.

For the demonstration example presented in detail in McDermott et al. (2011) we assume constant density and no deformation. To be able to evaluate the volume of the replacing fluid, the volume underneath the saturation surface needs to be calculated. Different analytical approximations can be considered for determining the location of the two phase flow front. Here for the demonstration of the method we apply the original 1D analytical solution derived by Buckley and Leverett (1941) for the replacement of one fluid with another in two phase flow.

The *Buckley* and *Leverett* solution is one of the simplest for two phase flow where capillary pressures are not considered to be causing any resistance to flow. The capillary pressure is, however, included indirectly in terms of allowing residual trapping by the consideration of a residual saturation for the calculation of the relative permeability functions below. Buckley and Leverett (1941) used relative permeabilities described by the functions given below (4.3.11) derived from their laboratory work.

$$k_{r1} = \frac{(S_1 - S_{1r})^2}{(1 - S_{1r} - S_{1r})^2} \quad \text{and} \quad k_{r2} = \frac{(1 - S_1 - S_{2r})^2}{(1 - S_{1r} - S_{2r})^2} \tag{4.3.11}$$

The *Buckley* and *Leverett* analytical solution of the saturation equation considering fractional flow functions is presented by Thorenz et al. (2002) as being

$$\Delta x = -\frac{u_{total}}{\phi} \frac{\partial \left(\frac{1}{1 + \frac{k_{r2}\mu_1}{\mu_2 k_{r1}}} \right)}{\partial S} \Delta t \tag{4.3.12}$$

From this equation it is possible to derive the saturation curve presented in Fig. (4.7a). The Eq. (4.3.12) has two possible saturations for one location. Using the equal area solution the actual location of the saturation front is determined by constructing a shock front whereby "Area 1" is equal to "Area 2" (see Fig. 4.7b).

In Fig. (4.7), the solution of the *Buckley and Leverett* equation has been normalised against the maximum distance *md* from the origin for the extension of the saturation front. Examining (4.3.12) it can be seen that the term $\frac{u_{total}}{\phi} \Delta t$ is a scaling term, and for the solution presented in Fig. (4.7) we set this to 1.

This means that it is possible for any combination of flow rates, porosity and time to be compared with the normalized analytical solution via a scaling factor. This fact is central to the application of this analytical solution.

The shape of the analytical solution from the origin to the saturation front can be approximated by a polynomial fitted to match the normalized analytical response (Fig. 4.7b). Therefore a standard response for the solution assuming constant

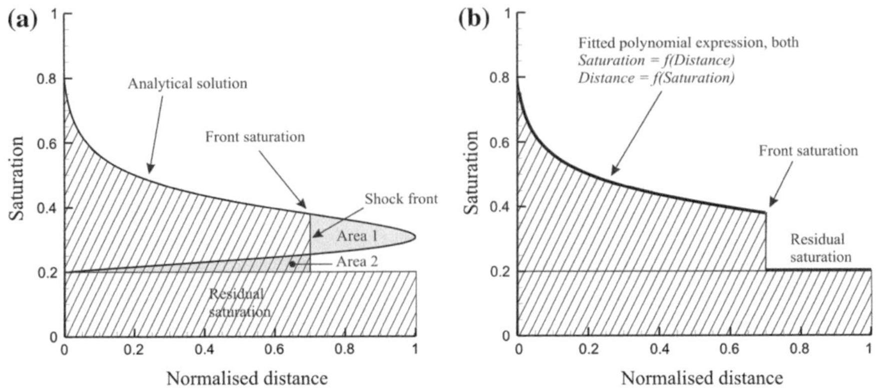

Fig. 4.7 **a** One-dimensional analytical solution for two phase flow and **b** the two-phase flow surface is approximated by a polynomial expression (McDermott et al. 2011)

material permeabilities and viscosities within an element may be evaluated by solving (4.3.12) and applying the appropriate scaling term. Depending on the saturation values of the nodes, the saturation front may be (1) present within the element, (2) have passed through the element, (3) not have reached the element. Each of these cases is handled individually and more details can be found in McDermott et al. (2011). For a detailed description of the comparison between the simplified 1D analytical solution for two phase flow by Buckley and Leverett (1941), and the different the FUG schemes see McDermott et al. (2011).

4.3.2.2 Demonstrative Example: Well Injection in a Heterogeneous Reservoir Rock

The proposed approach by McDermott et al. (2011) is used to simulate injection of supercritical CO_2 into a reservoir layer underlying a caprock. The fluid and material properties are also given in the above reference and are not repeated here.

The CO_2 spreads out laterally from the injection point, and forms channels as a result of the heterogeneity. This is demonstrated in Fig. 4.8, left, where the front tracking method can be seen to be providing sub-element scale information on the location of the saturation front. In Fig. 4.8 we also compare the results of a finite volume approach utilizing an identical physical model and the method presented by McDermott et al. (2011), the latter with and without front tracking. In this figure the front tracking location is presented, then removed for comparison to the finite volume (FV) approach. We note that the overall shape of the predicted radial flow patterns is similar, and many features can be cross referenced. The FUG-vT method predicts the formation of more discrete and higher saturated channels, while the Finite Volume scheme predicts more distribution in the saturating phase. This is due

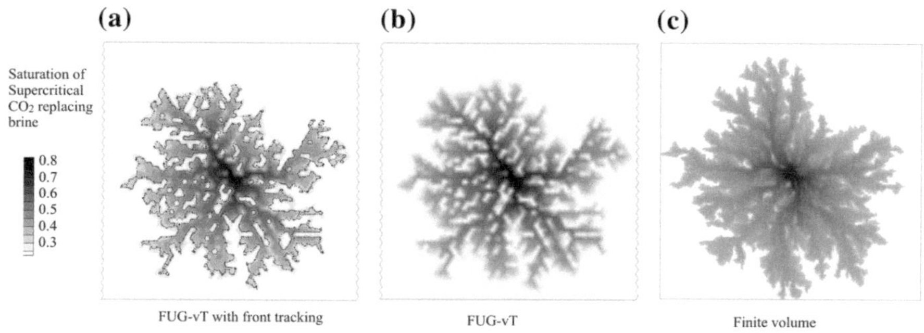

Fig. 4.8 Comparison of well injection of supercritical CO_2 in a heterogeneous reservoir rock. **a** FUG with front tracking, **b** FUG without front tracking and **c** a finite volume solution (McDermott et al. 2011)

in part to the differences in the numerical schemes, specifically concerning the tendency for the finite element approach to 'blur' permeability contrasts across elements as the fluxes are assigned to the nodes by integrating across the elements. This creates the possibility for channelled pathways that the finite volume approach would not include.

Strictly speaking the proposed model does not allow capillary trapping to occur, although the pattern of flow predicted would suggest this. The flow in the system is driven by the source term representing the well and pressure field developed by the well. Preferential flow as a consequence of the heterogeneous permeability field leads to the partial isolation of low permeability blocks within the flow system. This is exaggerated by the positive feedback caused by the relative permeability functions, i.e. the higher the saturation, the higher the permeability. Should capillary pressure also be included in the calculation, these low permeability areas would be even more sealed from the preferential flow channels. For more details of the practical application of the proposed mixed approach, advantages and disadvantages, see McDermott et al. (2011).

4.3.2.3 Effect of Dissolution on the Evolution of Viscous Fingering

Chuoke et al. (1959) pioneering work provides a theoretical analysis of the onset of viscous fingering in immiscible displacements in porous media by using the similarity between Hele-Shaw and porous media flows and ignoring the zone of partial saturation or volume concentration of the displacing fluid behind the front (capillary trapping) . In this type of approach, the existence of a sharp interface between the immiscible fluids is considered, corresponding to a macro average (representative elementary volume REV) of the interface at the micro-level between the immiscible fluids, in contrast with a volume of fluid formulation described in terms of saturation index (multiphase flow formulation), where a flow region of overlapping fluids is considering for $0 \leq S_w \leq 1$ without a specific defined definition of a fluid interface.

Sharp interface approach requires the definition of different equations of motion on each side of the interface and the tracking of the interface motion as the flow progresses, allowing direct consideration of interfacial processes by including them into the corresponding matching condition at the interface. In the case of a two dimensional immiscible displacement in a porous medium, fingers of width given by the characteristic scale $w \sim \left(\frac{k}{N_{Ca}}\right)^{1/2}$ found by the Hele–Shaw approximation, under-predict the experimental observations. This has led to the hypothesis of an effective surface tension γ , larger than the molecular surface tension and function of the wetting conditions, that varies with the large-scale curvature at the macro-scale (see Weitz et al. 1987). The use of a modified jump condition in terms of the effective surface tension is known as Chouke's boundary condition and the resulting interface instability analysis is referred to as Hele–Shaw–Chouke theory (for more details see Homsy 1987). As commented by Homsy, it is not clear how surface tension acting at the menisci in the pore space can provide a restoring force proportional to the macroscopic curvature. However, strong experimental evidence has been reported in the literature with excellent agreement between the experimentally observed characteristic macroscopic length scales and Hele–Shaw–Chouke predictions, by fitting the value of the effective surface tension. Most of the works dealing with immiscible displacement have been focused on two-dimensional cases. However, some works have also addressed the difference between two- and three-dimensional displacements (Xiaoping et al. 1997).

In the multiphase flow approach presented in the previous section, the variations in saturation in the overlapping region result in a gradual change of the mobility of both phases. This type of analysis is closely related to the stability of graded mobility process, see Gorell and Homsy (1983) and Hickernell and Yortsos (1986), where depending upon the mobility function a displacement that has an unfavourable viscosity ratio may still be linearly stable, even at infinite capillary number. Homsy's (1987), review article presents a very detailed analysis of these two types of approach, i.e. sharp front (Hele–Shaw–Chouke) and saturation index (multiphase flow). Besides, in the article he addresses the similarity between changes of mobility due to variation in saturation during immiscible displacements and due to changes in concentrations in miscible displacements.

In the sharp interface approach, the effect of capillary trapping of the wetting fluid (brine) on the evolution of CO_2 plume (non-wetting fluid) is not considered, which results in a local reduction of the difference between the mobility of the two phases and consequently affects the fingering evolution. Daripa and Pasa (2008) presented an analysis showing that the slowdown of instabilities due to capillarity is usually very rapid which makes the flow almost, but not entirely, stable. New sharp interface formulations and numerical solutions explicit consider the effect of dynamic wetting by including a nonlinear term in the pressure jump condition due to the trailing film left behind by the displaced fluid. Jackson et al. (2015a) used this type of approach to study the effect of capillary trapping on the nonlinear evolution of radial viscous fingering in a Hele–Shaw cell and observed that finger interaction is reduced and finger breaking mechanisms delayed but never fully inhibited, which allows the primary fingers to advance further into the domain before secondary fingers are generated, reducing the level of competition.

In this section we will present an analysis of the coupling effect between the surface dissolution at the interface between two immiscible fluids and the evolution of viscous fingering in a Hele–Shaw cell (for more details see Power et al. 2013). Dissolution processes will change the kinematic and dynamic conditions at the fluid interface of an immiscible displacement and consequently the dynamics of any interfacial instability. For simplicity in the present numerical solution a single-phase approximation is considered, this implies that the viscosity of the less viscous fluid is so small that it may be ignored. In the case of CO_2 sequestration, the viscosity of the CO_2 gas can be one or two orders of magnitude smaller than the brine's viscosity. Therefore, the type of single-phase formulation presented here only provides a leading order approximation of the complete asymptotic regular perturbation approach in terms of the viscosity ratio of the problem. The reader is referred to Jackson et al. (2015b) for an extension to a two phase formulation of the single phase approach presented here, where direct account of the viscos effects of the injected CO_2 and resident brine is considered.

The mathematical modeling of viscous fingering in Hele–Shaw cell is generally carried out by using a potential model formulation. Hele-Shaw flow takes place when a viscous fluid moves slowly between two fixed parallel plates, separated by a thin gap. The mean flow is two-dimensional, and the mean velocity components in the plane of the cell are related to the mean pressure by Darcy equations $u_i = -M\partial p/\partial x_i$ with $M = b^2/12\mu$ which is a function of the plate gap b and the fluid viscosity μ. For incompressible fluid of constant viscosity the conservation of mass implies that $\partial^2 p/\partial x_i^2 = 0$, i.e. the Laplace equation.

The boundary conditions at infinity are those for a growing bubble, i.e. constant pressure, while the boundary conditions at the fluid interface Γ are given by the kinematic and dynamic conditions obtained from the conservation of mass and momentum across the interface. In the evolution of the moving front, the patterns of the interfacial fingering are determined by the interface kinematic and dynamic matching condition. In the present case, the transfer velocity of the dissolved gas reduces the interface displacement velocity described by the kinematic matching condition, delaying the evolution of the fingering. On the other hand, the momentum flux across the interface, due to the CO_2 dissolution, modifies the dynamic matching condition at the interface with possible changes in the fingering pattern.

The interface matching conditions are obtained by considering the surface separating the two-phase flow (in reality the two flows are separated by a complex, narrow transitional layer, which is modeled here as a simple surface, see Edwards et al. 1991), which is represented here as a material pillbox straddling the moving and deforming interface between the two-phases, as shown in Fig. 4.9, where the pillbox encloses a small fluid volume in contact with the interface having an internal surface S and an interface surface E. At the pillbox, mass and momentum conservation are imposed and the corresponding kinematic and dynamic interface matching condition obtained by taking the limit when the surface S tends to the interface surface E (see Power at el. (2013) for more details). In contrast to cases without gas dissolution into the brine, where, as the interface develops fluid does not cross the interface, in the present case, CO_2 crosses the interface by dissolution to become part of the brine liquid. Under this condition, the sharp interface is represented as a moving driving surface, i.e. a surface at which density and velocity are discontinuous and mass and momentum are transferred across it (see Slattery 1999). In our case, the mass and momentum transfer is associated with the interface due to the dissolution process.

The detailed derivation of the approach is given in Power et al. (2013) and will not be repeated here. Where due to the dissolution of the CO_2 into the brine at the interface, two new terms are included in the interface matching condition; a mass transfer term, $\rho_g u_d$, in the kinematic condition and a momentum transfer, $\rho_g u_d^2$, in the dynamic condition with u_d being the dissolution rate velocity.

As previously commented in Sect. 4.3.1, at the fluid interface CO_2 will dissolve into the aqueous phase, where by the difference in density of the developed fluid mixture and the local brine a convective flow is induced, promoting gravitational instabilities and the formation of density fingering. According to the conservation of mass, the descending mixed fluid along the fingers will induce recirculation cells of brine fluid between the fingers with an associated fluid entrainment into the fingers from the surrounding brine. The entrained brine reduces the density difference and the CO_2 concentration at the interface diffusive boundary layer, resulting in enhancement of the dissolution process (convectively-enhanced dissolution) . Recent experimental and numerical studies by Kneafsey and Pruess (2009) and Neufeld et al. (2010) (see Sect. 4.3.1) have observed that the convectively-enhanced dissolution process tends to produce an almost constant dissolution flux after only a very short time, which will persist for a very long time, i.e. on the scale of years, corresponding to the constant values in Fig. 4.5, i.e. the horizontal lines in the figure for each Rayleigh number. Ward et al. (2014a, b), studied the effect of

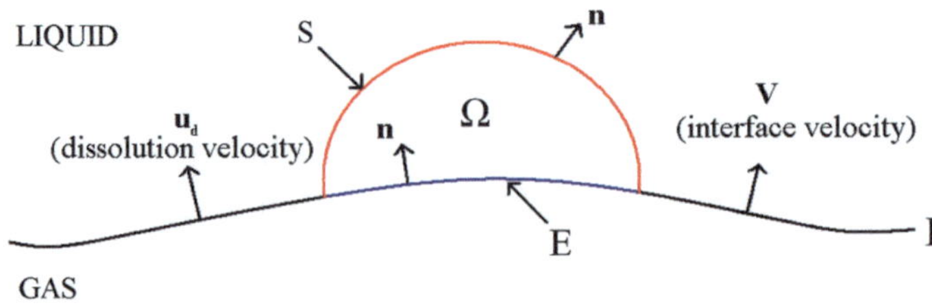

Fig. 4.9 Definition of pillbox at the fluid interface

chemical reactions on the convective enhanced dissolution process by considering that the solute concentration decays via a first-order chemical reaction, restricting the depth over which solute can penetrate the domain. Their simulations and analysis suggest that the time-averaged CO_2 uptake rate (dissolution flux) follows the classical scaling, namely being proportional to Ra (Rayleigh number), as observed in Fig. 4.5. However, the dissolution flux has a strong dependence on Ra. This provides evidence that the flux is restricted by the boundary layer, and that the plumes beneath accommodate this constraint. For convenience in this work, we will consider the CO_2 dissolution velocity to have a constant value, found from the dissolution flux estimates of Kneafsey and Pruess (2009) and Neufeld et al. (2010).

4.3.2.4 B-Spline Boundary Element Formulation (BEM) and Numerical Scheme

Since the interface between the two fluids experiences large deformations, and the correct determination of its shape is of foremost importance, the use of boundary integral representation (potential theory) is an attractive approach to find the numerical solution of the problem that has been successfully implemented before, see Degregoria and Schwartz (1986), Tosaka and Sugino (1994), Power (1994) and Zhao et al. (1995). On the other hand, in an immiscible displacement, surface tension plays a dominant role in the evolution of the viscous fingers, and so it is important that the shape of this interface is properly modeled in order that the local curvature at each interface point may be accurately evaluated. For this reason Power et al. (2013) employed boundary element formulation (BEM) model employing uniform cubic B-splines both for the geometry and the field variables is implemented, following Cabral et al. (1990) and Zhao et al. (1995).

In Power et al. (2013), the pressure field is expressed as a source potential (injection flux) plus a bounded perturbed field.

$$p = \hat{p} + \frac{Q}{2\pi M} \ln\left(\frac{r}{r_a}\right) \tag{4.3.13}$$

Therefore a boundary integral representational formula for the perturbed pressure evaluated at the fluid interface can be written in terms of the Laplace fundamental solution and corresponding flux, as (see Power and Wrobel 1995):

$$c(\xi)\hat{p}(\xi) = \int_{\Gamma} \hat{p}(\chi)q^*(\xi,\chi)d\Gamma_\chi - \int_{\Gamma} \hat{q}(\chi)p^*(\xi,\chi)d\Gamma_\chi + p_c, \tag{4.3.14}$$

where $c(\xi) = \theta/2\pi$, θ is the internal angle subtended at the source point ξ, $p^*(\xi, \chi)$ is the fundamental solution to the Laplace equation, $q^*(\xi, \chi)$ and $q(\xi)$ are the normal derivatives of $p^*(\xi, \chi)$ and $p(\xi)$, respectively. Here, p_c, is a constant value accounting for the no-flux condition of $\hat{p}$ across the fluid interface Γ, i.e.

$$\int_{\Gamma} \hat{q}(\chi) d\Gamma_\chi = 0 \tag{4.3.15}$$

By discretization of the surface boundary using standard boundary element approach, the above integral representation formula is converted in a linear matrix system of equation. In the discretization of the boundary Γ into N elements, the coordinates of each boundary point as well as the surface pressure and its normal derivative are described using B-splines functions. The local curvature at each point is evaluated using a fourth-order Lagrangian polynomial to minimize possible numerical oscillations. The unknown value of the interface pressure normal derivative, $\hat{q}$ is obtained after substituting the dynamic interface matching condition into the resulting matrix system. The value of $\hat{q}$ obtained is then substituted into the kinematic condition to find the interface displacement velocity for a given dissolution mass flux $\rho_g u_d$. In order to move the internal surface in time, a simple Euler approximation is used in integrating the surface normal displacement equation $\frac{ds}{dt} = V_n$, with V_n as the normal interface surface velocity.

The proposed B-spline BEM numerical scheme was verified by comparison its results with the evolution of the six tongues radial finger, without surface dissolution, studied experimentally by Paterson (1981), and numerically by Howison (1986) and Power (1994). The numerical results obtained for the fingering evolution, with the corresponding tip splitting, showed excellent agreement with previous numerical and experimental results, as well as analytical estimation of the finger's tip time evolution before the first splitting, showing asymptotical scale of $x \sim At^{0.6}$ (Almgren et al. 1993), for more details see Power et al. (2013).

The effect of dissolution is to erode the surface of the growing gas bubble. The effect is most pronounced in regions where the flow component normal to the bubble surface is small; most notably at the finger bases, where a stagnation point exists. Dissolution therefore works to thin the base of viscous fingers, eventually leading to their breaking at the base. These broken fingers move outwards due to convective forces in the surrounding fluid, and slowly normalise towards a circular shape due to the effect of surface tension, viscous sintering. With no further access to injected gas, the shed bubbles will eventually dissolve into the liquid. Power et al. (2013) reported a series of numerical examples showing the dissolution-driven breaking of viscous fingers.

We model the displacement of the brine of mobility $M = 5 \times 10^{-5}$ m^3 h/kg and density $p_l = 1.2 \times 10^3$ kg/m^3 by the CO_2 gas of lower mobility injected at a constant flux rate $Q = 0.5$ m^3/h. As previously reported by Kneafsey and Pruess (2009), in the case of CO_2-brine convectively enhances dissolution, a constant dissolution flux of the gas into the liquid is expected, here we used a constant rate $R = \rho_g u_d = 3.6 \times 10^{-3}$ kg/(m^2 h). Besides, according to Hele–Shaw–Chouke formulation, used in our simulations, an effective surface tension larger than the molecular surface tension needs to be considered in order to approximate the corresponding fingering evolution. The molecular surface tension of an open CO_2−brine interface varies due to temperature and pressure, but is of the order 0.02 to 0.07 kg m s^{-2}; see Bachu and Bennion (2009), here we used an effective surface tension $\gamma = 0.25$ kg m s^{-2}. Figure 4.10a shows the obtained fingers evolution for the above set of values corresponding to a CO_2-brine immiscible displacement when an initial irregular disturbance defined by $r = r_0 \cos((3/\sqrt{\pi})\theta^{3/2})$ is imposed where r is the radial distance from the origin.

As expected, a number of irregular fingering features are generated where it can be observed that the finger bases remain stationary, and fingers bifurcate when they grow to be sufficiently wide. Note that of the nine initial crests, only the fingers resulting three widest crests bifurcate immediately; the remaining fingers must grow and spread before the critical length scale is reached. Here the thinning of fingers due to the effect of dissolution is clearly visible at the bases, especially at the base of the higher-frequency disturbances.

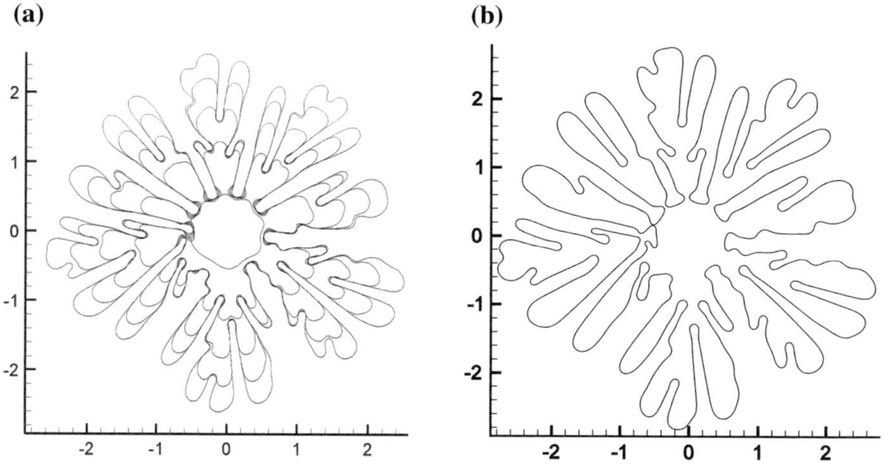

Fig. 4.10 **a** Fingering evolution before breaking, **b** fingering breaking (t = 22 h) by dissolution effect

The thinning of fingers at the base leads to breaking of the first finger at t = 22.0 h. These broken fingers rapidly re-grow towards the shed bubbles which will be split by the growing finger, further increasing the long-term rate of dissolution.

4.4 Example Case Studies

Auli Niemi, Zhibing Yang, Dorothee Rebscher, Bruno Figueiredo, Victor Vilarrasa, Jan Lennard Wolf and Franz May

In this chapter short example case studies are presented concerning modeling of real or proposed CO_2 injection sites. We start by exemplifying modeling of large scale storage sites by various methods and then proceed to examples addressing modeling of coupled hydro-chemical and hydro-mechanical processes.

4.4.1 Modeling of Large Scale Systems

Auli Niemi and Zhibing Yang

Predictions concerning the performance of large scale CO_2 storage systems involve, among other things, (1) estimating the induced pressure increase from the injection, to assure that it does not exceed the maximum allowed pressure tolerated by the formation and (2) estimating the CO_2 migration, to assure that the injected CO_2 does not get transported beyond allowable limits. When making such predictions, a number of uncertainties exist, ranging from uncertainties from data availability to the uncertainties arising from the use of different models and their underlying assumptions. To avoid errors arising from the use of specific models, Yang et al. (2015) and Tian et al. (2016) developed an approach to use a set of models of increasing complexity for modeling large scale sites and their performance in the long term. Yang et al. (2015) addressed the Dalders Monocline formation in the Baltic Sea, which site under the assumptions and injection scenarios tested showed to be pressure limited. Tian et al. (2016) in turn modeled the South Scania, Sweden, site, which under the investigated conditions and injection configurations was found to be migration limited.

The benefit of using models of increased complexity and accuracy is twofold. First, the use of several models obviously increases the level of confidence to model predictions, if several models with different simplifications and assumptions give similar results. Second, the use of models of increasing complexity can be useful

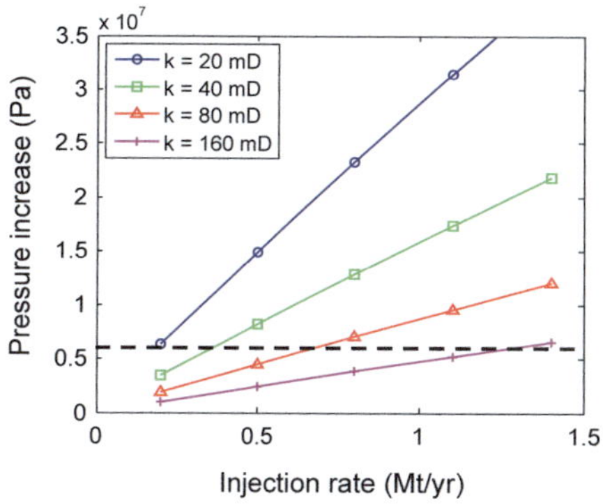

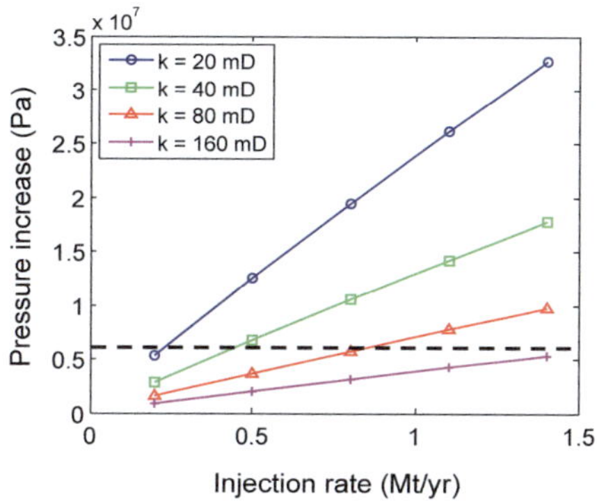

Fig. 4.11 Sensitivity of pressure increase to the permeability of the formation based on the preliminary simulations with a semi-analytical model. The distance from the injection well to the boundary is 5 km (*left panel*) and 25 m (*right panel*). The *dashed line* indicates the maximum allowed pressure increase (Yang et al. 2015)

when the simpler models are used to obtain first order-of-magnitude estimates and their results then used to guide the model scenarios for the more complex and computationally demanding models.

To exemplify, Yang et al. (2015) used the semi-analytical approach by Mathias et al. (2011) (see Sect. 4.1.3) to get the first estimates of the maximum allowed injection rate not to exceed the maximum allowed pressure and to carry out parameter sensitivity studies concerning the impact of variations in layer thickness, permeability, boundary conditions, etc. An example of the effect of permeability and distance to no-flow boundary (representing an impermeable fault) on pressure increase induced by a given injection rate is shown in Fig. 4.11. The next step was to use a more complicated but still somewhat simplified numerical model based on the vertical equilibrium assumption (e.g. Gasda et al. 2009, see also Sect. 4.1.2). This model, while still having simplifications in comparison to a 'full physics' high-fidelity model (Sect. 4.1.1), can give estimates of plume spreading, along with the information on pressure increase. The results from this model were then used to guide selecting the most detailed scenarios for the high-fidelity 'full-physics' modeling, carried out with the massively parallel TOUGH-MP simulator. Figure 4.12 shows example results of pressure plume evolution and spreading of the plume as simulated with the two models. Firstly, it can be seen that similar trends are observed in the results from the two models, regardless of the differences in their underlying assumptions. Secondly, it can be seen that the extent of the domain of the increased pressure is far more extensive than that of the CO_2 plume. For more discussions concerning the case study, the reader is referred to Yang et al. (2015) and for another example for a fundamentally different system to Tian et al. (2016).

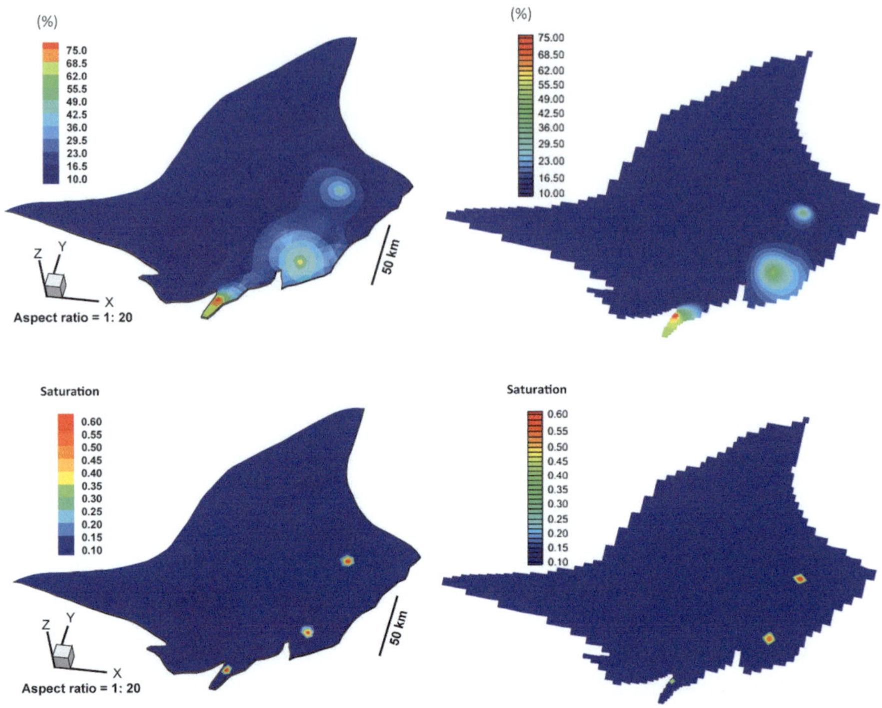

Fig. 4.12 Simulated overpressure distribution in percentage of the initial pressure (*upper panels*) as calculated with the VE model (*left*) and 3D TOUGH/MP (*right*) along with the simulated plume extent as expressed as depth averaged saturation distribution at 50 years after the start of injection (*lower panels*) with VE model (*left*) and TOUGH/MP (*right*). The injection rate is 0.5, 0.5 and 0.2 Mt/year from the three wells (Yang et al. 2015)

4.4.2 Modeling of Coupled Hydro-Geochemical Processes

Dorothee Rebscher, Jan Lennard Wolf and Franz May

Prominent computational tools for geochemical reaction modeling include CrunchFlow (Steefel 2009), Geochemist's Workbench (Bethke and Yeakel 2014), HST3D (Kipp 1997), together with PHREEQC forming PHAST (Parkhurst et al. 2010), HYTEC (van der Lee et al. 2003), HYDROGEOCHEM (Yeh and Tsai 2013), ORCHESTRA (Meeussen 2003), OpenGeoSys (Kolditz et al. 2012a) together with external geochemical solvers like IPHREEQC, open-source PFLOTRAN (Hammond et al. 2007; Lichtner et al. 2015), STOMP (White et al. 2012) with the reactive transport package ECKEChem (White and McGrail 2005), TOUGHREACT (Xu et al. 2014; Wei et al. 2015), and their derivatives or enhanced versions. Overviews of the capabilities of reactive transport codes in the subsurface are provided e.g. by Steefel et al. (2015), Zhang et al. (2012). Over the years, numerous program testing, benchmark tests, as well as case studies have been performed for a wide range of coupled THMC processes (MacQuarrie and Mayer 2005; Kolditz et al. 2012b, 2015). In recent years, increased attention has been paid to the subject of CO_2 storage in deep saline sandstone reservoirs and the associated geochemical reactions (e.g. Balashov et al. 2013).

In the following, we will present an example of coupled hydro-geochemical modeling to investigate the geochemical effects in a proposed small scale CO_2 injection experiment at the pilot injection site Heletz, Israel (Niemi et al. 2016), where an impurity gas (in this case SO_2) is included in the injected gas stream. In the case of impurity gases, the geochemical effects are of particular interest.

In general, the scope of such simulations is to enhance the understanding of flow and transport of impure CO_2 and brine, and to investigate the induced chemical reactions and associated changes in the reservoir, such as porosity, permeability, pH value, and mineral composition. Coupling of transport and chemistry provides insights into their interdependency and helps to test the design, including inhibiting the release of noxious H_2S to the surface, which may be caused by SO_2 disproportionation reactions in a CO_2–SO_2–H_2O mixture. Special attention should be given to potential impacts on groundwater. Decrease in pH values due to the entry of CO_2 and/or SO_2 into an aquatic solution and resulting changes of geochemical equilibriums can be expected, e.g. added stress on the environment caused by this acidification. Some geochemical conversions are certainly expected, for instance the formation of solid carbonate ensuring mineral trapping of CO_2, as a result of CO_2 dissolved in water reacting with the reservoir rock. Other implications caused by chemical reactions are also of concern, like the release of contaminants, mainly metals, during chemical reactions. These effects have to be considered site-specifically in risk assessments, preventive actions, and remediation plans.

TOUGHREACT (Xu et al. 2006) code and its parallel version V3.0-OMP is used for the modeling of the example case study. The multi-component, multiphase flow in the porous media is first calculated by the general-purpose numerical simulation program TOUGH2. This code simulates fluid flow in liquid and gaseous phases, and the transitions between the phases, occurring under pressure, viscous, and gravity forces according to Darcy's law (Pruess et al. 1999). As an equation of state module, the fluid property model ECO_2N is used, which was especially developed for geological sequestration of CO_2 in saline aquifers, handling both aqueous and CO_2-rich phases (Pruess 2005). Thermodynamic and thermo-physical properties of H_2O–NaCl–CO_2 mixtures are given for appropriate temperature, pressure, and salinity ranges, i.e. $10\ °C \leq T \leq 110\ °C$, $P \leq 60$ MPa, and salinity up to full halite saturation, thus covering the conditions of the storage reservoir at Heletz with a temperature of 66 °C, a pressure of 14.7 MPa, and a salinity of 0.055 (Niemi et al. 2016). Part of the TOUGH suite of codes is the TOUGHREACT code, a reactive fluid flow and geochemical transport simulator for investigating reactive chemistry in both aqueous and gaseous phases (Xu et al. 2006). The code is well established, successfully applied in numerous case studies and various code comparison studies, e.g. Steefel et al. (2015). The parallel version TOUGHREACT V3.0-OMP provides the option of including SO_2 as a trace gas, where SO_2 does not have to be injected as SO_2 dissolved in an additional saline fluid phase with the composition of the native brine. The latter more auxiliary method is often applied while using geochemical transport models, including the previous versions of TOUGHREACT.

At the Heletz pilot site (Niemi et al. 2016) a push-pull experiment with impurity gases is planned to be carried out. This is a typical example where complex coupled hydro-geochemical effects can be expected. The site geology, presented in detail in (Niemi et al. 2016), consists of sandstone reservoir layers with a cumulative thickness of approximately 20 m, located at depths of about 1650 m. Details of the modeling are explained in (Rebscher et al. 2015, 2014; Wolf et al., in press) and only some example results are presented here.

A basic 2D radial symmetric model on a field scale was constructed with a total of 3700 cells, representing three sandstone layers, adjacent by two shale layers. In the vertical, each of these five layers is portrayed by 3, 3, 9, 8, and 20 cells, respectively, their heights ranging layer dependent from about 0.36–1.4 m, with a cumulative height of 18 m. An incremental value of 1.06 for the horizontal grid sizes gives a higher discretization close to the injection well to obtain improved accuracy, resulting in horizontal cell sizes from 0.18 to 57 m with increasing distance to the well, covering 4000 m in total. Material parameters are based on Niemi et al. (2016) with permeabilities of about 10^{-13} and 10^{-15} m^2 for the sandstone and shale layers, respectively. Porosities vary from 10 to 20 %. The initial mineral composition of the sandstone is taken as follows: quartz is the prominent primary mineral with a volume fraction of almost 70 %. The primary carbonate of interest is calcite $CaCO_3$ with 3.7 %, secondary carbonates are ankerite, e.g. $CaFe(CO_3)_2$, dolomite $CaMg(CO_3)_2$, and siderite $FeCO_3$. Further primary minerals are orthoclase $KAlSi_3O_8$ and albite $NaAlSi_3O_8$ with 12 and 2.5 % and the clay minerals illite, e.g. $K_{0.85}Al_{2.85}Si_{3.15}O_{10}(OH)_2$, kaolinite $Al_2Si_2O_5(OH)_4$, and chlorite $(Mg, Fe)_3(Si, Al)_4O_{10}$, with 38.8, 32, and 14 % respectively. A secondary feldspar is anorthite $Ca(Al_2Si_2)O_8$.

The simulated injection scenario represents a potential schedule for injecting the binary mixture CO_2 to SO_2 in relation of 97 % to 3 %. The injection phase with a rate of approximately 0.28 kg/s for the first 100 h is followed by a relaxation phase of 144 h. During a shorter second injection phase, CO_2 saturated water is injected at a rate of 1 kg/s. All three phases together last 330 h, i.e. about 2 weeks. In the model, the injection occurs within the bottom cell of the sandstone layer A, located in the lower left corner of Fig. 4.13.

Both plots in this figure illustrate the situation after the recovery phase, i.e. after 244 h. It is evident that the area affected by the push–pull experiment is restricted to the vicinity of the injection point. As expected, the injection of the CO_2 to SO_2 mixture affects a strong decrease in pH value. Dissolving all the carbonate results with these initial mineral conditions in a slight decrease of porosity within a spatial range of about 3 m, because the precipitation of anhydrite features a higher molar volume compared to the one of dissolving calcite. Here, the impurity SO_2

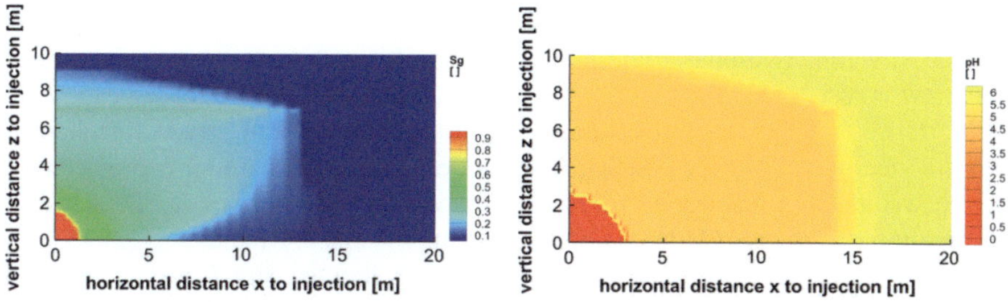

Fig. 4.13 Spatial distribution of CO_2 gas saturation (*left*) and Ph value (*right*) after the injection of 105 kg CO_2 to SO_2 in relation of 97 % to 3 % within 100 h, followed by a relaxation phase of 144 h; the diagrams show the whole vertical extent of 18 m of the five layer model, but only 20 m around the injection site in the horizontal direction (Rebscher et al. 2014)

preferentially dissolves into the aqueous phase due the higher solubility compared to CO_2. The higher aqueous SO_2 concentration and subsequently the higher reactivity zone is therefore constrained within the lower sandstone layer A, i.e. the lowest 7 m in Fig. 4.13. The resulting depleted supercritical CO_2 phase on the other hand reaches horizontal distances of about 15 and 10 m in the vertical, as can be seen in the spatial distributions of gas saturation S_g and pH value in Fig. 4.13. The carbonate dissolution induced by the decrease in pH value changes the porosity. Depending on the initial mineral composition of the rock, especially on the initial content of calcium in the minerals calcite ($CaCO_3$) and ankerite ($CaFe(CO_3)_2$), either a net increase or a net decrease in porosity, and hence in permeability, occurs (Wolf et al. in press).

4.4.3 Modeling of Coupled Hydro-Mechanical Systems

Bruno Figueiredo, Victor Vilarrasa and Auli Niemi

Pioneering work on modeling **coupled hydro-mechanical (HM)** processes has been carried out by e.g. Rutqvist and co-workers, including development of the TOUGH-FLAC code (Rutqvist 2011) that can handle such coupling. Examples of coupled hydro-mechanical modeling in CO_2 injection related applications include Rutqvist and Tsang (2002), Cappa and Rutqvist (2011, 2012), Castelletto et al. (2013), Goodarzi et al. (2015), Streit and Hillis (2004), Verdon et al. (2011), Vidal-Gilbert et al. (2010) and Vilarrasa et al. (2010b). As described in detail in Rutqvist (2011) in TOUGH-FLAC two established codes, namely TOUGH2 (Pruess et al. 1999) and FLAC 3D (Itasca 2012) that are linked to address the coupled hydro-mechanical processes for multiphase flow systems. TOUGH2 is used for solving multiphase flow and heat transport equations, and FLAC3D is used for solving geo-mechanical stress–strain equations. There are several other codes that can address the coupled hydro-mechanical problem as well, including some tailored towards two-phase/multiphase flow problems. A comprehensive recent overview is also given in Rutqvist (2011, 2012) and will not be repeated here.

In terms of geological storage of CO_2, the major hydro-mechanical (HM) concerns calling for coupled HM modeling are maintaining caprock integrity to avoid CO_2 leakage (Vilarrasa et al. 2010b; Rutqvist et al. 2007), fault reactivation due to shear failure (Cappa and Rutqvist 2011; Rinaldi et al. 2015), and potential induced seismicity and fluid leakage (Vilarrasa and Carrera 2015a, b; Zoback and Gorelick 2012, 2015; Juanes et al. 2012; Rutqvist et al. 2007; Rutqvist and Tsang 2005; Hawkes et al. 2004).

In the following we will exemplify, following Figueiredo et al. (2015), the coupled HM modeling through a case study for potential fault reactivation due to CO_2 injection. The study site is Heletz site in Israel (Niemi et al. 2016), which is a test site for scientific CO_2 injection field experiments. The potential reservoir for CO_2 injection consists of three high-permeability sandstone layers, named K, W and A, separated by impermeable shale layers and overlain by a thick caprock. The confinement formations, located above the caprock and below the dolomite layer, are constituted essentially by limestone. The storage formation is intersected by two pre-existing sub-vertical normal faults (F1 and F2) on two opposite sides of the injection point.

A 5 km $\times$ 2.5 km vertical cross section with 1 m thickness was developed within the framework of TOUGH-FLAC code (Fig. 4.14). The mesh constitutes of 15,000 elements with refined elements close to the faults and storage formations. An injection rate for storage of 10 kg/s—which is more than ten times the injection rate used in the field experiments, to exaggerate the effect—was used in the simulations.

The difference in the results obtained by considering the actual three-layer storage formation in comparison to an equivalent single-layer formation is analyzed. No fault reactivation was observed in either case. It was also found that for the two cases the pore pressure evolution is similar, but the differences in the evolution of CO_2 saturation are significant, the latter being attributed to the differences in CO_2 spreading in a single versus three-layer storage layers. Fault slip displacement and changes in permeability were found not to be significant as the plastic shear strains mainly occur in a fault section that is only about 10 m in length, corresponding to the thickness of the storage formation.

Three key parameters were identified as important for the fault zone hydromechanics in association with CO_2 injection and storage: the offset of the storage layers across faults, the permeability of the confining formations and the thickness of the storage formation. The first key parameter was found to lead to an increase in CO_2 leakage through the caprock (Fig. 4.15). The second and third key parameters were found to have a direct relation with the length of the fault section reactivated by shear failure.

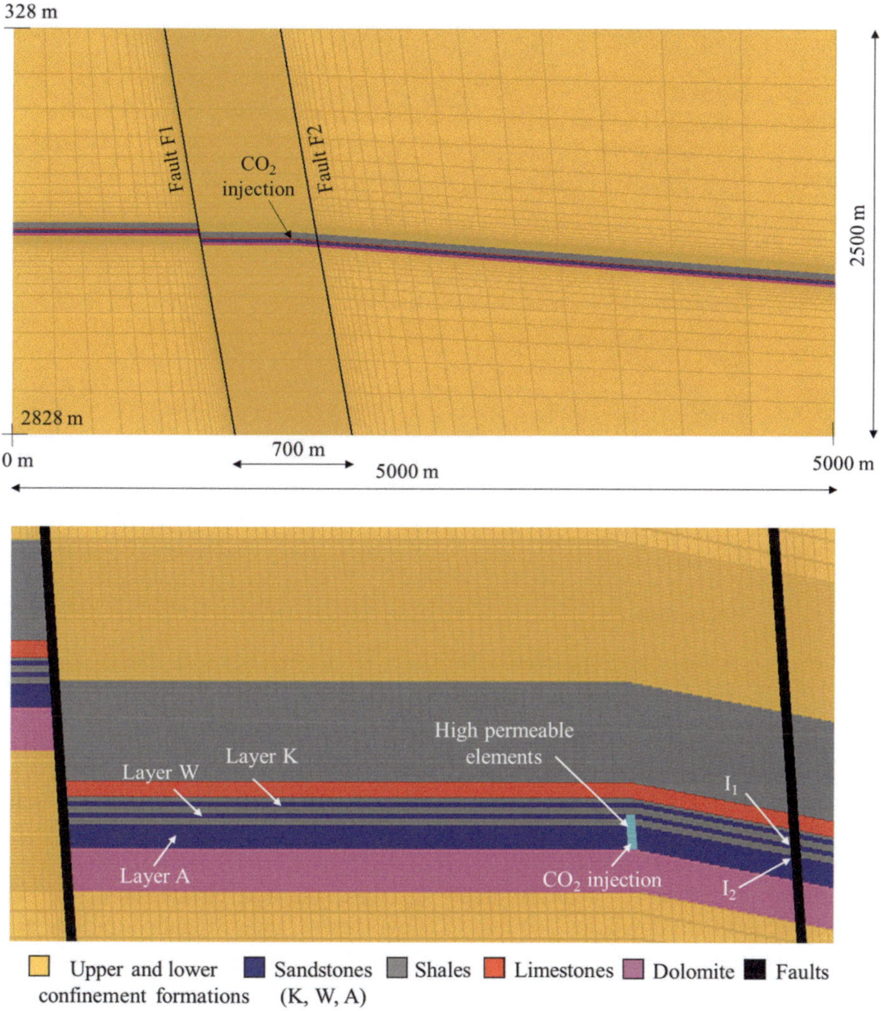

Fig. 4.14 Model mesh for the coupled simulations of CO_2 injection and faults slip (*above*) and a detail of the three-layer storage formation (the vertical scale has been expanded to display the storage details) (*below*) (Figueiredo et al. 2015)

Another example is the coupled **thermo-hydro-mechanical (THM) modeling** concerning liquid CO_2 injection presented in details in (Vilarrasa et al. 2013b, 2014). Most studies consider isothermal conditions. However, in general, CO_2 will reach the storage formation at a lower temperature than that corresponding to the geothermal gradient, especially at high flow rates (Paterson et al. 2008). A clear example of this can be found at e.g. In Salah, Algeria, where, even though CO_2 is injected at 35 °C at the surface, it reaches the storage formation at 50 °C (45 °C colder than the reservoir) (Bissell et al. 2011). Vilarrasa et al. (2013b) used

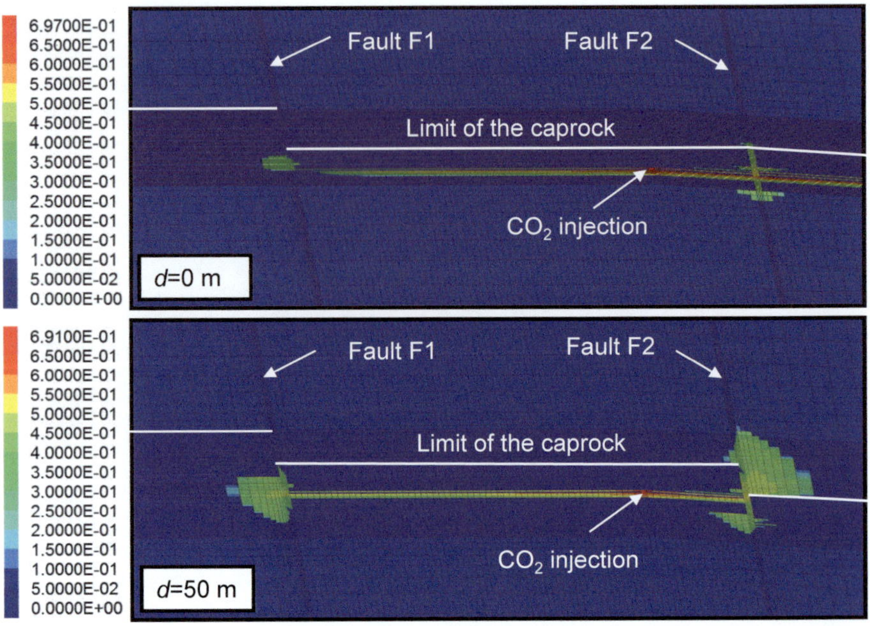

Fig. 4.15 CO_2 saturation (in %) obtained after 5 years of CO_2 injection with consideration of an offset d = 0 m (*above*) and 50 m (*below*) of the layers across the fault F2 (Figueiredo et al. 2015)

CODE_BRIGHT code for modeling the non-isothermal injection of 1 Mt/year of CO_2 at 20 °C in a reservoir with a mean temperature of 56 °C, to simulate liquid (cold and dense) CO_2 injection. Their simulation was coupled to mechanical deformation to address thermal stresses induced by the cold injection. Once liquid CO_2 enters into the reservoir, it heats up until thermal equilibrium with the geothermal gradient is reached, so that CO_2 evolves to supercritical conditions as it flows away from the well. As a result, the cold CO_2 region advances much behind the CO_2 plume interface. CO_2 remains in liquid conditions within the cold region, leading to a significantly higher density (around 900 kg/m^3) compared with the rest of the CO_2 plume (around 650 kg/m^3), where CO_2 stays in a supercritical state. This density difference affects overpressure as the denser fluid displaces a smaller amount of brine, thus inducing a slightly lower overpressure.

This lower overpressure is beneficial to maintain the geomechanical stability of the caprock and faults. However, cooling induces a thermal stress reduction that may lead to failure conditions (Gor et al. 2013). This thermal stress reduction is likely to induce shear failure conditions in the reservoir, which could explain part of the microseismic events detected at In Salah (Vilarrasa et al. 2015), or even the formation of hydraulic fractures in stiff reservoirs (Goodarzi et al. 2012). Nevertheless, the thermal stress reduction causes a stress redistribution around the cooled region leading to an increase of the horizontal stresses in the lower part of the caprock (Vilarrasa et al. 2013b). This stress redistribution tightens the caprock, reducing the risk of CO_2 leakage across it. Thus, the thermal stress reduction induced by cooling should not be feared, but care should be taken to avoid inducing excessive microseismicity.

References

Almgren R, Dai WS, Hakim V (1993) Scaling behavior in anisotropic Hele–Shaw flow. Phys Rev Lett 71:3461–3464

Andersen O, Gasda SE, Nilsen HM (2015) Vertically averaged equations with variable density for CO_2. Transp Porous Media 107:95–127

Bachu S, Bennion DB (2009) Interfacial tension between CO_2, freshwater, and brine in the range of pressure from (2 to 27) MPa, temperature from (20 to 125) °C, and water salinity from (0 to 334,000) mg L^{-1}. J Chem Eng Data 54:765–775

Balashov VN, Guthrie GD, Hakala JA, Lopano CL, Rimstidt JD, Brantley SL (2013) Predictive modeling of CO_2 sequestration in deep saline sandstone reservoirs: impacts of geochemical kinetics. Appl Geochem 30:41–56

Bethke C, Yeakel S (2014) Geochemist's workbench: release 10.0 reaction modeling guide. Aqueous Solutions LLC, Champaign

Bissell RC, Vasco DW, Atbi M, Hamdani M, Okwelegbe M, Goldwater MH (2011) A full field simulation of the in Salah gas production and CO_2 storage project using a coupled geo-mechanical and thermal fluid flow simulator. In: Energy procedia, 10th international conference on greenhouse gas control technologies, vol 4, pp 3290–3297

Bronstein IN, Semedjajew KA (1977) Taschenbuch der Mathematik (Handbook of mathematics). Harri Deutsch, Frankfurt

Buckley SE, Leverett MC (1941) Mechanism of fluid displacement in sands. Trans AIME 146:107–116

Cabral JJSP, Wrobel LC, Brebbia CA (1990) A BEM formulation using B-splines: I—uniform blending functions. Eng Anal Bound Elem 7:136–144

Cao H (2002) Development of techniques for general purpose simulators. Stanford University

Cappa F, Rutqvist J (2011) Impact of CO_2 geological sequestration on the nucleation of earthquakes. Geophys Res Lett 38:L17313

Cappa F, Rutqvist J (2012) Seismic rupture and ground accelerations induced by CO_2 injection in the shallow crust. Geophys J Int 190:1784–1789

Castelletto N, Gambolati G, Teatini P (2013) Geological CO_2 sequestration in multi-compartment reservoirs: geomechanical challenges. J Geophys Res Solid Earth 118:2417–2428

Cavanagh AJ, Haszeldine RS, Nazarian B (2015) The Sleipner CO_2 storage site: using a basin model to understand reservoir simulations of plume dynamics. First Break 33:61–68

Celia MA, Nordbotten JM (2009) Practical modeling approaches for geological storage of carbon dioxide. Ground Water 47:627–638

Chang KW, Hesse MA, Nicot J-P (2013) Reduction of lateral pressure propagation due to dissipation into ambient mudrocks during geological carbon dioxide storage. Water Resour Res 49:2573–2588

Chen Y, Durlofsky LJ, Gerritsen M, Wen XH (2003) A coupled local–global upscaling approach for simulating flow in highly heterogeneous formations. Adv Water Resour 26:1041–1060

Chuoke RL, van Meurs P, van der Poel C (1959) The instability of slow, immiscible, viscous liquid–liquid displacements in permeable media. Trans. AIME 216, pp 188–194

Cihan A, Zhou Q, Birkholzer JT (2011) Analytical solutions for pressure perturbation and fluid leakage through aquitards and wells in multilayered-aquifer systems. Water Resour Res 47: W10504

Class H, Ebigbo A, Helmig R, Dahle HK, Nordbotten JM, Celia MA, Audigane P, Darcis M, Ennis-King J, Fan Y, Flemisch B, Gasda SE, Jin M, Krug S, Labregere D, Beni AN, Pawar RJ, Sbai A, Thomas SG, Trenty L, Wei L (2009) A benchmark study on problems related to CO_2 storage in geologic formations. Comput Geosci 13:409–434

Daripa P, Pasa G (2008) Hydrodynamic stability of multi-layer Hele–Shaw flows. J Stat Mech Theory Exp 2008:P12005

de Marsily G (1986) Quantitative hydrogeology: groundwater hydrology for engineers. Academic Press, London

Degregoria AJ, Schwartz LW (1986) A boundary-integral method for two-phase displacement in Hele–Shaw cells. J Fluid Mech 164:383

Dentz M, Carrera J (2003) Effective dispersion in temporally fluctuating flow through a heterogeneous medium. Phys Rev E 68:36310

Dentz M, Tartakovsky DM (2008) Abrupt-interface solution for carbon dioxide injection into porous media. Transp Porous Media 79:15–27

Durlofsky LJ, Efendiev Y, Ginting V (2007) An adaptive local–global multiscale finite volume element method for two-phase flow simulations. Adv Water Resour 30:576–588

Edwards D, Brenner H, Wasan D (1991) Interfacial transport processes and rheology. Butterworth-Heinemann, Boston

Elenius MT, Nordbotten JM, Kalisch H (2014) Convective mixing influenced by the capillary transition zone. Comput Geosci 18:417–431

Emami-Meybodi H, Hassanzadeh H (2015) Two-phase convective mixing under a buoyant plume of CO_2 in deep saline aquifers. Adv Water Resour 76:55–71

Ferguson G, Woodbury AD (2005) Thermal sustainability of groundwater-source cooling in Winnipeg, Manitoba. Can Geotech J 42:1290–1301

Figueiredo B, Tsang C-F, Rutqvist J, Bensabat J, Niemi A (2015) Coupled hydro-mechanical processes and fault reactivation induced by CO_2 Injection in a three-layer storage formation. Int J Greenhouse Gas Control 39:432–448

Freifeld B, Zakim S, Pan L, Cutright B, Sheu M, Doughty C, Held T (2013) Geothermal energy production coupled with CCS: a field demonstration at the SECARB Cranfield Site, Cranfield, Mississippi, USA. Energy Procedia 37:6595–6603

Fu X, Cueto-Felgueroso L, Bolster D, Juanes R (2015) Rock dissolution patterns and geochemical shutdown of—brine–carbonate reactions during convective mixing in porous media. J Fluid Mech 764:296–315

Garcia JE, Pruess K (2003) Flow instabilities during injection of into saline aquifers. In: Presented at the TOUGH symposium, Berkeley, CA, USA

Gasda SE, Nordbotten JM, Celia MA (2009) Vertical equilibrium with sub-scale analytical methods for geological CO_2 sequestration. Comput Geosci 13:469–481

Gasda SE, Nordbotten JM, Celia MA (2011) Vertically averaged approaches for CO_2 migration with solubility trapping. Water Resour Res 47:W05528

Gasda SE, Nordbotten JM, Celia MA (2012) Application of simplified models to CO_2 migration and immobilization in large-scale geological systems. Int J Greenhouse Gas Control 9:72–84

Ghesmat K, Hassanzadeh H, Abedi J (2011) The impact of geochemistry on convective mixing in a gravitationally unstable diffusive boundary layer in porous media: CO_2 storage in saline aquifers. J Fluid Mech 673:480–512

Glimm J, Grove JW, Li X, Zhao N (1999) In: Chen G-Q, DiBenedetto E (eds) Simple front tracking, vol 238. Am. Math. Soc, Providence, pp 133–149

Goodarzi S, Settari A, Keith D (2012) Geomechanical modeling for CO_2 storage in Nisku aquifer in Wabamun Lake area in Canada. Int J Greenhouse Gas Control 10:113–122

Goodarzi S, Settari A, Zoback MD, Keith DW (2015) Optimization of a CO_2 storage project based on thermal, geomechanical and induced fracturing effects. J Pet Sci Eng 134:49–59

Gor GY, Elliot TR, Prévost JH (2013) Effects of thermal stresses on caprock integrity during CO_2 storage. Int J Greenhouse Gas Control 12:300–309

Gorell S, Homsy G (1983) A theory of the optimal policy of oil recovery by secondary displacement processes. SIAM J Appl Math 43:79–98

Häfner F, Sames D, Voigt H.-D (1992) Wärme- und Stofftransport. Springer-Lehrbuch, Berlin
Hammond G, Lichtner P, Lu C (2007) Subsurface multiphase flow and multicomponent reactive transport modeling using high-performance computing. J Phys Conf Ser 78:12025
Hassanzadeh H, Pooladi-Darvish M, Elsharkawy AM, Keith DW, Leonenko Y (2008) Predicting PVT data for CO_2–brine mixtures for black-oil simulation of CO_2 geological storage. Int J Greenhouse Gas Control 2:65–77
Hawkes CD, McLellan PJ, Zimmer U, Bachu S (2004) Geomechanical factors affecting geological storage of CO_2 in depleted oil and gas reservoirs: risks and mechanisms. In: Proceedings of gulf rocks 2004, presented at the sixth North America rock mechanics symposium (NARMS): rock mechanics across borders and disciplines, Houston, TX, USS
Helmig R (1997) Multiphase flow and transport processes in the subsurface: a contribution to the modeling of hydrosystems. Springer, Berlin, p 367
Hesse MA, Orr FM, Tchelepi HA (2008) Gravity currents with residual trapping. J Fluid Mech 611:35–60
Hickernell JF, Yortsos YC (1986) Linear stability of miscible displacement processes in porous media in the absence of dispersion. Stud Appl Math 74:93–115
Hidalgo JJ, Carrera J (2009) Effect of dispersion on the onset of convection during CO_2 sequestration. J Fluid Mech 640:441
Hidalgo JJ, Carrera J, Dentz M (2009a) Steady state heat transport in 3D heterogeneous porous media. Adv Water Resour 32:1206–1212
Hidalgo JJ, Carrera J, Medina A (2009b) Role of salt sources in density-dependent flow. Water Resour Res 45:W05503
Hidalgo, JJ, Dentz M, Cabeza Y, Carrera J (2015) Dissolution patterns and mixing dynamics in unstable reactive flow. Geophys Res Lett 42:6357–6364
Hidalgo JJ, Slooten LJ, Medina A, Carrera J (2005) A Newton–Raphson based code for seawater intrusion modelling and parameter estimation. In: Araguas L, Custodio E, Manzano M (eds) Groundwater and saline intrusion: selected papers from the 18th salt water in-trusion meeting, 18th SWIM, Cartagena, 2004. IGME, pp 111–120
Homsy GM (1987) Viscous fingering in porous media. Annu Rev Fluid Mech 19:271–311
Houseworth JE (2012) Matched boundary extrapolation solutions for CO_2 well-injection into a saline aquifer. Transp Porous Media 91:813–831
Howison SD (1986) Fingering in Hele–Shaw cells. J Fluid Mech 167:439–453
Huang AB, Chikhliwala ED, Yortsos YC (1984) Linear stability analysis of immiscible displacement including continuously changing mobility and capillary effects: part II—general basic flow profiles. SPE 13163, Society of Petroleum Engineers, Dallas, Tex
Huang H, Meakin P (2008) Three-dimensional simulation of liquid drop dynamics within unsaturated vertical Hele–Shaw cells. Water Resour Res 44:10
Huang P-M, Wang M-K, Chiu C-Y (2005) Soil mineral–organic matter–microbe interactions: impacts on biogeochemical processes and biodiversity in soils. Pedobiologia 49:609–635
Huang X, Bandilla KW, Celia MA (2014) Basin-scale modeling of CO_2 storage using models of varying complexity. Int J Greenhouse Gas Control 20:73–86
IPCC (2005) IPCC special report on carbon dioxide capture and storage. In: Prepared by working group III of the intergovernmental panel on climate change. Cambridge University Press, Cambridge, New York
Itasca (2012) FLAC3D, fast lagrangian analysis of continua in 3 dimensions. Version 5.0. Minneapolis, Minnesota
Jackson SJ, Stevens D, Giddings D, Power H (2015a) Dynamic-wetting effects in finite-mobility-ratio Hele–Shaw flow. Phys Rev E 92:23021
Jackson SJ, Stevens D, Power H, Giddings D (2015b) A boundary element method for the solution of finite mobility ratio immiscible displacement in a Hele–Shaw cell. Int J Numer Methods Fluids 78:521–551
Jessen K, Kovscek AR, Orr FM Jr (2005) Increasing CO_2 storage in oil recovery. Energy Convers Manag 46:293–311

Juanes R, Hager BH, Herzog HJ (2012) No geologic evidence that seismicity causes fault leakage that would render large-scale carbon capture and storage unsuccessful. Proc Natl Acad Sci USA 109:E3623 (author reply E3624)

Juanes R, MacMinn CW, Szulczewski ML (2010) The footprint of the CO_2 plume during carbon dioxide storage in saline aquifers: storage efficiency for capillary trapping at the basin scale. Transp Porous Media 82:19–30

Juanes R, Spiteri EJ, Orr FM, Blunt MJ (2006) Impact of relative permeability hysteresis on geological CO_2 storage. Water Resour Res 42:W12418

Kipp KL (1997) Guide to the revised heat and solute transport simulator: HST3D—version 2 (No. Water-Resources Investigations Report 97-4157). U.S. Geological Survey, Denver

Knabner P, Frolkovic P (1996) Consistent velocity approximation for finite volume or element discretizations of density driven flow in porous media. In: Aldama AA, Aparicio J, Brebbia CA, Gray WG, Herrera I, Pinder G (eds) Computational methods in subsurface flow and transport problems, computational methods in water resources XI. Computational Mechanics, Southamptonm, pp 93–100. Southampton

Kneafsey TJ, Pruess K (2009) Laboratory flow experiments for visualizing carbon dioxide-induced, density-driven brine convection. Transp Porous Media 82:123–139

Kolditz O (2001) Non-linear flow in fractured rock. Int J Numer Methods Heat Fluid Flow 11:547–575

Kolditz O, Bauer S, Bilke L, Böttcher N, Delfs JO, Fischer T, Görke UJ, Kalbacher T, Kosakowski G, McDermott CI, Park CH, Radu F, Rink K, Shao H, Shao HB, Sun F, Sun YY, Singh AK, Taron J, Walther M, Wang W, Watanabe N, Wu Y, Xie M, Xu W, Zehner B (2012a) OpenGeoSys: an open-source initiative for numerical simulation of thermo-hydro-mechanical/chemical (THM/C) processes in porous media. Environ Earth Sci 67:589–599

Kolditz O, Görke U-J, Shao H, Wang W (eds) (2012b) Thermo-hydro-mechanical-chemical processes in porous media. Lecture notes in computational science and engineering. Springer, Berlin

Kolditz O, Shao H, Wang W, Bauer S (eds) (2015) Thermo-hydro-mechanical-chemical processes in fractured porous media: modelling and benchmarking closed-form solutions, terrestrial environmental sciences. Springer, Cham

Kueper BH, McWhorter DB (1992) The use of macroscopic percolation theory to construct large-scale capillary pressure curves. Water Resour Res 28:2425–2436

Lake LW (1989) Enhanced oil recovery. Prentice Hall, New Jersey

Landman AJ, Schotting RJ (2007) Heat and brine transport in porous media: the Oberbeck–Boussinesq approximation revisited. Transp Porous Media 70:355–373

Lenormand R, Touboul E, Zarcone C (1988) Numerical models and experiments on immiscible displacements in porous media. J Fluid Mech 189:165–187

Lichtner PC, Hammond GE, Lu C, Karra S, Bisht G, Andre B, Mills R, Kumar J (2015) PFLOTRAN user manual, a massively parallel reactive flow and transport, model for describing surface and subsurface processes. Technical Report No. LA-UR-15-20403, Los Alamos National Laboratory, Los Alamos, NM, p 195

Lu C, Lichtner PC (2007) High resolution numerical investigation on the effect of convective instability on long term CO_2 storage in saline aquifers. J Phys Conf Ser 78:12042

MacMinn CW, Juanes R (2009) Post-injection spreading and trapping of CO_2 in saline aquifers: impact of the plume shape at the end of injection. Comput Geosci 13:483–491

MacMinn CW, Szulczewski ML, Juanes R (2011) CO_2 migration in saline aquifers. Part 2. Capillary and solubility trapping. J Fluid Mech 688:321–351

MacQuarrie KTB, Mayer KU (2005) Reactive transport modeling in fractured rock: a state-of-the-science review. Earth Sci Rev 72:189–227

Mathias SA, Gluyas JG, Martínez González, de Miguel GJ, Hosseini SA (2011) Role of partial miscibility on pressure buildup due to constant rate injection of CO_2 into closed and open brine aquifers. Water Resour Res 47:W12525

Mathias SA, Gluyas JG, Martínez González, de Miguel GJ, Bryant SL, Wilson D (2013) On relative permeability data uncertainty and CO_2 injectivity estimation for brine aquifers. Int J Greenhouse Gas Control 12:200–212

McDermott CI, Bond AE, Wang W, Kolditz O (2011) Front tracking using a hybrid analytical finite element approach for two-phase flow applied to supercritical CO_2. Transp Porous Media 90:545–573

Meakin P, Tartakovsky AM (2009) Modeling and simulation of pore-scale multiphase fluid flow and reactive transport in fractured and porous media. Rev Geophys 47:RG3002

Meeussen JCL (2003) ORCHESTRA: an object-oriented framework for implementing chemical equilibrium models. Environ Sci Technol 37:1175–1182

Neufeld JA, Hesse MA, Riaz A, Hallworth MA, Tchelepi HA, Huppert HE (2010) Convective dissolution of carbon dioxide in saline aquifers. Geophys Res Lett 37:L22404

Nicot J-P (2008) Evaluation of large-scale CO_2 storage on fresh-water sections of aquifers: an example from the Texas Gulf Coast Basin. Int J Greenhouse Gas Control 2:582–593

Niemi A, Bensabat J, Shtivelman V, Edlmann K, Gouze P, Luquot L, Hingerl F, Benson SM, Pezard PA, Rasmusson K, Liang T, Fagerlund F, Gendler M, Goldberg I, Tatomir A, Lange T, Sauter M, Freifeld B (2016) Heletz experimental site overview, characterization and data analysis for CO_2 injection and geological storage. Int J Greenhouse Gas Control 48:3–23

Niessner J, Hassanizadeh SM (2008) A model for two-phase flow in porous media including fluid–fluid interfacial area. Water Resour Res 44:W08439

Nilsen HM, Lie KA, Andersen O (2016) Robust simulation of sharp-interface models for fast estimation of CO_2. Comput Geosci 20:93–113

Nordbotten JM, Celia MA (2006) Similarity solutions for fluid injection into confined aquifers. J Fluid Mech 561:307–327

Nordbotten JM, Celia MA, Bachu S (2005) Injection and storage of CO_2 in deep saline aquifers: analytical solution for CO_2 plume evolution during injection. Transp Porous Media 58:339–360

Nordbotten JM, Flemisch B, Gasda SE, Nilsen HM, Fan Y, Pickup GE, Wiese B, Celia MA, Dahle HK, Eigestad GT, Pruess K (2012) Uncertainties in practical simulation of CO_2 storage. Int J Greenhouse Gas Control 9:234–242

Nordbotten JM, Michael AC (2011) Geological storage of CO_2: modeling approaches for large-scale simulation. Wiley, New Jersey

Obi EOI, Blunt MJ (2006) Streamline-based simulation of carbon dioxide storage in a North Sea aquifer. Water Resour Res 42:W03414

Olivella S, Gens Solé A, Carrera Ramírez J, Pérez Alonso, de Agreda E (1996) Numerical formulation for a simulator (CODE_BRIGHT) for the coupled analysis of saline media. Eng Comput 13:87–112

Parkhurst DL, Kipp KL, Charlton SR (2010) PHAST version 2-A program for simulating groundwater flow, solute transport, and multicomponent geochemical reactions. U.S. Geological Survey Techniques and Methods, 6-A35 (2010). Denver, Colo, p 25

Paterson L (1981) Radial fingering in a Hele Shaw cell. J Fluid Mech 113:513

Paterson L, Lu M, Connell L, Ennis-King JP (2008) Numerical modeling of pressure and temperature profiles including phase transitions in carbon dioxide wells. In: Presented at the SPE annual technical conference and exhibition, 21–24 Sept, Society of Petroleum Engineers, Denver, CO, USA

Pau GSH, Bell JB, Pruess K, Almgren AS, Lijewski MJ, Zhang K (2010) High-resolution simulation and characterization of density-driven flow in CO_2 storage in saline aquifers. Adv Water Resour 33:443–455

Power H (1994) The evolution of radial fingers at the interface between two viscous liquids. Eng Anal Bound Elem 14:297–304

Power H, Stevens D, Cliffe KA, Golin A (2013) A boundary element study of the effect of surface dissolution on the evolution of immiscible viscous fingering within a Hele–Shaw cell. Eng Anal Bound Elem 37:1318–1330

Power H, Wrobel LC (1995) Boundary integral methods in fluid mechanics. Computational Mechanics Publications, Southampton and Boston
Pruess K (2011) ECO2M: a TOUGH2 fluid property module for mixtures of water, NaCl, and CO_2, including super- and sub-critical conditions, and phase change between liquid and gaseous CO_2 (No. LBNL-4590E). Lawrence Berkeley National Laboratory, Berkeley
Pruess K (2005) ECO2N: a TOUGH2 fluid property module for mixtures of water, NaCl, and CO_2 (No. LBNL-57952). Lawrence Berkeley National Laboratory, Berkeley
Pruess K, García J, Kovscek T, Oldenburg C, Rutqvist J, Steefel C, Xu T (2004) Code intercomparison builds confidence in numerical simulation models for geologic disposal of CO_2. In: Energy, 6th international conference on greenhouse gas control technologies, vol 29, pp 1431–1444
Pruess K, Oldenburg CM, Moridis GJ (1999) TOUGH2 user's guide version 2 (No. LBNL–43134, 751729). Berkeley National Laboratory, Berkeley
Rasmusson M, Fagerlund F, Tsang Y, Rasmusson K, Niemi A (2015) Prerequisites for density-driven instabilities and convective mixing under broad geological CO_2 storage conditions. Adv Water Resour 84:136–151
Rebscher D, Wolf JL, Bensabat J, Niemi A (2015) Numerical simulations of the chemical impact of impurities on geological CO_2 storage—comparison between TOUGHREACT V2.0 and TOUGHREACT V3.0-OMP. In: Paper proceedings LBNL-190559, presented at the TOUGH symposium 2015, Lawrence, Berkeley National Laboratory, CA, USA, pp 493–500
Rebscher D, Wolf JL, Jung B, Bensabat J, Segev R, Niemi AP (2014) Effects of impurities in CO_2 spreading model development for field experiments in the framework of the CO2QUEST project. In: AGU fall meeting, abstr. 21
Riaz A, Hesse M, Tchelepi HA, Orr FM (2006) Onset of convection in a gravitationally unstable diffusive boundary layer in porous media. J Fluid Mech 548:87–111
Rinaldi AP, Vilarrasa V, Rutqvist J, Cappa F (2015) Fault reactivation during CO_2 sequestration: effects of well orientation on seismicity and leakage. Greenh Gas Sci Technol 5:645–656
Rutqvist J (2011) Status of the TOUGH-FLAC simulator and recent applications related to coupled fluid flow and crustal deformations. Comput. Geosci 37:739–750 (Transport of Unsaturated Groundwater and Heat Symposium 2009, TOUGH Symposium)
Rutqvist J (2012) The geomechanics of CO_2 storage in deep sedimentary formations. Geotech Geol Eng 30:525–551
Rutqvist J, Birkholzer J, Cappa F, Tsang C-F (2007) Estimating maximum sustainable injection pressure during geological sequestration of CO_2 using coupled fluid flow and geomechanical fault-slip analysis. Energy Convers Manag 48:1798–1807
Rutqvist J, Birkholzer JT, Tsang C-F (2008) Coupled reservoir–geomechanical analysis of the potential for tensile and shear failure associated with CO_2 injection in multilayered reservoir–caprock systems. Int J Rock Mech Min Sci 45:132–143
Rutqvist J, Tsang C-F (2005) Coupled hydromechanical effects of CO_2 injection. In: Apps JA, Tsang CF (eds) Developments in water science, underground injection science and technology. Elsevier, Amsterdam, pp 649–679
Rutqvist J, Tsang C-F (2002) A study of caprock hydromechanical changes associated with CO_2-injection into a brine formation. Environ Geol 42:296–305
Saaltink MW, Ayora C, Carrera J (1998) A mathematical formulation for reactive transport that eliminates mineral concentrations. Water Resour Res 34:1649–1656
Saaltink MW, Carrera J, Olivella S (2004) Mass balance errors when solving the convective form of the transport equation in transient flow problems. Water Resour Res 40:W05107
Schlumberger (2012) ECLIPSE technical description. Schlumberger, Sugar Land
Shamshiri H, Jafarpour B (2012) Controlled CO_2 injection into heterogeneous geologic formations for improved solubility and residual trapping. Water Resour Res 48:W02530
Singh VP, Cavanagh A, Hansen H, Nazarian B, Iding M, Ringrose PS (2010) Reservoir modeling of CO_2 plume behavior calibrated against monitoring data from Sleipner, Norway. In: Presented at the SPE annual technical conference and exhibition, Society of Petroleum Engineers, Florence, Italy

Slattery JC (1999) Advanced transport phenomena. Cambridge University Press, New York

Steefel CI (2009) CrunchFlow, software for modeling multicomponent reactive flow and transport, user's manual. Berkeley, Calif

Steefel CI, Appelo CAJ, Arora B, Jacques D, Kalbacher T, Kolditz O, Lagneau V, Lichtner PC, Mayer KU, Meeussen JCL, Molins S, Moulton D, Shao H, Šimůnek J, Spycher N, Yabusaki SB, Yeh GT (2015) Reactive transport codes for subsurface environmental simulation. Comput. Geosci 19:445–478

Streit JE, Hillis RR (2004) Estimating fault stability and sustainable fluid pressures for underground storage of CO_2 in porous rock. In: Energy, 6th international conference on greenhouse gas control technologies, vol 29, pp 1445–1456

Szulczewski ML, Hesse MA, Juanes R (2013) Carbon dioxide dissolution in structural and stratigraphic traps. J Fluid Mech 736:287–315

Thorenz C, Kosakowski G, Kolditz O, Berkowitz B (2002) An experimental and numerical investigation of saltwater movement in coupled saturated—partially saturated systems. Water Resour Res 38:1–5

Tian L, Yang Z, Fagerlund F, Niemi A (2016) Effects of permeability heterogeneity on CO_2 injectivity and storage efficiency coefficient. Greenh Gas Sci Technol 6:112–124

Tosaka N, Sugino R (1994) Boundary element analysis of moving boundary in Laplacian growth. In: Computational modeling of free and moving boundary problems II. Computational Mechanics Publication, Southampton

Tsang C-F (1991) Coupled hydromechanical-thermochemical processes in rock fractures. Rev Geophys 29:537–551

Unverdi SO, Tryggvason G (1992) A front-tracking method for viscous, incompressible, multi-fluid flows. J Comput Phys 100:25–37

van der Lee J, De Windt L, Lagneau V, Goblet P (2003) Module-oriented modeling of reactive transport with HYTEC. Comput Geosci 29:265–275

Verdon JP, Kendall J-M, White DJ, Angus DA (2011) Linking microseismic event observations with geomechanical models to minimise the risks of storing CO_2 in geological formations. Earth Planet Sci Lett 305:143–152

Vidal-Gilbert S, Tenthorey E, Dewhurst D, Ennis-King J, Van Ruth P, Hillis R (2010) Geomechanical analysis of the Naylor Field, Otway Basin, Australia: implications for CO_2 injection and storage. Int J Greenhouse Gas Control 4:827–839

Vilarrasa V, Bolster D, Dentz M, Olivella S, Carrera J (2010a) Effects of CO_2 compressibility on CO_2 storage in deep saline aquifers. Transp Porous Media 85:619–639

Vilarrasa V, Bolster D, Olivella S, Carrera J (2010b) Coupled hydromechanical modeling of CO_2 sequestration in deep saline aquifers. Int J Greenhouse Gas Control 2009(4):910–919

Vilarrasa V, Carrera J (2015a) Geologic carbon storage is unlikely to trigger large earthquakes and reactivate faults through which CO_2 could leak. Proc Natl Acad Sci USA 112:5938–5943

Vilarrasa V, Carrera J (2015b) Reply to Zoback and Gorelick: geologic carbon storage remains a safe strategy to significantly reduce CO_2 emissions. Proc Natl Acad Sci USA 112:E4511

Vilarrasa V, Carrera J, Bolster D, Dentz M (2013a) Semianalytical solution for plume shape and pressure evolution during CO_2 injection in deep saline formations. Transp Porous Media 97:43–65

Vilarrasa V, Olivella S, Carrera J, Rutqvist J (2014) Long term impacts of cold CO_2 injection on the caprock integrity. Int J Greenhouse Gas Control 24:1–13

Vilarrasa V, Rutqvist J, Rinaldi AP (2015) Thermal and capillary effects on the caprock mechanical stability at In Salah, Algeria. Greenh Gas Sci Technol 5:449–461

Vilarrasa V, Silva O, Carrera J, Olivella S (2013b) Liquid CO_2 injection for geological storage in deep saline aquifers. Int J Greenhouse Gas Control 14:84–96

Ward TJ, Cliffe KA, Jensen OE, Power H (2014a) Dissolution-driven porous-medium convection in the presence of chemical reaction. J Fluid Mech 747:316–349

Ward TJ, Jensen OE, Power H, Riley DS (2014b) High-Rayleigh-number convection of a reactive solute in a porous medium. J Fluid Mech 760:95–126

Wei X, Li W, Tian H, Li H, Xu H, Xu T (2015) THC-MP: high performance numerical simulation of reactive transport and multiphase flow in porous media. Comput Geosci 80:26–37
Weitz DA, Stokes JP, Ball RC, Kushnick AP (1987) Dynamic capillary-pressure in porous-media—origin of the viscous-fingering length scale. Phys Rev Lett 59:2967–2970
White MD, Bacon DH, McGrail BP, Watson DJ, White, SK, Zhang ZF (2012) Stomp subsurface transport over multiple phases: STOMP-CO_2 and Stomp-CO_2e Guide: version 1.0 (No. PNNL-21268). Pacific Northwest National Laboratory (PNNL), Richland
White MD, McGrail BP (2005) STOMP subsurface transport over multiple phases version 1.0 addendum: ECKEChem equilibrium-conservation-kinetic equation chemistry and reactive transport. United States Department of Energy, Washington, DC
Wilkinson D, Willemsen JF (1983) Invasion percolation: a new form of percolation theory. J Phys Math Gen 16:3365–3376
Wolf JL, Niemi A, Bensabat J, Rebscher D (in press) Benefits and restrictions of 2D reactive transport simulations of CO_2 and SO_2 co-injection into a saline aquifer using TOUGHREACT V3.0-OMP. Int J Greenhouse Gas Control
Xiaoping T, Juemin P, Junkai C (1997) Three dimensional simulation of unstable immiscible displacement in the porous medium. Appl Math Mech 18:81–89
Xu T, Sonnenthal E, Spycher N, Pruess K (2006) TOUGHREACT—a simulation program for non-isothermal multiphase reactive geochemical transport in variably saturated geologic media: applications to geothermal injectivity and CO_2 geological sequestration. Comput Geosci 32:145–165
Xu T, Sonnenthal E, Spycher N, Zheng L (2014) TOUGHREACT V3.0-OMP reference manual: a parallel simulation program for non-isothermal multiphase geochemical reactive transport. Berkeley, Calif
Yamamoto H, Zhang K, Karasaki K, Marui A, Uehara H, Nishikawa N (2009) Numerical investigation concerning the impact of CO_2 geologic storage on regional groundwater flow. Int J Greenhouse Gas Control 3:586–599
Yang Z, Niemi A, Tian L, Erlström M (2013a) Modelling of far-field pressure plumes for carbon dioxide sequestration. Energy Procedia, European Geosciences Union General Assembly 2013, EGUDivision Energy, Resources and the Environment, ERE, vol 40, pp 472–480
Yang Z, Tian L, Jung B, Joodaki S, Fagerlund F, Pasquali R, Vernon R, O'Neill N, Niemi A (2015) Assessing CO_2 storage capacity in the Dalders Monocline of the Baltic Sea Basin using dynamic models of varying complexity. Int J Greenhouse Gas Control 43:149–160
Yang Z, Tian L, Niemi A, Fagerlund F (2013b) Upscaling of the constitutive relationships for CO_2 migration in multimodal heterogeneous formations. Int J Greenhouse Gas Control 19:743–755
Yeh GT, Tsai CH (2013) HYDROGEOCHEM 6.0, a two-dimensional model of coupled fluid flow, thermal transport, HYDROGEOCHEMical transport, and geomechanics through multiple phase, systems version 6.0 (FACTM2D), theoretical basis and numerical approximation. Graduate Institute of Applied Geology, National Central University, Jhongli
Yortsos YC (1995) A theoretical analysis of vertical flow equilibrium. Transp Porous Media 18:107–129
Yortsos YC, Huang AB (1984) Linear stability analysis of immiscible displacement including continuously changing mobility and capillary effects: part I—simple basis flow profiles. Society of Petroleum Engineers, Dallas
Younes A, Ackerer P, Delay F (2010) Mixed finite elements for solving 2-D diffusion-type equations. Rev Geophys 48:rg1004
Zhang K, Wu YS, Pruess K (2008) User's guide for TOUGH2-MP—a massively parallel version of the TOUGH2 code (No. LBNL-315E). Lawrence Berkeley National Laboratory, Berkeley
Zhang L, Hou L, Wang L, Kan AT, Chen W, Tomson MB (2012) Transport of fullerene nanoparticles (nC60) in saturated sand and sandy soil: controlling factors and modeling. Environ Sci Technol 46:7230–7238
Zhao KXH, Wrobel LC, Power H (1995) Numerical simulation of viscous fingering using B-spline boundary elements. In: Wrobel LC, Brebbia CA (eds) Computational modeling of free and moving boundary problems II. Computational Mechanics, Southampton

Zhou Q, Birkholzer JT, Mehnert E, Lin Y-F, Zhang K (2010) Modeling basin- and plume-scale processes of CO_2 storage for full-scale deployment. Ground Water 48:494–514

Zoback MD, Gorelick SM (2015) To prevent earthquake triggering, pressure changes due to CO_2 injection need to be limited. Proc Natl Acad Sci 112:E4510

Zoback MD, Gorelick SM (2012) Earthquake triggering and large-scale geologic storage of carbon dioxide. Proc Natl Acad Sci 109:10164–10168

Chapter 5
Upscaling and Scale Effects

Marco Dentz, Jesus Carrera and Juan Hidalgo

Abstract Chapters 3 and 4 address the mathematical and numerical modeling of CO_2 geological storage. This chapter, in turn, focuses on a specific important aspect of modeling, namely that of scale effects and upscaling. The geological systems are heterogeneous, with heterogeneity occurring at various scales. This gives rise to what is commonly named the "scale effect". Certain process are critical at the scale of pores, while some of the effects of CO_2 injection may have an effect and need to be modeled at the scale of tens and even hundreds of kilometers. Furthermore, various processes may be important at different scales. This requires understanding and methods of linking processes over a span of the scales. This is the topic of the current chapter.

5.1 Scale Effects

The basic issue with upscaling may be illustrated with an example from (Bronstert et al. 2005). Watching a movie of boats in the ocean, alert spectators immediately and intuitively identify whether the boats were filmed in a pool or in the actual ocean. Pool waves may look as large as ocean ones, but they never look as foamy. The reason is that surface tension, which is an important force for small water bodies, becomes negligible when compared the inertial forces of true ocean waves. While it is evident that dominant processes may change with scale, the conventional continuum mechanics approach does not acknowledge it. The continuum mechanics approach is sound as long as the main processes remain unaltered. However, one of the points of

M. Dentz (✉) · J. Carrera · J. Hidalgo
Institute of Environmental Assessment and Water Research (IDAEA),
Spanish National Research Council (CSIC), Barcelona, Spain
e-mail: marco.dentz@csic.es

J. Carrera
e-mail: jesus.carrera.ramirez@gmail.com

J. Hidalgo
e-mail: juanj.hidalgo@idaea.csic.es

A. Niemi et al. (eds.), *Geological Storage of CO_2 in Deep Saline Formations*,
Theory and Applications of Transport in Porous Media 29,
DOI 10.1007/978-94-024-0996-3_5

this chapter is that such is rarely, if ever, the case for the processes involved in geological CO_2 storage. Variability is important not only because of the associated uncertainty, but also because large-scale behaviour of a spatially variable phenomenon may be significantly different from the small-scale behaviour. This means, changes in scale may (1) lead to changes in the effective parameters, (2) cause new processes to emerge, which often will imply (3) that the governing equations need to be changed. The term *scale effect* refers to any of these changes. Upscaling is a general term that refers to the procedures to derive these changes (in parameters, relevant processes, or governing equations) assuming that parameters, processes, and governing equations are known at a small scale. This chapter is motivated by the recognition that spatial and temporal variability is (1) very important, and (2) impossible to describe in detail. Therefore, scale effects should be expected in general.

Processes involved in CO_2 storage include: single and multiphase flow, solute transport, energy transport, chemical reactions and mechanical deformations. All of them suffer from scale effects. Single phase flow is governed by the flow equation, which basically entails fluid mass and momentum conservation. The latter is expressed by Darcy's law, whose only parameter is hydraulic conductivity. While the flow equation is broadly accepted, it is also well known that hydraulic conductivity displays significant scale effects, that is, its representative value grows with the size of the problem (Renard and de Marsily 1997; Sanchez-Vila et al. 2006). This implies that the hydraulic conductivity measured at one scale need not be equal to those described at a different, larger scale. Since measurements are often made at small scales (e.g., cm or m), whereas flow may be need at much larger (e.g., 10^2–10^3 m) scales, it is clear that upscaling is important.

The situation is more complex for transport. Traditionally continuum scale transport through homogeneous media has been represented by the equation (ADE) (Bear 1972):

$$\varphi \frac{\partial c(x,t)}{\partial t} + \mathbf{q} \cdot \nabla c(x,t) - D\nabla^2 c(x,t) = 0, \tag{5.1.1}$$

where φ is porosity, $c(x,t)$ represents the magnitude being transported per unit volume of fluid (e.g., concentration), $\mathbf{q}$ is the mean flux and D is the diffusion (dispersion) coefficient, which represents mixing caused by Brownian motion (or by fluctuations of the pore-scale fluid velocity around its mean).

Observed chemical transport in heterogeneous media does not behave as implied by (5.1.1) (Adams and Gelhar 1992; Berkowitz and Scher 1997; Carrera 1993; Kapoor et al. 1997; Neuman and Zhang 1990; Salamon et al. 2006; Steefel et al. 2005). For one thing both apparent dispersivity (Gelhar et al. 1992; Lallemand-Barres and Peaudecerf 1978) and porosity (Guimerà and Carrera 2000) Pdisplay marked scale effects. Dispersivity grows (almost) linearly with the scale of the problem, whereas porosity in fractured media grows with residence time. Differences are not restricted to transport parameters. Field observations differ from model predictions. It is therefore clear that the ADE is not adequate for simulating transport. A number of alternatives have been proposed in recent years. They will be discussed in Sect. 5.3.

However, the driving force in the most recent developments has been reactive transport. When dealing with reactive transport, it is convenient to distinguish between fast reactions, which tend to proceed in equilibrium, and slow reactions, whose rate depends on the distance to equilibrium, as quantified by saturation.

Rezaei et al. (2005) showed that the actual rate of equilibrium reactions in porous media is controlled by mixing. The interplay between transport and chemistry is non trivial. Specifically, performing geochemical calculations is needed, but does not suffice to predict where equilibrium reactions will take place, what will be their rate, or under which conditions will this rate be maximum. These outcomes are controlled by mixing. In fact, shortly thereafter, De Simoni et al. (2005) found a relatively simple expression to quantify mixing for Fickian dispersion. A result from these and many other works is that proper representation of reactive transport requires a proper representation of mixing. The large-scale ADE equates spreading and mixing in form of macrodispersion coefficients. While for a fluid at rest, or a hypothetical homogeneous porous medium, a diffusion or dispersion coefficient is an operational concept to quantify solute mixing, this is very different for transport in heterogeneous flow fields, as illustrated in Fig. 5.1. At practically relevant times, the macrodispersion concept quantifies the extension of the plume as opposed to the volume that is occupied by the solute as a consequence of mixing (Dentz et al. 2000; Kitanidis 1994). Thus, the ADE parameterized by a (temporally evolving) macrodispersion coefficient quantifies rather the extension of the plume than mixing, and is therefore not suited for the quantification of reactive transport.

One might argue that slow reactions, whose rate is controlled by local chemistry, should not suffer from scale effects. As it turns out, kinetic reaction rates observed in the field are much slower than what might be expected from laboratory measurements (White and Peterson 1990). This implies that local concentrations are not well represented in the ADE either (Lichtner and Kang 2007). In fact, most of the

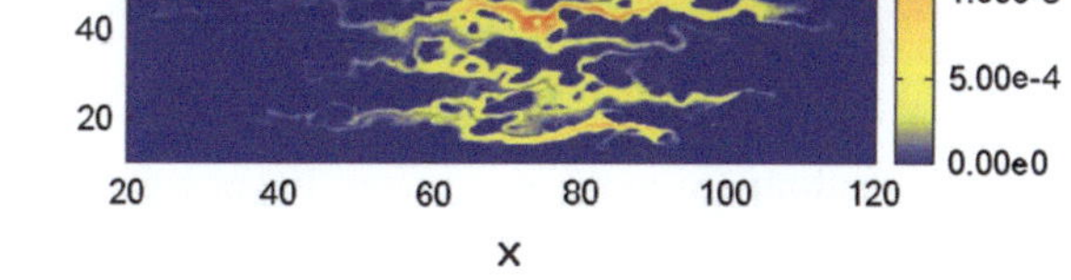

Fig. 5.1 Mixing refers to the rate at which different waters blend together. As such, it is controlled by local gradients and tends to destroy local irregularities in concentration. Spreading, on the other hand, tends to increase the extent of the plume or the overall width of an advancing front. They are linked, because irregular spreading tends to generate gradients perpendicular to the flow direction, but they are different concepts. Yet, both are equated in the ADE

alternatives to the ADE discussed in this chapter adopt non local formalisms. That is, transport processes at one point in space and time depend not only on the concentration field at that point, but also on its spatial variability far away and on its time history.

We find similar scale effects in for the description of multiphase in heterogeneous porous media. This may manifest on one hand, in scale effects in effective flow parameters such as intrinsic hydraulic conductivity, or scale effects in constitutive relationships such as capillary pressure saturation relations and relative permeability (Yang et al. 2013). On the other hand, heterogeneity effects the spreading of a displacement in a way similar as observed in the case of the miscible displacement of one fluid by the other, which can be described by the single phase flow macrodispersion coefficient described above. For two-phase flow the phenomenon is analogous (Langlo and Espedal 1994; Neuweiler et al. 2003; Bolster et al. 2009). Remarkably, for two-phase flow, this gives rise to a macrodispersive flux in the saturation equations for the two fluids. For highly heterogeneous porous media, one observes similar fluid retention phenomena as for solute transport in multicontinuum media (Di Donato et al. 2007; Geiger et al. 2013).

In all these applications, scale effects are caused by the interaction of spatial heterogeneity and the small scale flow and transport processes, which can be seen as collective phenomena that may be described by effective parameters, or require the constitution of flow and transport equations that are different from the ones on the local scale. In the following, we gives an description of various efforts to approach the upscaling problem, and quantify large scale flow and transport in heterogeneous porous formations. This description needs to be necessarily incomplete due to the vast amount of literature that has been dedicated to this important topic in the last 50 years.

5.2 Single Phase Flow

The following sections describe the upscaling of porous media flow from the pore to Darcy scale, and from the Darcy to the field scale.

5.2.1 Pore to Darcy Scale

On the pore scale, the medium is composed of void space that is available to the fluid and a solid phase that here is assume to be impermeable. Figure 5.2 shows a X-ray microtomography cross section of the heterogeneous pore-structure of a pure calcite limestone.

In the following, we briefly introduce into the description of single phase flow on this scale, and its upscaling to the continuum or Darcy scale, which is assumed to be much larger than a typical pore length scale.

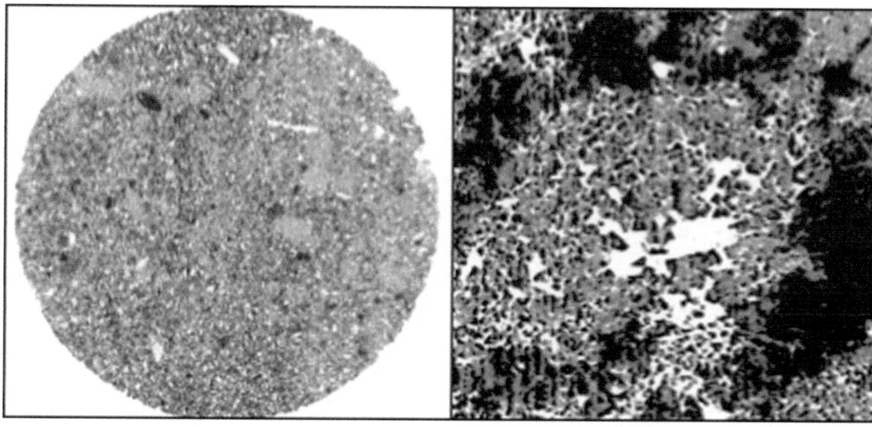

Fig. 5.2 X-ray microtomography cross section of a pure calcite limestone imaged at different resolutions. *Left* porosity map (diameter 1 cm; resolution 5 μm). *Right* processed image (600 × 600 μm, resolution 1 μm) showing the mobile domain (*in white*) and the micro-porosity distribution in the immobile domain formed by clusters of distinctly different size (*grey scale*). Zones in *black* are areas unconnected with the immobile domain (porosity lower than the percolation threshold, see also (Gouze et al. 2008)

The pore space available for fluid flow is denoted by Ω_f, the solid space by Ω_s. Their boundaries are denoted by $\partial\Omega_f$ and $\partial\Omega_s$. The space occupied by the bulk of the porous medium is denoted by $\Omega = \Omega_f \cup \Omega_s$. The solid grains are assumed to be impermeable. Flow in the pore space is governed by the Navier-Stokes equation

$$\rho_f \frac{\partial \mathbf{v}(\boldsymbol{x}, t)}{\partial t} + \mathbf{v}(\boldsymbol{x}, t) \cdot \nabla \mathbf{v}(\boldsymbol{x}, t) = \mu \nabla^2 \mathbf{v}(\boldsymbol{x}, t) - \nabla p(\boldsymbol{x}, t) + \rho_f g, \tag{5.2.1}$$

where ρ_f is the fluid density, $\mathbf{v}(\boldsymbol{x}, t)$ is the fluid velocity, μ is dynamic viscosity, $p(\boldsymbol{x}, t)$ is fluid pressure and g is the gravity acceleration. The first two terms on the left describe fluid acceleration, the first two terms on the right quantify the action of shear and stress on a fluid volume in terms of pressure (stress) and viscosity (shear), the last term quantifies body forces, which here is only the action of gravity.

Flow can be characterized by the dimensionless Reynolds number $Re = \ell v_c / v$ with ℓ a characteristic pore length scale, v_c a characteristic flow rate, and $v = \mu / \rho$ kinematic viscosity. The Reynolds number compares fluid inertia (ℓv_c) to vicous resistance (v). We assume that the fluid density is constant, which means that we focus on isothermal flow that is not affected by solute transport. Furthermore, for $Re = 1$ as is typically the case for porous media flows, inertia effects can be disregarded and thus, the Navier-Stokes equation simplifies to the Stokes equation

$$\rho_f \frac{\partial \mathbf{v}(\boldsymbol{x}, t)}{\partial t} = \mu \nabla^2 \mathbf{v}(\mathbf{x}) - \nabla p(\mathbf{x}) + \rho_f \mathbf{g}, \tag{5.2.2}$$

where $\mathbf{v}(\mathbf{x})$ is the pore velocity, μ is the fluid viscosity and $p(\mathbf{x})$ the fluid pressure, ρ_f is the fluid density and $\mathbf{g}$ is gravity acceleration. Since we assume that the fluid density is constant in time and space, fluid mass conservation implies $\nabla \cdot \mathbf{v}(\mathbf{x}) = 0$.

There are a series of approaches to solve the pore scale flow problem including Lattice-Boltzmann (Acharya et al. 2007; Kang et al. 2006; Willingham et al. 2008), smoothed particle hydrodynamics (Tartakovsky et al. 2007, 2009), computational fluid dynamics and pore network modeling (Li et al. 2006; Meile and Tuncay 2006), see also the recent review by Meakin and Tartakovsky (2009). In general these methods are computationally very costly and therefore often limited to relatively small flow domains or relatively simple pore geometries. If we are interested in the global fluxes, it is not necessary to know all the small scale details of the fluctuations of $\mathbf{v}(\boldsymbol{x}, t)$, but sufficient to determine the average fluid flow in a representative elementary volume (REV) V_r of the medium,

$$\mathbf{q}(\boldsymbol{x}, t) = \frac{1}{V_r} \int\limits_{V_r} d\mathbf{r} \mathbf{v}(\mathbf{x} + \mathbf{r}, t). \tag{5.2.3}$$

In order to illustrate the concept of the representative elementary volume, we consider the definition of porosity as done in the textbook by Bear (1972). Thus, we consider the volume of void space in a radius ℓ about a point x,

$$V_f(\mathbf{x}, \ell) = \int\limits_{V(\ell)} d\mathbf{r} \mathbf{I}(\mathbf{x} + \mathbf{r} \in \Omega_f), \tag{5.2.4}$$

where the indicator function $\mathbf{I}(\mathbf{r} \in \Omega_f)$ is 1 if the statement in its argument is true and 0 else, $V(\ell)$ is the bulk volume on the scale ℓ. The ratio $V_f(\ell)/V(\ell)$ fluctuates on a scale of the order of the pore diameter for small ℓ, where voids and grain are clearly distinguishable. For increasing scale ℓ, i.e., for ℓ larger than the characteristic pore size, this ratio converges to the constant volumetric porosity. The scale ℓ_r at which this transition happens defines the representative elementary volume $V_r = V(\ell_r)$. Notice that this concept requires that the pore space can be characterized by a characterstic length scale. This concept does not apply for media characterized by fractal pore size distributions because of the lack of a characteristic pore size.

The Stokes equation (5.2.2) can be upscaled, or averaged using the methods of Homogenization and volume averaging (Bear 1972; Hornung 1997; Whitaker 1986) for example. The average flow velocity $\mathbf{q}(x, t)$ then satisfies the equation (e.g. Bear 1972)

$$\mathbf{q}(\mathbf{x}, t) + \frac{\rho_f \mathbf{k}}{\varphi \mu} \frac{\partial \mathbf{q}(\boldsymbol{x}, t)}{\partial t} - \frac{\mu^2 \mathbf{k}}{\mu} \nabla \mathbf{q}(\boldsymbol{x}, t) = -\frac{\rho_f g \mathbf{k}}{\mu} \nabla h(\mathbf{x}, t), \tag{5.2.5}$$

where the permeability tensor **k** reflects the medium geometry and composition, μ^e is an effective viscosity. The hydraulic head is defined by $h(\mathbf{x},t) = p(\mathbf{x},t)/(\rho_f g) + z$, the porosity ϕ compares the pore volume to the bulk volume of the porous medium, $\phi = V_f/V$. The first terms on the left hand side comes from the visous term in (5.2.2), the second from the acceleration term. The third term represents shear losses at the fluid grain interfaces and was introduced by Brinkman (1949). The term on the right side summarizes pressure and body forces. Under certain conditions the evolution equation for $\mathbf{q}(\mathbf{x},t)$ may include memory terms that are non-local in time (e.g. Hornung 1997). Fluid acceleration typically occurs on such small time scales that the transient term in (5.2.5) can be disregarded. Also the so-called Brinkman term can typically be neglected, which gives the Darcy equation

$$\mathbf{q}(\boldsymbol{x},t) = -\mathbf{K}\nabla h(\mathbf{x},t), \mathbf{K} = \frac{\rho_f g \mathbf{k}}{\mu}, \tag{5.2.6}$$

where **K** is hydraulic conductivity. Conductivity **K** is not only a property of the porous medium, but through its dependence on fluid density and viscosity also of the fluid. Notice that it also depends on gravity acceleration g which makes it a characteristic of the planetary conditions.

Fluid mass conservation for a Darcy scale porous medium finally is expressed by the continuity equation (Bear 1972)

$$S_0 \frac{\partial h(\mathbf{x},t)}{\partial t} + \nabla \cdot [\mathbf{K}\nabla h(\mathbf{x},t)] = f(\mathbf{x},t), \tag{5.2.7}$$

where $f(\mathbf{x},t)$ represents the presence of volume sinks and sources. The specific volumetric storage $S_0 = \rho g[\alpha(1-\phi) + \beta\phi]$ expresses the compressibility of the medium through α and of the fluid through β. It quantifies the volume of water that is released from a unit volume of aquifer per unit decline in hydraulic head $h(x,t)$. In the following, we will assume that the hydraulic conductivity tensor is diagonal and isotropic such the $K_{ij} = K\delta_{ij}$.

Under the Dupuit assumption of predominantly horizontal flow, i.e., $h(\boldsymbol{x},t) = h(x,y,t)$, the integration of the flow equation over the thickness d_a of the aquifer gives (Bear 1972)

$$S\frac{\partial h(x,y,t)}{\partial t} + T\nabla^2 h(x,y,t) = F_v(x,y,t) + F_s(x,y,t), \tag{5.2.8}$$

where storativity $S = S_0 d_a$, transmissivity $T = K d_a$, $F_v(x,y,t)$ are the vertically integrated volume sources and sinks, and $F_s(x,y,t)$ are surface sources and sinks at the horizontal aquifer boundaries.

In the following we denote the Darcy scale also as continuum scale because it does not distinguish between solid grains and void space, but characterizes the physical medium properties in terms of the effective parameters hydraulic conductivity $\mathbf{K}$, porosity ϕ and specific volumetric storage S_0.

5.2.2 Darcy to Field Scale

The continuum-scale physical medium properties as quantified by the hydraulic conductivity K, porosity ϕ and specific storativity S_0 are in general spatially variable due to heterogeneity in the porous material. Hydraulic conductivity varies by 13 order of magnitude between different materials (Bear 1972). Variability in porosity and specific storage is typically much lower.

This spatial variability leads to large-scale flow behaviors that are quantitatively different from the ones observed at a local scale, where the medium can be assumed to be homogeneous. The local scale refers to a length scale that is shorter than the characteristic variability scale of the medium. The Darcy equation (5.2.6) and the mass conservation equation (5.2.7) in a heterogeneous medium are given by

$$\mathbf{q}(\boldsymbol{x}, t) = -K(\boldsymbol{x})\nabla h(\boldsymbol{x}, t) \tag{5.2.9}$$

$$S_0(\boldsymbol{x})\frac{\partial h(\boldsymbol{x}, t)}{\partial t} - \nabla \cdot \mathbf{q}(\boldsymbol{x}, t) = f(\boldsymbol{x}, t). \tag{5.2.10}$$

The detailed knowledge of the spatial variability of the physical medium properties in terms of $K(\boldsymbol{x})$ and $S_0(\boldsymbol{x})$, and the (numerical) solution of the local scale flow problem (5.2.10), can in principle quantify the observed flow behavior. However, the detailed characterization of the local scale medium fluctuations is in many practical applications not possible and also not desireable. As pointed out above for the upscaling from pore to Darcy scale, the characterization of large scale features of fluid flow does not require the detailed knowledge of the full local scale fluctuation behavior. Thus, coarse grained, averaged flow descriptions are required to quantify and explain observed large scale phenomena and to make predictions. As for the transition from the pore to the continuum scale discussed in the previous section, coarse graining and upscaling implies averaging of the Darcy equation (5.2.9).

In this context we distinguish between equilibrium and non-equilibrium approaches. Equilibrium approaches assume that large scale flow can be characterized by the Darcy equation characterized by an effective hydraulic conductivity. We call this an equilibrium approach because it assumes that flow at a coarse grained position $\boldsymbol{x}$ can be characterized by a single average value of hydraulic head. Non-equilibrium approaches average the flow Eq. (5.2.10) and arrive at flow equations that are characterized by memory kernels, which account for the fact that the support scale of the coarse flow description is not in local equilibrium.

In the following, we present approaches for the upscaling of the flow problem from Darcy to field scale. We focus hereby on the scale dependence of hydraulic conductivity Renard and de Marsily (1997), Sanchez-Vila et al. (2006), as well as non-equilibrium approaches for fluid flow in media with large parameter contrasts.

5.2.2.1 Steady Flow: Effective Hydraulic Conductivity

The characterization of large scale flow in terms of effective hydraulic conductivity assumes that an average Darcy flow velocity $\mathbf{q}(\boldsymbol{x})$ on the large scale obeys the Darcy law

$$\overline{\mathbf{q}}(\boldsymbol{x}) = -\mathbf{K}^e \nabla \overline{h}(\boldsymbol{x}), \tag{5.2.11}$$

where the effective hydraulic conductivity tensor $\mathbf{K}^e$ measures the average flux subject to a unit gradient of an average hydraulic conductivity $\overline{h}(\boldsymbol{x}, t)$. Equation (5.2.11) defines the effective hydraulic conductivity, which, in general, is a tensorial quantity. Directional dependence of large scale conductivity can be due to statistical anisotropy of the local hydraulic conductivity distribution, boundary distributions, or due to the domain geometry. Exact results for the effective hydraulic conductivity exist only for layered media, and two-dimensional porous media whose conductivities satisfy certain conditions.

Stratified Media

A heterogeneous porous medium whose conductivity values are organized in strata of equal thickness d and thus depend only on the z-direction, $K(x) = K(z)$ The medium can be characterized by the sequence of conductivity values $\{K_n\}$ in the strata, where the subscript n determines the spatial position within the medium. We consider the case of steady flow characterized by $\nabla \cdot \mathbf{q}(x) = 0$.

Let us consider now the effective flow behavior in such a medium. First, we consider flow in z-direction, perpendicular to the direction of stratification, which implies that flow is aligned with the direction perpendicular to the stratification. The flow Eq. (5.2.10) now becomes $\frac{\partial}{\partial z} K(z) \frac{\partial h(z,t)}{\partial z} = 0$. The solution for hydraulic conductivity for constant head boundary conditions is given by integration as $h(z) = h_0 - K_H G \int_0^z dz' q/K(z')$, where $G = \frac{\partial \overline{h}(z)}{\partial z} = (h_w - h_0)/L$ is the large scale hydraulic gradient, h_0 and h_w are the hydraulic heads at inlet and outlet; the vertical extension of the domain is $w = Nd$; $K_H = (w/d)(\sum_n K_n^{-1})^{-1}$ is the harmonic average of the specific sequence of conductivity values. The Darcy equation can now be written in terms of the global hydraulic head $\overline{h}(z)$ and K_H as

$$q = -K_H \frac{\partial \overline{h}(z)}{\partial z}. \tag{5.2.12}$$

Notice that the flow problem is completely defined by the global hydraulic gradient and the harmonic mean of the conductivity values within the flow domain. Details such as the exact sequence $\{K_n\}$ do not play a role.

We consider now flow in x-direction, aligned with the direction of stratification. In this case, the flow Eq. (5.2.10) becomes $\frac{\partial}{\partial x} K(z) \frac{\partial h(x,t)}{\partial x} = 0$. For constant head boundary conditions at $x = 0$ and $x = L$, the solution for the hydraulic head is independent of the z-direction. Thus the Darcy equation reads as $q(z) = -K(z)\partial h(x)/\partial x$, and we obtain for the global flow $\overline{q} = w^{-1} \int_0^w dzq(z)$ the effective Darcy equation

$$\overline{q} = -K_A \frac{\partial h(x)}{\partial x}, \tag{5.2.13}$$

where $K_A = (d/w) \sum_n K_n$ is the arithmetic mean over the values in the conductivity sequence $\{K_n\}$. Again, the details of the sequence are not of importance.

Stochastic Modeling

The stochastic modeling approach interprets this sequence as a stochastic process in space. This means the conductivities K_n that form the sequence $\{K_n\}$ are assumed to be random variables characterized by a certain distribution $p_K(K)$. The stochastic process, i.e., the random sequence $\{K_n\}$ is characterized by the joint distribution of conductivity values K_n. Notice that each realization of this stochastic process, this means, each random sequence of conductivity values, defines an aquifer realization. In many geological media, the distribution of hydraulic conductivity $p_K(K)$ is found to follow approximately a log-normal distribution (Renard and de Marsily 1997; Sanchez-Vila et al. 2006). In fact, the sequence $\{K_n\}$ is typically modeled as a multi-lognormally distributed stochastic process, which implies that $\{Y_n = \ln(K_n)\}$ is a multi-Gaussian distributed stochastic process. This means that the joint distribution of the Y_n is a multi-Gaussian distribution. Thus, the stochastic process $\{Y_n\}$ can be characterized by its mean $\overline{Y_n}$ and covariance function $\overline{(Y_i - \overline{Y_i})(Y_j - \overline{Y_j})} = C_{Y,ij}$. It is typically assumed that the process is stationary, which implies that the mean $\overline{Y_n} = \overline{Y}$, i.e., it does not depend on the position within the medium, and that the covariance $C_{Y,ij} = C_{Y,i-j}$, i.e., it depends only on the relative distance between the strata. The variance of log-conductivity is given by $\sigma_Y^2 = C_{Y,ii}$. Processes with this property are termed stationary processes.

The stochastic modeling approach substitutes now the spatial harmonic and arithmetic averages by their respective ensemble averages $\overline{K}_H = \left(\overline{1/K}\right)^{-1}$ and $\overline{K}_A = \overline{K}$. The spatial and ensemble average quantities are in general not equal. The spatial average K_A is in general an average over a finite number of strata, while the stochastic average in principle implies an average over an initinite number of realizations. The process $\{K_n\}$ is called ergodic, if the infinite space limit of the arithmetic average and its ensemble average coincide,

$$\lim_{w \to \infty} \frac{d}{w} \sum_{n=1}^{w/d} K_n = \overline{K}_A. \tag{5.2.14}$$

This implies that the heterogeneity features present in a single medium realization are representative of the ensemble of aquifers. Thus, in an ergodic medium, the stochastic average conductivity may be used to predict the effective hydraulic conductivity to be used to describe flow on a large scale. Large scale in this particular example means, large compared to the characteristic size of a stratum. Here we are primarily interested in ergodic processes because we want to use the stochastic approach to make predictions on flow and transport in heterogeneous media.

In general, hydraulic conductivity and other physical and chemical medium characteristics are continuous functions of the spatial position, $K(x)$. A more detailed account on the stochastic modeling approach in hydrogeology can be found in the textbook by Dagan (1989), Gelhar (1993) and Rubin (2003).

Two-Dimensional Isotropic Media

We consider the exactly solvable case of flow in isotropic two-dimensional media presented in Matheron (1967). Hydraulic conductivity $K(x)$ is modeled as a stationary and ergodic random field with the property that $K(x)$ and $K(x)^{-1}$ obey the same statistics. This is the case if $K(x) = \exp[Y(x)]$ is multi-lognormally distributed with istropic correlation properties, i.e., $Y(x) = \ln[K(x)]$ is multi-normally distributed. Cleary, the distribution of $K(x)^{-1} = \exp[-Y(x)]$ is again lognormally distributed. The following derivation uses the duality argument of Keller (1964), see also the paper by Dean et al. (2007).

The Darcy velocity is assumed to the divergence-free, i.e., $\nabla \cdot \mathbf{q}(x) = 0$. Therefore, it can be represented in terms of a streamfunction $\psi(x)$ as $\mathbf{q}(x) = -\mathbf{e}_3 \times \nabla\psi'(x)$, where $\mathbf{e}_3$ denotes the unit vector perpendicular to the two-dimensional flow plain. Darcy's law (5.2.9) relates conductivity $h(x)$ and hydraulic conductivity $K(x)$ to the streamfunction $\psi(x)$ as $K(x)\nabla h(x) = \mathbf{e}_3 \times \nabla\psi(x)$. A vectorial multiplication of this equation from the left with $\mathbf{e}_3$ gives a dual relation between the streamfunction $\psi(x)$ and the inverse hydraulic conductivity $K(x)$ with the hydraulic head, $K(x)^{-1}\nabla\psi(x) = -\mathbf{e}_3 \times \nabla h(x)$. By defining now $K'(x) = K_0^2/K(x)$, such that the distributions of $K'(x)$ and $K(x)$ are identical, and further, defining $\psi'(x) = \psi(x)/K_0$, we can write the following dual system of equations for $\psi'(x)$ and $h(x)$,

$$K(x)\nabla h(x) = -K_0\mathbf{e}_3 \times \nabla\psi'(x), \quad K'(x)\nabla\psi'(x) = K_0\mathbf{e}_3 \times \nabla h(x). \tag{5.2.15}$$

As $K(x)$ and $K'(x)$ are statistically identical, we obtain for both averages $\overline{K(x)\nabla h(x)} = K^e\nabla\overline{h}(x)$ and $\overline{K'(x)\nabla\psi(x)} = K^e\nabla\overline{\psi}(x)$, respectively. As such, we obtain by averaging (16) the relation $K^e\nabla\overline{h}(x) = -K_0\mathbf{e}_3 \times \nabla\overline{\psi'}(x)$. Vectorial multiplication from the right by $\mathbf{e}_3$ gives $K_0\nabla\overline{\psi'}(x) = K^e\mathbf{e}_3 \times \nabla\overline{h}(x)$. This relation is now compared to the average of the second equation in (16), $K^e\nabla\overline{\psi'}(x) = K_0\mathbf{e}_3 \times \nabla\overline{h}(x)$. Thus, we obtain directly that $K^e = K_0$. For the multi-lognormal random conductivity field with isotropic correlation function, it is easy to verify that $K_0 = K_G$, the geometric mean conductivity, such that $K^e = K_G$.

Three-Dimensional Isotropic Media

The problem of finding effective hydraulic conductivity in three-dimensional heterogeneous porous media has been pursued in a systematic way by using stochastic modeling. In the following, we want to briefly outline the basic idea following the method presented in the paper by Gutjahr et al. (1978).

As above, hydraulic conductivity here is represented by a stationary isotropic multi-lognormally distributed spatial random field. The basic methodology consists in seeking an approximation for the effective hydraulic conductivity K^e of first order in the variance σ_Y^2 of log-hydraulic conductivity $Y(x)$. To this end, $Y(x)$ is separated in its ensemble mean value $\overline{Y}$ and fluctuations $Y'(x) = Y(x) - \overline{Y}$ about it. Along the same lines, hydraulic head $h(x)$ is decomposed into its mean value $\overline{h}(x) = \overline{h}(x)$ and fluctuations $h'(x) = h(x) - \overline{h}(x)$ about it. It is furthermore assumed that the average hydraulic gradient is constant $\nabla\overline{h}(x) = G_H\mathbf{e}_1$ and aligned with the one-direction of the coordinate system as a consequence of constant head boundary conditions. Using this decomposition in (5.2.9) and discarding terms that are of order higher than 2 in the fluctuating quantities, we obtain for the average Darcy flow

$$\overline{\mathbf{q}}(x) = -K_G G_H \mathbf{e}_1 - K_G\left[\frac{\sigma_Y^2}{2} + \overline{Y'(x)\nabla h'(x)}\right], \tag{5.2.16}$$

where we used the expansion $K(x) = K_G\left[1 + Y'(x) + Y'(x)^2/2 + cldots\right]$ and the fact that by definition $\overline{Y'(x)} = \overline{h'(x)} = 0$. Notice that the lowest order term in (5.2.16) is identical to the exact result for two-dimensional isotropic media derived in the precious section.

In order to evaluate (5.2.16) consistently in σ_Y^2, it remains to determine an expression for $h'(x)$ linear in $Y'(x)$. To this end, we consider the flow equation for $h(x)$ in steady state, which is obtained by taking the divergence of (5.2.9) as $\nabla^2 h(x) + \nabla Y(x)\cdot\nabla h(x) = 0$. Using the decompositions of $h(x)$ and $Y(x)$ in mean and fluctuations in the steady state flow equation, we obtain for $h'(x)$ the following equation

$$\nabla^2 h'(x) = -\frac{\partial Y'(x)}{\partial x_1} G_H, \tag{5.2.17}$$

where we disregard contributions that are quadratic in the fluctuations. This Poisson equation can be solved, for example by the method of Green's functions such that $h'(x) = \int dx' Y'(x') g(x - x') G_H$, where an infinite flow domain is assumed. The Green's function satisfies $\nabla^2 g(x - x') = -\delta(x - x')$ and is given by $g(x) = 1/(4\pi|x|)$. Inserting the expression for $h'(x)$ into (5.2.16) and solving the remaining integrals gives

$$\overline{\mathbf{q}}(x) = -K_G\left[1+\sigma_Y^2\left(\frac{1}{2}-\frac{1}{3}\right)\right]G_H. \tag{5.2.18}$$

such that the effective conductivity is identified to be $K^e = K_G\left[1+\sigma_Y^2\left(\frac{1}{2}-\frac{1}{3}\right)\right]$. Notice that we used the dimensionality of space only when specifiying the Green's function to solve (5.2.17). The same perturbation calculation holds also for $d=1$ and $d=2$ dimensions. The general perturbation expression for the effective hydraulic conductivity in d spatial dimensions is given by $K^e = K_G\left[1+\sigma_Y^2\left(\frac{1}{2}-\frac{1}{d}\right)\right]$.

Notice that the arithmetic average of $K(x)$ is given by $K_A = K_G\exp(\sigma_Y^2/2)$, the harmonic average by $K_H = K_G\exp(-\sigma_Y^2/2)$. In analogy, starting from the above perturbation approximation for K^e it was conjectured (Gelhar and Axness 1983) that $K^e = K_G\exp[\sigma_Y^2\left(\frac{1}{2}-\frac{1}{d}\right)]$, which in the case of $d=1$ reduces to the exact result $K^e = K_H$ and for $d=2$ to the exact result $K^e = K_G$. In general it is found (Matheron 1967) that the effective conductivity is bounded between the harmonic and arithmetic mean conductivities, $\overline{K}_H \le K^e \le \overline{K}_A$.

Anisotropic Media

Statistically anisotropic media are characterized by directional dependence of the correlation length of the fluctuations of hydraulic conductivity. Notice that for anisotropic media, as pointed out above, effective dispersion is actually a tensorial quantity. The stochastic perturbative methdology is identical to the one reviewed in the previous paragraph, with the difference that here the covariance of the loghydraulic conductivity fluctuations is directionally depended. Using this methodology, Gelhar and Axness (1983) derived explicit expressions for the hydraulic conductivity tensor in $d=3$ spatial dimensions. Specifically, for a mean flow aligned with the bedding, i.e., aligned with the direction of stratification of the aquifer, and isotropic correlation length $\lambda_1 = \lambda_2 = \lambda_h > \lambda_3$ in the horizontal direction, these authors find that the effective hydraulic conductivity tensor is diagonal with isotropic conductivity in the horizontal, $K_{11}^e = K_{22}^e = K_h$, and a different value $K_{33}^e = K_v \neq K_h$ in the vertical. For this case, the results obtained by Gelhar and Axness (1983) can be written as Sanchez-Vila et al. (2006)

$$K_h = K_G\left(1+\frac{\chi}{2}\sigma_Y^2\right), \quad K_v = K_G\left[1-\left(\frac{1}{2}-\chi\right)\sigma_Y^2\right], \tag{5.2.19}$$

where χ is defined by

$$\chi = \frac{\rho^2}{\rho^2-1}\left[1-\frac{\arctan\left(\sqrt{\rho^2-1}\right)}{\sqrt{\rho^2-1}}\right], \quad \rho = \frac{\lambda_h}{\lambda_v}. \tag{5.2.20}$$

These expressions reduce to the ones for three-dimensional isotropic media presented above in the limit $\rho \to 1$.

Comprehensive reviews of approaches to determine the effective hydraulic conductivity for anisoptropic bounded and infinite porous media as well as for non-stationary conductivity fields can be found in the papers by Renard and de Marsily (1997) and Sanchez-Vila et al. (2006).

Further Remarks

It remains to remark that effective or equivalent single phase flow properties in heterogeneous porous media have been intensely studied over the past 50 years, and the brief account given above is all but complete.

In the literature (Sanchez-Vila et al. 2006; Wen and Gómez-Hernández 1996) a distinction is made between effective and equivalent hydraulic conductivities. The term effective hydraulic conductivity is defined in an ensemble sense in the context of a stochastic aquifer model. Thus it relates ensemble average flux to an average head gradient. The equivalent hydraulic conductivity is defined in a single aquifer by spatial averaging. It relates the spatially averaged Darcy flux with the spatially average hydraulic gradient. Thus, equivalent hydraulic conductivities are representative of a certain region of the medium or medium block. We touched on this issue when reviewing equivalent conductivites in stratified media, and their representation in a stochastic model. Equivalent and effective hydraulic conductivities in this sense may be equated if the underlying medium or region of the medium has ergodic properties.

The necessity to define equivalent hydraulic conductivities arises in the problem of the representation of point measurements of hydraulic conductivity in a coarse numerical grid, the size of which, however, may not fulfil the ergodicity conditions required to equal ensemble and spatial average. The coarse-grained medium representation characterized by spatially variable (from block to block) equivalent conductivity is considered a large scale aquifer equivalent. Specifically, this equivalent medium should be characterized by the same mean flow as the fine scale medium. This requirement poses some consistency conditions for the equivalent conductivity values. They cannot be considered an intrinisc property of the medium block, but depend on the conductivities of the surrounding blocks; equivalent conductivies are in general non-local quantities (Neuman and Orr 1993). Wen and Gómez-Hernández (1996), Renard and de Marsily (1997) and Sanchez-Vila et al. (2006) provide comprehensive review of approaches, methods and techniques for the determination of equivalent hydraulic conductivies.

The heterogeneous nature of a natural medium leads on one hand to large scale flow properties that are different from the local scale ones, which are quantified in the context of this section by effective and equivalent hydraulic conductivities. These heterogeneities, on the other hand, induce uncertainty on these averages, due to the incomplete knowledge of the details of the particular medium under consideration. The representation of the heterogeneous porous medium as a realization of an ensemble of media provides the framework for the quantification of uncertainty on hydraulic heads and equivalent conductivities (Dagan 1986; Neuman and Orr 1993). Winter et al. (2003) focus on the quantification of uncertainty in highly heterogeneous geological media that are composed of domains of very distinct hydraulic properties. These authors consider heterogeneity in both the geometry of these domains as well as in the physical medium properties of each domain.

In the following section, we will briefly review the dual porosity approach for the modeling of flow in media with with sharp contrasts in hydraulic conductivity.

5.2.2.2 Unsteady Flow: Dual Continuum Media

In this section, we consider the situation that the local equilibrium condition invoked above is not valid, this means that hydraulic head is not in local equilibrium on the support scale of the spatially averaged model. Such conditions are typical for fractured media characterized by sharp contrasts between the hydraulic properties in the fracture and matrix domains.

Barenblatt et al. (1960) suggested a dual continuum model for flow in fractured media illustrated schematically in Fig. 5.3. Both fracture and matrix domain are represented by Darcy scale porous media characterized by distinct hydraulic condutivities, K_f and K_m, respectively, and specific storage coefficients, S_f and S_m, respectively. The volume fraction φ_f of the fracture domain is typically much smaller than the volume fraction φ_m of the matrix domain, $\varphi_f = \varphi_m$. Notice that here the subscripts f and m refer to the fracture and matrix domains, respectively. The basis idea is to capture the large scale behavior by averaging the flow problem over a representative elementary volume of the medium but respect the disparity in the hydraulic properties in the two continua. Due to this disparity the heads in fracture and matrix may be very different.

Similar as for the upscaling from pore to Darcy scale, a REV here is defined as the averaging volume, for which the ratios of fracture to bulk volume and matrix to bulk volume converge to the characteristic constant values φ_f and φ_m. This requires that a characteristic matrix length scale exists. The REV represent the support

Fig. 5.3 Sketch of a double continuum medium inspired by the geometry of the Bristol channel (Geiger et al. 2013)

volume of the large scale flow description, this means, the coordinate x in the large scale flow model represents a portion of the detailed medium description of the size of the REV.

Instead of defining a single average hydraulic head representative for the REV, the dual continuum approach defines intrinsic averages over the heads in the fracture and matrix domains as

$$h_f(x,t) = \frac{1}{V_f}\int\limits_{\Omega_f} d\mathbf{r} h(x+\mathbf{r},t), h_m(x,t) = \frac{1}{V_m}\int\limits_{\Omega_m} d\mathbf{r} h(x+\mathbf{r},t), \tag{5.2.21}$$

where Ω_f and Ω_m represent the fracture and matrix portions of the REV and V_f and V_m their respective volumina. The large scale flow behavior can then be described by the following coupled set of equations

$$S_f\phi_f\frac{\partial h_f(x,t)}{\partial t} - K_f\phi_f\nabla^2 h_f(x,t) = \Gamma(x,t), \tag{5.2.22}$$

$$S_m\phi_m\frac{\partial h_m(x,t)}{\partial t} - K_m\phi_m\nabla^2 h_m(x,t) = -\Gamma(x,t), \tag{5.2.23}$$

where the source term $\Gamma(x,t)$ quantifies mass transfer between the fracture and matrix continua. It can be determined by the requirement of head and flux continuum at the interface between the fracture and matrix continua. The total hydraulic head $h_t(x,t)$ in the REV is given by $h_t(x,t) = \varphi_f h_f(x,t) + \varphi_m h_m(x,t)$.

The exchange term $\Gamma(x,t)$ is estimated in Barenblatt et al. (1960) as

$$\Gamma(x,t) = -\omega\left[h_f(x,t) - h_m(x,t)\right], \tag{5.2.24}$$

where the mass transfer rate ω is related to the matrix conductivity K_m and the characteristic block size dimensions. If the permeability in the matrix is significantly smaller than the permeability in the fracture, $K_m \ll K_f$, the direct contributions of the matrix to the macoscopic flow may be disregarded, i.e., $K_m = 0$ in (5.2.23). In this case the matrix continuum contributes to the large scale flow behavior in an indirect way and acts as a reservoir that exchanges mass with the fracture continuum. The exchange term $\Gamma(x,t)$ can be then expressed in terms of the hydraulic head in the fracture continuum as

$$\Gamma(x,t) = -\varphi_m S_m \frac{d}{dt}\int\limits_0^t dt' g_m(t-t')h_f(x,t') + \varphi_m S_m g_m(t) h_{m0}, \tag{5.2.25}$$

where the memory function $g_m(t) = \omega' \exp(-\omega' t)$ with $\omega' = \omega/(\varphi_m S_m)$, and h_{m0} is the initial head in the matrix. Notice that the specific form (5.2.24) assumes that the hydraulic head in the fracture is in quasi equilibrium. If this is not given, the memory function $g_m(t)$ is in general obtained by solving the flow problem in the fine scale matrix domain with fixed head $h_f(x,t)$ at the fracture–matrix interface. Furthermore, if the matrix continuum is characterized by a distribution of characteristic matrix scales and matrix conductivies, this means, it is composed of a set of multiple continua, the source term $\Gamma(t)$ is given by the sum of the contributions of each of the different continua as $\Gamma(t) = \sum_i \varphi_{m,i} S_{m_i} g_{m,i}(t)$ with $\varphi_{m,i}$ and $S_{m,i}$ the volume fraction and specific storage of the ith matrix continuum, and $g_{m,i}(t)$ the memory function that quantifies the mass exchange between the fracture and ith matrix continuum.

5.3 Solute Transport

The following sections give a brief overview on transport upscaling from the pore to the field scale, and discusses large scale transport modeling approaches. Before this, we want to briefly discuss some of the assumptions that are related to continuum scale transport descriptions based on the well known advection-dispersion equation.

First, the concentration $c(x,t)$ of a dissolved substance is defined over a support volume, or representative elementary volume $V_r : \ell^d$, over which concentration is assumed to be constant. The dimensionality of space is denoted by d, ℓ is a characteristic local scale. This means, it is assumed that a smallest support volume exists within which concentration gradients are essentially zero. Concentration can then be defined as

$$c(x,t) = \frac{M(x,t)}{V_r}, \qquad (5.3.1)$$

where $M(x,t)$ is the amount of solute contained in the volume V_r at the position x at time t. It is further assumed that concentration changes, or perturbations of the solute distribution over the support scale are accommodated, or relaxed on a time scale that is much smaller than the observation time scale T_{obs}. The physical relaxation mechanism on the local scale is diffusion or local dispersion. Thus, the characteristic relaxation time scale is given by $\tau_D = \ell^2/D$, with D a diffusion or dispersion coefficient. If the condition $\tau_D = T_{obs}$ is fulfilled, concentration changes in a given time interval Δt at a given position x only depend on the fluxes to and from the actual position within the time interval Δt. Under these conditions solute transport may be modeled by advective and dispersive mass transfer in terms of the advection-dispersion equation

$$\frac{\partial c(x,t)}{\partial t} + \nabla \cdot \mathbf{u} c(x,t) - D\nabla^2 c(x,t) = 0, \qquad (5.3.2)$$

where **u** is the transport velocity. Depending on the size of the support volume, or representative elementary volume, and the medium and flow properties, the above conditions on uniqueness of concentration values on the local scale, and mass transfer properties, may not be fulfilled. In the following sections, we will discuss these issues and related scale effects on conservative transport from pore to Darcy and from Darcy to field scale.

5.3.1 Pore to Darcy Scale

Solute transport in the fluid portion Ω_f of a porous medium can be described by the advection-diffusion equation (Bear 1972)

$$\frac{\partial C(\mathbf{x},t)}{\partial t} + \nabla \cdot [\mathbf{v}(\mathbf{x})C(\mathbf{x},t) - D\nabla C(\mathbf{x},t)] = 0 \tag{5.3.3}$$

where $C(\mathbf{x},t)$ is the solute concentration, $\mathbf{v}(\mathbf{x})$ is the Stokes velocity, and D is the effective molecular diffusion coefficient. The solid grains are assumed to be impermeable to both flow and transport. The transport problem (5.3.3) can also be formulated in an equivalent Lagrangian framwork in terms of the equation of motion of the position $\mathbf{x}(t)$ of solute particles. These are given by Risken (1996)

$$\frac{d\mathbf{x}(t)}{dt} = \mathbf{v}[\mathbf{x}(t)] + \sqrt{2D}\xi(t), \tag{5.3.4}$$

where $\xi(t)$ denotes a Gaussian white with zero mean and unit variance, which models the erratic motion of the solute particle due to diffusion.

The transport domain may be quite complex depending on the pore geometry, and as a consequence the flow field $\mathbf{v}(\mathbf{x})$ may be very variable. The complexity in the flow field as well as in the boundary conditions, make the pore scale flow and transport problem challenging, in terms of characterization and in terms of the actual solution of flow and transport due to the high number of degrees of freedom. For practical applications, however, it is desirable to have a transport description in terms of a few effective flow and transport parameters. As for the flow problem in the previous section, this requires averaging, or coarse graining of the pore-scale transport equation (5.3.3).

Taylor Dispersion

We first consider the example of transport in a single pore that is idealized here for a two-dimensional medium by a channel with constant aperture $2a$. This example serves to illustrate a series of issues related to the upscaling of transport in heterogeneous media in general.

Due to the particular geometry, the flow velocity is aligned with the channel and depends only on the position along the channel cross-section. The solution of the flow Eq. (5.2.2) in steady state for a two-dimensional channel is given by the Hagen-Poiseuille profile

$$v(x_2) = v_0\left[1 - \left(\frac{x_2}{a}\right)^2\right], \quad v_0 = \frac{-\Delta p}{\mu a^2 L}, \tag{5.3.5}$$

where Δp is the pressure drop along the length L of the channel. The mean flow velocity over the channel cross-section is given by $v_m = 2v_0/3$. Transport in the channel is described by the advection dispersion equation

$$\frac{\partial C(\mathbf{x}, t)}{\partial t} + v(x_2)\frac{\partial C(\mathbf{x}, t)}{\partial x_1} - D\nabla^2 C(\mathbf{x}, t) = 0. \tag{5.3.6}$$

This transport problem can be understood in terms of its characteristic length and time scales. The characteristic diffusion length over a time t is given by $\ell_D(t) = \sqrt{2Dt}$, while the characteristic advection length is given by $\ell_v = v_m t$. The characteristic scale $\tau_{vD} = 2D/v_m^2$ marks the time at which the two length scales are equal. The second characteristic time scale marks the time for complete diffusive mixing over the channel cross-section, $\tau_D = 2a^2/D$. For advection dominated transport, the two time scales are clearly separated, $\tau_{vD} = \tau_D$. The relative importance between adection and diffusion is measured by the Péclet number, which here is defined by $Pe = av_m/D = \sqrt{\tau_D/\tau_{vD}}$.

In the early time regime for $t \ll \tau_{vD}$, diffusive mass transfer is more efficient than advective, and therefore transport is essentially described by diffusion. In the late time regime for $t \gg \tau_D$ the solute is uniformly distributed over the channel cross-section and is essentially one-dimensional. In this late time regime, the solute distribution can be fully characterized in terms of the vertically averaged concentration

$$c(x_1, t) = \frac{1}{2a}\int_{-a}^{a} dx_2 C(\mathbf{x}, t). \tag{5.3.7}$$

For large Peclet numbers, the longitudinal solute dispersion is in general much larger than given by molecular diffusion D. The non-uniform vertical velocity cross-section (5.3.5) leads to rapid solute transport in the center of the channel and slower transport at the channel boundaries. This leads to stretching of an initial line source as illustrated in Fig. 5.4, and thus enhanced, purely advective solute spreading. This advective spreading induces vertical concentration gradients and thus vertical mass transfer, by which the solute eventually samples the vertical flow heterogeneity.

At times $t \gg \tau_D$, the solute has sampled the full velocity spectrum. In this regime an effective or equivalent dispersion coefficient can be defined in terms of the characteristic spreading distance during the time τ_D. The characteristic

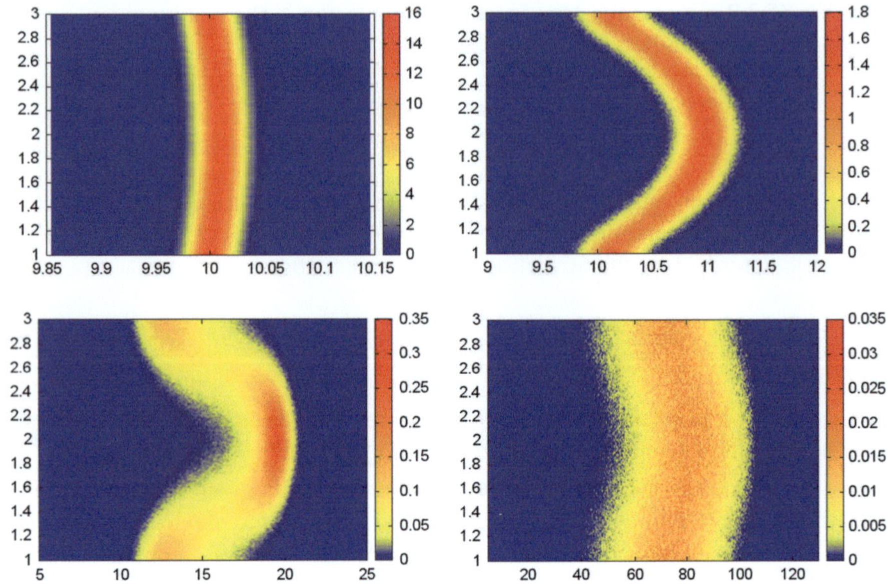

Fig. 5.4 Solute plumes evolving from a line injection for $D = 10^2$, $u_0 = 1$, at times 10^{-2}, 1, 10, and 100

spreading distance is given by $\delta v \tau_D$, where δv are the characteristic velocity fuctuations experienced by the solute. They are of the order of the mean velocity, $\delta v : v_m$. Thus, one obtains for the effective dispersion coefficient the estimate $D^e : v_m^2 \tau_D^2 / \tau_D = v_m^2 a^2 / D$, which reads for the non-dimensional dispersion coefficient as $D^e/D : Pe^2$. The exact value of the Taylor dispersion coefficient for channel flow is $D^e = \frac{2}{105} v_m^2 a^2 / D$, and for flow in a radial pipe, $D^e = \frac{1}{48} v_m^2 a^2 / D$, where here a is the pipe radius. The Taylor dispersion is inversely proportional to the molecular diffusion coefficients. The smaller molecular diffusion, the more time indvidual solute packets spend at different velocities, and thus, the larger the horizontal spread. Increasing diffusion reduces this contrast and therefore the Taylor dispersion coefficient. In the asymptotc time regime $t \gg \tau_D$, the evolution of the solute distribution is essentially one-dimensional and can be characterized by the advection-dispersion equation

$$\frac{\partial c(x_1,t)}{\partial t} + v_m \frac{\partial c(x_1,t)}{\partial x_1} - D^e \frac{\partial^2 c(x_1,t)}{\partial x_1^2} = 0. \tag{5.3.8}$$

Notice that the simplified representation of the advective terms, compared to (5.3.6) is compensated by an increased dispersion coefficient. The impact of velocity fluctuations on large scale solute transport is represented in terms of an effective dispersion coefficient. This concept is frequently used for large scale transport modeling in geophysical and turbulent fluid flows, as well as for the modeling of transport in heterogeneous porous media, as we will see in the following.

Before, that, let us consider briefly the dispersion behavior in the intermediate time regime $\tau_{vD} = t = \tau_D$. In this time regime, the solute is not in vertical equilibrium, this means concentration cannot be characterized by a single average value $c(x_1, t)$ at a longitudinal position x_1. Vertically averaged concentration profiles in this time regime are characterized by steep leading edge and a long trailing tail, see Fig. 5.5. Such spatial features of a large scale solute plume, are often termed anomalous because they do not adjust themselves to advection-dispersion models such as the asymptotic Eq. (5.3.8). This behavior is also reflected in the behavior of the apparent longitudinal dispersion coefficient, which is defined as half the rate of change of the longitudinal second centered moment of the $C(\mathbf{x}, t)$ as Aris (1956)

$$D^a(t) = \frac{1}{2}\frac{d}{dt}\left\{ \int_{-\infty}^{\infty} dx_1 \int_{-a}^{a} dx_2 x_1^2 C(\mathbf{x}, t) - \left[\int_{-\infty}^{\infty} dx_1 \int_{-a}^{a} dx_2 x_1^2 C(\mathbf{x}, t) \right]^2 \right\}. \quad (5.3.9)$$

In the asymptotic regime for $t \gg \tau_D$, it converges toward the constant D^e, while in the intermediate regimes it evolves from the local diffusion coefficient towards the asymptotic Taylor dispersion coefficient, see Fig. 5.6. Such features, evolution of

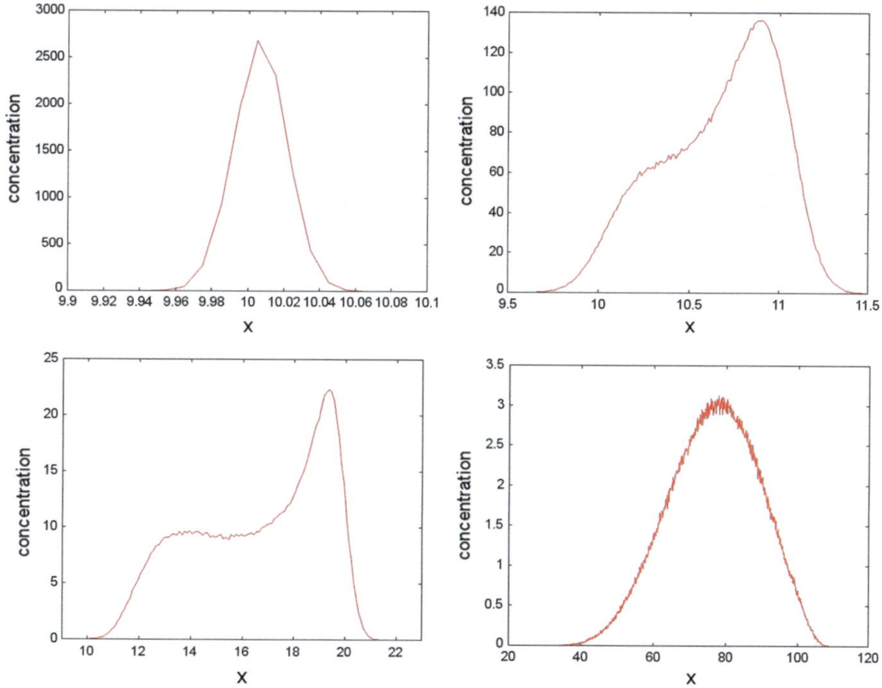

Fig. 5.5 Vertically integrated solute concentration at times 10^{-2}, 1, 10, and 100

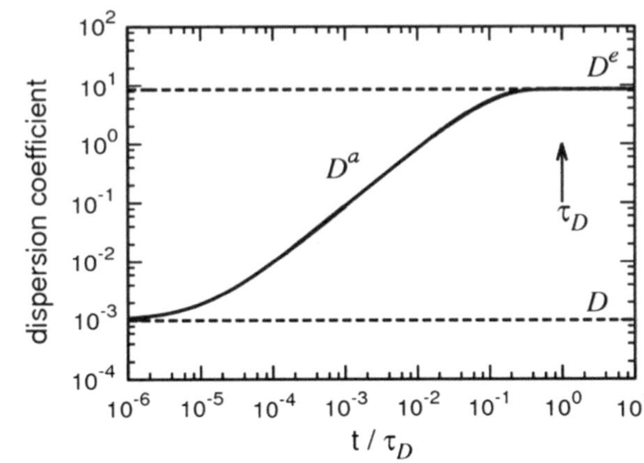

Fig. 5.6 Temporal behavior of the apparent dispersion coefficient for D = 10^{-3}

apparent dispersion, tails and leading edges in solute distributions, are the result of incomplete mixing on the support scale of the upscaled model, which here is the channel cross-section. Similar features are found in heterogeneous porous media on larger scales, as discussed in the introductory section and below. Another point can be made here. The basic mechanisms leading to anomalous behaviors, namely spatial and/or temporal fluctuations, are the same mechamisms that ultimateley lead to scale effects in effective transport coefficients, if they exist.

Hydrodynamic Dispersion

In the previous section, we considered transport in a single pore. For a porous medium, which can be seen as a network of pores, the dispersion behavior is generally different. Here the characteristic length scale is given by the typical pore length ℓ_p. Together with the average flow velocity v_m and diffusion, we can define the Péclet number $Pe = v_m \ell_p / D$, which compares the advection time scale $\tau_v = \ell_p / v_m$ and the diffusion time $\tau_D = \ell_p^2 / D$ over a pore length. Notice that the pore length marks the correlation scale for the fluctuations of the pore velocity $\mathbf{v}(\mathbf{x})$. Thus, we can obtain a rough estimate for the effect of pore-scale velocity fluctuations on the dispersion behavior of a solute. For large Peclet numbers, this means for advection dominated transport, the typical spread of the solute distribution after a few pore length will be given by square distance traveled during the advection time, $(v_m \tau_v)^2$ per advection time τ_v, which gives $D^e : v_m \ell_p$, and accordingly for the dimensionless dispersion coefficient $D^e / D : Pe$. Notice the difference between the scaling here and the scaling of the Taylor dispersion coefficient above with Pe^2.

In general, it is found in experiments and pore-scale flow and transport simulations (Bijeljic and Blunt 2006; Pfannkuch 1963; Sahimi 1995) that the longitudinal hydrodynamic dispersion coefficient depends non-linearly on the Péclet number. For $Pe < 1$, this means for diffusion dominated scenarios, $D^e / D \leq 1$ because diffusion is restricted by the solid matrix. For increasing $Pe > 1$, the longitudinal hydrodynamic dispersion coefficient evolves as $D^e / D : Pe^{1.2}$. Then for

$Pe > 400$, this means in advection dominated scenarios, is behaves as $D^e/D : Pe$, as motivated above.

More rigorous treatments on the determination of hydrodynamic dispersion coefficients and the upscaling of transport from the pore to the Darcy scale use volume averaging (e.g. Whitaker 1999) and homogenization theory (e.g. Hornung 1997), moment methods (Brenner 1980), as well as numerical porescale models (e.g. Meakin and Tartakovsky 2009).

Upscaling the pore-scale transport problem (5.3.3) implies deriving the evolution equation for the volume averaged concentration

$$c(\mathbf{x}, t) = \frac{1}{V_f} \int\limits_{\Omega_f} d\mathbf{r} C(\mathbf{x} + \mathbf{r}, t), \tag{5.3.10}$$

where V_f denotes the fluid volume within the representative elementary volume V_r. Volume averaging and homogenization theory derive and advection dispersion equation for Darcy scale transport that is given by Bear (1972)

$$\varphi \frac{\partial c(\mathbf{x}, t)}{\partial t} + \nabla \cdot [\mathbf{q} c(\mathbf{x}, t) - \mathbf{D} \nabla c(\mathbf{x}, t)] = 0, \tag{5.3.11}$$

where φ is porosity, $\mathbf{q}$ is the Darcy velocity given by (5.2.9), and $\mathbf{D}$ is the hydrodynamic dispersion tensor.

Further Remarks

It needs to be noticed that approaches that employ hydrodynamic dispersion coefficients assume that the support volume, the REV V_r, of the average, Darcy scale transport description is well mixed. This means that the solute concentration can be uniquely defined at each position $\mathbf{x}$ in the coarse grained medium by it local average value. The dominant mechanism for concentration homogenization on the support scale is diffusion. This means that the times for which Eq. (5.3.11) is valid need to be much larger than the diffusion time scale τ_D (Brenner 1980), or in general the homogenization time scale over the support volume. This condition may not be met in media that are characterized by dead end pores and permeable solid grains that are accessible for solute diffusion. In such scenarios, the homogenization time scale may be much larger than the one estimated for diffusion in the pore space. Even though the medium properties such as porosity may be homogeneous on the REV scale, REV scale transport may deviate from the behavior predicted by the advection dispersion Eq. (5.3.11). These transport situation can be accounted for by multi-continuum approaches such as the ones proposed in Quintard and Whitaker (1994) and Lichtner and Kang (2007).

5.3.2 Darcy to Field Scale

As discussed in Sect. 5.2.2, Darcy scale heterogeneity in the physical medium properties induce spatial variability porosity and hydraulic conductivity. The latter induces spatial variability in the Darcy flow velocity through (5.2.9), and therefore also in the velocity depend hydrodynamic dispersion tensor. In addition, temporal fluctuation in the flow conditions may induce time dependence in the Darcy velocity and hydrodynamic dispersion. Thus, solute transport in a Darcy scale heterogeneous porous medium is expressed by the advection-dispersion equation

$$\phi(\mathbf{x})\frac{\partial c(\mathbf{x},t)}{\partial t} + \nabla \cdot [\mathbf{q}(\mathbf{x},t)c_i(\mathbf{x},t) - \mathbf{D}(\mathbf{x},t)\nabla c_i(\mathbf{x},t)] = 0 \qquad (5.3.12)$$

As for porescale transport (5.3.3), also the Darcy scale transport problem (5.3.12) can be formulated in terms of the Lagrangian trajectories $\mathbf{x}(t)$ of "solute particles" (e.g. Dagan 1989; Risken 1984). The equation of motion for Darcy-scale solute particles is given by the Langevin equation (e.g. Delay et al. 2005; Kinzelbach 1988; LaBolle et al. 1996; Salamon et al. 2006)

$$\frac{\mathrm{d}\mathbf{x}_i(t)}{\mathrm{d}t} = \frac{\mathbf{q}[\mathbf{x}_i(t),t] - \nabla \cdot \mathbf{D}[\mathbf{x}_i(t),t]}{\phi[\mathbf{x}_i(t)]} + \sqrt{\frac{2\mathbf{D}[\mathbf{x}_i(t),t]}{\phi[\mathbf{x}_i(t)]}} \cdot \xi(t), \qquad (5.3.13)$$

where $\xi(t)$ now is a Gaussian white noise characterized by zero mean and correlation $\xi_i(t)\xi_j(t') = \delta_{ij}\delta(t-t')$. It needs to be emphasized that the support scale of this description is the same as the one for (5.3.12). That is, the fact that here transport is represented in terms of solute particles does not imply that this description resolves spatial scales that are smaller than the support scale of the continuum description.

The continuum-scale physical medium properties as quantified by the hydraulic conductivity may in general vary on many scales. This spatial variability leads to large-scale transport behavior that is quantitatively different from its local scale counterpart. A manifestation of this scale behavior is the increase of solute dispersion with increasing scale (e.g. Gelhar et al. 1992; Lallemand-Barres and Peaudecerf 1978) as quantified by macrodispersion coefficients (e.g. Dagan 1984; Gelhar and Axness 1983; Neuman et al. 1987). Other manifestations of spatial heterogeneity are so-called non-Fickian or "anomalous" transport features such as the tailing of breakthrough curves towards long times, and spatial solute distributions characterized by forward or backward tails, and the non-linear evolution of the spatial variance of the concentration distribution with time (Berkowitz et al. 2006; Dentz et al. 2011b; Neuman and Tartakovsky 2009).

One may argue that the detailed knowledge of the spatial variability of the physical medium properties and the (numerical) solution of the local scale flow and transport problems (5.2.9) and (5.3.12), allows to quantify observed large scale transport behaviors. However, the detailed characterization of the local scale medium fluctuations is in many practical applications not possible and numerical transport simulators need to operate on coarse support scales due to limitations of computer power. But more than this, the observed "anomalous" large scale transport behaviors can be seen as collective phenomena that result from the interplay of medium heterogeneity on one hand and local scale flow and transport processes on the other. Thus, from a practical point of view, it is essential to have coarse grained, average models at hand to quantify, explain and predict observed large scale transport phenomena. Methodologies to quantify spatial heterogeneity and its impact on transport include stochastic averaging (e.g. Dagan 1984; Gelhar and Axness 1983; Neuman et al. 1987; Rubin 2003) volume averaging (e.g. Whitaker 1999), homogenization theory (Hornung 1997).

The macrodispersion approach (e.g. Gelhar and Axness 1983) models effective transport by the same dynamical model as local scale transport. In this modeling framework, the evolution of the (ensemble) mean concentration $\overline{c}(\mathbf{x}, t)$ is given by

$$\overline{\phi}\frac{\partial \overline{c}(\mathbf{x}, t)}{\partial t} + \overline{\mathbf{q}} \cdot \nabla \overline{c}(\mathbf{x}, t) - \nabla \cdot \mathbf{D}^m \nabla \overline{c}(\mathbf{x}, t) = 0, \tag{5.3.14}$$

with $\overline{\phi}$ average porosity, $\overline{\mathbf{q}}$ the average Darcy velocity and $\mathbf{D}^m$ the macrodispersion tensor. Macrodispersion describes the action of flow heterogeneity on enhanced solute spreading, along the same lines as Taylor dispersion quantifies the impact of flow variability along a pipe cross-section, and hydrodynamic dispersion quantifies the effect of pore scale flow variability. In the context of transport in geophysical flows, the phenomenon of enhanced contaminant dispersion is described by eddy diffusivity.

Approaches to account for the impact of medium heterogeneity on the large scale transport behavior may be roughly divided into two groups. The first group quantifies the large scale behavior in terms of effective transport parameters such as macrodispersion coefficients, effective porosity and effective retardation coefficients, for example. It is often observed that effective transport parameters evolve with the time, or with travel distance of the transported solute. This gives an indication that the ADE based transport dynamics may be invalidated, along with the observation of non-Fickian large scale transport behaviors. As already mentioned at the end of the previous section, the validity of the advection-dispersion model (5.3.12) on the Darcy scale is conditional to the requirement of homogeneity of the REV, which may be invalidated depending on the spatial scales of the underlying heterogeneity and the mass transfer time scales. This is more so when passing from the Darcy to the field scale, where homogenization scales may be much larger than the observation time scales.

Thus, the second groups seeks to cast the emerging non-Fickian transport dynamics in large scale equations and transport laws. Such models are often based on so-called non-local process models. This means that the change of concentration at a given time is influenced by the history of the transport process. This notion may be illustrated with the example of a double porosity medium characterized by a part that is mobile and one that is immobile with respect to advection. At a given point the average solute concentration is composed of a contribution from the mobile continuum that may pass fast and one of the immobile continuum that may have already been trapped for some time. Thus transport is determined by its history. This is also manifest in the double porosity flow model that defines two hydraulic heads at one point of the coarse grained flow domain.

5.3.2.1 Effective Transport Models

Heterogeneity induced large scale transport dynamics have been quantified in a series of approaches. Starting with the macrodispersion approach that represents transport by the large scale ADE (5.3.14) over blockeffective macrodispersion approaches to spatially and temporally non-local transport models.

Block-Effective Macrodispersion

The block-dispersion approach accounts for the fact that the macrodispersion concept is only of limited applicability for practically relevant time and length scales. Instead, the medium is coarse grained into regions of characteristic size λ_c. While macrodispersion coefficients integrate the full spectrum of variabiliy of the flow fluctuations on spreading, block-effective dispersion coefficients quantify the flow fluctuations on a scale that is smaller than the coarse grainging scale λ_c. The coarse graining scale λ_c may be seen as the size of the grid block in a numerical model. While λ_c-scale fluctuations are accounted for explicitely in the numerical model, sub-scale fluctuations are quantified by block-effective macrodispersion coefficients (Beckie et al. 1996; Rubin et al. 1999). As the coarse graining scale increases, flow variability is transferred from advection to block-effective macrodispersion. The coarse graining scale may also been seen as the homogenization length scale of the physical model, or in other words as averaging scale at which concentration can be represented by a single average value.

Multirate Mass Transfer (MRMT) Approach

The MRMT (Carrera et al. 1998; Haggerty and Gorelick 1995) approach, or multicontinuum approach is based on the observation that non-Fickian transport in heterogeneous medium can be caused by the contrast of fast (advective) transport in regions of high flow velocity and slow transport in regions characterized by small flow velocities. This mechanism is modeled in the MRMT approach by dividing the medium into a mobile continuum, in which transport is due to advection and dispersion, and a set of immobile regions, in which transport can be due to diffusion and slow advection. The mobile and immobile regions are connected through linear mass transfer, which allows to quantify the medium heterogeneity in terms of the distribution of typical mass transfer times (e.g. Cunningham et al. 1997; Gouze et al. 2008; Willmann et al. 2008). This modeling framework is similiar in nature to the dual porosity approach for unsteady flow in fractured media. Notice that the behavior of the large scale concentration is given the average over the suite of local concentration values in the mobile and immobile continua. Thus, this approach represents the non-Fickian average transport behavior, and, explicitly, concentration variability at the support scale.

Continuous Time Random Walk (CTRW) Approach
Similar as the MRMT approach, the CTRW framework (Berkowitz et al. 2006) is based on the observation that medium heterogeneity gives rise to a spectrum of characteristic mass transfer time scales. This means, that for a given characteristic transport distance, the mass transfer times scales vary due to the spatial heterogeneity of the medium. The CTRW approach generalizes classical random walk (RW) models, such as (5.3.4) and (5.3.13). Classical RWs model solute transport in terms of random particle transitions in space during a constant time increment, that is, the particle trajectory is a stochastic process. CTRW generalizes this approach by introducing a distributed time increment that accounts for the variability of mass transfer time scales. This means, in a CTRW not only the spatial but also the time increment is modeled as a stochastic process. The impact of spatial heterogeneity is quantified by the joint distribution of transition length and times (e.g. Berkowitz and Scher 1997; Dentz and Castro 2009; Le Borgne et al. 2008). It has been shown that the MRMT and CTRW approaches are equivalent under some conditions (Dentz and Berkowitz 2003).

Moment Equations and Projector Formalism
Moment equations represent the governing equations for the statistical (ensemble) moments of the solute concentration such as the mean concentration and the mean squared concentration. This approach is based on a stochastic interpretation of the Darcy-scale ADE (5.3.10). This method yields spatio-temporally non-local equations for the mean concentration, concentration variance, and in principle also for the higher order ensemble concentration moments (e.g. Morales-Casique et al. 2006). The resulting moment equations are characterized by memory kernels which in principle can be related to the statistics of the underlying heterogeneity.

The projector formalism (e.g. Cushman et al. 2002; Cushman and Ginn 1993) determines the ensemble moments of concentration by stochastic averaging of the concentration distribution in terms of the particle trajectories given by (5.3.11). Like the moment equation approach, the projector formalism yields effective equations for the ensemble concentration moments that are characterized by memory kernel that are related to the heterogeneity statistics.

Fractional Advection-Dispersion Equations
Fractional-advection dispersion equations (e.g. Benson et al. 2000; Cushman and Ginn 2000) generalize the ADE to the effect of introducing fractional-order space and time derivatives (West et al. 2003). Fractional-order space and time derivatives of a given function can be seen convolutions of this function with a power-law kernel function (in space or time). Time-fractional ADEs can be shown to be equivalent to CTRWs characterized by power-law transition time distributions (Metzler and Klafter 2000) and MRMT (Schumer et al. 2003). Space-fractional ADEs may be related Levy flight random walk model, this means RWs characterized by a power-law distribution of space increments (Meerschaert et al. 2002; Metzler and Klafter 2000). The phenomenological basis is again the realization that non-Fickian transport may be caused by broad distribution of mass transfer time scales.

Stochastic-Convective Streamtube Approach

The stochastic streamtube approach models the heterogeneous medium as a set of individual streamtubes that are characterized by different, but typically constant flow velocities (e.g. Cirpka and Kitanidis 2000; Dagan and Bresler 1979; Ginn et al. 1995). Flow heterogeneity is accounted for statistically by the distribution of flow velocities across the different streamtubes. Transport in a single streamtube is given by the one-dimensional advection-dispersion equation characterized by constant transport parameters. The average concentration is then given by the sum of the partial concentration in the different streamtubes weighted by the distribution of streamtube velocities.

5.3.2.2 Mixing

In the previous section, we focused on models to describe average transport in heterogeneous media. We mentioned that non-Fickian transport can be caused by the notion of incomplete mixing, this means concentration variability on the support volume. In this section, we focus on mixing processes in heterogeneous media. Mixing is the process that leads to homogenization of heterogeneous concentration distributions, dilution of concentrated solution, and is a precondition for chemical reactions as it puts segregated chemical species into contact.

In homogeneous media under uniform and constant flow conditions, solute mixing is due to local scale dispersion. For heterogeneous media this is different. The spatial heterogeneity of the field leads to stretching and folding of the solute plume. At times that are smaller than the mass transfer time over a typical heterogeneity scale, these mechanisms increase the solute spread but not the mixing of the solute (e.g. Kitanidis 1994). Thus, for heterogeneous media, the processes of spreading and mixing need to be separated. Both processes are of course closely coupled. The concentration contrasts that are generated by the spread of the solute enhance mass transfer due to diffusion and local dispersion and thus lead to enhanced mixing. In the following we will briefly describe measures for these processes in terms of effective dispersion coefficients and concentration variability.

Effective Dispersion Coefficients

Motivated by the fact that mixing in homogeneous media is fully quantified in terms of diffusion and dispersion, macroscale dispersion may be considered as a large scale mixing process. However, we have already seen, when discussing transport in a channel that the Taylor dispersion coefficient describes solute transport only at asymptotically large times, this means at times much larger than the diffusion time over the channel cross-section.

Macroscopic dispersion coefficients can be defined by the method of moments (e.g. Aris 1956) in terms of the rate of growth of the average variance of the spatial solute distribution. Average here is understood to be a stochastic average. The ensemble variance is quantified by the second centered moment of the average concentration $\overline{c}(\mathbf{x}, t)$

$$\kappa_{ij}^{ens}(t) = \int d\mathbf{x} x_i x_j \bar{c}(\mathbf{x}, t) - \int d\mathbf{x} x_i \bar{c}(\mathbf{x}, t) \int d\mathbf{x} x_j \bar{c}(\mathbf{x}, t), \quad (5.3.15)$$

and its temporal change is defined as the ensemble dispersion coefficients

$$D_{ij}^{ens}(t) = \frac{1}{2} \frac{d\kappa_{ij}^{ens}(t)}{dt}. \quad (5.3.16)$$

The macrodispersion coefficients in (5.3.14) are given by the long time limits $D_{ij}^m = lim_{t \to \infty} D_{ij}^{ens}(t)$ if they exist. For transport in stratified media with infinite transverse extension, this is not the case (e.g. Matheron and De Marsily 1980) for example.

Using stochastic modeling, Gelhar and Axness (1983) obtained for the longitudinal macrodispersion coefficient in the limit of large Péclet numbers

$$D_L^m = \sigma_Y^2 \bar{q} \lambda, \quad (5.3.17)$$

where σ_Y^2 is the variance of the log-hydraulic conductivity field and λ its correlation length and $\bar{q}$ the average flow velocity. This result was obtained using a first-order perturbation expansion in σ_Y^2. In this approximation, the transverse dispersion coefficient turns out to be of the order of the local dispersion coefficient. Notice that the basic dependence on correlation length and mean velocity is the same as the one obtained for hydrodynamic dispersion above because the underlying physical processes are the same.

The ensemble dispersion coefficients (5.3.16) evolve in time (Dagan 1988) on the advection time scale $\tau_q = \lambda / \bar{q}$. Ensemble dispersion measures the spreading of the average plume. As illustrated in Fig. 5.7, in addition to the spreading of the individual plumes, it quantifies an artificial spreading effect due to variability in the center of mass positions between different plume realizations.

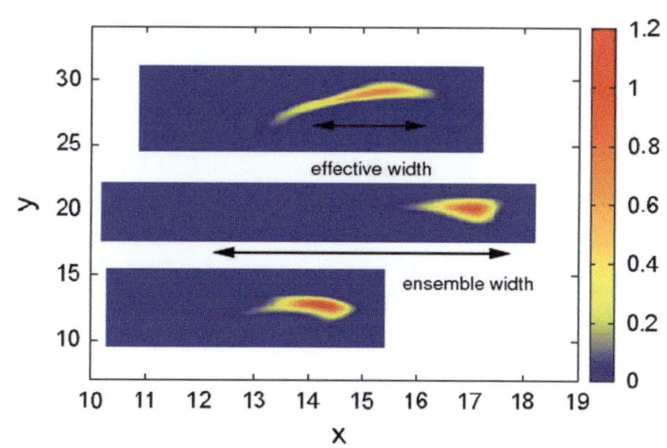

Fig. 5.7 Illustration of the difference between the ensemble and effective measures for the plume width. The figures shows three plumes evolving from point-injections along the line at $x_1 = 10$ at time $t = 5\tau_q$ in three different realizations of a heterogeneous flow field

The spatial variance of the solute distribution in a single realization is given by

$$\kappa_{ij}(t) = \int \mathrm{d}\mathbf{x} x_i x_j c(\mathbf{x}, t) - \int \mathrm{d}\mathbf{x} \int \mathrm{d}\mathbf{x}' x_i x'_j c(\mathbf{x}, t) c(\mathbf{x}', t). \tag{5.3.18}$$

The effective spatial variance of the concentration distribution is defined by the ensemble average over the variance in each realization, $\kappa_{ij}^{eff}(t) = \overline{\kappa_{ij}(t)}$, as illustrated in Fig. 5.8. The difference between the ensemble and effective variances is exactly the center of mass fluctuations between realizations (Batchelor 1949, 1952; Kitanidis 1988). Thus, the effective dispersion coefficient, defined as half the temporal rate of change of the effective width

$$D_{ij}^{eff}(t) = \frac{1}{2}\frac{d\kappa^{eff}(t)}{dt} \tag{5.3.19}$$

measures the average spreading behavior in a single aquifer realization.

The evolutions of the longitudinal ensemble and effective dispersion coefficients for a solute plume that evolves from a point injection can be approximated for $Pe \gg 1$ and $\sigma_Y^2 < 1$ by Dentz et al. (2000)

$$D_{11}^{ens} = \sqrt{\frac{\pi}{2}}\sigma_Y^2 \bar{q}\lambda \operatorname{erf}(t/\tau_q), D_{11}^{eff} = \sqrt{\frac{\pi}{2}}\sigma_Y^2 \bar{q}\lambda \left[1 - (1 + 4t/\tau_D)^{-\frac{d-1}{2}}\right]. \tag{5.3.20}$$

Their temporal evolution is illustrated in Fig. 5.8. The effective dispersion coefficients evolve on the dispersion time scale $\tau_D = \lambda^2/D$, which measures the time for dispersive mass transfer over a correlation length of the medium. This reflects the physical mechanisms that eventually lead to macroscopic dispersion for a point source. At early times, smaller than τ_q, the solute has not experienced flow heterogeneity because the plume has not yet traveled over a correlation distance of the flow field. For times larger than τ_q, the plume is exposed to flow heterogeneity.

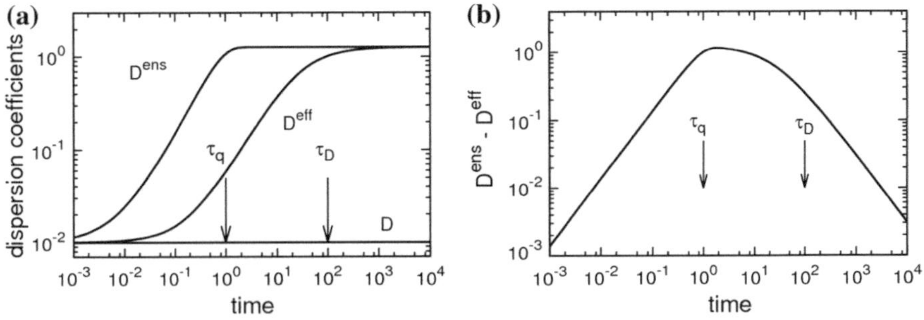

Fig. 5.8 Illustration of **a** longitudinal effective versus ensemble dispersion coefficients and **b** growth rate of center of mass fluctuations

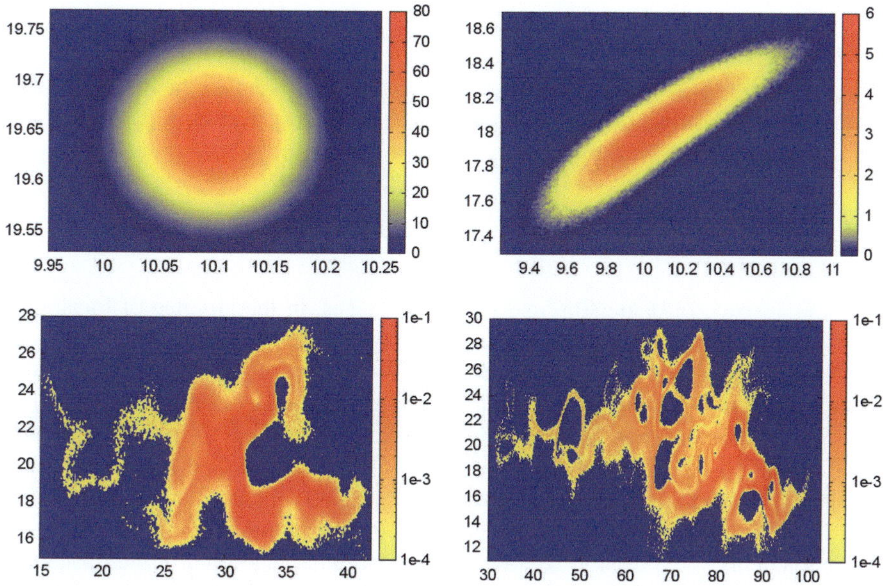

Fig. 5.9 Solute plumes evolving from a point injection in a heterogeneous flow field for $Pe = 10^2$ at times $10^{-1}\tau_q$, τ_q, $20\tau_q$ and $50\tau_q$

It starts to be deformed by the flow field, and concentration gradients are created due to the stretching and compression of the material elements the solute plume is composed of. This is illustrated in Fig. 5.9. This figure illustrates also that the effective variance κ_{11}^{eff} measures, at intermediate times $\tau_q < t < \tau_D$, the rate by which the plume spreads due to advective heterogeneity rather than actual mixing. However, an equivalent Gaussian plume characterized by the effective dispersion coefficient overestimates the actual solute dilution because it overestimates the actual volume occupied by the solute. In the following, we will discuss measures to describe the process of mixing and the dilution state of a system.

Mixing and Dilution

The dilution index was introduced by Kitanidis (1994) in order to describe the actual dilution state of a transport system. It is based on the entropy of the normalized concentration distribution $c(\mathbf{x}, t)$, which is defined by

$$H(t) = -\int d\mathbf{x} c(\mathbf{x}, t) \ln[c(\mathbf{x}, t)]. \quad (5.3.21)$$

$H(t)$ is a measure for the amount of space the solute may occupy, or more precisely, it quantifies the logarithm of the volume accessed by the solute. Thus, exponential of the entropy $H(t)$

$$E(t) = \exp[H(t)]. \tag{5.3.22}$$

Is a measure for the volume occupied by solute. $E(t)$ is called the dilution index. Kitanidis (1994) shows that the temporal change of the system entropy for a $c(x, t)$ that satisfies (38) is given by

$$\frac{dH(t)}{dt} = \frac{1}{\phi} \int \frac{d\mathbf{x}}{c(\mathbf{x}, t)} [\nabla c(\mathbf{x}, t) \mathbf{D} \nabla c(\mathbf{x}, t)]. \tag{5.3.23}$$

In the absence of dispersion and diffusion, that is, in the absence of local scale mixing mechanisms, the entropy is given by the one of the initial system, $H(t) = H_0$ and the dilution index accordingly, $E(t) = E_0$. This means in the absence of dispersion, the entropy of the system remains constant. This is evident because the flow fields under consideration are volume conserving because of $\nabla \cdot \mathbf{q}(x) = 0$. Notice also that entropy $H(t)$ in an infinite domain is maximized by a Gaussian distribution.

The expression in the square brackets in (5.3.23) represents the mechanism leading to dilution, namely the attenuation of concentration gradients by local dispersion. This is expression is closely related to what is known in the turbulence community as the scalar dissipation. The scalar dissipation rate describes the velocity by which concentration variability is destroyed. Concentration variability is measured in terms of the evolution of the global concentration variance σ_c^2

$$\sigma_c^2(t) = \int d^d x \overline{c'(\mathbf{x}, t)^2}, \tag{5.3.24}$$

where the concentration fluctuations are $c'(\mathbf{x}, t) = c(\mathbf{x}, t) - \overline{c}(\mathbf{x}, t)$. Based on (5.3.12), one obtains an evolution equation for the concentration variance (Kapoor and Kitanidis 1998)

$$\frac{d\sigma_c^2(t)}{\mathrm{d}t} = -\frac{2}{\phi} \int \mathrm{d}^d x \left[\overline{\nabla c'(\mathbf{x}, t) \mathbf{D} \nabla c'(\mathbf{x}, t)} - \nabla \overline{c}(\mathbf{x}, t) \mathbf{D}^m \nabla \overline{c}(\mathbf{x}, t) \right], \tag{5.3.25}$$

This relation can be seen as a balance equation for concentration variability. The second term on the right hand side is quantified terms of the macrodispersion tensor and the average concentration. It is a source terms and represents the creation of concentration variance due to spreading. The first term is the scalar dissipation rate. It is a sink terms that represents the desctruction of concentration variability due local scale dispersion.

Notice that (5.3.25) is not closed because it depends on the local concentration fluctuations. The interaction by exchange with the mean (IEM) model (Villermaux and Devillon 1972) approximates the scalar dissipation rate by an expression linear in σ_c^2 so that the first term on the right of (5.3.25) describes first order degradation. Other closure approximation employed in the turbulence literature can be found in

Pope (2001). It turns out that the simple IEM closure cannot describe the decay of concentration variance in porous media (de Dreuzy et al. 2012).

We discussed above that the notions of incomplete mixing and non-Fickian transport are closely related. Non-Fickian transport can be seen as a consequence of the fact that the macroscopic support volume is not well mixed. Le Borgne et al. (2011) investigated the evolution of the scale at which concentration variability is negligible. Intuitively one would assume this mixing scale evolves diffusively according to $\sqrt{Dt}$, which is the case in a homogeneous medium. For strongly heterogeneous media, however, the action of the heterogeneous flow field distorts the plume and counteracts the homogenization by local dispersion. As a consequence of this competition, the mixing scale evolves slower than in a homogeneous medium, which indicates persistent non-Fickian transport in heterogeneous media.

Most effective models summarized above describe transport in terms of the average concentrations. As we have seen here, the mixing process is related to the creation and attenuation of local scale concentration gradients, this means, it is a local process. Thus, the correct quantification of mixing in a large scale transport model requires a measure for local scale concentration variability. This relates to another issue intrinsic to spatial heterogeneity, namely uncertainty. Incomplete knowledge on the fluctuation details of the heterogeneous medium induce uncertainty on the concentration values. Incomplete mixing and uncertainty can be quantified in terms of the probability density function (PDF), $p_c(c, \mathbf{x}, t)$ of concentration at a given spatial position of the coarse grained medium (Lichtner and Tartakovsky 2003; Pope 2001). The distribution of concentration PDFs may be determined using Monte-Carlo simulations in a stochastic framework or by local spatial sampling of concentration values. (Semi)-analytical approaches are based on PDF equations, this means evolution equations for $p_c(c, \mathbf{x}, t)$, or so-called mapping approaches.

The knowledge of the local concentration variability and a correct representation of mixing are of particular importance for the modeling of reactive transport in heterogeneous media because chemical reactions are intrinsically local phenomena that rely on mass transfer to bring the chemical species into contact.

5.4 Reactive Transport

Reactive transport is in general modeled by the combination of a conservative transport equation that accounts for the evolution of the concentration due to physical mass transfer mechanisms and a source term, the reaction rate, which accounts for chemical reactions. In the previous section, we discussed conservative transport models at pore, Darcy and field scales. We emphasized that homogeneity of concentration at the relevant support scale plays an important role, which can be related to characteristic mass transfer time scales over the support volume. In the context of reactive transport, additional characteristic times emerge, the characteristic reaction scales τ_r.

Chemical reaction rate laws are typically determined in well mixed environments. This means there are no mass transfer limitations, or, in other words, concentration changes due to physical mass transfer are equilibrated on time scales much smaller than the times scales τ_r for changes in the species concentrations due to chemical reactions. The relation of the characteristic physical mass transfer time scales τ_m on the support scale and the reaction time scale τ_r is measured by the Damköhler number

$$Da_{mic} = \frac{\tau_m}{\tau_r}. \tag{5.4.1}$$

Thus, for a microscopic Damköhler number $Da_{mic} = 1$, the evolution of the concentration of a given species i can be described by the rate law

$$\frac{dc_i(t)}{dt} = r[\{c_j(t)\}], \tag{5.4.2}$$

with $r[\{c_j(t)\}]$ the reaction rate that in general depends on the concentrations of all reacting species. As mentioned above, the formulation (5.4.2) requires that the support volume on which the species concentration $c_i(t)$ is defined is well mixed at any time. This implies that concentration can be defined locally by a single value, and that concentration changes on the support scale occur on time scales smaller than the reaction time scales. Under these conditions, reactive transport can be described by the advection-dispersion reaction equation (ADRE)

$$\frac{\partial c_i(x,t)}{\partial t} + \nabla \cdot \mathbf{u} c_i(x,t) - D\nabla^2 c_i(x,t) = r[\{c_j(x,t)\}]. \tag{5.4.3}$$

As we have seen in the previous section, for transport in heterogeneous media, uniqueness of concentration on the support scale is not always given, again depending on the medium heterogeneity and the associated mass transfer time scales. In the following, we discuss these issues in relation to reactive transport modeling from the pore to the Darcy and Darcy to the field scale.

5.4.1 Pore to Darcy Scale

At the pore scale, the evolution of the concentration $C_i(\mathbf{x}, t)$ of an aqueous species i can be expressed by a combination of the ADE (5.3.3) and suitably chosen source terms and boundary conditions,

$$\frac{\partial C_i(\mathbf{x},t)}{\partial t} + \nabla \cdot [\mathbf{v}(\mathbf{x})C_i(\mathbf{x},t) - D\nabla C_i(\mathbf{x},t)] = \sum_j v_{ij} r_j[\{C_n(\mathbf{x},t)\}]. \qquad (5.4.4)$$

We assume that the molecular diffusion coefficient D is the same for all species. The notation $\{C_n(\mathbf{x},t)\}$ indicates the set of all species concentrations. The reaction rate for the jth homogeneous reaction is denoted by $r_j[\{C_n(\mathbf{x},t)\}]$ with the v_{ij} the stoichiometric coefficients. As outlined above this formulation assumes that mass transfer on the microscopic support scale can be disregarded. The reaction rate on the right side represents fast and slow chemical reactions in the fluid phase. The terms fast and slow now refer to the typical reaction time scales compared to the diffusion time scale τ_D over a characteristic medium length scale, which are compared by the Damköhler number Da_j

$$Da_j = \frac{\tau_D}{\tau_{r,j}} \qquad (5.4.5)$$

with $\tau_{r,j}$ the reaction time scale of the j-th reaction. Notice that the Damköhler number defined here is different from Da_{mic} defined above, which refers to mass transfer on the microscopic support volume. Fast chemical reaction, this means reactions j for which $Da_j \gg 1$, react almost instantenously upon a concentration change due to physical mass transfer. Such reactions are intrinsically mixing-limited.

For fast reversible reactions, indexed in the following by f, the species concentrations are related by the mass action law (e.g. Saaltink et al. 1998; Steefel and MacQuarrie 1996)

$$\prod_i [\gamma_i c_i(\mathbf{x},t)]^{v_{fi}} = K_f, \qquad (5.4.6)$$

where K_f is the equilibrium constant of the fth equilibrium reaction. The γ_i are the activity coefficients of the ith species, which depend in general on all species concentrations. Here we assume them to be constant. The reaction rates for $r_f[\{C_n(\mathbf{x},t)\}]$ can in principle be approximated by their limit for $Da_f \to \infty$ (Sanchez-Vila et al. 2007)

$$r_f^e(\mathbf{x},t) = \lim_{Da_f \to \infty} r_f[\{C_n(\mathbf{x},t)\}]. \qquad (5.4.7)$$

where $r_f^e(\mathbf{x},t)$ is the equilibrium reaction rate of the fth reaction. It can be expressed in terms of the transport and mixing dynamics of the conservative transport system (De Simoni et al. 2005, 2007). The right side of (5.4.4) can now be written as

$$\sum_j v_{ij} r_j[\{C_n(\mathbf{x},t)\}] = \sum_f v_{if} r_f^e(\mathbf{x},t) + \sum_s v_{is} r_s[\{C_n(\mathbf{x},t)\}]. \qquad (5.4.8)$$

where we indexed the slow reactions by s.

Reactions between the fluid and solid phases take place at the fluid-solid boundaries where minerals are present. We focus here on precipitation-dissolution reactions. The boundary conditions at the fluid solid interface can then be written as (e.g. Lichtner and Kang 2007; Whitaker 1999)

$$-\mathbf{n}_{fs}(\mathbf{x}) \cdot D\nabla C_i(\mathbf{x},t)|_{\mathbf{x}\in\partial\Omega_s} = -\sum_m \nu_{im}\alpha_m(\mathbf{x})j_m[\{C_n(\mathbf{x},t)\}]|_{\mathbf{x}\in\partial\Omega_s}, \qquad (5.4.9)$$

where $\mathbf{n}_{fs}(\mathbf{x})$ is the unit normal vector on $\partial\Omega_s$ pointing from the fluid into the solid phase; $\alpha_m(\mathbf{x})$ is the relative mineral surface area. The relative mineral surface area is non-zero only at those locations on the solid fluid interface where the mineral is present. The reaction flux of the mth mineral at its surface is denoted by $j_m[\{C_n(\mathbf{x},t)\}]$. It depends on the aqueous concentrations at the solid surface. A discussion of the boundary conditions at the fluid-solid interface can be found in Knabner et al. (1995).

Based on transition state theory (Hänggi et al. 1990) the reaction flux $j_m[\{C_n(\mathbf{x},t)\}]$ at the solid-fluid boundary may be modeled as (e.g. Lichtner and Kang 2007)

$$j_m[\{C_n(\mathbf{x},t)\}]|_{\mathbf{x}\in\Omega_s} = k_m(1-\Omega_m(\mathbf{x},t)), \qquad (5.4.10)$$

where k_m is the kinetic rate constant. The saturation state $\Omega_m(\mathbf{x},t)$ is given by

$$\Omega_m(\mathbf{x},t) = K_m^{-1}\prod_j\left[\gamma_j C_j(\mathbf{x},t)\right]^{\nu_{jm}}, \mathbf{x}\in\Omega_s. \qquad (5.4.11)$$

K_m is the equilibrium constant of the mth mineral reaction, ν_{jm} are the stoichiometric coefficients and γ_j denotes the activity coefficient of the jth aqueous species. The activity coefficients relate the species activities to their molal concentration. For mineral species, it is equal to one.

For practical applications of reactive transport modeling on the Darcy and field scale, the full characterization of the physical and chemical pore scale medium properties and the (numerical) solution of the porescale reactive flow and transport problems on macrosopic length scale are virtually impossible. As for the problems of single phase flow and conservative transport, upscaled reactive transport models are required which depend only on a few effective flow, transport and reaction parameters. Reactive transport is affected by physical heterogeneity which affects solute mixing and thus conservative mass transfer time scales as well as chemical heterogeneity, for example in the specific reactive surface areas.

Mathematical approaches to reactive transport upscaling from the pore to the Darcy scale has used volume averaging (Edwards et al. 1993; Kechagia et al. 2002; Lichtner and Kang 2007; Quintard and Whitaker 1994), Homogenization theory (Mikelić et al. 2006), pore-network modeling (Li et al. 2006; Meile and Tuncay 2006) and smoothed particle hydrodynamics (Battiato et al. 2009; Tartakovsky et al. 2008). Traditionally, Darcy scale transport is quantified by an ADRE characterized by effective transport and reaction parameters which can be obtained from volume averaging of the porescale reactive transport problem (5.4.4). The evolution equation for the volume averaged species concentration $c_i(\mathbf{x}, t)$ is then given by Lichtner (1985), Quintard and Whitaker (1994)

$$\varphi(\mathbf{x})\frac{\partial c_i(\mathbf{x},t)}{\partial t} + [1-\varphi(\mathbf{x})]\frac{\partial s_i(\mathbf{x},t)}{\partial t} + \nabla \cdot [\mathbf{q}(\mathbf{x},t)c_i(\mathbf{x},t) - \mathbf{D}(\mathbf{x},t)\nabla c_i(\mathbf{x},t)]$$
$$= \phi(\mathbf{x})\sum_i v_{ij}r_j[\{c_n(\mathbf{x},t)\}] + r_i^{(s)}[\mathbf{x},\{c_n(\mathbf{x},t)\}], \tag{5.4.12}$$

The bulk reaction rate for the fluid solid reactions $r_i^{(s)}(\mathbf{x}, t)$ is given by

$$r_i^{(s)}[\mathbf{x},\{c_n(\mathbf{x},t)\}] = \sum_m v_{im}a_m(\mathbf{x})k_m(1-\Omega_m[\{c_n(\mathbf{x},t)\}]), \tag{5.4.13}$$

with $a_m(\mathbf{x})$ the bulk reactive surface area. Notice that heterogeneous reactions are represented, like homogeneous reactions, through an average source term.

Recall that a reactive transport description in terms of the ADRE assumes that concentration is unique on the support scale V_r and that mass transfer is much faster than the characteristic reaction time scales. As series of studies have investigated the validity of the ADRE (5.4.12) using volume averaging (Kechagia et al. 2002), pore-network simulations (Li et al. 2006; Meile and Tuncay 2006) and smoothed particle hydrodynamics (Tartakovsky et al. 2008). These works find consistently discrepancies between reaction rates obtained at the pore scale and their Darcy scale representation (5.4.12). Battiato et al. (2009) analyzed the conditions under which the ADRE provides a valid description of Darcy scale reactive transport, theoretically, using volume averaging, and numerically, using smoothed particle hydrodynamics. These authors analyze systematically the physical and chemical conditions under which the macroscale description in terms of an ADRE is valid ($Da = 1$), and identify the regimes for which a hybrid description is necessary that represents both concentration variability on the support scale as well as the bulk Darcy scale transport properties. To this end, Lichtner and Kang (2007) present a reactive multicontinuum approach that is based on the conceptualization of the medium as consisting of mutiple interaction continua. As in the dual porosity and multirate mass transfer models discussed above, this description quantifies the behavior of the average bulk concentration and represents explicitly concentration heterogeneity at the support scale. Battiato et al. (2011) develop a (numerical) hybrid method that consists of a coupled system of equations that represents (i) the evolution of the mean concentration through a macroscale reactive transport equation on a coarse grid, (ii) mass transfer between the coarse grid and a fine grid at nodes where the conditions for the macroscale description are not met, (iii) fine scale reactive transport equation at the hybrid node supported with continuum and boundary conditions.

As outlined at the beginning of this section, kinetic rate laws are often determined in the laboratory in well-mixed environments, either batch or flowthrough reactors. Under such conditions, mass transfer limitations and possible localization of geochemical reactions as observed in in-situ measurements do not arise (e.g. Li et al. 2006; Steefel et al. 2005). Li et al. (2008) conducted a thorough experimental and numerical study on the scale dependence of mineral dissolution rates and analyzed the role of concentration gradients in single pores and fractures. These findings again challenge the assumption of well-mixedness necessary for the Darcy-scale reactive transport description (5.4.12) to be valid.

5.4.2 Darcy to Field Scale

The straightforward generalization of the ADRE approach from the Darcy to the field scale, represents large scale reactive transport by the combination of the macoscale ADE (5.3.14) and average reaction rates that are formulated in terms of the mean species concentrations

$$\overline{\varphi}\frac{\partial \overline{c}_i(\mathbf{x},t)}{\partial t} + \overline{\mathbf{q}} \cdot \nabla \overline{c}_i(\mathbf{x},t) - \nabla \cdot \mathbf{D}^m \nabla \overline{c}_i(\mathbf{x},t) = \overline{\varphi}\,\overline{r}[\{\overline{c}_n\}] \tag{5.4.14}$$

We encounter here the same conceptual issues as in the transition from the pore to the Darcy scale. The average reactive transport description (5.4.14) is valid if the species concentrations are uniform on the macroscopic REV, and mass transfer times on the REV scale are much smaller than the characteristic reaction scale. If these conditions are not fulfilled, the ADRE is only of limited applicability. In fact, it has been found that macroscale models of reactive transport misrepresent reaction rates in heterogeneous media (e.g. Ginn et al. 1995; Gramling et al. 2002; Kapoor et al. 1997; MacQuarrie and Sudicky 1990; Molz and Widdowson 1988). As pointed out above, macrodispersion coefficients overestimate solute dispersion and simulate a homogeneity of the transport system that is not real at practically relevant times (Kapoor and Kitanidis 1998; Kitanidis 1994).

Recall that non-Fickian conservative transport observed at the macroscale is related to the fact that concentration is not unique at the macroscopic support scale, or in other words to mass transfer limitations on the support scale. As a consequence, the evolution of average concentration shows memory effects and is quantified by non-local average transport models. Similar scale phenomena have been found for reactive transport. The impact of chemical and physical heterogeneity may lead to large scale reaction rate laws different from those at the local scale. For example, it has been found that reactive transport under heterogenous local equilibrium adsorption properties may be represented on the large scale as a kinetic sorption process (Dentz and Berkowitz 2003; Dentz and Castro 2009; Espinoza and Valocchi 1997).

One may now be tempted to improve the macroscale ADRE along the same lines as for conservative transport, and combine a non-local effective transport model for the average species concentration with reaction rate laws formulated again in terms of the average species concentrations. Even though such a formulation represents the impact of heterogeneity on macroscale transport, it does not represent explicitly the concentration variability at the support scale. Chemical reactions, just like solute mixing, are intrinsically local phenomena, and depend in general on the local scale species concentrations and their gradients.

Thus, also for crossing from the Darcy to the field scale, hybrid models are required that represent the mean behavior of the species concentrations on one hand and that can model explicitly variability in the local species concentrations on the other. As outlined above and in Sect. 5.3.2, both the multicontinuum approach (e.g. Donado et al. 2009; Lichtner and Kang 2007; Liu et al. 2008) and the stochastic-convective approach (e.g. Ginn 2001) have these attributes and have therefore been used for the modeling of multispecies reactive transport in Darcy-scale heterogenous media.

The multicontinuum approach identifies a connected mobile primary continuum and a suite of secondary continua characterized statistically by a distribution of physical and chemical characteristics, as for example typical mass transfer times, porosity, reactive surface areas and sorption coefficients. The mobile continuum is characterized by a volume averaged mobile concentration that communicates with the fine scale secondary continua through first-order or diffusive mass transfer. Reactive transport is solved for the mobile and each immobile subdomain separately and the average concentrations and reaction rates are determined in terms of the local species concentrations (Dentz et al. 2011a; Lichtner and Kang 2007; Willmann et al. 2010).

As discussed above, the stochastic-convective model distinguishes solute concentrations between the individual streamtubes, and thus represents small scale concentration variability. In this approach reactive transport is solved for each individual streamtube (e.g. Cirpka and Kitanidis 2000; Ginn 2001; Seeboonruang and Ginn 2006a, b), and therefore local concentration fluctuations can be quantified in the overall reaction behavior. Notice, however, that the streamtubes are typically not connected such that transverse mixing between streamtubes is not represented. Therefore, global mixing may actually be underestimated because concentration gradients between streamtubes cannot be attenuated and therefore may impact on the prediction of the global reaction behavior. Thus, while the macrodispersion approch may be considered a maximum dilution model (Kitanidis 1994), the stochastic-convective approach represents a minimal mixing model (Robinson and Viswanathan 2003; Seeboonruang and Ginn 2006a).

5.5 Multiphase Flow

This section discusses scale effects in multiphase flow through porous media. We focus on the simultaneous flow of two immiscible fluids in a rigid porous matrix. Fluid flow is characterized by the fluid-solid and fluid-fluid interfaces. The presence of interfaces between the fluids leads to phenomena different from the ones observed in single phase flow.

On the pore scale, flow can be described by the Stokes equation (5.2.2) for each of the two fluids. At the fluid-fluid interface, the flow velocities are continuous, while the fluid pressures in each phase may be different (Marle 1981), sustained by the surface tension σ at the interface between the two fluids. For two fluids at rest in a capillary, the interaction between the fluids and the solid walls gives the pressure difference at equilibrium by the Young-Laplace equation as

$$\Delta p = \sigma\left(\frac{1}{r_1} + \frac{1}{r_2}\right)\cos(\theta), \tag{5.4.15}$$

where r_1 and r_2 are the curvature radii of the respective fluids. In a cylindrical capillary, the radii $r_1 = r_2 = r$ are equal to the capillary diameter. The contact angle θ $(0 < \theta < \pi)$ is defined as the angle between the capillary wall and the interface between the two fluids. It depends on the surface tensions between the liquids and the liquids and the solid, which are related through the Young-Dupré equation (Bear 1972; Sahimi 1995, 2011). For $\theta < \pi/2$, the fluid from whose side the angle is measured, is called wetting, for $\theta > \pi/2$ it is called non-wetting. As discussed in Bear (1972), contact angle and interfacial tension depend in general on the direction the interface is moving (capillary hysteresis).

As for single phase flow, coarse graining of the pore-scale flow problem is required to derive effective flow descriptions for practically relevant temporal and spatial scales. This has been undertaken by volume averaging (Hassanizadeh and Gray 1990), homogenization theory (Bourgeat 1997), and numerical pore scale flow modeling (Meakin and Tartakovsky 2009). The Darcy scale two-phase flow problem then is described by an extension of the Darcy equation for each fluid

$$\mathbf{q}_\alpha = -\frac{k}{\mu_\alpha} k_{r,\alpha}(\nabla p_\alpha - \rho_\alpha \mathbf{g}), \tag{5.4.16}$$

where the subscript α denotes the fluid phase, ρ_α its density, p_α its pressure, μ_α its viscosity, and $\mathbf{g}$ is the downward pointing vector of gravity acceleration. Mass conservation for each fluid is described by

$$\varphi \frac{\partial S_\alpha}{\partial t} + \nabla \cdot \mathbf{q}_\alpha = 0, \tag{5.4.17}$$

where S_α denotes the saturation, or volumetric fraction of fluid phase α in the medium; both fluids are assumed to be incompressible such that $\rho_\alpha =$ constant.

Here we have $\sum_\alpha S_\alpha = 1$. As pointed out in Bear (1972), this system of equations was proposed by Wyckoff and Botset (1936) and Muskat (1937) as a generalization of the single phase flow description in Darcy scale porous media. Equations (5.4.16)–(5.4.17) are complemented by constitutive relationships that relate the pressures in each of the two fluids, and describe the relative saturation $k_{r,\alpha}$ for each fluid. In the simultaneous flow of two fluids, the pore space available for one fluid is limited due to the presence of the other fluid, which therefore reduces its permeability. With this reasoning, relative permeability $k_{r,\alpha}$ is related to fluid saturation S_α. An extensive discussion on the concept of relative permeability and the dependence on fluid saturation can be found in Bear (1972).

Capillary pressure $p_c = p_{nw} - p_w$ denotes the difference between the pressures in the wetting and non-wetting fluid phases. This concept is motivated by the pore-scale interfacial pressure difference (5.4.15) between the two fluids. Capillary pressure is in general related to fluid saturation, $p_c = p_c(S_\alpha)$. As outlined on Marle (1981), for a conic capillary the interface is spherical and fluid saturation may be uniquely related to the interfacial pressure. This is not true anymore for porous media, which are characterized by a complex pore geometry and a distribution of pore length scales. Capillary hysteresis and pore-scale heterogeneity lead to hysteresis effects in the capillary pressure-saturation relationship, which depends on whether the wetting fluid displaces the non-wetting (imbibition curve) or vice versa (drainage curve). Furthermore, notice that the Young-Laplace equation (5.4.15) describes the conditions at the interface at equilibrium. Thus, the simple $p_c(S_\alpha)$ relationship may not be sufficient to describe dynamic phenomena (Cueto-Felgueroso and Juanes 2012; Das and Mirzaei 2012). Thus, several authors proposed dynamic relationships between the phase pressures and saturations of the type (see Dahle et al. 2005 and references therein).

$$p_{nw} - p_w = p_c(S_\alpha) + F\left(\frac{\partial S_\alpha}{\partial t}, S_\alpha\right). \tag{5.4.18}$$

Marle (1981) discusses that (multi-)phase flow in porous media is mainly based on phenomenological macroscale laws that describe fluid flow at the Darcy scale based on macrocopic quantities such as fluid saturations and phase pressures. These laws then are verified or challenged by experiments. This macroscopic approach is contrasted to the microscopic approch that ideally understands porous media flow based on the *elementary laws of visous flow in a pore and from the laws of capillarity*. At the time, *no serious study [had] yet been undertaken of the flow of several fluids*. To date, volume averaging and homogenization theory as well as pore-scale flow simulations have shed new light on the foundations of the Darcy-scale multiphase flow description. Yet many open questions remain (Cueto-Felgueroso and Juanes 2012; Das and Hassanizadeh 2005).

In the remainder of this section, we will focus on the quantification of large scale two-phase flow in Darcy-scale porous media. We will rely on the description of two-phase flow through Eqs. (5.4.16) and (5.4.17) complemented by the static capillary pressure-saturation relationship $p_{nw} - p_w = p_c(S_\alpha)$, and saturation dependent relative permeability $k_{r,\alpha}(S_\alpha)$. We first focus on the large scale description of two-phase flow in moderately heterogeneous porous media. We present an approach based on stochastic modeling of spatial heterogeneity, which render a non-local large scale flow model. The front roughening due to spatial heterogeneity can described here in terms of a flow parameter similar to the macrodispersion coefficients discussed above. For transport in strongly heterogeneous media with sharps contrasts in hydraulic conductivity such as fractured media, we review multicontinuum and multirate mass transfer approaches. We then consider the derivation of large scale two-phase flow models through dimensional reduction. Finally, we give an overview over the upscaling of convective mixing in porous media.

5.5.1 Macrodispersion Two-Phase Flow Model

Spatial heterogeneities can lead to an increase in the interfacial area between two fluids, and thus have an impact on the dissolution efficiency between fluid phases. The spreading of the saturation distribution in a homogeneous and a heterogeneous medium is illustrated in Fig. 5.10. While the homogeneous model is characterized

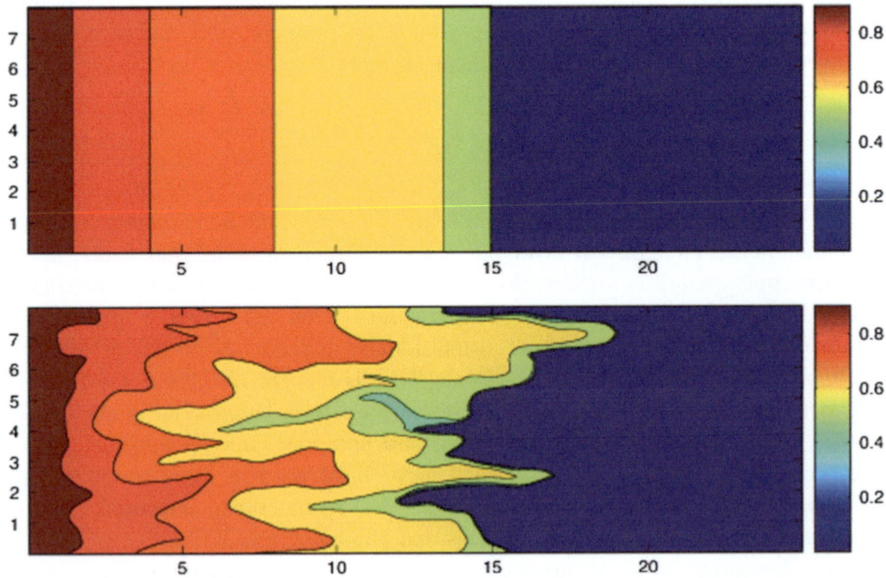

Fig. 5.10 Saturation distributions in *top* a homogeneous medium, and *bottom* a heterogeneous medium. The displacing fluid is injected at the left boundary at constant rate. Gravity and capillary effects are disregarded (after Bolster et al. 2009)

by a constant saturation value along a cross-section, the heterogeneous model is characterized by a distribution of saturation values, and with larger interface length due to heterogeneity-induced front roughening (Noetinger et al. 2004).

As discussed in Sect. 5.3.2, dispersive transition zones in solute transport problems have typically been characterized by spatial moments. Due to the qualitative similarity of heterogeneity-induced front roughening, similar approaches have been applied to two-phase flow. Cvetkovic and Dagan (1996) and Dagan and Cvetkovic (1996) used Lagrangian perturbation theory in a stochastic modeling approach in order to determine the averaged cumulative recovery of the displacing fluid and the spatial moments of the fluid distribution. They found that spatial heterogeneities cause a dispersive growth of the second moment. Also Zhang and Tchelepi (1999) observed a dispersion effect for horizontal immiscible displacement. Langlo and Espedal (1995) quantified the heterogeneity-induced front dispersion with an effective dispersion coefficient. Their approach was extended by Neuweiler et al. (2003) to quantify the dispersion coefficient analytically, by Bolster et al. (2009) to include temporal fluctuations in the flow field, and Bolster et al. (2011) to account for the impact of gravity and heterogeneity on front spreading. Panfilow and Floriat (2004) use a homogenization approach to quantify the front dispersion in a periodic medium.

In the following, we outline the stochastic-perturbative approach to the quantification of large scale flow in moderately heterogeneous media. As in the context of single phase flow and transport, see Sects. 5.2.2 and 5.3.2, the heterogeneous medium is modeled as a realization of an ensemble of random media. A large scale flow equation can then be derived by ensemble averaging of the flow problem. Perturbation theory is used to relate the local scale fluctuations of hydraulic conductivity to the large scale flow formulation. Medium and fluid are incompressible so that porosity ϕ and phase densities ρ_α are constant.

Quantities that refer to the wetting and non-wetting phases, in the following are marked by the subscript w and nw, respectivley. Incompressibility and mass conservation imply that the divergence of the total specific discharge $\mathbf{Q} = \sum_\alpha q_\alpha$ is zero, $\nabla \cdot \mathbf{Q} = 0$. Eliminating $\mathbf{q}_{nw}$ from Eq. (68) in favor of $\mathbf{Q}$ by using (67) gives the two-phase flow problem in the fractional flow formulation (Bear 1972) for the non-wetting phase as

$$\frac{\partial S_{nw}}{\partial t} + \nabla \cdot \left[\mathbf{Q} f_{nw}(S_{nw}) - \frac{k \Delta\rho_{nw} \mathbf{g}}{\mu_w} g_{nw}(S_{nw}) \right] - \nabla \cdot \left[\frac{k}{\mu_w} g_{nw}(S_{nw}) \nabla p_c(S_{nw}) \right] = 0, \tag{5.5.1}$$

where $\Delta\rho_{nw} = \rho_w - \rho_{nw}$. We set $\phi = 1$ for simplicity (which is equivalent to rescaling time). The fractional flow function $f_{nw}(S_{nw})$ and modified fractional flow function $g_{nw}(S_{nw})$ are defined by

$$f_{nw}(S_{nw}) = \frac{\lambda_{nw}(S_{nw})}{\lambda_{nw}(S_{nw}) + \lambda_w(S_{nw})}, g_{nw}(S_{nw}) = k_{r,w}(S_{nw})f_{nw}(S_{nw}). \tag{5.5.2}$$

where the phase mobilites are defined by $\lambda_\alpha = k_{r,\alpha}/\mu_\alpha$. Notice that Eq. (5.5.1) may be considered a non-linear advection-dispersion equation for the phase separation, in which the second term on the left side represents non-linear advection, the second term non-linear dispersion.

For simplicity, we disregard gravity and set capillary pressure constant in the following. Equation (5.5.1) then simplifies to the Buckley-Leverett equation

$$\frac{\partial S_{nw}}{\partial t} + \mathbf{Q} \cdot \nabla f_{nw}(S_{nw}) = 0, \tag{5.5.3}$$

For constant capillary pressure, the total flow velocity $\mathbf{Q}$ satisfies the Darcy equation

$$\mathbf{Q} = -k\Lambda(S_{nw})\nabla p_{nw}, \Lambda(S_{nw}) = \frac{k_{r,nw}(S_{nw})}{\mu_{nw}} + \frac{k_{r,w}(S_{nw})}{\mu_w}. \tag{5.5.4}$$

We focus on the steady displacement of the wetting by the non-wetting as illustrated in Fig. 5.10. The total flux at the inflow boundary is prescribed as $\mathbf{Q} = Q_0\mathbf{e}_1$, where $\mathbf{e}_1$ is the unit-vector in 1-direction. It can be decomposed into its imposed mean and fluctuations about it, $\mathbf{Q}(x) = Q_0\mathbf{e}_1 + \mathbf{Q}'$. Substituting the flux decomposition into (72), the local scale saturation equation reads as

$$\frac{\partial S_{nw}(x,t)}{\partial t} + Q_0\frac{\partial f_{nw}(S_{nw})}{\partial x_1} + \mathbf{Q}' \cdot \nabla f(S_{nw}) = 0. \tag{5.5.5}$$

In analogy to the decomposition of total discharge and permeability, we assume that the saturation can be decomposed into its ensemble average $\overline{S}_{nw}$ and fluctuations about it $S_{nw} = \overline{S}_{nw} + S'_{nw}$. Furthermore, assuming that the saturation fluctuations are small, the fractional flow function can be expanded about its mean as $f(S_{nw}) = f(\overline{S}_{nw}) + \frac{df(\overline{S}_{nw})}{d\overline{S}_{nw}}S'_{nw} + \ldots$ Using these decompositions in (5.5.5) and subsequently performing the ensemble average gives the large scale flow equation

$$\frac{\partial \overline{S}_{nw}}{\partial t} + Q_0\frac{\partial f_{nw}(\overline{S}_{nw})}{\partial x_1} = -\nabla \cdot \overline{\mathbf{Q}'S'_{nw}}\frac{df(\overline{S}_{nw})}{d\overline{S}_{nw}}. \tag{5.5.6}$$

The flux term on the right side depends on the local scale fluctuation of the total flow velocity and phase saturation, and therefore it is not closed. However, by subtracting the average equation (5.5.6) from the local scale equation (5.5.5) and disregarding contributions that are quadratic or higher order in the fluctuations, one obtains the following equation for the fluctuations of the phase saturations

$$\frac{\partial S'_{nw}}{\partial t} + Q_0 \frac{\partial}{\partial x_1} \frac{df_{nw}(\overline{S}_{nw})}{d\overline{S}_{nw}} S'_{nw} = -\mathbf{Q}' \cdot \nabla f_{nw}(\overline{S}_{nw}). \tag{5.5.7}$$

This is a linear, closed form equation for S'_{nw} that is based on disregarding higher-order fluctuations of flow velocity and saturation. Therefore, the large scale flow equation (5.5.6) is closed perturbatively here. Notice, that in the previous section, the Dupuit approximation has been used to close the vertically averaged large scale flow problem.

The linear equation (5.5.7) can be solved in terms of the associated Green's function $G(x, t|x', t')$, which solves (5.5.7) for the initial condition $G(x, t|x', t') = \delta(x - x')$. Thus, we obtain for S'_{nw} the expression

$$S'_{nw} = -\int_0^t dt' \int d\mathbf{x}' G(\mathbf{x}, t|\mathbf{x}', t') \mathbf{Q}'(x') \cdot \nabla' f\left[\overline{S}_{nw}(\mathbf{x}', t')\right], \tag{5.5.8}$$

Inserting (5.5.8) into (5.5.6), gives a closed form large scale equation for the average saturation that can be written as the spatially and temporally non-local non-linear advection-dispersion equation

$$\frac{\partial \overline{S}_{nw}}{\partial t} + Q_0 \frac{\partial f(\overline{S}_{nw})}{\partial x_1} - \nabla \cdot \int_0^t dt' \int d\mathbf{x}' \frac{df(\overline{S})}{d\overline{S}_{nw}(\mathbf{x}, t)} \mathbf{k}(\mathbf{x}, t|\mathbf{x}', t') \nabla' f\left[\overline{S}_{nw}(\mathbf{x}', t')\right] = 0. \tag{5.5.9}$$

The impact of the local scale flow fluctuations, and therefore the impact of heterogeneity on large scale flow, is encoded in the kernel function

$$k_{ij}(x, t|x', t') = \overline{Q'_i(x) G(x, t|x', t') Q'_j(x')}. \tag{5.5.10}$$

Equation (5.5.9) has a similar structure as (5.5.1). However, while in (5.5.1) the (non-local) dispersive flux term originates from the presence of capillary pressure, here it comes from spatial heterogeneity on the Darcy scale. As outlined previously, non-local fluxes typically occur when averaging. The non-linear character of the two-phase problem is preserved during the upscaling excercise. However, in order to close the large scale flow problem, the non-linearity is quasi-decoupled in (5.5.7) by using perturbation theory, so that S' can be related to the large scale saturation through the Green's function G. Explicit expressions for G can be found in Neuweiler et al. (2003) and Bolster et al. (2009).

In order to quantify the dispersion of the displacement front, Neuweiler et al. (2003) equate the non-local flux term on the right side of (5.5.9) to

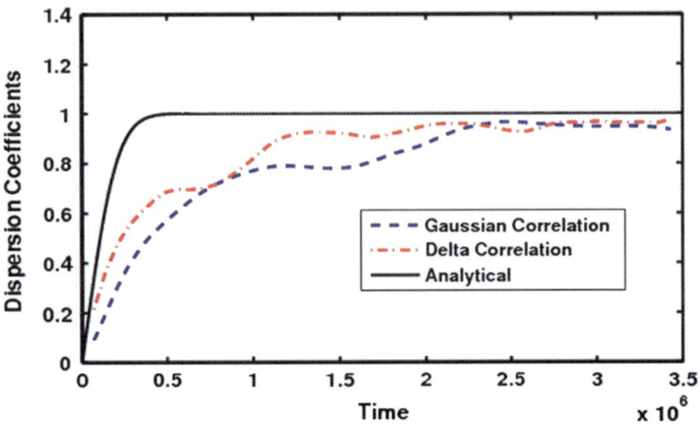

Fig. 5.11 Temporal behavior of the dispersion coefficient (5.5.12) and the evolution of the front width obtained from the numerical solution of the heterogeneous two-phase problem for a Gaussian and Delta autocorrelation functions with $\sigma_Y^2 = 0.5$, normalized by the long-time asymptotic value of (5.5.12) from (after Bolster et al. 2009)

$$\int_0^t dt' \int d\mathbf{x}' \frac{df(\overline{S}_{nw})}{d\overline{S}_{nw}(x,t)} \mathbf{k}(x,t|x',t') \nabla' f\left[\overline{S}_{nw}(x',t')\right] \equiv \mathbf{D}^*(t) \nabla f\left[\overline{S}_{nw}(x,t)\right], \quad (5.5.11)$$

which defines the dispersion tensor $\mathbf{D}^*(t)$. A perturbation theory calculation (Bolster et al. 2009; Neuweiler et al. 2003) then gives the following compact expression for the macrodispersion coefficient in flow direction

$$D_{11}^* = \sigma_Y^2 Q_0^2 \int_0^{Q_0 t} \mathrm{d}x_1 C_{11}(x_1), \quad (5.5.12)$$

where C_{11} denotes the correlation function of Q_1'. It may be approximated by the corelation function for single phase flow as outlined in Neuweiler et al. (2003) and Bolster et al. (2009). Figure 5.11 illustrates the evolution of the front spreading obtained from numerical solution of the two-phase flow problem (5.5.3) and the analytical expression (5.5.12).

In summary, the stochastic perturbative approach yields a large scale two-phase flow equation that is characterized by a non-local and non-linear dispersive flux term. The impact of heterogeneity in permeability leads to a roughening of the interface between the two fluids that can be described by an effective dispersion coefficient. It needs to be noted that dispersion here does not describe mixing, or blending, of the two immiscible fluids, but purely heterogeneity-induced spreading.

Thus, the dispersion coefficient and the related interface width serve rather as bulk measure for the interface length.

5.5.2 *Multicontinuum Two-Phase Flow Model*

The multicontinuum approach for two-phase flow in highly heterogeneous media is in line with the ones reported in Sects. 5.2.2 and 5.3.2 for single phase flow and transport, respectively. We consider a highly heterogeneous medium of the type illustrated in Fig. 5.11 with clearly delineated mobile and immobile flow regions.

The invading fluid now is the wetting and the displaced the non-wetting phase. We formulate the two-phase flow problem in terms of the saturation of the wetting phase, which here is denoted by $S = S_w$. The two-phase flow problem in the fractional flow formulation reads in terms of the wetting fluid phase as

$$\phi\frac{\partial S}{\partial t} + \nabla\cdot\left[\mathbf{Q}f_w(S) - \frac{k\Delta\rho_w\mathbf{g}}{\mu_{nw}}g_w(S)\right] + \nabla\cdot\left[\frac{k}{\mu_{nw}}g_w(S)\nabla p_c(S)\right] = 0, \quad (5.5.13)$$

where $\Delta\rho_w = \rho_{nw} - \rho_w$. The fractional flow function $f_w(S)$ and modified fractional flow function $g_w(S)$ for the wetting phase are defined by

$$f_w(S) = \frac{\lambda_w(S)}{\lambda_w(S) + \lambda_{nw}(S)}, g_w(S) = k_{r,w}(S)f_w(S). \quad (5.5.14)$$

It is assumed that flow within the matrix blocks is mainly capillary dominated, this means viscous and gravity forces are subdominant in the immobile regions. Parameters, parameter functions and observables that refer to the fracture domain in the following are marked by subscript *f*, their matrix counterpart by subscript *m*. Thus, the flow equation (5.5.13) in the fracture domain reads as

$$\phi_f\frac{\partial S_f}{\partial t} + \nabla\cdot\left[\mathbf{Q}_f f_w(S_f) - \frac{k_f\Delta\rho_w\mathbf{g}}{\mu_{nw}}g_w(S_f)\right] + \nabla\cdot\left[\frac{k_f}{\mu_{nw}}g_w(S_f)\nabla p_{c,f}(S_f)\right] = 0, \quad (5.5.15)$$

For the matrix domain, one obtains from (5.5.13) by disregarding viscous and and gravity forces

$$\phi_m\frac{\partial S_m}{\partial t} + \nabla\cdot\left[\frac{k_m}{\mu_{nw}}g_w(S_m)\nabla p_{c,m}(S_m)\right] = 0. \quad (5.5.16)$$

The conditions at the interface are continuity of flux and pressure, which are expressed by

$$k_f g_w(S_f)\mathbf{n}_{fm} \cdot \nabla p_{c,f}(S_f) = k_m g_w(S_m)\mathbf{n}_{fm} \cdot \nabla p_{c,m}(S_m), \tag{5.5.17}$$

$$p_{c,f}(S_f) = p_{c,m}(S_m) \tag{5.5.18}$$

for $\mathbf{x} \in \partial\Omega_{fm}$, where $\partial\Omega_{fm}$ denotes the fracture-matrix interface, and $\mathbf{n}_{fm}$ the unit vector perpendicular to the interface.

Using a multiscale expansion, Tecklenburg et al. (2013) derive the following large scale flow model for the (volume) averaged saturation $\overline{S}_f$

$$\varphi_f \phi_f \frac{\partial \overline{S}_f}{\partial t} + \nabla \cdot \left[\varphi_f \overline{\mathbf{Q}}_f f_w(\overline{S}_f) - \frac{\varphi_f k_f \Delta\rho_w \mathbf{g}}{\mu_{nw}} g_w(S_f)\right] + \nabla \cdot \left[\frac{\varphi_f k_f}{\mu_{nw}} g_w(\overline{S}_f) \nabla p_{c,f}(\overline{S}_f)\right]$$
$$- \nabla \cdot \phi_f \mathbf{D}^* \nabla f_w(\overline{S}_f) = -\phi_m \varphi_m \frac{\partial \overline{S}_m}{\partial t}, \tag{5.5.19}$$

where $\overline{\mathbf{Q}}_f$ is the average total flow in the facture domain. As in Sect. 5.2.2, φ_f and φ_m denote the volume fractions of the fracture and matrix domains. The macrodispersion tensor $\mathbf{D}^*$ can be identified with the one defined in (5.5.11). The average saturation $\overline{S}_m$ in the matrix domain is given by the volume average of S_m, which is determined from (5.5.16) with the boundary condition

$$p_{c,m}(S_m) = p_{c,f}(\overline{S}_f) \tag{5.5.20}$$

for $\mathbf{y} \in \partial\Omega_m$; $\partial\Omega_m$ denotes the boundary of the matrix domain, and $\mathbf{y}$ the position vector in the coordinate system attached to the matrix domain.

The closed system of equations constituted by Eqs. (5.5.19) and (5.5.16) with the boundary condition (5.5.20) constitute the multicontinuum large scale flow model for the average fracture saturation $\overline{S}_f$. Note that the resulting flow model depends on the non-linear diffusion properties between the mobile and immobile zones as expressed by (5.5.16) and (5.5.20). Thus, we seek to simplify the capillary diffusion problem (5.5.16) thus and relate it back to the single phase flow problem. To this end, we expand (5.5.16) in the form

$$\varphi_m \frac{\partial S_m}{\partial t} - \nabla_y \cdot \left[\mathcal{D}'_c \nabla_y S_m\right] = 0, \mathcal{D}'_c = \frac{k_m}{\mu_{nw}} g_w(S_m) \frac{dp_{c,m}(S_m)}{dS_m}, \tag{5.5.21}$$

where $\mathcal{D}'_c$. represents the non-linear capillary diffusion coefficient (McWhorter and Sunada 1990). We approximate $\mathcal{D}'_c$ by a suitably chosen constant value such that $\mathcal{D}'_c = \mathcal{D}_c =$ constant (Tecklenburg et al. 2013). This approximation may appear quite drastic. However, we are not so much interested in the distribution of saturation inside the immobile region, as in the average matrix saturation $\overline{S}_m$. Schmid and Geiger (2012) determined the characteristic time scale τ_c for capillary diffusion into a matrix block of typical size ℓ_m in terms of the capillary flow properties within the matrix. Thus, the constant average capillary diffusion coefficient can is estimated as $\mathcal{D}_c = \ell_m^2/(2\tau_c)$. With this simplification, (5.5.21) reduces to a linear diffusion equation that can be solved by the method of Green's functions. As a matter of fact, the immobile flow problem is similar to the corresponding matrix problems for single phase flow transport, with the difference that the diffusion coefficient $\mathcal{D}_c$ here is determined from the capillary flow properties in the matrix, and that the matrix saturation at the boundary of the matrix domain is given implicitly in terms of the average fracture saturation by (5.5.20).

For a uniform initial saturation S_{m0} inside the matrix, S_m can be determined in terms of the Green's function that solves the diffusion problem (5.5.21) in the matrix domain for a unit Delta pulse on the matrix boundary. The case of heterogeneous distributions is detailed in Tecklenburg et al. (2013). Finally, one obtains for the average saturation in the immobile zones

$$\overline{S}_m = S_{m0} - S_{m0} \int_0^t dt' g_m(t') + \int_0^t dt' g_m(t - t') S_{mb}(t'). \tag{5.5.22}$$

where we defined the boundary saturation S_{mb} from (5.5.20) as $S_{mb} = p_{c,m}^{-1}[p_{c,f}(\overline{S}_f)]$. Inserting this expression into (5.5.19) gives a single equation large scale flow model (Geiger et al. 2013; Tecklenburg et al. 2013).

$$\begin{aligned} &\varphi_f \phi_f \frac{\partial \overline{S}_f}{\partial t} + \nabla \cdot \left[\varphi_f \mathbf{Q} f_w(\overline{S}_f) - \frac{\varphi_f k_f \Delta\rho_w \mathbf{g}}{\mu_{nw}} g_w(S_f) \right] \\ &+ \nabla \cdot \left[\frac{\varphi_f k_f}{\mu_{nw}} g_w(\overline{S}_f) \nabla p_{c,f}(\overline{S}_f) \right] - \nabla \cdot \varphi_f \mathbf{D}^* \nabla f_w(\overline{S}_f) = \Gamma(t), \end{aligned} \tag{5.5.23}$$

where the source term $\Gamma(t)$ is given by

$$\Gamma(t) = -\varphi_m \phi_m \frac{d}{dt} \int_0^t dt' g_m(t - t') S_{mb}(t') + \varphi_m \phi_m S_{m0} g_m(t). \tag{5.5.24}$$

In the absence of capillary and gravity forces in the fracture, and for a linear fractional flow function $f_w(\overline{S}_w) = \overline{S}_w$, the large scale flow model is equal to the corresponding MRMT solute transport model.

The large scale model (5.5.23) and (5.5.24) can be readily generalized to account for a distribution of immobile continua characterized by different geometries and hydraulic properties. In this case, the source term $\Gamma(t)$ is generalized to $\Gamma(t) = \sum_i \varphi_{m,i} \phi_{m,i} g_{m,i}(t)$, where $\varphi_{m,i}$ and $\phi_{m,i}$ are volume fraction and porosity of the ith immobile continuum. The memory functions $g_{m,i}(t)$ quantify the mass exchange between the ith immobile continuum and the mobile continuum.

Multicontinuum models have been used for the large scale modeling of unsaturated flow and for the simulation of water-oil diplacement for simulating oil recovery from fractured reservoirs. For the modeling of unsaturated flow, Šimůnek et al. (2003) and Gerke (2006) distinguish multi-permeability and multi-porosity approaches. In the former, all the continua, characterized by different hydraulic properties, are mobile, and connected through a mass transfer term, while the latter distinguish a mobile and a set of immobile continua. In the oil and gas community such models are termed dual-permeability and dual-porosity models, respectively (Kazemi et al. 1976). Di Donato et al. (2007) present a multirate mass transfer model that has certain similarities with the one described above. These authors consider first-order mass transfer between the fracture and the matrix, similar to the one reported in Sect. 5.2.2. It is assumed, however, that the mass flux between the fracture and matrix domains is independent on the fracture saturation, but depends only on the difference between the maximum matrix saturation and the actual saturation.

5.5.3 Vertically Integrated Models

Vertically integrated large scale models seek to simplify the full two-phase flow problem and express the large scale flow problem in terms of the vertically integrated observables. This implies the projection, or compression of the original dynamic equation, and subsequent modeling of flux terms that are in general depend on the small scale details of the flow problems. This is a well-known problem in flow and transport upscaling. The macrodispersion approach reported above for the two-phase flow problem, and in Sect. 5.3.2 for transport, models the non-local flux terms by the macrodispersion closure $\overline{\mathbf{u}'c} = \mathbf{D}^m \nabla \overline{c}$, where $\mathbf{u}'$ is a velocity fluctuation, $\mathbf{D}^m$ a macroscale dispersion tensor, and $\overline{c}$ an average concentration. The multicontinuum approaches for flow and transport detailed above and in Sects. 5.2.2 and 5.3.2 represents the fluxes between the small and large scales through a memory function that is determined by the solution of the small scale flow or transport problem.

Large scale averaged models do not resolve variability along the vertical, which, however, is needed to model the fluxes and relate them to the large scale variables. Nordbotten and Celia (2012) provide an approach based on compression and reconstruction of features, which we want to briefly summarize in the following. These authors employ an heterogeneous multiscale method, which provides a methodological framework to relate the incomplete large scale, or coarse scale model to the small scale, or fine scale model and thus close the coarse scale description. This approach defines a compression operator $\mathcal{C}$, that projects the fine scale on the coarse scale variables, and a reconstruction operator R that approximates the fine scale variable based on the coarse variable. Consistency requires that $\mathcal{C}\mathcal{R} = \mathrm{II}$ with II the identity operator. The reverse is in general not true, $\mathcal{C}\mathcal{R} \neq \mathrm{II}$

because the reconstruction in general provides an approximation to the true fine scale variable.

The compression operator $\mathcal{C}$ may represent a spatial integration, or projection of a characteristic value of the small scale variable. The former applies in the definition of coarse scale phase saturation

$$\overline{S}_\alpha = \mathcal{C}_{S_\alpha} S_\alpha = \int_{\zeta_B}^{\zeta_T} dx_3 S_\alpha. \tag{5.5.25}$$

In the following, the subscript marks the variable to which compression and reconstruction are applied. The reconstructed saturation is denoted by

$$\hat{S}_\alpha = \mathcal{R}_{S_\alpha} \overline{S}_\alpha. \tag{5.5.26}$$

For the compression of pressure the second option applies. As emphasized by Nordbotten and Celia (2012), pressure is not an additive quantity, and it is in general not a independent on the vertical position. It is assumed for the following that the Dupuit assumption is valid, this means that flow is predominantly horizontal, $\mathbf{e}_3 \cdot \mathbf{q}_\alpha \approx 0$. Coarse phase pressure is defined by projection as

$$\overline{p}_\alpha = \mathcal{C}_{p_\alpha} p_\alpha = p_\alpha (x_3 = \zeta_0) \tag{5.5.27}$$

with ζ_0 a reference height. The reconstruction operator $\mathcal{R}_{p_\alpha}$ is obtained immediately as

$$\mathcal{R}_{p_\alpha} \overline{p}_\alpha = \overline{p}_\alpha - \rho_\alpha g (x_3 - \zeta_0). \tag{5.5.28}$$

For simplicity, we consider a domain with horizontal bottom and top boundaries at $x_3 = \zeta_B$ and $x_3 = \zeta_T$. Vertical averaging of (5.4.17) gives

$$\varphi \overline{S}_\alpha + \nabla_{||} \cdot \overline{\mathbf{q}}_\alpha = 0, \tag{5.5.29}$$

where zero flow is assumed at the horizontal boundaries, and porosity ϕ is assumed to be constant. The coarse phase saturations satisfy $\sum_\alpha \overline{S}_\alpha = 1$. The subscript $||$ denotes the horizontal direction. The fine scale flux $\mathbf{q}_\alpha$ is given by (5.4.16). Notice that (5.5.29) is not closed because $\overline{\mathbf{q}}_\alpha$ depends on the fine scale saturations through relative permeability $k_{r,\alpha}$. The equation can be closed subject to a suitable saturation reconstruction, such that the fine scale saturation can be related to the coarse saturation via (5.5.26). Nordbotten and Celia (2012) derive a reconstruction operator based on the Dupuit reconstruction (5.5.28) and the saturation dependence of capillary pressure. Based on a unique relation between capillary pressure and saturation, one may write

$$S_\alpha = p_c^{-1}(p_{nw} - p_w). \tag{5.5.30}$$

Using the Dupuit reconstruction (5.5.28), the pressure difference between the non-wetting and wetting fluids can be approximated by $p_{nw} - p_w \approx \overline{p}_{nw} - \overline{p}_w - (\rho_{nw} - \rho_w)g(x_3 - \zeta_0)$. This gives the reconstructed saturation as

$$\hat{S}_\alpha = p_c^{-1}[\overline{p}_{nw} - \overline{p}_w - (\rho_{nw} - \rho_w)g(x_3 - \zeta_0)]. \tag{5.5.31}$$

The compression of the reconstructed phase saturation $\hat{S}_\alpha$ gives the coarse saturation

$$\overline{S}_\alpha = \mathcal{C}_{S_\alpha}\hat{S}_\alpha = \mathcal{C}_{S_\alpha}p_c^{-1}[\overline{p}_{nw} - \overline{p}_w - (\rho_{nw} - \rho_w)g(x_3 - \zeta_0)]. \tag{5.5.31}$$

This follows from the consistence requirement mentioned above. The coarse saturation can be shown (Nordbotten and Celia 2012) to be monotone and invertible in the pressure difference $\overline{p}_{nw} - \overline{p}_w$, such that a coarse capillary pressure can be defined as

$$\overline{p}_c(\overline{S}_\alpha) = \overline{p}_{nw} - \overline{p}_w. \tag{5.5.32}$$

Thus, capillary pressure can be reconstructed as a function of the coarse phase saturation as

$$\hat{p}_c(\hat{S}_\alpha) = \overline{p}_c(\overline{S}_\alpha) - (\rho_{nw} - \rho_w)g(x_3 - \zeta_0), \tag{5.5.33}$$

and therefore, the reconstruction of the phase saturation $\hat{S}_\alpha$ can now be related to the coarse phase saturation through

$$\hat{S}_\alpha = \mathcal{C}_{S_\alpha}\overline{S}_\alpha \equiv p_c^{-1}\left[\overline{p}_c(\overline{S}_\alpha) - (\rho_{nw} - \rho_w)g(x_3 - \zeta_0)\right]. \tag{5.5.34}$$

The coarse grained Darcy flux $\overline{q}_\alpha$ can then be written in terms of the coarse grained variables as Nordbotten and Celia (2012)

$$\overline{q}_\alpha = \int_{\zeta_B}^{\zeta_T} dx_3 \mathbf{e}_\parallel \cdot \mathbf{q}_\alpha = \int_{\zeta_B}^{\zeta_T} dx_3 k \frac{k_{r,\alpha}(\mathcal{R}_{S_\alpha}\overline{S}_\alpha)}{\mu_\alpha} \nabla_\parallel \mathcal{R}_{p_\alpha}\overline{p}_\alpha. \tag{5.5.35}$$

One may now define the integrated permeability and coarse scale mobilities as

$$\overline{k} = \int_{\zeta_B}^{\zeta_T} dx_3 k, \overline{\lambda}_\alpha(\overline{S}_\alpha) = \overline{k}^{-1} \int_{\zeta_b}^{\zeta_T} dx_3 k \frac{k_{r,\alpha}(\mathcal{R}_{S_\alpha}\overline{S}_\alpha)}{\mu_\alpha}. \tag{5.5.36}$$

Thus, the equation for the coarse saturation $\overline{S}_\alpha$ now reads as

$$\phi\frac{\partial \overline{S}_\alpha}{\partial t} - \nabla_{\|} \cdot \left[\overline{k}\, \overline{\lambda}_\alpha(\overline{S}_\alpha)\nabla_{\|}\overline{p}_\alpha\right] = 0. \tag{5.5.37}$$

This equation, together with the coarse capillary pressure $\overline{p}_c(\overline{S}_\alpha)$ constitute a closed large scale flow model for the coarse phase saturations. This particular model assumes vertical equilibrium. Nordbotten and Celia (2012) discuss simplified models derived from this approach, and also consider generalizations that relax the assumptions of vertical equilibrium, and immiscibility, and incompressibility of the two fluids.

5.5.4 Convective Mixing

The previous sections described modeling approaches to quantify the large scale features of the flow of two immiscible fluids. In this section, we consider the mixing of the phases due to dissolution at the interface and the subsequent occurrence of convective instabilities. Dissolution of CO_2 in the formation water, and possible subsequent geochemical reactions are mechanisms that increase the storage safety. After injection, the buoyant CO_2 will spread and migrate laterally as a gravity current relative to the denser ambient brine. As CO_2 migrates, dissolution takes place in the interphase with the brine. The CO_2-brine mixture is denser than either of the pure fluids which results in an unstable density stratification. This configuration of fluids triggers a Rayleigh-Bénard hydrodynamic instability known as convective mixing that enhances the dissolution of CO_2.

Onset of convection

The time for the onset of convection is a critical factor in the CO_2 dissolution problem. At the critical time convection starts to dominate over diffusion and dissolution grows. The dissolution of CO_2 on top of the formation water leads to an unstable boundary-layer problem analogous to the one found when a fluid is overlaid by a cold boundary (Rees et al. 2008). Following this analogy, CO_2 dissolution is conceptualized in a semi-infinite domain with an impervious top boundary, at which concentration is prescribed equal to CO_2 solubility. The brine is assumed incompressible, the Boussinesq approximation valid and the porous medium homogeneous.

Under these assumptions, the critical time $t_c \propto D\left(\frac{\varphi\mu}{k\Delta\rho g}\right)^2$ (Hassanzadeh et al. 2007; Riaz et al. 2006). The coefficient of proportionality depends on the specific stability analysis method. Dominant mode analysis (Ben et al. 2002; Riaz et al. 2006) gives the largest coefficients, and methods which use a white noise initial condition (Caltagirone 1980; Ennis-King and Paterson 2005; Foster 1965) the lowest. It is found (Slim and Ramakrishnan 2010) that instabilities only happen for Rayleigh-Darcy numbers R larger than 32.5. The obtained critical time is layer-thickness dependent for 32.5 < R < 75. This limitation disappears if the underlying fluid is allowed to cross the top boundary. All the stability analysis methods overestimate the time for the onset of convection Slim and Ramakrishnan (2010) which can be attributed to the use of physically unrealistic perturbations (Daniel et al. 2013; Tilton et al. 2013).

The time for the onset of convection is also very sensitive to the conceptual model. More realistic models that include heterogeneity in the permeability field (Rapaka et al. 2009; Xu et al. 2006) or hydrodynamic dispersion (Hidalgo and Carrera 2009) predict shorter times for the convection to appear. However, when chemical reactions that consume carbon are taken into account (Andres and Cardoso 2011; Ghesmat et al. 2011), the time for the onset of convection increases because the boundary layer stabilizes.

Convective mixing and characteristic scales

Once convection has developed, the flow is characterized by a fingering pattern as illustrated in Fig. 5.12. The dissolution rate during this regime is approximately constant and larger than the one due to purely diffusive dissolution (Kneafsey and Pruess 2009). A heuristic analysis of the system suggests that the dissolution flux should be scale free, that is, independent of the Rayleigh number *Ra* (Hidalgo and Carrera 2009; Slim and Ramakrishnan 2010). Numerical simulations in 2D (Hidalgo and Carrera 2009; Riaz et al. 2006) and 3D (Pau et al. 2010) support this hypothesis. However, experimental data obtained using analogue fluid systems suggest a $Ra^{4/5}$ scaling of the dissolution fluxes (Backhaus et al. 2011; Neufeld et al. 2010). The scaling of the dissolution fluxes can be studied through the evolution of the scalar dissipation rate, a magnitude often easier to compute.

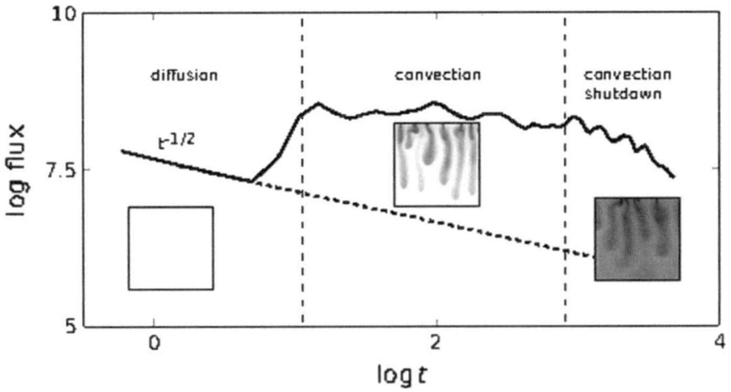

Fig. 5.12 Log dissolution flux versus log-time in arbitrary dimensionless units displaying the three main stages of CO_2 dissolution. At the begining CO_2 dissolves into brine through diffusion. The flux decays as $t^{-1/2}$. After the onset of convection the dissolution flux increases sharply. Convection dominated flux remains constant until the underlying brine saturates and the CO_2 flux is shutdown. The three *insets* show snapshots of the CO_2-brine system representative of the three regimes

This method shows that, when the Darcy-Boussinesq model is assumed, there is no dependence of the dissolution flux on *Ra* (Hidalgo et al. 2012). Similar results have been obtained for 3D simulations (Fu et al. 2013). Therefore the observed non-linear scaling has yet to be explained using effects not present in the traditional model.

Dissolution flux plays a major role in the large scale dynamics of CO_2 migration. As the CO_2 dissolves in the system, the buoyant gravity current slows down and eventually is totally arrested (MacMinn and Juanes 2013). The behavior of the convection-driven dissolution fluxes presents an intermittent behavior linked to finger growth and merging (Slim et al. 2013). Ultimately, the dissolution flux decreases as the system becomes saturated (Hidalgo et al. 2013). The nature of the convection shutdown depends on the system considered. For 2D box models the flux decays as $\approx t^{-(n+1)/n}$ where n is an exponent that characterizes the dependence of density with the concentration (Hewitt et al. 2013). However if the dynamics of the CO_2 gravity current is considered flux, decays first diffusively and later sub-diffusively (Szulczewski et al. 2013).

Heterogeneity and chemical reactions have an impact on dissolution. The presence of horizontal obstacles in the aquifer decreases the magnitud of dissolution fluxes (Elenius and Gasda 2013). The magnitude of the reduction depends on the size and distribution of the barriers. Moderate chemical reactions inhibit finger growing but favor tip splitting. This enhances convection and increases the dissolution flux. With increasing efficiency of chemical reactions, finger formation is suppressed and dissolution is controlled by diffusion (Andres and Cardoso 2011).

References

Acharya RC, Valocchi AJ, Werth CJ, Willingham TW (2007) Pore-scale simulation of dispersion and reaction along a transverse mixing zone in two-dimensional porous media. Water Resour Res 43:W10435

Adams EE, Gelhar LW (1992) Field study of dispersion in a heterogeneous aquifer: 2. Spatial moments analysis. Water Resour Res 28:3293–3307

Andres JTH, Cardoso SSS (2011) Onset of convection in a porous medium in the presence of chemical reaction. Phys Rev E 83:46312

Aris R (1956) On the dispersion of a solute in a fluid flowing through a tube. Proc R Soc Lond Math Phys Eng Sci 235:67–77

Backhaus S, Turitsyn K, Ecke RE (2011) Convective instability and mass transport of diffusion layers in a Hele-Shaw geometry. Phys Rev Lett 106:104501

Barenblatt GI, Zheltov IP, Kochina IN (1960) Basic concepts in the theory of seepage of homogeneous liquids in fissured rocks [strata]. J Appl Math Mech 24:1286–1303

Batchelor G (1949) Diffusion in a field of homogeneous turbulence. I. Eulerian analysis. Aust J Chem 2:437–450

Batchelor GK (1952) Diffusion in a field of homogeneous turbulence. Math Proc Camb Philos Soc 48:345–362

Battiato I, Tartakovsky DM, Tartakovsky AM, Scheibe T (2009) On breakdown of macroscopic models of mixing-controlled heterogeneous reactions in porous media. Adv Water Resour 32:1664–1673

Battiato I, Tartakovsky DM, Tartakovsky AM, Scheibe TD (2011) Hybrid models of reactive transport in porous and fractured media. Adv Water Resour New Comput Meth Softw Tools 34:1140–1150

Bear J (1972) Dynamics of fluids in porous media. Elsevier, New York

Beckie R, Aldama AA, Wood EF (1996) Modeling the large-scale dynamics of saturated groundwater flow using spatial-filtering theory: 1. Theoretical development. Water Resour Res 32:1269–1280

Ben Y, Demekhin EA, Chang H-C (2002) A spectral theory for small-amplitude miscible fingering. Phys Fluids (1994-Present) 14:999–1010

Benson DA, Wheatcraft SW, Meerschaert MM (2000) Application of a fractional advection-dispersion equation. Water Resour Res 36:1403–1412

Berkowitz B, Scher H (1997) Anomalous transport in random fracture networks. Phys Rev Lett 79:4038–4041

Berkowitz B, Cortis A, Dentz M, Scher H (2006) Modeling non-Fickian transport in geological formations as a continuous time random walk. Rev Geophys 44:RG2003. doi:10.1029/2005RG000178

Bijeljic B, Blunt MJ (2006) Pore-scale modeling and continuous time random walk analysis of dispersion in porous media. Water Resour Res 42:W01202

Bolster D, Dentz M, Carrera J (2009) Effective two-phase flow in heterogeneous media under temporal pressure fluctuations. Water Resour Res 45:W05408

Bolster D, Neuweiler I, Dentz M, Carrera J (2011) The impact of buoyancy on front spreading in heterogeneous porous media in two-phase immiscible flow. Water Resour Res 47:W02508

Bourgeat A (1997) Two-Phase Flow. In: Hornung U (ed) Homogenization and porous media, interdisciplinary applied mathematics. Springer, New York, pp 95–127

Brenner H (1980) Dispersion resulting from flow through spatially periodic porous media. Philos Trans R Soc Lond 297(1430):81–133

Brinkman HC (1949) A calculation of the viscous force exerted by a flowing fluid on a dense swarm of particles. Appl Sci Res 1:27–34

Bronstert A, Carrera J, Leavesley G, Mölders N (2005) Scale issues. In: Bronstert A, Carrera J, Kabat P, Lütkemeier S (eds) Coupled models for the hydrological cycle integrating atmosphere, biosphere and pedosphere. Springer, pp 21–43

Caltagirone J-P (1980) Stability of a saturated porous layer subject to a sudden rise in surface temperature: comparison between the linear and energy methods. Q J Mech Appl Math 33:47–58

Carrera J (1993) Chemistry and migration of actinides and fission products an overview of uncertainties in modelling groundwater solute transport. J Contam Hydrol 13:23–48

Carrera J, Sánchez-Vila X, Benet I, Medina A, Galarza G, Guimerà J (1998) On matrix diffusion: formulations, solution methods and qualitative effects. Hydrogeol J 6:178–190. doi:10.1007/s100400050143

Cirpka OA, Kitanidis PK (2000) An advective-dispersive stream tube approach for the transfer of conservative-tracer data to reactive transport. Water Resour Res 36:1209–1220. doi:10.1029/1999WR900355

Cueto-Felgueroso L, Juanes R (2012) Macroscopic phase-field model of partial wetting: bubbles in a capillary tube. Phys Rev Lett 108:144502

Cunningham JA, Werth CJ, Reinhard M, Roberts PV (1997) Effects of grain-scale mass transfer on the transport of volatile organics through sediments: 1. Model development. Water Resour Res 33:2713–2726

Cushman JH, Ginn TR (1993) Nonlocal dispersion in media with continuously evolving scales of heterogeneity. Transp Porous Media 13:123–138

Cushman JH, Ginn TR (2000) Fractional advection-dispersion equation: a classical mass balance with convolution-Fickian Flux. Water Resour Res 36:3763–3766

Cushman JH, Bennethum LS, Hu BX (2002) A primer on upscaling tools for porous media. Adv Water Resour 25:1043–1067

Cvetkovic V, Dagan G (1996) Reactive transport and immiscible flow in geological media. ii. applications. Proc R Soc Lond A, 452:202–328

Dagan G (1984) Solute transport in heterogeneous porous formations. J Fluid Mech 145:151–177
Dagan G (1986) Statistical theory of groundwater flow and transport: pore to laboratory, laboratory to formation, and formation to regional scale. Water Resour Res 22:120S–134S
Dagan G (1988) Time-dependent macrodispersion for solute transport in anisotropic heterogeneous aquifers. Water Resour Res 24:1491–1500
Dagan G (1989) Flow and transport in porous formations. Springer, Berlin
Dagan G, Bresler E (1979) Solute dispersion in unsaturated heterogeneous soil at field scale: I. Theory1. Soil Sci Soc Am J 43:461
Dagan G, Cvetkovic V (1996) Reactive transport and immiscible flow in geological media. I. General theory. Proc R Soc Lond Math Phys Eng Sci 452:285–301
Dahle HK, Celia MA, Hassanizadeh SM (2005) Bundle-of-tubes model for calculating dynamic effects in the capillary-pressure–saturation relationship. Transp Porous Media 58:5–22
Daniel D, Tilton N, Riaz A (2013) Optimal perturbations of gravitationally unstable, transient boundary layers in porous media. J Fluid Mech 727:456–487
Das DB, Hassanizadeh SM (2005) Upscaling multiphase flow in porous media. Springer, Berlin
Das DB, Mirzaei M (2012) Dynamic effects in capillary pressure relationships for two-phase flow in porous media: experiments and numerical analyses. AIChE J 58:3891–3903
de Dreuzy J-R, Carrera J, Dentz M, Le Borgne T (2012) Time evolution of mixing in heterogeneous porous media. Water Resour Res 48:W06511
De Simoni M, Carrera J, Sánchez-Vila X, Guadagnini A (2005) A procedure for the solution of multicomponent reactive transport problems. Water Resour Res 41:W11410
De Simoni M, Sanchez-Vila X, Carrera J, Saaltink MW (2007) A mixing ratios-based formulation for multicomponent reactive transport. Water Resour Res 43:W07419
Dean DS, Drummond IT, Horgan RR (2007) Effective transport properties for diffusion in random media. J Stat Mech Theory Exp 2007:P07013
Delay F, Ackerer P, Danquigny C (2005) Simulating solute transport in porous or fractured formations using random walk particle tracking. Vadose Zone J 4:360
Dentz M, Berkowitz B (2003) Transport behavior of a passive solute in continuous time random walks and multirate mass transfer. Water Resour Res, 39(5):1111
Dentz M, Berkowitz B (2005) Exact effective transport dynamics in a one-dimensional random environment. Phys Rev E 72:31110
Dentz M, Castro A (2009) Effective transport dynamics in porous media with heterogeneous retardation properties. Geophys Res Lett 36:L03403
Dentz M, Kinzelbach H, Attinger S, Kinzelbach W (2000) Temporal behavior of a solute cloud in a heterogeneous porous medium: 1. Point-like injection. Water Resour Res 36:3591–3604
Dentz M, Gouze P, Carrera J (2011a) Effective non-local reaction kinetics for transport in physically and chemically heterogeneous media. J Contam Hydrol 120–121:222–236
Dentz M, Le Borgne T, Englert A, Bijeljic B (2011b) Mixing, spreading and reaction in heterogeneous media: a brief review. J Contam Hydrol 120–121:1–17
Di Donato G, Lu H, Tavassoli Z, Blunt MJ (2007) Multirate-transfer dual-porosity modeling of gravity drainage and imbibition. SPE J. 12:77–88
Donado LD, Sanchez-Vila X, Dentz M, Carrera J, Bolster D (2009) Multicomponent reactive transport in multicontinuum media. Water Resour Res 45:W11402
Edwards DA, Shapiro M, Brenner H (1993) Dispersion and reaction in two-dimensional model porous media. Phys Fluids A, 5:837–848
Elenius MT, Gasda SE (2013) Convective mixing in formations with horizontal barriers. Adv Water Resour 62(Part C):499–510
Ennis-King J, Paterson L (2005) Role of convective mixing in the long-term storage of carbon dioxide in deep saline formations. Soc Petrol Eng 10(03):349–356
Espinoza C, Valocchi AJ (1997) Stochastic analysis of one-dimensional transport of kinetically adsorbing solutes in chemically heterogeneous aquifers. Water Resour Res 33:2429–2445
Foster TD (1965) Stability of a homogeneous fluid cooled uniformly from above. Phys Fluids 1958–1988(8):1249–1257

Fu X, Cueto-Felgueroso L, Juanes R (2013) Pattern formation and coarsening dynamics in three-dimensional convective mixing in porous media. Philos Trans R Soc Lond Math Phys Eng Sci 371:20120355

Geiger S, Dentz M, Neuweiler I (2013) A novel multi-rate dual-porosity model for improved simulation of fractured and multiporosity reservoirs. SPE J 18:670–684

Gelhar LW (1993) Stochastic subsurface hydrology. Prentice-Hall, Upper Saddle River

Gelhar LW, Axness CL (1983) Three-dimensional stochastic analysis of macrodispersion in aquifers. Water Resour Res 19:161–180

Gelhar LW, Welty C, Rehfeldt KR (1992) A critical review of data on field-scale dispersion in aquifers. Water Resour Res 28:1955–1974

Gerke HH (2006) Preferential flow descriptions for structured soils. J Plant Nutr Soil Sci 169:382–400

Ghesmat K, Hassanzadeh H, Abedi J (2011) The impact of geochemistry on convective mixing in a gravitationally unstable diffusive boundary layer in porous media: CO_2 storage in saline aquifers. J Fluid Mech 673:480–512

Ginn TR (2001) Stochastic–convective transport with nonlinear reactions and mixing: finite streamtube ensemble formulation for multicomponent reaction systems with intra-streamtube dispersion. J Contam Hydrol 47:1–28

Ginn TR, Simmons CS, Wood BD (1995) Stochastic-convective transport with nonlinear reaction: biodegradation with microbial growth. Water Resour Res 31:2689–2700

Gouze P, Melean Y, Le Borgne T, Dentz M, Carrera J (2008) Non-Fickian dispersion in porous media explained by heterogeneous microscale matrix diffusion. Water Resour Res 44:W11416

Gramling CM, Harvey CF, Meigs LC (2002) Reactive transport in porous media: a comparison of model prediction with laboratory visualization. Environ Sci Technol 36:2508–2514

Guimerà J, Carrera J (2000) A comparison of hydraulic and transport parameters measured in low-permeability fractured media. J Contam Hydrol 41:261–281

Gutjahr AL, Gelhar LW, Bakr AA, MacMillan JR (1978) Stochastic analysis of spatial variability in subsurface flows: 2. Evaluation and application. Water Resour Res 14:953–959

Haggerty R, Gorelick SM (1995) Multiple-rate mass transfer for modeling diffusion and surface reactions in media with pore-scale heterogeneity. Water Resour Res 31:2383–2400

Hänggi P, Talkner P, Borkovec M (1990) Reaction-rate theory: fifty years after Kramers. Rev Mod Phys 62:251–341

Hassanizadeh SM, Gray WG (1990) Mechanics and thermodynamics of multiphase flow in porous media including interphase boundaries. Adv Water Resour 13:169–186

Hassanzadeh H, Pooladi-Darvish M, Keith DW (2007) Scaling behavior of convective mixing, with application to geological storage of CO_2. AIChE J 53:1121–1131

Hewitt DR, Neufeld JA, Lister JR (2013) Convective shutdown in a porous medium at high Rayleigh number. J Fluid Mech 719:551–586

Hidalgo JJ, Carrera J (2009) Effect of dispersion on the onset of convection during CO_2 sequestration. J Fluid Mech 640:441–452

Hidalgo JJ, Fe J, Cueto-Felgueroso L, Juanes R (2012) Scaling of convective mixing in porous media. Phys Rev Lett 109:264503

Hidalgo JJ, MacMinn CW, Juanes R (2013) Dynamics of convective dissolution from a migrating current of carbon dioxide. Adv Water Resour 62(Part C):511–519

Hornung U (1997) Homogenization and porous media. Springer, New York

Kang Q, Lichtner PC, Zhang D (2006) Lattice Boltzmann pore-scale model for multicomponent reactive transport in porous media. J Geophys Res Solid Earth 111:B05203

Kapoor V, Kitanidis PK (1998) Concentration fluctuations and dilution in aquifers. Water Resour Res 34:1181–1193

Kapoor V, Gelhar LW, Miralles-Wilhelm F (1997) Bimolecular second-order reactions in spatially varying flows: segregation induced scale-dependent transformation rates. Water Resour Res 33:527–536

Kazemi H, Merrill LS, Porterfield KL, Zeman PR (1976) Numerical simulation of water–oil flow in naturally fractured reservoirs. Soc Pet Eng J. 16:317–326

Kechagia PE, Tsimpanogiannis IN, Yortsos YC, Lichtner PC (2002) On the upscaling of reaction-transport processes in porous media with fast or finite kinetics. Chem Eng Sci 57:2565–2577

Keller JB (1964) A theorem on the conductivity of a composite medium. J Math Phys 5:548–549

Kinzelbach W (1988) The random walk method in pollutant transport simulation. In: Custodio E, Gurgui A, Ferreira JPL (eds) Groundwater flow and quality modelling, NATO ASI series. Springer, Amsterdam, pp 227–245

Kitanidis PK (1988) Hydrologic research: The U.S.—Japan experience prediction by the method of moments of transport in a heterogeneous formation. J Hydrol 102:453–473

Kitanidis PK (1994) The concept of the dilution Index. Water Resour Res 30:2011–2026

Knabner P, van Duijn CJ, Hengst S (1995) An analysis of crystal dissolution fronts in flows through porous media. Part 1: compatible boundary conditions. Adv Water Resour 18:171–185

Kneafsey TJ, Pruess K (2009) Laboratory flow experiments for visualizing carbon dioxide-induced, density-driven brine convection. Transp Porous Media 82:123–139. doi:10.1007/s11242-009-9482-2

LaBolle EM, Fogg GE, Tompson AFB (1996) Random-walk simulation of transport in heterogeneous porous media: local mass-conservation problem and implementation methods. Water Resour Res 32:583–593

Lallemand-Barres A, Peaudecerf P (1978) Recherche des relations entre la valeur de la dispersivité macroscopique d'un milieu aquifer, ses autres caracteristiques et les conditions de mesure. Bull BRGM 3:227–287

Langlo P, Espedal MS (1994) Macrodispersion for two-phase, immiscible flow in porous media. Adv Water Resour 17:297–316

Langlo P, Espedal M (1995) Macrodispersion for two-phase, immiscible flow in porous media. Adv Water Resour 17:297–316

Le Borgne T, Dentz M, Carrera J (2008) Lagrangian statistical model for transport in highly heterogeneous velocity fields. Phys Rev Lett 101:90601

Le Borgne T, Dentz M, Davy P, Bolster D, Carrera J, de Dreuzy J-R, Bour O (2011) Persistence of incomplete mixing: a key to anomalous transport. Phys Rev E 84:15301

Li L, Peters CA, Celia MA (2006) Upscaling geochemical reaction rates using pore-scale network modeling. Adv Water Resour 29:1351–1370

Li L, Steefel CI, Yang L (2008) Scale dependence of mineral dissolution rates within single pores and fractures. Geochim Cosmochim Acta 72:360–377

Lichtner PC (1985) Continuum model for simultaneous chemical reactions and mass transport in hydrothermal systems. Geochim Cosmochim Acta 49:779–800

Lichtner PC, Kang Q (2007) Upscaling pore-scale reactive transport equations using a multiscale continuum formulation. Water Resour Res 43:W12S15

Lichtner PC, Tartakovsky DM (2003) Stochastic analysis of effective rate constant for heterogeneous reactions. Stoch Environ Res Risk Assess 17:419–429

Liu C, Zachara JM, Qafoku NP, Wang Z (2008) Scale-dependent desorption of uranium from contaminated subsurface sediments. Water Resour Res 44:W08413

MacMinn CW, Juanes R (2013) Buoyant currents arrested by convective dissolution. Geophys Res Lett 40:2017–2022. doi:10.1002/grl.50473

MacQuarrie KTB, Sudicky EA (1990) Simulation of biodegradable organic contaminants in groundwater: 2. Plume behavior in uniform and random flow fields. Water Resour Res 26:223–239

Marle CM (1981) Multiphase flow in porous media. Gulf Publishing Co, Houston

Matheron G (1967) Composition des perméabilités en milieu poreux héterogène. Méthode de Schwydler et règles de pondération. Revue de l'Institute Français du Petrole 443–466

Matheron G, De Marsily G (1980) Is transport in porous media always diffusive? A counterexample. Water Resour Res 16:901–917

McWhorter DB, Sunada DK (1990) Exact integral solutions for two-phase flow. Water Resour Res 26:399–413

Meakin P, Tartakovsky AM (2009) Modeling and simulation of pore-scale multiphase fluid flow and reactive transport in fractured and porous media. Rev Geophys 47:RG3002
Meerschaert MM, Benson DA, Scheffler H-P, Becker-Kern P (2002) Governing equations and solutions of anomalous random walk limits. Phys Rev E 66:60102
Meile C, Tuncay K (2006) Scale dependence of reaction rates in porous media. Adv Water Resour 29:62–71
Metzler R, Klafter J (2000) The random walk's guide to anomalous diffusion: a fractional dynamics approach. Phys Rep 339:1–77
Mikelić A, Devigne V, van Duijn C (2006) Rigorous upscaling of the reactive flow through a pore, under dominant Peclet and Damkohler numbers. SIAM J Math Anal 38:1262–1287
Molz FJ, Widdowson MA (1988) Internal inconsistencies in dispersion-dominated models that incorporate chemical and microbial kinetics. Water Resour Res 24:615–619
Morales-Casique E, Neuman SP, Guadagnini A (2006) Non-local and localized analyses of non-reactive solute transport in bounded randomly heterogeneous porous media: theoretical framework. Adv Water Resour 29:1238–1255
Muskat M (1937) The flow of homogeneous fluids through porous media: soil science. McGraw-Hill, New York
Neufeld JA, Hesse MA, Riaz A, Hallworth MA, Tchelepi HA, Huppert HE (2010) Convective dissolution of carbon dioxide in saline aquifers. Geophys Res Lett 37:L22404. doi:10.1029/2010GL044728
Neuman SP, Orr S (1993) Prediction of steady state flow in nonuniform geologic media by conditional moments: exact nonlocal formalism, effective conductivities, and weak approximation. Water Resour Res 29:341–364
Neuman SP, Tartakovsky DM (2009) Perspective on theories of non-Fickian transport in heterogeneous media. Adv Water Resour 32:670–680
Neuman SP, Zhang Y-K (1990) A quasi-linear theory of non-Fickian and Fickian subsurface dispersion: 1. Theoretical analysis with application to isotropic media. Water Resour Res 26:887–902
Neuman SP, Winter CL, Newman CM (1987) Stochastic theory of field-scale fickian dispersion in anisotropic porous media. Water Resour Res 23:453–466
Neuweiler I, Attinger S, Kinzelbach W, King P (2003) Large scale mixing for immiscible displacement in heterogeneous porous media. Transp Porous Media 51:287–314
Noetinger B, Artus V, Ricard L (2004) Dynamics of the water–oil front for two-phase, immiscible flow in heterogeneous porous media. 2: Isotropic media. Transp Porous Media 56:305–328
Nordbotten JM, Celia MA (2012) Geological storage of CO_2: modeling approaches for large-scale simulation. Wiley, New York
Panfilow M, Floriat S (2004) Nonlinear two-phase mixing in heterogeneous porous media. Transp Porous Media 57:347–375
Pau GSH, Bell JB, Pruess K, Almgren AS, Lijewski MJ, Zhang K (2010) High-resolution simulation and characterization of density-driven flow in CO_2 storage in saline aquifers. Adv Water Resour 33:443–455. doi:10.1016/j.advwatres.2010.01.009
Pfannkuch H-O (1963) Contribution a l'etude des d placement de fluides miscible dans un milieu poreux. Revue de l'Institut français du pétrole 18:215
Pope SB (2001) Turbulent flows. Meas Sci Technol 12:2020
Quintard M, Whitaker S (1994) Convection, dispersion, and interfacial transport of contaminants: homogeneous porous media. Adv Water Resour 17:221–239
Rapaka S, Pawar RJ, Stauffer PH, Zhang D, Chen S (2009) Onset of convection over a transient base-state in anisotropic and layered porous media. J Fluid Mech 641:227–244
Rees DAS, Selim A, Ennis-King JP (2008) The instability of unsteady boundary layers in porous media. In: Vadász P (ed) Emerging topics in heat and mass transfer in porous media, theory and applications of transport in porous media. Springer, Amsterdam, pp 85–110
Renard P, de Marsily G (1997) Calculating equivalent permeability: a review. Adv Water Resour 20:253–278. doi:10.1016/S0309-1708(96)00050-4

Rezaei M, Sanz E, Raeisi E, Ayora C, Vázquez-Suñé E, Carrera J (2005) Reactive transport modeling of calcite dissolution in the fresh-salt water mixing zone. J Hydrol 311:282–298

Riaz A, Hesse M, Tchelepi HA, Orr FM (2006) Onset of convection in a gravitationally unstable diffusive boundary layer in porous media. J Fluid Mech 548:87–111. doi:10.1017/S0022112005007494

Risken PDH (1984). The Fokker-Planck equation, Springer Series in Synergetics. Springer, Berlin, pp 63–95

Risken H (1996) The Fokker-planck equation. Springer, Heidelberg, New York

Robinson BA, Viswanathan HS (2003) Application of the theory of micromixing to groundwater reactive transport models. Water Resour Res 39:1313

Rubin Y (2003) Applied stochastic hydrogeology. Oxford University Press, New York

Rubin Y, Sun A, Maxwell R, Bellin A (1999) The concept of block-effective macrodispersivity and a unified approach for grid-scale and plume-scale-dependent transport. J Fluid Mech 395:161–180

Saaltink MW, Ayora C, Carrera J (1998) A mathematical formulation for reactive transport that eliminates mineral concentrations. Water Resour Res 34:1649–1656

Sahimi M (1995) Flow and transport in porous media and fractured rock. VCH, Weinheim

Sahimi M (2011) Flow and transport in porous media and fractured rock: from classical methods to modern approaches. Wiley, New York

Salamon P, Fernàndez-Garcia D, Gómez-Hernández JJ (2006) A review and numerical assessment of the random walk particle tracking method. J Contam Hydrol 87:277–305

Sanchez-Vila X, Guadagnini A, Carrera J (2006) Representative hydraulic conductivities in saturated groundwater flow. Rev Geophys 44:RG3002

Sanchez-Vila X, Dentz M, Donado LD (2007) Transport-controlled reaction rates under local non-equilibrium conditions. Geophys Res Lett 34:L10404

Schmid KS, Geiger S (2012) Universal scaling of spontaneous imbibition for water-wet systems. Water Resour Res 48:W03507

Schumer R, Benson DA, Meerschaert MM, Baeumer B (2003) Fractal mobile/immobile solute transport. Water Resour Res 39:1296

Seeboonruang U, Ginn TR (2006a) Upscaling heterogeneity in aquifer reactivity via exposure-time concept: forward model. J Contam Hydrol 84:127–154

Seeboonruang U, Ginn TR (2006b) Upscaling heterogeneity in aquifer reactivity via the exposure-time concept: inverse model. J Contam Hydrol 84:155–177. doi:10.1016/j.jconhyd.2005.12.010

Šimůnek J, Jarvis NJ, van Genuchten MT, Gärdenäs A (2003) Review and comparison of models for describing non-equilibrium and preferential flow and transport in the vadose zone. J Hydrol 272:14–35

Slim AC, Ramakrishnan TS (2010) Onset and cessation of time-dependent, dissolution-driven convection in porous media. Phys Fluids 22:124103

Slim AC, Bandi MM, Miller JC, Mahadevan L (2013) Dissolution-driven convection in a Hele–Shaw cell. Phys Fluids (1994-Present) 25:24101

Steefel CI, MacQuarrie KTB (1996) Approaches to modeling of reactive transport in porous media. Rev Miner Geochem 34:85–129

Steefel CI, DePaolo DJ, Lichtner PC (2005) Reactive transport modeling: an essential tool and a new research approach for the Earth sciences. Earth Planet Sci Lett 240:539–558

Szulczewski ML, Hesse MA, Juanes R (2013) Carbon dioxide dissolution in structural and stratigraphic traps. J Fluid Mech 736:287–315. doi:10.1017/jfm.2013.511

Tartakovsky AM, Meakin P, Scheibe TD, Eichler West RM (2007) Simulations of reactive transport and precipitation with smoothed particle hydrodynamics. J Comput Phys 222:654–672

Tartakovsky AM, Redden G, Lichtner PC, Scheibe TD, Meakin P (2008) Mixing-induced precipitation: experimental study and multiscale numerical analysis. Water Resour Res 44: W06S04

Tartakovsky AM, Tartakovsky GD, Scheibe TD (2009) Effects of incomplete mixing on multicomponent reactive transport. Adv Water Resour 32:1674–1679

Tecklenburg J, Neuweiler I, Dentz M, Carrera J, Geiger S, Abramowski C, Silva O (2013) A non-local two-phase flow model for immiscible displacement in highly heterogeneous porous media and its parametrization. Adv Water Resour 62(Part C):475–487

Tilton N, Daniel D, Riaz A (2013) The initial transient period of gravitationally unstable diffusive boundary layers developing in porous media. Phys Fluids (1994-Present) 25:92107

Villermaux J, Devillon JC (1972) Représentation de la coalescence et de la redispersion des domaines de ségrégation dans un fluide par un modèle d'interaction phénom. In: Proceedings of 2nd international symposium chemistry reaction engineering. Elsevier, New York

Wen X-H, Gómez-Hernández JJ (1996) Upscaling hydraulic conductivities in heterogeneous media: an overview. J Hydrol 183:9–32

West B, Bologna M, Grigolini P (2003) Physics of fractal operators. Springer, Berlin

Whitaker S (1986) Flow in porous media I: a theoretical derivation of Darcy's law. Transp Porous Media 1:3–25

Whitaker S (1999) The method of volume averaging. Kluwer, Berlin

White A, Peterson M (1990) Role of reactive-surface-area characterization in geochemical kinetic models. In: Chemical modeling of aqueous systems II. American chemical society, pp 461–475

Willingham TW, Werth CJ, Valocchi AJ (2008) Evaluation of the effects of porous media structure on mixing-controlled reactions using pore-scale modeling and micromodel experiments. Environ Sci Technol 42:3185–3193

Willmann M, Carrera J, Sánchez-Vila X (2008) Transport upscaling in heterogeneous aquifers: what physical parameters control memory functions? Water Resour Res 44:W12437

Willmann M, Carrera J, Sanchez-Vila X, Silva O, Dentz M (2010) Coupling of mass transfer and reactive transport for nonlinear reactions in heterogeneous media. Water Resour Res 46: W07512

Winter CL, Tartakovsky DM, Guadagnini A (2003) Moment differential equations for flow in highly heterogeneous porous media. Surv Geophys 24:81–106

Wyckoff RD, Botset HG (1936) The flow of gas-liquid mixtures through unconsolidated sands. J Appl Phys 7:325–345

Xu X, Chen S, Zhang D (2006) Convective stability analysis of the long-term storage of carbon dioxide in deep saline aquifers. Adv Water Resour 29:397–407. doi:10.1016/j.advwatres.2005.05.008

Yang Z, Niemi A, Fagerlund F, Illangasekare T (2013) Two-phase flow in rough-walled fractures: comparison of continuum and invasion-percolation models. Water Resour Res 49:993–1002

Zhang D, Tchelepi H (1999) Stochastic analysis of immiscible two-phase flow in heterogeneous media. SPE J 4:380–388

Chapter 6
Laboratory Experiments

Philippe Gouze, Katriona Edlmann, Christopher Ian McDermott and Linda Luquot

Abstract The hydro-thermo-mechanical and chemical properties of reservoir rocks and the surrounding sealing units are important data for assessing the performance of a CO_2 storage. Laboratory measurements on rock samples are the first method to assess these properties and evaluate the reservoir injectivity and storage potential. Beyond standard techniques, this chapter also presents state of the art laboratory experiments capable of reproducing the in situ conditions during CO_2 injection. In addition, these methods are also used to investigate the coupling between the hydro-thermo-mechanical and chemical properties.

6.1 Introduction

The hydro-thermo-mechanical and chemical properties of reservoir rocks and the surrounding sealing units are crucial data for assessing the performance of a CO_2 storage. Laboratory measurements on rock samples are the first method to assess these properties and evaluate the reservoir injectivity and storage potential. Then laboratory experiments reproducing the conditions of CO_2 injection can be performed for characterizing the mechanisms of alteration of these properties and for

P. Gouze (✉)
Géosciences, Université de Montpellier, CNRS, 34095 Montpellier Cedex 5, France
e-mail: philippe.gouze@univ-montp2.fr

K. Edlmann
School of Geosciences, University of Edinburgh, Edinburgh, UK
e-mail: katriona.edlmann@ed.ac.uk

C.I. McDermott
School of Geoscience, Edinburgh Collaborative of Subsurface Science and Engineering (ECOSSE), University of Edinburgh, Edinburgh, UK
e-mail: cmcdermo@staffmail.ed.ac.uk

L. Luquot
Spanish National Research Council (IDAEA-CSIC), 08034 Barcelona, Spain
e-mail: linda.luquot@iqac.csic.es

A. Niemi et al. (eds.), *Geological Storage of CO_2 in Deep Saline Formations*,
Theory and Applications of Transport in Porous Media 29,
DOI 10.1007/978-94-024-0996-3_6

determining the input parameters for predictive modeling. This chapter presents a panel of methods and procedures of laboratory testing of rock and fluid samples that can be applied during the exploration phase when boreholes are drilled and core samples retrieved.

The properties of the geomaterials forming the storage system—the reservoir and the caprock—prior to the CO_2 injection are usually determined in the laboratory using samples cored while drilling the wells or samples taken from quarries or other cored sites considered to be similar. Also, the changes of these properties, triggered by the injection of CO_2, can be assessed by laboratory experiments reproducing the dynamic processes taking place during the different stages of the storage and by studying geological systems that have been exposed to natural CO_2 flooding. The well cement annulus that plays a key role in the storage confinement is also an important component of the storage system.

The methods applied for measuring the properties of the storage system within the framework of CO_2 sequestration projects are to a large extent similar to those routinely performed in oil and gas industry during the exploration of hydrocarbon reservoirs. For instance, the characterization of the reservoir rock involves standard investigations including petrographic characterization as well as the measurement of hydrodynamic and mechanical properties. Basic properties such as porosity, thermal conductivity, permeability and most of the mechanical properties require a set of specific measurement tools that are relatively simple and similarly implemented in many academic and industrial research institutes. Standard laboratory measurements are also employed to determine fluid properties such as viscosity and solubility for mixtures involving several components and phases. Conversely, the evaluation of properties such as relative permeability (for parameterizing the flow of two or several fluid phases) and mineral reactivity requires more complex experimental protocols. They involve several steps and often rely on empirical model and on the knowledge of some other properties related to the pore structure, rock composition and fluid composition. Nevertheless these measurements follow procedures that are quite well established and widely used, albeit each team may operate alternative protocols.

Beside these basic measurements, CO_2 underground storage triggered additional challenges and promoted the development of new laboratory tools. Research activities related to the injection of CO_2 in reservoir, and consequently the development of specific laboratory tools and protocols, were first initiated in the frame of the development of enhanced oil and gas recovery using CO_2. The specificity of CO_2 underground storage compared to standard oil and gas industrial activities is twofold. First CO_2 storage involves filling geological reservoirs with a fluid that is potentially chemically aggressive and may cause strong and often irreversible alterations of the storage system hydrodynamic and mechanical properties. Second, the steric, thermic and dynamic effects, including buoyancy forces, caused by the massive injection of supercritical CO_2 ($scCO_2$) may alter the mechanical properties of the reservoir and cause confinement failure. Thus, the characterization of the mechanical and chemical mechanisms that can alter the properties of the confinement is a critical issue for CO_2 underground storage. Specifically, the development of fractures in the well cement and in the caprock layers, and the potentially induced chemical alteration processes in case of leakage of CO_2-rich brine and $scCO_2$-brine mixtures, must be investigated in detail using laboratory experiments because it is impractical to be determined at field scale.

Laboratory experiments reproducing the complex dynamic of the storage system have been developed mainly during the last decade. The main objective of these experiments is to investigate the coupled thermo-hydro-mechanical and chemical (THMC) mechanisms that are involved during the injection of the CO_2 into the reservoir. New experimental rigs allowing the simulation of the temperature, pressure, fluid composition and fluid flow conditions while measuring the effects on the change in the petrophysical, hydrodynamic and mechanical properties have been designed and constructed, for instance within the frame of the MUSTANG project. These tools have been used not only to explore the behavior of the reservoir when invaded by $scCO_2$ or CO_2—enriched brine, but also (1) to simulate leakages through fractured caprocks or damaged well cement annulus, (2) to study the efficiency of the injection of remediation fluids (healing reagents) and (3) to measure the hydro-mechanical coupling parameters during caprock fracturing.

The following sections will give an overview of the different laboratory techniques and experiments that are essential not only for parameterizing the models, but also to characterize the outcomes of complex processes that could not be easily studied at depth in the course of field tests. Yet, it is important to keep in mind that property measurements and mass transfer experiments performed in the laboratory used samples of some cubic centimeters in volume and therefore the representativeness of the results relies on the quality of the sample. The representativeness of the sample hinges on two main factors: the spatial heterogeneity of the studied geological unit and the care taken during sampling in order to minimize the alteration of the properties. The pertinent choice of the samples as well as the number of samples and measurements needed to characterize a property or a process is a critical step of the experimental protocol.

6.2 Measuring Hydrodynamical Properties

CO_2 storage in geological formations involves numerous processes at a broad range of scales. For example, at large scale and far from the injection well, fluid pressure buildup will be sensitive to large scale permeability (Birkholzer et al. 2012), whereas, at local scale, the CO_2 and brine relative permeability will control the dynamic of the system (Juanes et al. 2006). With time, CO_2 will partially dissolve into the water. This dissolution will be controlled by hydrodynamic parameters; initially by molecular diffusion and then accelerated by dispersion (Hidalgo and Carrera 2009; Riaz et al. 2006). Consequently, it is essential to evaluate all the hydrodynamical properties of the reservoir to model (1) the CO_2 dissolution rate into the water; (2) the CO_2 plume localization and (3) the fraction of CO_2 in water depending on the structural characteristics of the storage system and on the dynamic conditions enforced by the CO_2 injection.

These measurements are quite similar to those routinely performed for evaluating oil and gas reservoirs and specifically to assess the performance of enhanced oil recovery using CO_2. For instance, the theoretical aspects related to measurements of porosity, absolute permeability, electrical properties (used to evaluate tortuosity and formation factor), thermal and chemical diffusion, fluid saturation, interfacial tension, wettability, capillary pressure and relative permeability have been described in detail in several books (Dandekar 2013; Sahimi 2011; Pinder and Gray 2008). In this section we will presents briefly the objectives and methods for measuring the hydrodynamic properties of the reservoir rocks, the caprocks and the cements and then focus on their use in the frame of CO_2 underground storage where these properties may evolve with time due to chemical reactions. For instance mineral dissolution and precipitation can strongly modify pore connectivity and consequently (relative) permeability, but also the diffusivity, the reactive surface area and virtually all the properties of the material (Guarracino et al. 2014; Gouze and Luquot 2011; Luquot and Gouze 2009; Navarre-Sitchler et al. 2009; Noiriel et al. 2009).

6.2.1 Porosity and Structural Parameters

6.2.1.1 Porosity and Tortuosity

Porosity denotes the volume fraction of void in a rock. Usually there are *interconnected pores* and *isolated pores*. The former designates the fraction of the porosity in which fluid(s) can flow if a pressure gradient is applied. The value of connected porosity ϕ and its geometrical properties are of primary interest for characterizing hydrodynamic processes. *Connected porosity* is routinely measured using gas (helium or nitrogen) porosimeter. This is a fast and cheap method where the Boyle's law ($p_1 V_1 = p_2 V_2$) is used to determine the porosity from pressurizing at pressure p_1 a tank of known volume V_1 containing a rock core of known total volume V_T, then opening this tank to a second tank of known volume V_2 and measuring p_2. The connected porosity is:

$$\phi = V_T^{-1}[1 - V_1 + ((p_2 V_2)/(p_2 - p_1))] \tag{6.2.1}$$

Another standard method for investigating the porosity and the pore structure is *mercury intrusion porosimetry* (MIP) which provides precious information on the distribution of the pore access (Giesche 2006). This method consists of injecting mercury (i.e. a non-wetting liquid) at increasing pressure into a rock core. This corresponds to a drainage experiment (see Sect. 6.2.3). The porosity is known from the total volume injected fluid while the size of the pore access (pore throat) is inferred from the pressure p needed to force the liquid into capillaries-like throats of radius r against the surface tension force using the Young-Laplace equation $r(p) = 2\gamma \mathrm{Cos}\alpha / p$, where γ is the interfacial tension and α the contact angle between the rock and the air. MIP is an invasive method and the sample cannot be used for other purposes after measurement, but it is a unique method for determining throat diameter ranging over five orders of magnitude from few nanometers to hundreds of micrometers.

In addition to the value of the porosity, the geometry of the pore network (pores and throats) controls the flow of single or multi-phase fluids. There are several (complementary) approaches for measuring the average geometrical properties of the pore space. The most studied ones are certainly the tortuosity factor and the specific surface.

Tortuosity is a key parameter for characterizing macroscopically flow and transport properties in rocks. There are many models for tortuosity estimation based on the underlying physical process involved. The geometrical tortuosity τ_G denotes the sample-averaged ratio of the length of the mean trajectory between two points to the lineal distance between these two points. The mean trajectory is the curve that lies at equal distance to the pore walls. Tortuosity is clearly related to the degree of connection of the pore network such as the permeability is and thus a lot of research work has been devoted to relate permeability to porosity and tortuosity (see Rashid et al. 2015 for a critical review of these models). Albeit the definition of τ_G is simple, measuring the geometrical tortuosity of porous media requires defining first the geometrical skeleton of the pore space and consequently requires 3D imaging of rock samples.

Other definitions of the tortuosity can be useful. For instance electrical tortuosity (Sect. 6.2.1) and diffusional tortuosity (Sect. 6.3.1) denote the factor needed to relate the effective diffusion (of electrical potential or a solute respectively) to the diffusion that would be measured for a porosity of 1, i.e. in the fluid phase without rock. The widely used terming of electrical and diffusional tortuosity may be confusing; generally speaking they are geometrical factors which denote the average tortuosity of the displacement of electrons or tracers respectively. In simple structures such as bundles or networks of tubes displaying smooth changes in diameter, the electrical and diffusional tortuosity factors denotes the sample-scale geometrical tortuosity of diffusion paths. However, in more complex structures such as in carbonate reservoirs or even more in tight rocks such as forming the reservoir caprocks (tight carbonates, marl and claystones) the diffusional (or electrical) tortuosity factors encompasses constrictivity effects that are usually impossible to measure independently.

6.2.1.2 Electrical Tortuosity

Electrical conductivity σ_e of a water-saturated rock of known porosity ϕ is widely used to infer tortuosity using the generalized Archie's law:

$$\sigma_o/\sigma_e = \phi^{-m} \tag{6.2.2}$$

where σ_o denotes the conductivity of the water or the brine filling the connected porosity of the rock and m is a constant that depends on the geometry of the pores.

The ratio σ_o/σ_e is called the formation factor (F) because it typifies a given reservoir formation. Values of F are generally lower than about 100 for reservoir rocks while the formation factor values of caprocks are very high and denote the complexity and the high tortuosity of the claystone, tight carbonate and marl. Rashid et al. (2015) reported values ranging from 23 to 2565 with an average value of 313.

If the rock contains a low fraction (i.e. less than around 15 %) of nonconductive phases such as oil or CO_2, the porosity (ϕ) in the Archie's law denotes the fraction of the pore saturated with the brine. Thus the electrical conductivity is a method for measuring residual oil or gas saturation but requires knowing the value of m measured at brine saturation. The exponent m, called the cementation exponent, increases as the connectedness of the pore network decreases. For reservoir rocks, the value of m ranges from 1.5 to 2.5 for sandstones but can reach values as large as 5 for carbonates (Glover 2009). While ϕ and m are characteristic properties of the rock explicitly given by Eq. 6.2, electrical conductivity measurements are habitually performed to extract the electrical tortuosity τ factor. Various models have been proposed in the literature to link the electrical tortuosity with the formation factor, e.g. $F = \tau^n/\phi$, (Clennell 1997), but the usual definition of the electrical tortuosity assumes $n = 1$ and therefore τ is explicitly obtained from the electrical conductivity measurement of the water saturated sample. Accordingly, the tortuosity τ is ubiquitously related to the porosity: $\tau = \phi^{(1-m)}$.

Electrical conductivity measurements are quite easy to implement and there is a huge amount of literature discussing the technical and methodological aspects of this technique. The fundamental step is the measure of the sample conductivity using the Ohm's law, $\sigma_e = I/(S\Delta V)(L/S)$ by imposing a difference of potential ΔV between the two edges of a sample of section S and length L and measuring the current density I. Yet, there are two main limitations. First, the presence of significant fractions of clay mineral triggers non-negligible surface conductivity (σ_s) mechanisms at the pore-rock interface. In this case Archie's law must be modified, $\sigma_e = \sigma_s + \sigma_0/F$ and it is necessary to perform the measurements at different values of σ_0 for evaluating F. The second difficulty is the possible alteration of the rock sample by the brine used for the measurements specifically in carbonates and clay-rich sandstones. In any case it is recommended to use solutions that are in thermodynamic equilibrium with the rock-forming mineral to minimize fluid-rock reactions.

It is worth noticing that there are several different types of notation for the relation between the formation factor and the porosity. For instance it is often proposed the following relationship $F = \alpha/\phi^m$, where m is the cementation factor and α is named cementation intercept, lithology factor or sometimes improperly designated as the tortuosity factor. But this approach is not valid because this would indicate that $F = \alpha$ for $\phi = 1$ or in other words that $\sigma_0 = \alpha\sigma_e$ whereas by definition $\sigma_0 = \sigma_e$ in the absence of solid.

6.2.1.3 Specific Surface

The pore-mineral interface area is a critical property of reservoir rock for modeling chemical reactions such as dissolution reactions triggered by the acidification of the brine during CO_2 storage. Knowing the geometry of the pore surface is also useful for modeling relative permeability, capillary pressure and wettability mechanisms (see Sect. 6.2.3). The gas sorption method is often used for evaluating the area of the surface of the pores connected to the sample boundaries. A given mass of inert gas is injected in a sample and the amount of gas adsorbed is measured by volumetric, mass-flux or the most often pressure methods at one or several temperatures (isotherms). The raw data are processed following the BET method (Brunauer et al. 1938) in order to extract the specific surface area, i.e. surface area per unit mass. The interface between the pore and the solid phases can be also evaluated by processing microscopic images of thin sections or X-ray 3D microtomographic images (see Sect. 6.2.1). In both cases the calculated surface area is controlled by the optical resolution of the imaging technique and a fraction of the surface roughness is not resolved while BET method probes the solid surface area with a nanometer scale resolution.

6.2.1.4 Properties Extracted from X-Ray Micro-tomography Techniques

Since a decade X-ray micro-tomography (XRMT) has become a highly regarded mean for assessing pore structure geometry. Form the 3D X-ray absorption images the first step is to produce binary images where the solid and the void phases are identified. From these binary images many parameters, such as the connected and unconnected porosity, the surface to volume ratio and the geometrical tortuosity of the connected porosity can be extracted easily (Nikolaidis and Pitas 2001). Other morphological descriptors of the pore structure such as proposed in Torquato (2002) can be implemented for measuring the distribution of some characteristic of the pore size. For instance Lu and Torquato (1993) discussed the calculation and the use the chord-length distribution and the free-path distribution functions and of their momentum for macroscopically characterizing the microstructures in relation with transport problems.

XRMT images are often used to extract the porosity skeleton (Fig. 6.1). The skeleton denotes the one-dimensional reduction, centered inside the porosity body, of the connected porosity that keeps some relevant geometrical and topological features of the pore properties (Siddiqi and Pizer 2008; Lee et al. 1994). It is a powerful tool for analyzing large microtomography images of porous media (e.g. Lindquist and Venkatarangan 1999). The skeleton can be defined in tip voxels having exactly one neighbor, regular voxels having exactly two neighbors and node voxels having more than two neighbors. Tip, regular and node voxels are usually associated with porosity dead-ends, throats and pores respectively. More generally the fraction of each of these types of skeleton voxels is an interesting indicator of

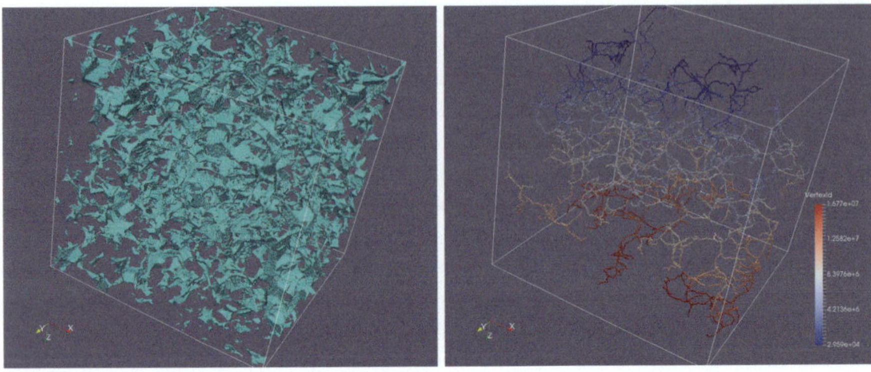

Fig. 6.1 *Left* Connected porosity of a 2563 voxel sample of Fontainebleau sandstone (voxel side size 5.06 μm) and *right* the corresponding skeleton

the overall complexity of the pore space structure. For instance, Luquot et al. (2014) used this method for identifying and characterizing the change of the porosity geometry triggered by the injection of CO_2-rich brine into reservoir rock samples. While the skeleton is obtained the radius of the largest inscribed spheres centered at each voxel forming the skeleton can be computed using for instance the Euclidean distance transform algorithm introduced by Meijster et al. (2002). This enables to derive the pore size distribution in the sample and detect the preferential flow paths.

More evolved data processing can be applied to evaluate the pore size distribution and the connectivity as well as the thermal and chemical diffusion coefficient and the diffusional tortuosity by solving the Laplace equation using for example random walk methods (Dentz et al. 2012; Sen 2004). More computation-demanding simulations, using for instance Finite Volume or Lattice Boltzmann methods (e.g. Gharbi 2014; Pan et al. 2001; Succi 2001) can be performed using the XRMT binarized images in order to calculate the permeability by solving the Navier-Stokes equation (e.g. Gharbi 2014), the hydrodynamic dispersion (e.g. Gjetvaj et al. 2015) and the elastic mechanical properties (e.g. Wojtacki et al. 2015). However, the investigated volume is inversely proportional to the complexity of the mathematical problem to be solved due to computation limitations. For example the determination of the diffusional tortuosity can be typically performed on images containing many thousands of pores, while permeability calculation is limited to sub-volumes of few hundreds of pores and mechanical properties to sub-volumes of few tens of pores. For comparison, electrical conductivity measurements, permeability measurements and mechanical tests performed in the laboratory on typical 1–2 inch diameter cores sample give properties averaged over 10^5–10^7 pores.

Moreover it is worth noticing that the representativeness of the results is highly dependent on the imaging quality and resolution and even more importantly on the accuracy of the image binarization or segmentation (e.g. Schlüter et al. 2014). Nevertheless the possibilities offered by both the imaging equipment and the numerical processing of data are continuously improved and the so called *numerical core analysis* is now widely used, specifically because it is a non-invasive technique. Because of this remarkable advantage, XRMT can be used to monitor the changes of these properties with time in the course of dynamic experiments involving the displacement of immiscible phases or chemical rock alteration. For instance XRMT imaging has been successfully used to measure the phase saturation during imbibition and drainage experiments performed for evaluating relative permeability (Silin et al. 2011; Perrin and Benson 2010). The direct imaging of the spatial distribution of the phases (ex: CO_2 and brine) during such experiment is a powerful tool for understanding, for instance, hysteresis mechanisms or to tackle experimental bias such as induced by the finite size of the sample when performing imbibition and drainage experiments. Similarly, XRMT has been used intensively for measuring the change of porosity and other hydrodynamic properties caused by the dissolution and the precipitation of minerals linked to by the chemical disequilibrium triggered by the CO_2 injection (see Sect. 6.4).

6.2.2 Hydrodynamical Properties for Single Fluid Flow

6.2.2.1 Permeability

Intrinsic (or absolute) permeability is a fundamental property which is needed for parametrizing reservoir models. The widely use technique for determining permeability from core samples is steady state flow tests using aqueous liquids or gas. When using an incompressible liquid such as water, the fluid is injected using a constant flow rate Q [$m^3\ s^{-1}$] through the sample and the pressure drop ΔP [Pa] between the inlet and the outlet of the sample is measured. Permeability k [m^2] is evaluated from applying Darcy's law:

$$k = (4\mu QL)/(\pi d^2 \Delta P) \tag{6.2.3}$$

where L and d denote the length and the diameter of the core sample and is μ the viscosity [Pa s]. Note that μ depends on the pressure, temperature and the composition of the fluid. Water permeability is generally an accurate method, specifically when performing the measurement for different flow rate. Doing so allows verifying that the applied flow rates are in the range where Darcy's law is valid by checking the linear relation between k and ΔP. However the use of water can trigger chemical reactions with the rock-forming mineral and consequently alter the estimated value of k (Al-Bulushi et al. 2012) and makes the measurement a destructive one. Therefore it is recommended to use water previously thermodynamically equilibrated with the rock-forming minerals. Some permeameters allows the application of axial and radial confining pressures to reproduce the mechanical constrains at depth.

Conversely, constant pressure injection while measuring the flow rate, for instance using a mass flow meter, is performed when using gas (usually nitrogen or helium). The advantages of gas permeametry are that gas does not alter the rock and the measurements are easy and fast. Nevertheless, the relation between the measured values of Q and k cannot be modeled by Darcy's law but require models that take into account the low density and the compressibility of the gas. The standard approach for processing the measured values of ΔP and Q in order to evaluate k involves the following steps (Jones and Roszelle 1978):

- Measure Q for four or five different values of the inlet and outlet gas pressures (P_{in} and P_{out});
- Evaluate the so called *gas permeability* k_G using for instance the Scheidegger's law (Wu et al. 1998; Scheidegger 1974): $k_G = (8\mu QLP_{out})/(\pi d^2(P_{in}^2 - P_{out}^2))$. The value of k_G depends on the pressure and to lesser extent to the type of gas used;
- Calculate the mean gas pressure P_{av} in the core for each measurement and plot the gas permeability against the inverse of the mean gas pressure in order to evaluate the coefficient b of the relation $k_G = k(1 + (b/P_{av}))$ that is known as the Klinkenberg's correction for the pressure-dependent slip effect of the gas at the pore-solid interface (Klinkenberg 1941). The value of the intrinsic permeability is obtained by evaluating the value of k_G at the origin: $k = k_G$ when $(1/P_{av}) \rightarrow 0$, i.e. extrapolating the measurement for an infinite pressure for which gases behave as a liquid.

The accuracy of the method depends strongly on the effective linearity of the measured values of the couple k_G and $1/P_{av}$. Often the values are very badly aligned which indicates that the law (Scheidegger's law) used to evaluate k_G is not appropriate or highlight large inaccuracy of the measurements (Bloomfield and Williams 1995). This technique can be used for rocks with permeability ranging from about 0.1 mD to 10 D.

Similar values of permeability can be measured using the so called probe-permeameter or micro-permeameter. This technique consists in measuring the pressure decay while forcing a flow of gas through a small injection tip pressed and sealed against the surface of a rock sample. By construction the method is applied to unconfined samples and the measured value may different from the value at the reservoir pressure specifically for claystones. Yet, this is a fast and simple method that can be used directly on the field to evaluate the permeability of cores (Halvorsen and Hurst 1990).

Measuring the low values of permeability of the tight rocks (i.e. permeability smaller than 0.1 mD down to about 10^{-5} mD) forming the reservoir caprocks is not a routine task. Such measurement cannot be performed using steady-state gas flow techniques because the duration required to reach steady state is too long. Non-steady-state gas flow techniques applied to cores confined at reservoir pressure can be used. The method consists in monitoring at both edge of the sample, the pressure change caused by a pressure pulse, a pressure increment or an oscillating pressure perturbation applied at on edge of the sample using a piston pump. The permeability as well as the storativity is calculated using models that require a precise knowledge of the sample size and porosity (Metwally and Sondergeld 2011; Fischer 1992). For extremely low permeability rocks, i.e. ranging from 10^{-5} to 10^{-6} mD, transient gas flow techniques can be applied to crushed-rock and consists in monitoring the pressure decay following a pressure pulse into a closed container containing crushed-rock particles of size less than 1 mm (Cui et al. 2009).

6.2.2.2 Diffusivity

Diffusion is a ubiquitous mechanism. Mass transfer by diffusion is the dominant process in low permeability materials such as claystones and cements, but is also essential in reservoir where advection is very low (i.e. for low value of the Peclet number, see Sect. 6.4.1), for instance far from the injection well.

The effective diffusion coefficient or diffusivity coefficient is routinely measured at laboratory scale using specific diffusion cell, such as displayed in Fig. 6.2a, where the rock sample is sandwiched between two reservoirs, one of them containing the solute tracer while the other one contains tracer-free water (Tinker 1969). As diffusion is a slow process, the thickness of sample is usually small (few centimeters) in order to make the measurement tractable. Even though, diffusion measurement in tight rocks requires monitoring the tracer concentration for several tens of days. Whereas the principle is simple its set up is not straightforward specifically for permeable media (reservoir rocks) for which avoiding advection due to pressure gradient or buoyancy effects is challenging. Measuring the effective diffusion in claystone or other tight rocks displaying very low pore size porosity requires some great care as well. For instance, tracer molecule or ion of different sizes can probe different portions of the porosity and ion exclusion phenomenon

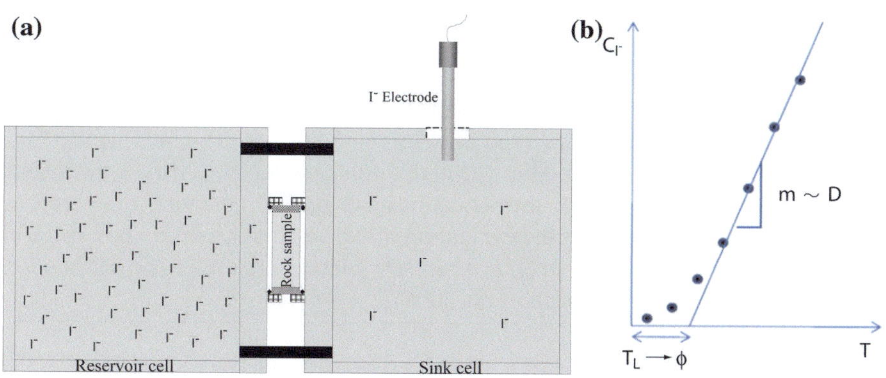

Fig. 6.2 **a** Diffusion cell apparatus for steady-state diffusion measurement. **b** Concentration versus time in the outlet reservoir displaying the linear increase of concentration occurring when the mass flux of tracer is steady state

can occurs. For example, such a phenomenon prevents chlorine and iodine anion from having access to the same porosity as a water tracer like tritiated water (HTO) or deuterium oxide (HDO). Furthermore, the intensity of the exclusion mechanism usually depends on the ionic strength of the water saturating the pores (Van Loon et al. 2003).

The simple way for measuring the breakthrough curve in the outlet reservoir is to install electrodes that measure cation or anion concentration. Nevertheless, the use of specific tracers (e.g. organic or radioactive) or mixture of tracer may exclude the use of an in situ monitoring probe and therefore implies sampling the outlet reservoir fluid. The calculation of the effective diffusion coefficient D_e is based on Fick's second law assuming one-dimensional transport geometry:

$$\phi \frac{\partial C}{\partial t} = D_e \frac{\partial^2 C}{\partial x^2} \tag{6.2.4}$$

where C is the tracer concentration [mol m^{-3}], t the time [s] and D_e the effective diffusion coefficient [m^2 s^{-1}]. Usually the measurements are performed with the initial condition $C(0 \leq x \leq L, t = 0) = 0$, where L denotes the sample of length.

Using reservoirs of large volume allows assuming that there is a finite interval of time during which the concentration boundary can be considered as constant $C(x = 0, t > 0) = C_i$ with C_i the initial tracer concentration in the inlet reservoir and $C(x = L, t > 0) = 0$ if there is no tracer in the outlet reservoir initially, while the flux of tracer Φ is constant $\Phi = -D_e \partial C/\partial x = D_e C_i/L$ (Crank 1975; van Brakel and Heertjes 1974). For this period of time, the concentration of tracer in the outlet reservoir, C_o, increases linearly with time $C_o(t) = D_e C_i S t / V_o L$ (Fig. 6.2b), where V_o denotes the outlet reservoir volume and S the surface of the sample in contact with the reservoir. Accordingly, the effective diffusion coefficient is given by:

$$D_e = \frac{V_o L}{C_i S} \frac{\partial C_o(t)}{\partial t} \tag{6.2.5}$$

Analytical solutions of the complete transient diffusion problem (Eq. 6.2.4), including the change of concentration in the reservoir and the change in the reservoir volume if samplings are performed, can be obtained in the Laplace domain and then numerically inversed in order to be fitted to the experimental data. In this case both the porosity and the effective diffusion coefficient can be evaluated. Also it is worth noticing that the formal equivalence between the Fick's law and the Ohm's law (Sect. 6.2.1) allows the evaluation of the value of D_e from knowledge of the electrical tortuosity: $D_e = \phi D_0/\tau$, where D_0 denotes the molecular diffusion of the tracer in water (e.g. Garrouch et al. 2001).

6.2.2.3 Dispersion

Evaluating the coefficient of dispersion of the Fickian advection-dispersion equation (see Chap. 3) is not straight forward because this coefficient is strongly dependent on the measurement scale. Considering samples of characteristic size and time the tracer need to cross a rock sample of few centimeters long for usual flow rates (<1 md^{-1}), the asymptotic dispersion is usually not reached and therefore is not possible to determine the Fickian dispersion coefficient (Gjetvaj et al. 2015; Cortis and Berkowitz 2004; Levy and Berkowitz 2003; Berkowitz et al. 2000). Still, dispersion mechanisms can be studied by performing tracer tests that are similar to those used for diffusivity (Sect. 6.2.2) but applying a constant flow rate at one edge of the sample. While the steady state flow in sample is reached, a pulse of tracer is injected as close as possible to the sample inlet. The main technical issues are (i) the injection of a pulse of tracer of very short duration while minimizing the flow perturbation, (ii) avoiding spurious dispersion before the tracer enter the sample and (iii) assuring a homogeneous concentration at the sample inlet surface. Thus the volume of the traced fluid to be injected must be small and the concentration of the tracer sufficiently high to allows the measurement of the effluent tracer concentration over, at least, 3 orders of magnitude in concentration, but sufficiently low to avoid density effects (Tenchine and Gouze 2005). Furthermore, the tracer must not interact with the rock-forming minerals, i.e. avoiding sorption and chemical reactions such as dissolution and precipitation. For these reasons fluorescent dies are often used because their low reactivity and because optical techniques using high sensitivity sensors allows measuring a large range of tracer concentration values down to values as low as 10^{-10} mol of tracer per mole of water.

There are two methods to perform tracer tests. The standard method, named flow-through tracer test, consists in injecting the tracer at one side of the core and measure the breakthrough curve at the opposite side of the core. However, it can be useful to perform push-pull experiments in order to better characterize the long-time behavior of the dispersion mechanisms and compare the reversible part of the dispersion due to the hydrodynamic spreading of the tracer to the irreversible part of the tracer dispersion triggered by diffusion (Gouze et al. 2008).

The equipment displayed in Fig. 6.3 is designed to allow both flow-through and push-pull experiments without any change in the circuit when changing experiment type. Water is pumped into the circuit with two piston pumps, passes through tracer injection point, sample core and out of the system. Arrows indicate the direction of flow for different experiment stages. Blue arrows indicate flow direction in the branches of the circuit where flow direction is always the same, green arrows indicate flow direction during a flow-through experiment and during the pull phase of push-pull experiment, while red arrows indicate flow direction during push phase of push-pull experiment. The TELOG sensor (ICARE Lab. Montpellier) is a high resolution optical for measuring fluorescent dye implementing pulsed light sources and a high sensitivity photomultiplier for measuring tracer concentration from 10^{-6} to 10^{-10} mol of tracer per mole of water.

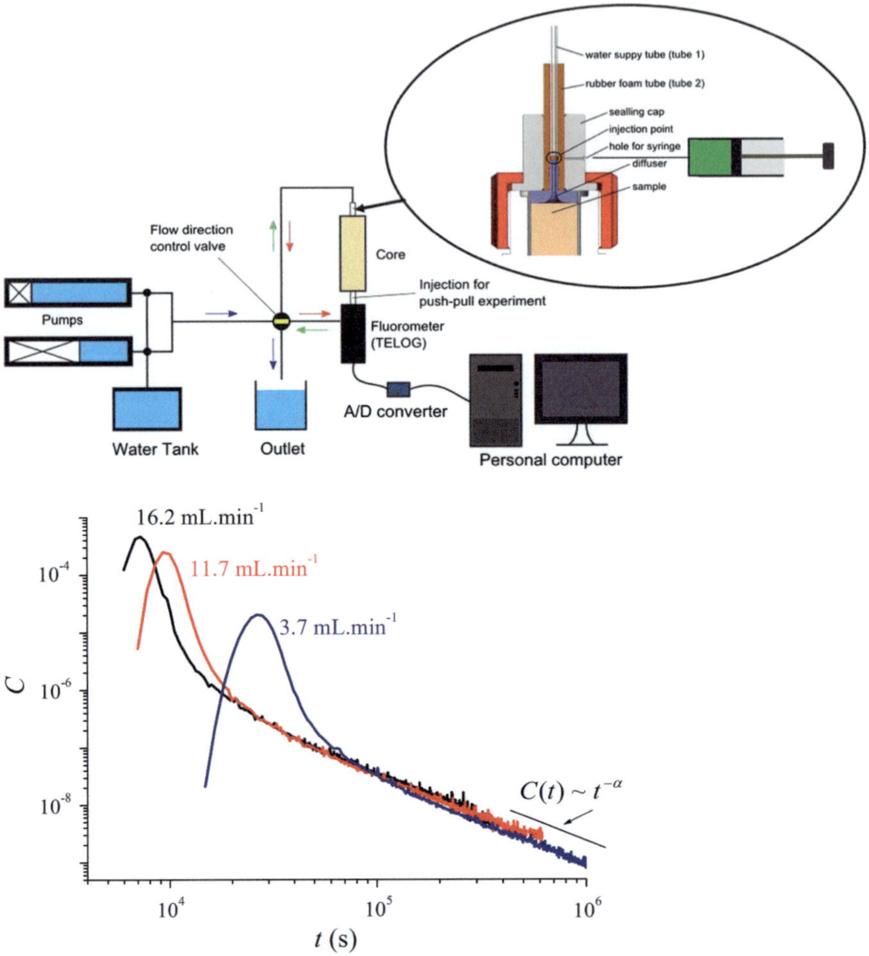

Fig. 6.3 *Top* Schematic diagram of the tracer test equipment layout with arrows indicating flow direction. *Blue arrows*-permanent flow direction; *green arrows*-flow direction for flow through experiments and the-pull phase of push-pull experiments; *red arrows*-flow direction during the push phase of push-pull experiments. *Bottom* Example of breakthrough curves (concentration versus time) measured during a passive tracer dispersion experiments in carbonate cores of diameter 90 mm and length 557 mm for different flow rates. Details of the experiments are given in Gouze et al. (2009)

For flow-through experiment configuration the water arrives at the top of the sample. The given volume of tracer is injected with a micro syringe to precisely dose the tracer volume and control the duration of the injection. Then, the solute passes through the sample and the outlet concentration is measured by the TELOG sensor. For the *push* phase of a push-pull experiment, flow direction valve is placed in position to direct flow at the bottom of cell. There is another injection point just between the TELOG and sample. Tracer is push for a period of time depending on flow rate, permeability and length of sample in order to investigate as much as possible of the porous media without any tracer leaving from the sample at the other side. At the end of push period the control valve is switched and the water starts to flow from the top of the sample like during flow through experiment. In order to emphasize dispersion due to diffusion a resting time (no flow) between the push phase and the pull phase can be done.

Usually, the measured breakthrough curves (BTC), i.e. the tracer concentration in the effluent versus time, display strongly asymmetric shapes with long tails (i.e., for times larger than the time corresponding to the advective peak arrival t_a) that appears to decrease more or less as a power law of time $C(t \gg t_a) \sim t^{-\beta}$, and indicates an apparently infinite variance. These are characteristics of non Fickian dispersion and the interpretation of these data must be done in the framework of non-Fickian models. Non-Fickian dispersion properties, their origin and their relation to the geological heterogeneity are the source of debates. Authors have explored different approaches for better modeling pre-asymptotic processes that result from (long-range) spatial correlation of geological structures and consequently of fluid velocities. Often these models leads to non-local in time transport equations that are solved using random walk approaches such as Continuous Time Random Walk (Berkowitz et al. 2000) or Time Domain Random Walk (Russian et al. 2016). The results of dispersion experiments are useful for parametrizing the models presented in Chap. 5, such as based on the Multirate Mass Transfer, Continuous Time Random Walk or Fractional Advection-Dispersion Equations approaches.

6.2.3 Measuring Capillary Pressure and Relative Permeability for $scCO_2$-Brine Systems

The objective of this section is to present the methods for measuring the relative permeability versus saturation data that are required for applying the extended or generalized Darcy law to biphasic flow (see Chap. 3 and Sect. 5.5). Two and three non-miscible phases flow in porous rocks involves complex mechanisms (e.g. Sahimi 2011) and thus measuring relative permeability requires demanding laboratory and data processing works.

In the following we will assume that the porosity of rock is saturated with a mixture of two immiscible fluids: CO_2 and brine. The relative permeability $k_i^{(r)}(S_i)$ is a dimensionless measure of the effective permeability of phase i ($i = C$ for CO_2 or b for brine) for a saturation (S_i) and is defined as $k_i^{(r)}(S_i) = k_i^{(e)}(S_i)/k^*$ where $k_i^{(e)}(S_i)$ denotes the effective permeability of phase i and k^* the reference permeability (in m^2). The choice of reference permeability is not critical as soon as it is consistent with the model in which the values of $k_i^{(r)}$ will be implemented, but often the intrinsic permeability is used or alternatively the CO_2 permeability measured at irreducible brine saturation. Similarly, relative permeability of CO_2 and brine are usually expressed as the function of the brine saturation $k_i^{(r)}(S_b)$.

The (relative) displacement of CO_2 or brine is controlled by a range of physical and chemical factors that yield viscous forces and capillary forces. Their respective control on the hydrodynamics of the fluid pair is scaled by the capillary number $Ca = \mu v/\sigma$, where μ, v and σ denote the average dynamic viscosity [Pa s], the average fluid velocity [m s^{-1}] and the surface or interfacial tension between the two fluid phases in [N m^{-1} or Pa m]. Usually, capillary forces dominate in natural reservoir flow (Hilfer and Øren 1996; $Ca < 10^{-8}$). However, in the vicinity of the injection well the value of Ca increases as v increases.

6.2.3.1 Wettability and Capillary Pressure

Capillary forces arise from the interactions between the fluid pair and the rock-forming minerals in contact with the fluids. When two fluid phases are present, the wettability denotes the contact angle θ_{fs} of the fluid f with the solid phase s. Wettability results from surface tension due to attractive forces between the fluid molecules and the solid surface while cohesive forces takes place within each of the fluid phase and create an interfacial tension σ. If $\theta_{fs} < 90°$ for a given fluid phase and a given solid phase, then this fluid preferentially covers the solid phase and it is named the wetting fluid. Like for interfacial tension, wettability depends on pressure and temperature. Under typical reservoir conditions most of the rock-forming minerals such as quartz, feldspar, calcite, dolomite and clay minerals are brine wet and the CO_2 as a gas or a supercritical phase is the non-wetting fluid. Examples of CO_2 wettability values for reservoir and seal rocks as well as discussions of the potential wettability alteration due to the flushing of rocks by CO_2 are given in Iglauer et al. (2015), Chiquet et al. (2007), Zhu et al. (2011) and Wang et al. (2012).

When one of the fluid pair displaces the other one, capillary forces triggers differential pressure between them that is named the capillary pressure P_c (in Pa). P_c is defined as the pressure difference between the non-wetting phase and the wetting phase. At equilibrium, i.e. without any external forces such as an advective fluid flow, the Young–Laplace equation gives the capillary pressure as the function of the interfacial tension σ, the wetting angle θ of the brine on the surface of the capillary and the curvature radii of the respective fluids. In a capillary of the radius r the Young–Laplace equation is written $P_c = (2\sigma \mathrm{Cos}\theta_{fs})/r$. These parameters are defined at pore scale, but macroscopic equivalent values are needed for describing multiphase flow in porous media. At macroscopic scale P_c is primarily controlled by the extension of the interface area between the fluid pair and the solid and thus depends on both the fraction of each of the phase saturating the pore space and the porosity topology. Evidently pores in rocks are not capillary-shaped and pore and throat diameters are (highly) variable. From the seminal works of Leverett (1941) and Bear (1972) there have been many different derivations of the capillary pressure in porous media using geometrical and, or, thermodynamic considerations (see for example Hassanizadeh and Gray 1993). These model are useful for instance to evaluate the pore size distribution from capillary pressure measurements such as mentioned in Sect. 6.2.1.

Sample scale capillary pressure curves are typically measured by performing drainage experiments which consists in forcing the displacement of a wetting fluid by a non-wetting fluid, using centrifuge methods or mercury (Hg) injection porosimetry (MIP). For the later, the pressure needed to inject Hg in the vacuumed sample gives the Hg/air capillary pressure $P_c^{(Hg/a)}$ versus the (cumulative) volume of Hg injected at room temperature. These data can be extrapolated to CO_2/brine capillary pressure $P_c^{(C/b)}$ at relevant temperature (and pressure) using the relation given by Washburn (1921):

$$\frac{P_c^{(C/b)}}{P_c^{(Hg/a)}} = \frac{\sigma^{(C/b)} \cos \theta^{(C/b)}}{\sigma^{(Hg/a)} \cos \theta^{(Hg/a)}} \tag{6.2.6}$$

The values of the interfacial tension can be found in the literature for the given fluid, pressure and temperature ($\sigma^{(C/b)} \cong 3 \times 10^{-2}$ N m^{-1} in reservoir conditions and $\sigma^{(Hg/a)} \cong 0.5$ N m^{-1} at room temperature; e.g. Georgiadis et al. 2010), but values of the contact angle for the system CO_2/brine/rock under consideration are often not known and usually it is assumed that $\theta^{(C/b)} = \theta^{(Hg/a)}$ (Hingerl et al. 2016; Krevor et al. 2012; Pentland et al. 2011). The capillary pressure$P_c^{(C/b)}$ versus brine saturation S_b data are commonly fitted using the Brooks-Corey model (Brooks and Corey 1964):

$$P_c^{(C/b)} = P_{\min}^{(C/b)} \left(\frac{1 - S_b^{(0)}}{S_b - S_b^{(0)}} \right)^{1/A} \tag{6.2.7}$$

where $P_{\min}^{(C/b)}$, S_b and $S_b^{(0)}$ denote the minimum pressure needed for the CO_2 to penetrate the sample, the actual brine saturation and the residual brine saturation, respectively. $S_b^{(0)}$ is also often termed critical brine saturation or irreducible brine saturation, and denotes the maximum brine saturation at which the brine remains immobile while the sample is flushed with CO_2. In Eq. 6.2.7 A is a fitting (positive) parameter named the pore size distribution index. $P_{\min}^{(C/b)}$ can be regarded as a fitting parameter or can be evaluated from drainage experiment such as described below. Note that Eq. 6.2.7 denotes a simple monotonic decreasing function $P_c^{(C/b)}(S_b)$ and consequently it is sometimes impossible to fit experimental data with the Brooks-Corey equation (e.g. see an example in Krevor et al. 2012).

Drainage experiments using brine and CO_2, as well as imbibition experiments (flushing CO_2 with brine) can be performed in order to measure the $P_c^{(C/b)}$ versus

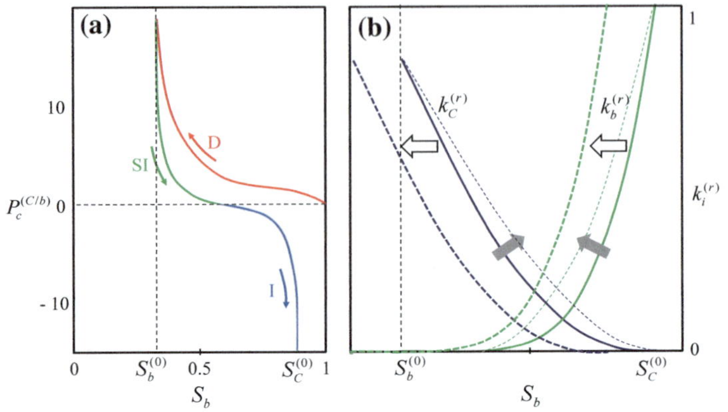

Fig. 6.4 **a** Schematic representation of capillary pressure versus brine saturation. Label D, SI and I denotes the forced drainage, spontaneous imbibition and forced imbibition stage respectively. **b** Schematic representation of relative permeability curves for CO_2 and brine versus brine saturation for a given value of the capillary number and wettability. *Plain arrows* indicate the expected change of the shape of relative permeability curves when the value of the capillary number increases. Lower wettability will tend to shift the curves toward lower values of the brine saturation

S_b. These experiments are also used to measure relative permeability $k_i^{(r)}$ versus S_i ($i = CO_2$ or brine) as it will be explained in the next section (Sect. 6.2.3). Figure 6.4a displays a typical capillary pressure curve $P_c^{(C/b)}$ versus S_b. Starting from fully brine saturated rock sample, the primary process (noted D) is the forced drainage of the brine by the CO_2 which ends when only residual brine saturation $S_b^{(0)}$ is reached and gives the maximum capillary pressure. The second stage (noted SI) displays the spontaneous imbibition where the brine invades the rock due to capillary forces until $P_c^{(C/b)} = 0$, then the third process (noted I) is the forced imbibition for which the brine displaces the CO_2. Eventually the brine saturation reaches an asymptotic value corresponding to the residual CO_2 saturation $S_C^{(0)}$.

6.2.3.2 Experimental Evaluation of the CO_2 and Brine Relative Permeability

Several methods have been proposed for the calculation of relative permeability from capillary pressure data, stating from the seminal works of Purcell (1949), then discussed for instance in Li and Horne (2002). Most of these models relate directly on the evaluation of the pore size distribution derived that can be derived from MIP (Brooks and Corey 1966) or centrifuge experiments (Pinter and Bodi 2012). The calculation of relative permeability from capillary pressure versus saturation data have been largely used to unsaturated flow of water in soil applications using models relating capillary pressure to saturation such as proposed by Brooks and Corey (1964), Durner (1994), Kosugi (1996), Seki (2007) and van Genuchten (1980). Applications to the flow of oil and brine in reservoir rocks are discussed in Honarpour et al. (1986) and a review of these methods can be found in Li and Horne (2006). The *op. cit.* authors concluded that these methods give acceptable results for drainage but are less accurate for imbibition.

Specific experiments using the brine and CO_2 at pressure and temperature of the targeted reservoir are often preferred to extrapolations from mercury injection or centrifuge methods using brine and air at room conditions. The basic procedure for measuring relative permeability is to perform imbibition (displacing CO_2 with brine) and drainage (displacing brine with CO_2) experiments or injecting mixture of CO_2 and brine into a rock sample. The effective permeability must be measured over the largest-as-possible range of fluid saturations in order to construct the relative permeability curves. Fluid displacement experiments, i.e. drainage or imbibition performed at different intrusion pressure, are named transient methods (Berg et al. 2013; Johnson et al. 1959) while simultaneous injection of the two phases at given fractions are named steady state flow methods (Virnovsky et al. 1995; Dake 1978).

As already mentioned, the transient flow of two immiscible fluids depends on the sample-scale capillary pressure (and consequently on the geometry of the pore network), on the wettability of the fluids for this given rock sample, and on the interfacial tension between the CO_2 and the brine. These complex mechanisms together with the thermodynamic properties of the fluids hinder the measurement procedure. Hereafter are listed some major issues related to the experimental appraisal of relative permeability. Wettability depends on the brine composition (e.g. some solute components can act as surfactants) and furthermore both wettability and interfacial tension depend on temperature (e.g. Hamouda et al. 2008). Thus, it is recommended to perform the measurements at temperature close to that of the reservoir using brine of composition corresponding to the thermodynamically equilibrated reservoir brine. For CO_2 underground storage applications, the CO_2 is in supercritical state and furthermore the brine and the CO_2 must be thermodynamically equilibrated under experimental pressure and temperature conditions prior to be pumped into the sample in order to avoid drying of the residual brine saturation, specifically for drainage experiments. This implies that a fraction of CO_2 dissolves in the brine and creates an acidic fluid that can promote the dissolution of minerals such as carbonates (see Sect. 6.4.4). Also, relative permeability experiments usually start from cleaned cores and a specific attention must be played to the cleaning protocol which may alter the wettability of the surface of some minerals. Furthermore, capillary effects occurring at the sample edges where the fluid(s) are injected and produced can hamper the measurement if not treated properly using adapted experimental techniques and data processing (Huang and Honarpour 1998). These mechanisms are named capillary end-effects and can be the source of high uncertainty in the results if not recognized and corrected appropriately. Finally it is important to mention that hysteresis mechanisms triggered by the existence of residual phase fraction and multiple metastable configurations of the (pore-scale) distribution of the phases depending on the saturation history (Cueto-Felgueroso and Juanes 2016; Moebius and Or 2012; Morrow 1970; Haines 1930) may require performing both drainage and imbibition measurements at different flow rates (e.g. Pini and Benson 2013). Finally, it is recommended to apply a differential pressure much smaller than the average operating pressure in order to minimize errors on the volume calculations of the CO_2 which is a compressible fluid at the pressure and temperature of the reservoirs. All together these different issues make the measurement of relative permeability challenging and induce the use sophisticated equipment, experimental protocols and data processing. Therefore, the procedures are noticeably variable from one research center to another with specific technical and theoretical adaptations to improve the measurement accuracy for the different types of rocks and reservoir conditions. For example the multiple-core method known also as the Penn State method, consists in sandwiching the rock sample with two other rock samples to reduce capillary end-effects at the inlet and outlet of the central sample during steady state experiments where two fluid phases are pumped through the sample at constant flow rates. In any case, relative permeability experiments are time-consuming, e.g. from several hours for high permeability rocks to weeks for the low permeability rocks.

The unsteady state method inherits from capillary pressure measurements and is the most used method for measuring relative permeability because it is usually noticeably faster than the steady state method. For CO_2 storage applications the optimal experimental protocol includes a drainage stage followed by an imbibition stage. The core sample is initially saturated with brine under vacuum, then brine is injected at different pressure in order to measure the intrinsic permeability k using Darcy law. Subsequently, CO_2 is pumped at constant flow rate or at constant pressure while the pressure drop across the sample or the total flow rate as well as the fraction of CO_2 and brine produced are monitored. Note that applying constant pressure minimizes the source of error due to the compressibility of the CO_2 phase. CO_2 is injected until the fraction of brine produced at the sample outlet is zero and allows evaluating $S_b^{(0)}$ the residual brine saturation and $k_C^e(S_b^{(0)})$ the effective permeability of CO_2 at the residual brine saturation. Then a similar procedure is applied but with the injection of the brine. When the fraction of CO_2 produced at the outlet reach zero one obtains the value of the permeability of the brine at the residual CO_2 saturation $k_C^{(e)}(S_b^{(0)})$. The residual saturation of CO_2, $S_b^{(0)}$ is an important parameter because pore scale residual trapping of CO_2 (as micrometer-sized bubbles hold by capillary forces in the rock pores) is a key mechanism for prevents the CO_2 from leaking back to the surface.

The value of the flow rate must be set according to the objective of the study. For CO_2 storage application it is important to obtain relative permeability curves for conditions corresponding to low value of Ca in order to model the hydrodynamics of the reservoir far from the injection location (usually $Ca < 10^{-8}$). However high flow rates are usually preferred for minimizing the impact of capillary end-effects that may take place at the sample edges that are in contact with the single phase fluid. Yet high flow rate may trigger the displacement of fine particles that may be present into the pores and consequently compromise the measurement. To determine the appropriate flow rate it is recommended to perform few preliminary experiments at increasing values of the flow rate in order to determine when capillary end-effects are negligible while keeping in the range of value that are representative of the targeted reservoir. Together with the complexity of the data processing that will be discussed below, the main disadvantage of unsteady state methods, where a fluid displaces another fluid of distinctly different thermodynamic properties, is that this configuration can promote flow localization (fingering) mechanisms caused by the interplay of gravity and viscosity segregation effects and the porosity heterogeneities. Thus, the phase with the higher mobility (i.e. the CO_2) may bypass regions of low permeability. In this case the measured relative permeability will not be representative of the entire sample (Chang et al. 1997) and furthermore not representative of the reservoir because these localization mechanisms are also amplified by the finite-size nature of the sample.

In the course of drainage-imbibition experiments capillary pressure in the sample is space and time distributed and there are several methods (often named data reduction methods) used to process the experimentally measured time-resolved data (i.e. the flow rate, the cumulative injected volume, the phase fraction produced and the pressure drop ΔP) and evaluate the effective and relative permeability values versus saturation. For high flow rate experiments the most used processing methods are those proposed by Welge (1952), Johnson et al. (1959) and Jones and Roszelle (1978). These methods are built from assumption-based simplifications of the Buckley-Leverett model of two phases displacement; the main assumption being to neglect capillary pressure effects. Eliminating inaccuracies caused by the capillary end-effect, specifically for low flow rate experiments, has been investigated by several authors (e.g. Ramakrishnan and Cappiello 1991; Virnovsky et al. 1995). As a general description, these methods need an evaluation the phase saturation at the edge of the sample using empirical relationships requiring the calculation of the fractional flow that is defined as the derivative of the produced flow rate of the drained phase by the flow rate of the injected phase. Then the relative permeability is expressed as the function of fractional flow, the viscosity and other model-specific parameters through iterative (or history matching) methods using trial-and-error simulations (e.g. Zhang et al. 2012). The derivation of these different models is complex and several softwares and graphical methods have been developed to assist the processing of the experimental data, most of them being specifically adapted to a given experimental protocol developed by a given research center. While unsteady state method is fast relative to the steady sate method, it is often practically impossible to perform measurement for the low end of the brine saturation values.

Conversely, the steady state method requires more complex equipment and more importantly involves longer experimental durations, but allows simple data processing and often produce more reproducible and potentially accurate results. The principle of the steady state method is to measure the effective or relative permeability of the two fluids at different saturation values when the capillary pressure along the sample is negligible, i.e. the fraction CO_2 and brine is the same along the sample. The duration of the experiment is typically ten times longer than for the high flow rate unsteady state experiments and makes this method restricted to high permeability rocks. The first steps of the experiment are similar to those described for the unsteady state method for drainage: the core sample is saturated with brine under vacuum, and then brine is injected to measure the intrinsic permeability k. Next, brine and CO_2 are co-injected starting from low value of the CO_2/brine ratio. It is usually required to pump a volume of mixture corresponding to 4–6 times the sample pore volume before steady state is reached, i.e. when the phase fraction of the produced mixture and the pressure drop are constant. This procedure is repeated for increasing the CO_2/brine ratio until 100 % CO_2 is injected and the value of the residual brine saturation and the permeability of the CO_2 at residual brine saturation $k_C^{(e)}(S_b^{(0)})$ are measured. Note that some rocks displays fast growing capillary pressure when the brine saturation decreases and thus it is difficult to avoid saturation gradient along the sample when injecting the highest CO_2/brine ratios because high pressure is needed to overpass the high interfacial tension between CO_2 and brine and the low viscosity of the CO_2; this is an open issue discussed for instance in Levine et al. (2014), Pini and Benson (2013) and Krevor et al. (2012). The same procedure can be repeated but with decreasing the CO_2/brine ratio to simulate an imbibition process, until 100 % brine is injected and the value of the residual CO_2 saturation and the permeability of the brine at residual CO_2 saturation $k_C^{(e)}(S_b^{(0)})$ are measured.

Albeit the difficulties of measuring the relative permeability of CO_2 at residual brine saturation mentioned above, the processing of the experimental data is quite simple because of the steady state nature of the biphasic flow in the rock. Indeed at steady state the saturation of each phase and the capillary pressure are constant along the core and the pressure drop for each of the phase equal the total pressure drop ΔP. Thus, the relative permeability of phase i (i = CO_2 or brine) at each saturation stage is given by the generalized Darcy's law: $k_i^{(r)} \cong (\mu_i L Q_i)/(kA\Delta P)$ where L and A are the length and the edges surface of the sample and μ_i and integration of the heterogeneityQ_i denotes the viscosity and the flow rate (in $m^3\ s^{-1}$) of the phase i. Capillary end effects steady state methods can be as significant as for the unsteady state method, (Kamath et al. 1995), but can be handled using adequate experimental protocol such as proposed by Pini and Benson (2013). Some results comparison between these two methods for CO_2-brine systems are given in Mathias et al. (2013) and Burnside and Naylor (2014). Figure 6.4b displays a typical relative permeability curve $k_i^{(r)}$ versus S_b where the effects of capillary number Ca and wettability are outlined.

For both steady and unsteady state methods, X-ray attenuation can be used to determine the distribution of the phase within the core sample. Usually X-ray absorbents, such as NaBr, are added to the brine to increase the contrast between the phases. It is specifically useful for verifying that no gradient of saturation exits along the sample for steady state experiments and for determining the amplitude of the capillary pressure end-effects (Berg et al. 2013). Imagery methods are specifically helpful for obtaining relative permeability data in low-permeability rocks (Zhang et al. 2014) where the effect of capillary pressure are dominant due to tightness of the pores and throats and the capillary end-effect significant (Akin and Kovscek 1999). X-ray tomography, which allows a 3D characterization of the phase distribution, is also a very promising technique to help processing both the unsteady state and the steady state permeability experiments, giving crucial information on the displacement of the front and properties of the interfaces between the phases, but also to study the CO_2 trapping mechanisms as the function of the properties of the pore network (Andrew et al. 2014; Hingerl et al. 2016; Iglauer et al. 2011; Krevor et al. 2012; Rahman et al. 2016).

Once the experimental values of CO_2 and brine relative permeability versus brine saturation are obtained, heuristic models are usually employed to interpret the relative permeability curves and extrapolate them to brine saturation values not achievable experimentally. Several models have been proposed starting from the simple formulation proposed by Corey (1954) where the two fitting parameters are the exponents n and m of the following power law equations:

$$k_b^{(r)} = k_b^{(0)}\left[S_b^*\right]^n \tag{6.2.8}$$

and

$$k_C^{(r)} = k_C^{(0)}\left[1 - S_b^*\right]^m \tag{6.2.9}$$

with S_b^* is the normalized water saturation defined as

$$S_b^* = \left(S_b - S_b^{(0)}\right)/\left(1 - S_b^{(0)} - S_C^{(0)}\right) \tag{6.2.10}$$

Other models with higher degree of freedom are often required to model the data. For examples the exponential models proposed by Chierici (1984) involves two fitting parameters per phase while the more recent LET model (Lomeland et al. 2005; Ebeltoft et al. 2014) involves three fitting parameters per phase (I_i, L_i and E_i):

$$k_b^{(r)} = \frac{k_b^{(0)}\left[S_b^*\right]^{L_b}}{\left[S_b^*\right]^{L_b} + E_b\left[1 - S_b^*\right]^{T_b}} \tag{6.2.11}$$

$$k_C^{(r)} = \frac{k_C^{(0)}\left[1 - S_b^*\right]^{L_C}}{\left[1 - S_b^*\right]^{L_C} + E_b\left[S_b^*\right]^{T_C}} \tag{6.2.12}$$

A comparative analysis of the performance and associated uncertainty of these models using steady state relative permeability measurements can be found in Moghadasi et al. (2015).

6.3 Experiments for Solid Matrix Mechanical Properties

This section summaries state-of-the-art experimental techniques for solid matrix mechanical testing, followed by a discussion on implications and remaining issues for mechanical properties within the context of CO_2 storage. The experimental equipment and conventional methods to determine static elastic moduli, dynamic elastic moduli and strength parameters will be only briefly outlined as they have been described in several publications (e.g. Jaeger et al. 2007) and used for decades.

Experimental measurements of rock mechanical properties determine the elastic moduli and strength parameters. The elastic moduli define the rocks ability to resist and recover from deformations produced by applied stress. These are primarily represented by the modulus of elasticity (E) which is a measure of the stiffness of the sample, i.e. the samples resistance against being compressed by a uniaxial stress and Poisson's ratio (v) which is a measure of the lateral expansion relative to longitudinal contraction. The strength parameters define the rocks resistance to deformation, defining how well cemented the rock is. They are primarily represented by the uniaxial compressive strength, triaxial stress factor, cohesion and angle of internal friction. The uniaxial compressive strength (Co) is the maximum stress the rock sample can withstand. If shear failure occurs the greatest shearing stress always occurs on the planes that contain the σ_2 axis which means in practice fractures generally form at an angle between 45° and 30° to the principal axes. With reference to Mohr-Coulombs work, the shearing stress can be related to the concept of internal friction, which suggests that at failure the relationship between the magnitude of shear stress (τ) and normal stress (σ_n) is:

$$|\tau| = So + \mu\sigma_n| \tag{6.3.1}$$

where So is the inherent shear stress of the rock, termed cohesion and μ is the coefficient of internal friction. The angle of internal friction (φ) is related to the coefficient of internal friction μ by:

$$\text{Tan}\,\varphi = \mu \tag{6.3.2}$$

The triaxial stress factor (k) is expressed in terms of the principal stresses (σ_1 and σ_3) which generated the normal and shear stresses (σ_n and τ) at failure and is directly related to the angle of internal friction (φ) by:

$$k = (1 + \text{Sin}\,\varphi)/(1 - \text{Sin}\,\varphi) \tag{6.3.3}$$

The same basic test facility is used for both elastic deformation and strength parameter determination. The samples are loaded into a pressure vessel and axial stress is transferred to the ends of the rock sample via hardened steel platens within a pressure vessel held within a servo controlled stiff testing machine. Radial stress (confining pressure) is transmitted to the rock sample via oil retained in the annular space between a rubber sleeve surrounding the sample and the body of the pressure vessel and applied through hydraulic fluids generated by a servo controlled pressure intensifier. Strain gauges applied to the surface of the rock sample allow axial and radial strain changes during testing to be recorded. For conventional rock mechanical testing the Hoek-Franklin triaxial cell is used which allows a stress state of $\sigma_1 > \sigma_2 = \sigma_3$ to be realized (Hoek and Franklin 1968). The true in-situ stress state of $\sigma_1 > \sigma_2 > \sigma_3$ cannot be achieved in a Hoek cell, however the true triaxial pressure vessel, (Smart et al. 1999) has an array of 24 trapped tubes where selective pressurisation of the tubes enables differential radial stresses to be generated, while axial stresses are applied as conventionally, through steel plates.

6.3.1 *Static Elastic Moduli Testing*

Samples are loaded hydrostatically (the axial and radial confining stress are the same) in increasing increments, once each hydrostatic stress level is reached the axial stress is increased and decreased by approximately 3 kN to induce vertical and horizontal strain. The axial (σ_a) and radial (σ_r) strains are measured by the strain gauges. The modulus of elasticity (E), is calculated as the ratio of the change in axial stress (σ_a) to the change in axial strain (ε_a) and Poisson's ratio (v) is calculated from the ratio of the accompanying change in radial strain (ε_r) to the change in axial strain (ε_a).

The velocity of elastic waves in a rock is a function of its density and elasticity, therefore the mechanical parameters of modulus of elasticity (E), Poisson's ratio (v), shear modulus (G), bulk modulus (K) and compressibility (Cb) can also be determined if the bulk density (ρ_b) of the rock is known. Elastic properties derived from acoustic measurements are known as dynamic elastic moduli. There is a wide range of experimental evidences to show that static and dynamic moduli of rocks are different, (Fjar et al. 1992). In general the static elastic modulus for dry rocks is less that the dynamic modulus. The most common reason for the described is ascribed to the presence of microcracks.

6.3.2 *Strength Parameter Testing*

A basic compressive strength test involves loading a sample to failure at a constant value of confining pressure. This results in a single pair of maximum and minimum principle stresses and the determination of stress at failure, i.e. the uniaxial compressive strength. This straightforward test does not facilitate the determination of any other strength parameters which require the construction of a Mohr's failure envelope, generated from several Mohr's circles determined on the same sample at different confining pressures. The uniaxial compressive strength (UCS) can be calculated by dividing the load at failure by the cross sectional area of the sample.

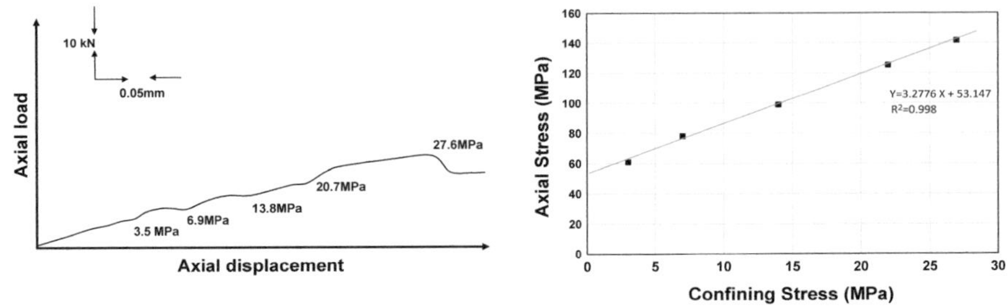

Fig. 6.5 Example multi failure axial stress/confining stress plot

From the multi-failure axial stress/confining stress results, a linear function is fitted to the data, which can be expressed in terms of the principle stress as follows (Fig. 6.5):

$$\sigma_a = UCS + \sigma_r k \tag{6.3.4}$$

where σ_a is the axial stress, σ_r the radial (confining) stress, UCS the uniaxial compressive stress and k the triaxial stress factor. Cohesion (So) is calculated from:

$$So = UCS/(2\sqrt{k}) \tag{6.3.5}$$

and the angle of internal friction (φ) is calculated from Eq. 6.3.3.

Along with conventional rock mechanical testing procedures, there are a number of further mechanical tests that enhance our understanding of the mechanical properties of storage reservoir and cap rocks:

- Acoustic emission (AE) describes the sound waves produced when a material undergoes stress as a result of an external force. Acoustic emission sensors detect stress waves motion that cause a local dynamic material displacement and convert this displacement to an electrical signal. Acoustic emissions can be used to detect plastic deformation indicator and crack propagation rate.
- Dynamic rock mechanical properties low and ultra-low (static) frequency experiments can be conducted to increase accuracy of the dynamic moduli measurements. This is because both P-wave and S-wave velocities increase as frequency increases from low frequency to ultrasonic frequencies. Ultrasonic data can overestimate velocities at lower frequencies, changing Young's modulus (less so for Poisson's Ratio).
- Compressibility is a measure of the relative volume change of a fluid or solid as a response to a stress change. The compression measurement procedure requires the servo hydraulic controlled stiff compression rig to have a double acting attenuator, with an integral displacement transducer (LVDT) incorporated in the actuator providing an electrical signal proportional to the piston rod displacement, this facilitates the determination of the sample compressibility.

6.3.3 Implications and Remaining Issues for CCS

All the experimental methods described above will provide mechanical property input data for the reservoir model, however they are snapshot measurements and it is important to remember that the CO_2 injection and storage system is a dynamic system that changes over time. As injected CO_2 changes the in-situ stress state, temperature, fluid density and formation water pH the fluid/rock/chemical interactions instigate changes in the porous network and in turn the elastic and strength properties of the rocks. It is important to understand how these mechanical properties evolve over time within the storage system.

6.3.3.1 Rock Mechanical Changes on Exposure to CO_2

Reservoir rock will not fail due to fluid flow alone, but as a result of stresses acting in the near well bore area. These stresses are caused by the pressure difference between the formation fluids and wellbore, fluid frictional forces, the reservoir stress state, thermal stresses and changing chemistry. When the magnitude of the combined forces exceeds the strength of the formation, the rock will fail. The key to wellbore stability is that the stresses acting on the rock surrounding the wellbore must not exceed the strength of the rock. There are a number of wellbore stability (sanding) predictors that require the rock mechanical input data of uniaxial compressive strength, dynamic shear modulus, poisons ratio, bulk modulus, modulus of elasticity, cohesion and angle of internal friction, Table 6.1.

Although the required rock strength and required mechanical input data can be experimentally determined before injection by the testing methods described above, they do not take account of the changing nature of the mechanical parameters as CO_2 injection progresses and the fluid/rock/chemical interactions instigate changes in the porous network and in turn the elastic and strength properties of the rocks. An empirical relationship must be developed between the reservoir and cap rock and its change in mechanical properties due to CO_2 exposure over time under in-situ conditions. There is no single downhole tool that can directly measure the rock

Table 6.1 Inputs in stability prediction models

Wellbore stability prediction model	Calculation—required mechanical inputs
Stein and Hilchie (1972)	Dynamic shear modulus
Stein et al. (1974)	Dynamic shear modulus
Tixier et al. (1975)	Poisson's ratio, shear modulus and bulk modulus
Coates and Denoo (1980)	Poisson's ratio, modulus of elasticity, shear modulus and bulk modulus
Risnes et al. (1982)	Cohesion and angle of internal friction
Morita et al. (1987)	Poisson's ratio, modulus of elasticity
Weingarten and Perkins (1992)	Cohesion and angle of internal friction
Sarda et al. (1993)	Effective stress, uniaxial compressive strength

elastic and strength properties of the in-situ reservoir rocks and fewer to capture the changing in mechanical properties with CO_2 injection.

Mechanical properties can be determined by the following downhole methods:

- Deformability tests that apply load to the rock surface exposed on the borehole wall and measure the resulting deformation. This can derive static mechanical properties, but this is limited to the modulus of elasticity and can only be conducted before cement casing.
- Acoustic travel times and bulk density from downhole logging, can be applied to empirical relationships relating compressional wave velocity, shear wave velocity and bulk density to the dynamic elastic parameters of modulus of elasticity, Poisson's ratio, bulk compressibility and shear modulus.
- Application of empirical correlations of porosity, composition and texture to calculate mechanical properties, (Edlmann et al. 1998).
- Indirect measurements contributing to the understanding of the mechanical nature of the reservoir such as borehole integrity from the calliper log, lithology determination and fracture identification from core, logging and wellbore imagery.

In addition to poor down-hole measurement of the mechanical properties there is little understanding of how the changes in fluid and matrix properties with the injection of CO_2 will impact on these measurements and these issues need addresses.

6.3.3.2 Remaining Mechanical Issues for Underground CO_2 Storage

There are limited studies into the effect of CO_2 injection on the mechanical properties of reservoir and caprock and the important highlights of these investigations are:

- Oikawa et al. (2008) conducted tri-axial experiments using Berea sandstone under simulated geological storage conditions and found the Modulus of Elasticity and Poisson's ratio to be suppressed slightly in the presence of CO_2.
- To undertake rock physics modeling of CO_2 bearing rocks the fluid substitution model is normally used to predict rock modulus change with a change in pore fluids, Hossain (2012), centered on Gassman's equation (Gassman 1951) input parameters are the effective bulk modulus of the rock matrix the shear modulus of the CO_2 bearing rock and the porosity. However these are input as static values and the impact of how the CO_2 bearing fluid may change these properties are not included but may have an important influence over time.
- The thermal, hydrodynamic and chemical process described in Chaps. 3 and 4 will contribute to altering the mechanical properties of the reservoir and cap rock during the CO_2 injection and storage process. Understanding the dynamic relationship between CO_2 exposure and mechanical properties influenced by mineralogy, grain size, porosity, cement depositional environment, etc. is a crucial area of research.

6.4 Fluid-Rock Interactions and Properties Changes

This section gives an overview of the experimental techniques used for investigating the mass transfers between the water-CO_2 mixture flowing in the pore space and the rock-forming minerals. Standard techniques using static or flooded reactors (*batch reactor*) dedicated to measure mineral solubility values and speciation will be only briefly outlined because they have been described in several publications and used for decades for applications ranging for instance from subsurface pollution, ore-deposit studies and hydrothermal processes (Sect. 6.4.2). Conversely we will focus on specific equipment dedicated to reproduce both the hydrodynamic and thermodynamic conditions expected during CO_2 storage operations (Sect. 6.4.3). More specifically, this type of a laboratory apparatus is designed to reproduce the pressure, temperature, fluid flow and chemical conditions. Accordingly, this kind of equipment, which is also used for studying other dynamic reactive geological settings, such as hydrothermal systems, are often called *flow-through rock* (*or reaction-percolation*) *experimental bench* or for specific application to CO_2 geological storage: *CO_2 sequestration evaluation flow system*. In the following we will use the acronym FTS (flow-through system) for simplification.

The experimental protocols attached to FTS aim at measuring the chemical fluxes and the changes of the rock properties, i.e. the change of the petrophysical, hydrodynamical and mechanical characteristics induced by the alteration of the solid fraction of the porous or fractured geomaterial under consideration. Some companies propose this type of equipment derived from those used in oil-industry for simulating multiphase flow or testing enhanced oil recovery techniques, the injection of $scCO_2$ being one of these techniques. Yet, experimental tools fully dedicated to study CO_2 underground storage are often designed and operated by academic research centers for specific objectives and therefore might display different technical characteristics. Nevertheless, the concepts are similar and we will outline the general technological characteristics and illustrate the use of these tools for studying mass transfer processes occurring into the reservoir during CO_2 injection, but also for studying scenarios of leakage into fractured caproks and well cement annulus (Sect. 6.4.4).

The rationale (Sect. 6.4.1) for developing specific equipment and performing such experiments coupling flow and reaction emerges from the complexity of the thermo-hydro-chemical mechanisms controlling the alteration of the rocks in the context of the forced injection of potentially highly reactive water-CO_2 mixture into potentially heterogeneous geomaterials. As a consequence modeling such mechanisms at reservoir scale is still challenging and requires not only a pertinent parameterization of the continuum approach-based modeling codes, but also to determine the validity of the underlying concepts and the accuracy of the numerical tools. Uncertainties concern essentially the integration of the heterogeneity into the macroscopic models (Chap. 5), the verification of the validity of the hypothesis used to derive the continuum approach formulation (Chap. 3) and the measurement of the effective parameters, such as the reactive surface area for example.

6.4.1 Rationale

The injection of CO_2 into deep reservoirs leads to large chemical disequilibrium and consequently can cause noticeable mass exchanges by mineral dissolution and precipitation that in turn may change strongly and often irreversibly the hydrodynamical and mechanical properties of the reservoirs. This is particularly important in the vicinity of the injection well where both the disequilibrium between the pore-fluid enriched in CO_2 and the rock-forming minerals and the fluid velocity (i.e. the renewing of the reactants) are large. Well cement and caprock can also encounter dissolution and precipitation in case of leakages through fractures either induced by overloading pressure during injection or reactivated by tectonic events.

In such conditions some of the usual assumptions supporting the validity of the macroscopic or continuum scale mass conservation equations may not be satisfied. For instance, the full mixing assumption, implying that the concentration is homogeneous at the scale of the support volume, used to derive the continuum scale advection-dispersion-reaction equation that model solute transport in porous media (see Chap. 3) may be invalid. For these dynamic systems displaying far-from-equilibrium conditions, not only pore scale concentration gradients may take place but mass fluxes controlling the return to equilibrium are non-linearly related to local concentration. Self-organized dissolution features, that tend to noticeably increase the hydrodynamic and mechanical properties heterogeneity, are expected. In other words, CO_2 injection in deep reservoir can trigger conditions in the reservoir, caprock and well cement for which the continuum scale (at which the pore scale structure and mass transfer mechanisms are averaged) approach may not preserve some aspects of the dynamics at the micro-scale. Yet the system will ubiquitously tends to return to equilibrium and therefore a very large range of conditions of flow, solute transport and dissolution-precipitation reactions are expected along the flow path from the injection well to the far field. Figure 6.6 illustrates typical trends of the Peclet (Pe_λ) and Damköhler (Da_λ) numbers as the function of the distance to the injection point. These two parameters are classically used to evaluate the hydrodynamic and hydrochemical conditions respectively. The Peclet (Pe_λ) number for a support volume of characteristic distance λ is conventionally written

$$Pe_\lambda = V\lambda / D \tag{6.4.1}$$

where V [m s^{-1}] is the average velocity and D [m^2 s^{-1}] denotes the dispersion or the diffusion coefficient depending on the studied scale. The Damköhler number (Da_λ) denotes the ratio of the characteristic time of the reactant renewing process at the reactive surface of the mineral to the characteristic time of reaction. Different formulations can be found in the literature depending on whether the renewing of the reactant is controlled by diffusion or advection, i.e. depending on the value of Pe. Here we propose a generalized definition of the Damköhler number:

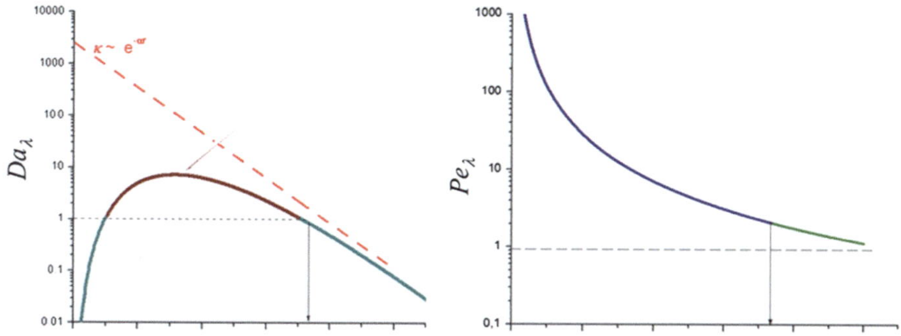

Fig. 6.6 Peclet (Pe_λ) and Damköhler (Da_λ) numbers as the function of the distance (r) from the injection point. Because the fluid velocity is proportional to the inverse of the distance in radially divergent flow, $Pe_\lambda \propto 1/r$. Conversely, the Da values may display more complex ternds as the function of r. Here we assume for matter of illustration that κ decreases exponentially with the distance to the injection ($k \propto e^{-\alpha r}$) but in reality the expression of $\kappa(r)$ is the result of complex coupling between the hydrodynamic transport and the multi-component reactions. Nevertheless it is expected that $\kappa(r)$ is a monotonic decreasing function of r because the system tends to equilibrium (i.e. $\kappa = 0$) far from the injection

$$Da_\lambda = \kappa\lambda / V\left(1 + Pe^{-1}\right) \tag{6.4.2}$$

where κ is the effective reaction kinetic velocity [s^{-1}]. The expression of κ is generally derived from the Transition State Theory (see Sect. 6.4.2). Figure 6.6 shows that a large range of conditions combining different values of Pe_λ and Da_λ can be expected. Close to the injection, large flow rate acts to maintain the system far from equilibrium whereas thermodynamical equilibrium is reached at a certain distance from the injection. Between these two extremes situations mass transfers locally dominated by diffusion, dispersion, advection or kinetically controlled reactions may occur. Similarly, a large range of hydrodynamic and hydro chemical conditions can be expected in fractures crossing caprock layers or well cement annulus depending on the differential pressure acting at the fracture boundaries and on the aperture.

A large number of experimental and theoretical studies, not specifically applied to CO_2 storage, has been devoted to study the influence of dissolution processes on the physical and chemical properties of porous media (e.g. Noiriel et al. 2004; Hoefner and Fogler 1988; Rege and Fogler 1987; Schechter and Gidley 1969). For instance, leading mechanisms such as pore coalescence (Schechter and Gidley 1969) and the formation of highly conductive flow channels (Carroll et al. 2013; Golfier et al. 2002; Renard et al. 1998; Daccord et al. 1993) involve particularly complex feedback effects. Yet, as a matter of fact the large number of studies related to CO_2 storage has boosted the construction of experimental apparatus specifically designed for using brine and CO_2 mixture for a large range of pressure and temperature. An overview of the equipment and studies is proposed in Sects. 6.4.2 and 6.4.3.

6.4.2 *Batch Experiments*

This section outlines the batch experiments undertaken to investigate the rock—CO_2 fluid interactions, the theory of which has been presented in Chap. 3. A brief summary of batch reactor and reaction kinetic theory is presented followed by descriptions of experimental batch equipment and analysis techniques used for instance within the MUSTANG project to characterize the fluid–rock interactions processes.

6.4.2.1 Batch Reactors

A batch reactor is a heated vessel where reactants are placed and reactions are allowed to proceed for a given time. It is designed for recurrent sampling of the fluid for analysis. The concentration and temperature are assumed to be uniform (sometimes the vessel is equipped with a stirrer) and all elements spend the same time in the reactor to ensure the same residence time. From a thermodynamic standpoint, a batch reactor represents a closed system. The system will tend to thermodynamic equilibrium with time. Batch reactors are typically used to study reaction kinetics under controlled conditions. It can be used to find reaction rate constants, activation energy and to determine the order of the reactions.

6.4.2.2 Rate Law Basics

For nearly all forward reactions, the rate is proportional to the product of the concentrations of the reactants, each raised to some power. For the general reaction:

$$aA + bB \leftrightarrow cC + dD \tag{6.4.3}$$

The instantaneous sample-averaged mass-transfer rate R [mol m^{-3} s^{-1}] can be modeled using the transition state theory (Lasaga 1998):

$$R = r(1 - I)^{-m} \tag{6.4.4}$$

where r is the effective rate constant and I the saturation index, $I = IAP/K_{eq}$, with IAP the Ion Activity Product and K_{eq} the equilibrium constant. The higher the r value is, the faster the reaction proceeds. The effective rate is usually defined as the product of the intrinsic kinetic constant r' [mol m^{-2} s^{-1}] that denotes the net flux of the reactant, by the reactive surface area σ [m^2 m^{-3}]. For CO_2 storage conditions, values of r' determined from batch experiments for different dissolution and precipitation reactions can be found in Pokrovsky et al. (2009) and references herein.

Two types of analysis are usually involved for determining the rate law for a given reaction:

- Integral method—A reaction rate is assumed and the initial ratios of concentration are known. If the final molar concentrations are plotted against time and the data fits a straight line, then the assumption of second order kinetics is true and the rate constant may be calculated from the slope of the line.
- Differential method—A reaction rate order is assumed, concentration versus time data from the batch experiments are plotted and tangents drawn at various points. The slopes of these tangents are the rate of reaction at these concentrations.

Chemical reactions are often classified according to their kinetics order:

- Zero order—the reaction rate is independent of the reactant concentration values. A change in temperature is the only factor that can change the rate of a zero order reaction.
- First order in the reactant—The reaction rate is proportional to the concentration of one reactant, e.g. radioactive decay.
- Second order in the reactant—the reaction rate is proportional to the product of the concentration of two reactants, or to the square of the concentration of a single reactant. Many dissolution and precipitation reactions are second order reactions.

6.4.2.3 Equipment Overview

Different batch reactor systems have been designed to investigate a wide variety of rock-CO_2-brine interactions under a large range of in-situ pressure and temperature conditions up to 20 MPa and 200 °C respectively. The experiments involve brine and/or CO_2 and powdered (to provide the maximum surface area for reaction) single minerals or multi-mineral rock, rock chips or rock samples. Similar experiments can be performed to study the reactions with well cement materials.

The reactor itself can be as simple as a tube (stainless steel, titanium or hastelloy) equipped at one end by a CO_2 inlet with close off valve and at the other end a 60 micron filter and a pressure release valve set at a given pressure. The tube is wrapped by a heating tape with a thermostat. More sophisticated types of reactor are often used to improve the sampling or the control of the fluid composition. For instance a reaction vessel within a pressure vessel such as presented in Fig. 6.7 can be used perform single phase experiment (only CO_2-rich brine and rock) and allow the sampling of the brine in the course of the experiment without modifying the pressure.

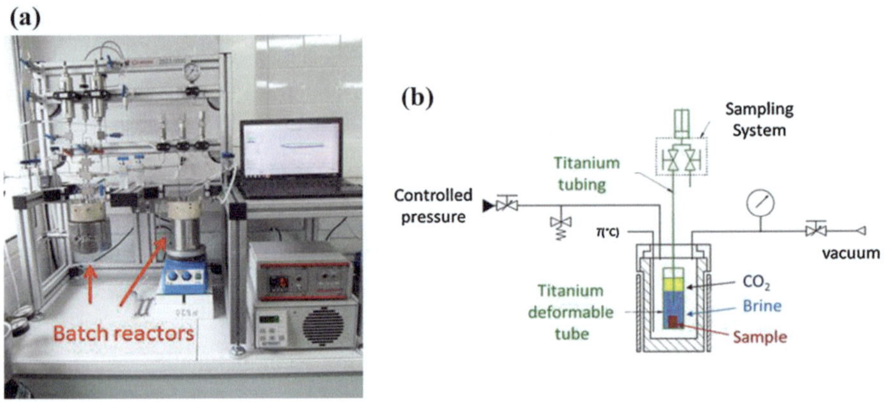

Fig. 6.7 High temperature, high pressure brine and CO_2 batch equipment. *Left* ICARE Lab stirred and non-stirred reactors. *Right* Reactor equipped with deformable titanium jacket

6.4.2.4 Analysis Techniques

There are a number of analytical techniques for measuring the fluids, gases and rock chemistry. The fluid analysis is divided into fluid element and gas composition analysis:

Element Concentrations

Major elements are routinely measured using ICP-AES (inductively Coupled Plasma Atomic Emission Spectrometry). A plasma source is used to dissociate the sample into its constituent atoms or ions exciting them to a higher energy level. They return to their ground state by emitting photons of a characteristic wavelength depending on the element present. The light is recorded by an optical spectrometer which when calibrated against standards provides a quantitative analysis of the sample.

Minor and trace elements are measured with ICP-MS (Inductively Coupled Plasma Mass Spectrometry). A plasma source is used in the same way as the ICP-AES, however the emitted ions are extracted through a series of cones into a mass spectrometer, usually a quadrupole. The ions are separated on the basis of their mass-to-charge ratio and a detector receives an ion signal proportional to the concentration.

Titration method is also often used, for example to measure the alkalinity and the chlorine content. Chemical equilibrium is a function of the concentrations of the fluids in equilibrium. The equilibrium constant value can be determined if any one of these concentrations can be measured. The concentration of the fluid in question is measured for a series of solutions with known analytical concentrations of the reactants. Titration is performed with one or more reactants in the titration vessel and one or more reactants in the burette. Knowing the analytical concentrations of reactants initially in the reaction vessel and in the burette, all analytical concentrations can be derived as a function of the volume (or mass) of titrant added.

Gas Compositions

In situ Raman analysis: A spectroscopic technique based on inelastic scattering of monochromatic light, usually from a laser source. Inelastic scattering means the frequency of photons changes on interaction with a sample and the frequency of the re-emitted photons are shifted in comparison with the original monochromatic frequency called the Raman Effect. Which when calibrated against standards provides a quantitative analysis of the sample.

Infra-red analysis: Infrared spectroscopy exploits the fact that molecules absorb specific frequencies that are characteristic of their structure. These absorptions are resonant frequencies, i.e. the frequency of the absorbed radiation matches the transition energy of the bond or group that vibrates. The exact frequency at which a given vibration occurs is determined by the strengths of the bonds involved and the mass of the component atoms

Gas chromatography: Gas chromatography is a method of separating the components of a solution and measuring their relative quantities. A sample is rapidly heated and vaporized to separate the components by distributing the sample between two phases: a stationary phase and a mobile phase. The mobile phase is a chemically inert gas that serves to carry the molecules of the sample through a heated column. Sample components are separated based on their boiling points and relative affinity for the stationary phase, which is most often a viscous liquid (wax) within the column. The higher a component's affinity for the stationary phase, the slower it comes off the column. The components are then detected and represented as peaks on a chromatogram.

6.4.2.5 Rock Mineralogy Analysis

The rock mineral analysis can be undertaken on powdered samples or using thin sections of solid samples. Whole rock analysis is not actually a complete analysis of the rock, but describes a specific method of ore preparation followed by analysis for eleven common rock-forming elements. The procedure involves a lithium borate fusion process on a small portion of pulverized solid material. The values for the rock-forming elements are then expressed as their more common oxide compound.

Powder is used also for X ray diffraction (XRD) for identifying mineral types. XRD is based on Bragg's Law, where the atomic faces of crystals cause an incident beam of X-ray beams to interfere with one another as the leave that crystal at certain angles of incidence (θ):

$$n\lambda = 2d \sin\theta \tag{6.4.5}$$

where d is the distance between atomic layers and (λ) is the wavelength of the incident X-ray beam and n in an integer. XRD measurements are based on observing the scattered intensity of an X-ray beam hitting a sample as a function of incident and scattered angle, polarization, and wavelength or energy. X-ray Powder diffraction facilitates qualitative identification of minerals in all rock types, measure the average spacing between layers of atoms, determine the orientation of a single crystal and determine the crystal structure.

X-ray fluorescence (element quantification): X-ray fluorescence (XRF) is when a sample is exposed to X-rays of high energy. As the X-ray (or photon) strikes an atom (or a molecule) in the sample, energy is absorbed by the atom. If the energy is high enough, a core electron is ejected out of its atomic orbital. An electron from an outer shell then drops into the unoccupied orbital, to fill the hole left behind. This transition gives off an X-ray of fixed, characteristic energy that can be detected by a fluorescence detector. The energy needed to eject a core electron is characteristic of each element, and so is the energy emitted by the transition.

Thin sections are used for microscopy observation, specifically using electronic microscopes. Scanning Electron Microscopy (chemical observation and structural properties): A tungsten filament source produces a stable and high current electron beam. When the electron beam interacts with the sample, the electrons lose energy by repeated random scattering and absorption within a teardrop-shaped volume of the specimen known as the interaction volume. The size of the interaction volume depends on the electron's landing energy, the atomic number of the specimen and the specimen's density. EDS (energy dispersive) X-ray analysis: The PGT Spirit EDS X-ray analysis system allows qualitative and quantitative mineral analysis as well as X-ray imaging of samples. The number and energy of the X-rays emitted from a specimen are measured by an energy-dispersive spectrometer. As the energy of the X-rays are characteristics of the difference in energy between the two shells, and of the atomic structure of the element from which they were emitted, this allows the elemental composition of the specimen to be measured.

6.4.3 Flow-Through Experiments

A flow-through experiment consists in injecting a fluid at a controlled flow rate through a sample of permeable rock, usually of cylindrical shape, called a core. The fluid can be a single phase fluid, such as liquid water containing dissolved ions and gases or a mixture of phases. The permeable sample can be a core of porous reservoir rock or a core containing natural fractures or fractures made artificially to mimic natural hydraulic discontinuities. The main goal of these dynamic experiments involving reactive fluids is to determine the time-elapsed change of the sample properties at macroscopic scale (i.e. core scale) such as permeability and porosity and to measure effective reaction rates in dynamic conditions. Moreover, these experiments can be associated with X-ray microtomography in order to observe directly the spatial distribution of the mass transfers with a resolution of some microns. The cores are usually imaged before and after the experiment. X-ray non-absorbent confinement cells can also be used to perform recurrent imaging and tackle simultaneously the changes of the pore structure and the absolute displacement of the different phases in the porosity. Nevertheless, the use of this technique is presently limited by several factors such as the duration of the experiment, the possibility of working at elevated pressure and temperature and the duration of the scan (largely dependent on the X-ray source flux and the sample size) during which it must be assumed that mass transfers are negligible. In any case, the sample size must be decided in order to fit the field of view of the camera; higher is the targeted resolution, smaller is the core diameter. In most of the cases core diameter (*D*) ranges from few millimeters to few centimeters and the geometrical resolution ranges from 0.5 to 20 microns depending on the applied imaging techniques.

The duration of the experiments typically ranges from some hours to weeks. For example, the alteration of a carbonate rich reservoir core flowed by a CO_2-saturated brine at 50–90 °C can be studied in few days, while the same conditions applied to the study of sandstones or fractured cements and claystones may require weeks.

6.4.3.1 FTS Equipment

As a general rule, the system includes motorized pumps equipped with displacement encoders which allow an accurate control of the inflow rate, a confinement cell were the sample is installed, heated and pressurized and a back pressure system allowing to control the outlet pressure while allowing the fluid to be withdrawn from the circuit and sampled for chemical analysis. A schematic representation of a flow-through apparatus is given in Fig. 6.8.

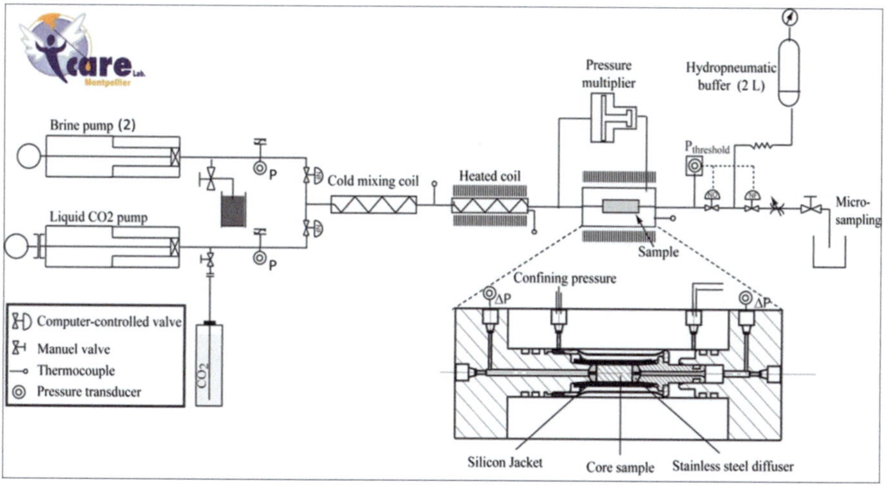

Fig. 6.8 Schematic representation of the ICARE Lab CO_2 sequestration evaluation flow system (model ICARE 1, design R. Leprovost, U. Montpellier)

A standard circuit requires up to 20 valves, most of them being usually electro-pneumatic valves (EPV); their opening and closing being mastered by the computer program which also control the pump flow rate.

The back pressure system is probably the most critical part of the FTS and different techniques can be used as it will be explained below. The second critical issue for reproducing the range of reservoir conditions is the flow rate. Modern computer controlled piston pump can deliver flow for a large range of values, but the main issue is often the capability of controlling the very low flow rates required to mimic in situ condition in small sized samples. Delivering low flow rate implicates the use of pumps having small pistons and high-technology motors and position encoders. Going down to flow rate of less than about 10^{-11} m^3 s^{-1} ($\sim 10^{-3}$ mL min^{-1}) requires specialized equipment using high-end pumps and a precise control of the temperature.

Others issues to be considered when designing such experimental bench for temperature up to 150 °C include the corrosive nature of the fluid (usually low pH brines) and the need of accurate temperature control and adapted circuit geometry when mixtures of scCO_2-water (or brine) are involved. Usually corrosion resistant alloys such as Hastelloy- C276 are used at least for the most exposed parts such as the pumps, the valves, and the pressure sensor membranes. The use of accurate and well calibrated pressure sensors is also required in order to obtain accurate values of the permeability over 2 or possibly 3 orders of magnitude. For some applications it is beneficial to set the pressure gradient instead of the flow rate. In this case the flow rate is measured from the displacement of the pump piston or counting the motor axe rotation using optical or electronic encoders.

Inlet Pumps: A combination of two piston pumps is required to obtain a constant flow rate for injection volumes larger that the volume of one pump. In this configuration, one of them pumps the liquid through the sample, while the other one refills from and external fluid tank. This procedure is controlled by a microprocessor installed in the pump electronic control card or directly by a computer. The objective of these experiments being to inject brines and CO_2 through the sample for a large and controlled range of partial pressure values, the question of mixing the CO_2 with the brine is a central issue. Two cases can be distinguished.

If the objective is to pump brine in which the partial pressure is lower than the total pressure, then the pumped fluid is a single phase. In this case the issue is to dissolve CO_2 into the brine in order to obtain a given constant CO_2 partial pressure whatever the total pressure is. Two main techniques can be applied. First the CO_2-brine mixture can be prepared in a tank by pressurizing the volume of brine required for the entire experiment with CO_2 at a pressure corresponding to the targeted CO_2 partial pressure. For values larger than the pressure of commercial CO_2 bottles (about 6 Mpa), the use of a pressure booster withdrawing and compressing liquid CO_2 from a bottle equipped with a plunger is necessary. This mixture will be used to refill the piston pumps. An alternative solution consists in adding to the FTS a piston pump refilled with liquid CO_2. The mixture will be obtained by running the brine and CO_2 pumps at the flow rates corresponding to the fraction of CO_2 needed to obtain the desired CO_2 partial pressure. As the volume of CO_2 to pump is only few percent of the volume of brine, the pump dedicated to the liquid CO_2 must have a minimal flow rate about two orders of magnitude smaller than the minimum flow rate needed for the experiment. This is the limiting factor of this technique (the minimum flow rate is here conditioned by the CO_2 pump characteristics) which conversely allows changing easily the CO_2 partial pressure in the course of the experiment.

For experiments requiring the injection of scCO_2-brine mixture, i.e. the injection of a biphasic fluid, an additional supercritical CO_2 pump is required. Such pump consists in a cooled pressure booster that pumps liquid CO_2 from a plunger-equipped bottle and a heating system allowing the pressurized liquid CO_2 to reach the supercritical phase. Note that the differential density between liquid and supercritical CO_2 must be taken into account for calculating the effective flow rate. The use of scCO_2 in FTS will be discussed in more detail below.

Back Pressure: Back pressure is necessary to reproduce the conditions at depth. It is also required for avoiding degassing of the CO_2 in the circuit; the total pressure must be higher or equal to the partial pressure of CO_2 to perform single phase experiments. Mainly three techniques can be used to control the total pressure of the outlet circuit. All of them produce pressure fluctuations but with distinctly different amplitudes and frequencies. As the objective is to maintain a pressure as constant as possible, all the different systems described below must be associate with a pressure damper that usually consists in a cylinder with a free mobile piston separating the flowing fluid from a volume of pressured gas. If the pressure fluctuations are small, then the free-piston cylinder can be replaced by a membrane.

The simpler one consists in using constant pressure valves. However this type of low-cost tools may be not sufficient for maintaining a sufficiently constant pressure at low and high flow rates. These valves are also temperature sensitive. As a result, large amplitude, low frequency pressure fluctuations are expected. Moreover freezing of the valve can occur when large amount of supercritical or liquid CO_2 have to be released by the valve. Micro-particles, for instance produced by the rock dissolution, may prevent the full closing of the valve and therefore may jeopardize irreversibly the experiment.

The second system that was experienced in the frame of the MUSTANG project is the use of a combination of two EPV in series, separated by a tube of limited volume, that alternatively open and close in order to deliver a flow rate that adapts to the pressure at the outlet of the sample while keeping it in a constant range of few tenths of MPa. Using a pressure damper containing 1 L of pressured gas is usually sufficient to obtain fluctuations of less than 0.1 MPa. This system is efficient and relatively cheap, but the valves are aging very fast because of their frequent activation (usually several hundred of operation for a standard experiment).

The third system consists in using back pressure piston pumps; ideally two piston pumps acting similarly to the dual injection pumps described above. These pumps will be programmed to maintain the desired pressure by continuously adapting the pistons displacement. Yet, the use of a single piston pump can be preferred for decreasing costs. In this case, a pressure damper containing about 1L of pressurized gas will allow pressure fluctuation to be less than 0.1 Mpa.

Sampling: sampling for cations analysis can be performed after degassing at the outlet of the circuit. However it is often necessary to sample the pressurized brine mixture or brine-CO_2 mixture, for evaluating the remaining CO_2 partial pressure, for instance. High pressure syringe with a screw-moving piston can be used. A simple technique for evaluating the CO_2 partial pressure is to equip the syringe with a micrometric screw gauge and a pressure sensor (or manometer) and measure the volume of gas released from the adiabatic decompression of the sample. The partial pressure (P_{CO_2}) is:

$$P_{CO_2}(P,T,s) = P_e \times P_d \times V_{dg}/R \times T \times V_L \times mCO_2(P,T,s) \tag{6.4.6}$$

where P_e and P_d denotes the pressure of the experiment and the pressure after decompression, V_{dg} and V denotes the volume of gas after decompression and the volume of the fluid sampled, R is the gas constant, T is the temperature and mCO_2 is solubility of CO_2 (Duan et al. 2006; Sun and Dubessy 2010).

Core Sample Holder: There are several designs of sample holder, named also confinement cell. Several providers propose standard designs such as Hassler type or Hoek type core holders adapted to different core sample sizes. Research laboratories are also developing core holders/confinement cells, usually to fit non-standard sample size. Whereas these different types of cells differ from each other in terms of technical details impacting mainly the protocol for the sample installation, the principle is similar for all the sample holder types. The cylinder of rock is installed into a flexible tube (or jacket) that can be made of different substance such as silicon, Teflon or vulcanized materials and then positioned in between two flow diffusers such as displayed in Fig. 6.8. This configuration allows applying an external pressure on the jacket ensuring that the fluid will not flow around de sample but is distributed equally over the sample diameter. This applied pressure is also used to reproduce de confinement pressure at depth (radial confinement pressure). Axial pressure mimicking the effect of the overburden along the sample axis can also be setup (Fig. 6.8). The axial and radial pressure can be operated by external piston pumps, gas tanks or made proportional to the inlet flow pressure using a pressure multiplier such as shown in Fig. 6.8. This later solution has two advantages; first it is unexpansive compared to the purchase of piston pumps and second it ensures that the confinement pressure increase (or alternatively decrease) proportionally to the fluid pressure during the operation of pressure loading (or unloading) at the beginning and end of the experiment, avoiding deviatoric stresses that may damage the sample irreversibly. In general it is sufficient to apply axial and radial pressures of about 110–115 % of the pore pressure for ensuring both a total sealing of the sample-jacket interface and reproduce natural pressure configurations.

Control of the Composition of the Inlet Fluid: Controlling the composition of the inlet fluid (brine) is important to reproduce the natural conditions. The ionic content can be set easily, but controlling the partial pressure of the gas requires some attention. This aspect, often disregarded in flow-through experiments, is important for example to reproduce anoxic conditions such as almost always the case in reservoirs, or even impose partial pressure of secondary gases such as methane of hydrogen to reproduce reducing conditions. A specific device has been developed in the frame of the MUSTANG project in order to make a controlled preconditioning of the brine, before being mixed with the CO_2. This device, called *ICARE Lab Fluid Preconditioner*, permits removing or adding dissolved gases or tracers in the fluid and keeping this volume of fluid (up to 5 L) under a pressure of 0.8 MPa during the duration of the experiment. The piston pumps will be refilled from this pressurized tank.

Use of scCO_2: The conceptual design made by the different functional parts of a flow-through bench presented above (synthetized by the schema given in Fig. 6.8) is designed primarily to single phase flow experiments for which the partial pressure of CO_2 can be set from zero up to the minimum value of the pressure in the circuit, i.e. the pressure set by the back pressure system. However, one may want to simulate the flow of pure scCO_2 or mixture of scCO_2 and CO_2-saturated brine. In this case one must install a scCO_2 pump that is made of three components. The central part of the pump is a pressure booster that withdraws liquid CO_2 from a CO_2 bottle equipped with a plunger. To ensure a perfect functioning, the booster pump is cooled down to few Celsius (generally 4 or 5 °C) by a cooling system. On the high pressure side of the pressure booster the CO_2 is heated to the desired temperature (above 32 °C) to produce scCO_2. In order to avoid any scCO_2 trapping in the circuit and ensure a perfect mixing the scCO_2 must be injected as close as possible to the sample through a diffuser.

6.4.3.2 Measurements and Data Analysis

The two main measurements allowed by flow-through experiment are the monitoring of the pressure drop in the sample and the recurrent sampling and composition analysis of the outlet fluid. The fluid analyses are performed using the different methods presented in Sect. 6.4.2.4. The pressure drop is measured by a differential pressure sensor recording the pressure difference $\Delta P(t)$ between the inlet and the outlet of the sample. Permeability $k(t)$ is obtained from the differential pressure using the Darcy law (Eq. 6.2.3) for porous media.

In case of fractured low permeability rocks (usually $<10^{-19}$ m^2 for cements and claystones), the permeability and fracture aperture $a_h(t)$ are estimated using the cubic law approximation:

$$a_h(t) = \sqrt[3]{\frac{12\mu Q L}{w \Delta P(t)}} \tag{6.4.7}$$

and

$$k(t) = \frac{a_h^2(t)}{12} \tag{6.4.8}$$

where w is the width of the fracture.

The time-resolved porosity of the sample cannot be measured directly, but in certain cases the change of porosity $(\partial\phi/\partial t)$ can be evaluated from the fluid composition. The simple case is when the only chemical process is the dissolution of a single mineral. Let assume that the rock sample is made of Magnesium-rich calcite $Ca_\alpha Mg_\beta CO_3$, where and α and β denotes the fraction of Ca and Mg respectively in the calcite. The volume of dissolved calcite $V(t)$ is:

$$V(t) = \upsilon Q \int_{t'=0}^{t'=t} \alpha \Delta C_{\mathrm{Ca}}(t') + \beta \Delta C_{\mathrm{Mg}}(t') dt' \tag{6.4.9}$$

where υ is the calcite molar volume and ΔC_{Ca} and ΔC_{Mg} denotes the difference of calcium and magnesium concentration respectively between the outlet and the inlet fluids. The sample-scale porosity change of porosity is given by:

$$\frac{\partial \varphi(t)}{\partial t} = \frac{1}{V_s}\frac{\partial V(t)}{\partial t} \tag{6.4.10}$$

where V_s is the total sample volume (i.e. $V_s = \pi L D^2/4$). To evaluate the time-resolve porosity $\phi(t)$ one must integrate Eq. (6.4.10) which requires the knowledge of the value the porosity at a given time t, usually at $t = 0$. This can be achieved by non-invasive laboratory measurement using a gas porosimeter for example, or calculated from the analyses of XMT images.

When the dissolution process involves several minerals sharing the same cations or when both dissolution and precipitation occurs simultaneously, the mass balance (Eq. 6.4.9) is much difficult to calculate. However this can be completed if both the composition of the involved minerals and the stoichiometry of the reactions are known. Accurate chemical analysis of the rock sample combining different techniques such as chemical imaging using electronic microscopy, X-ray diffraction and bulk analysis may provide the necessary data. Then, if the value $V_i(t)$ for all the dissolved and precipitated minerals i can be calculated, then the change of porosity $(\partial\phi/\partial t)$ is obtained by Eq. (6.4.10) where the second member $\partial V(t)/\partial t$ is replaced by the sum $\sum_i \partial V/\partial t$.

6.4.4 *Examples of Experimental Studies of Fluid-Rock Interactions*

6.4.4.1 Experimental Investigation into the Sealing Capacity of Naturally Fractured Shale Caprocks Under scCO_2 Flow

A high pressure and temperature CO_2 flow rig was designed and built to investigate the sealing properties of naturally fractured caprocks under in-situ conditions and supercritical CO_2 and gaseous CO_2 flow (Edlmann et al. 2013; McCraw et al. 2016). One unfractured (sample B-c) and two naturally fractured caprock 38 mm diameter core samples (Samples B-a and B-b) were obtained from the North Sea East Brae field at a depth of circa 4 km, Fig. 6.9.

In addition to the flow results, petrophysical and mineralogical data was collected from the caprock matrix and fracture surface before and after exposure to CO_2. Porosity is around 17 %, the mean pore radius around 900 nM and X-Ray Diffraction (XRD) investigations of both the caprock matrix and fracture fill material show that quartz (43 %) and illite (≈25 %) are the primary minerals that make up the Kimmeridge clay caprock, with minor chlorite (8 %) and kaolinite (8 %).

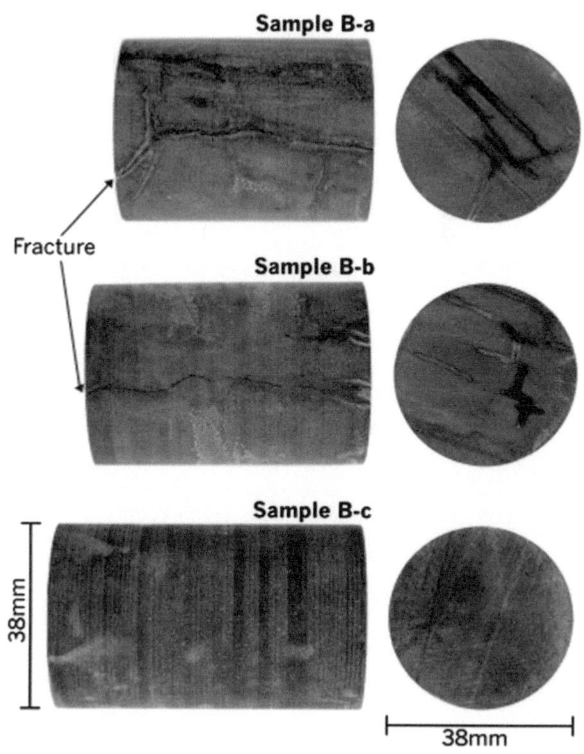

Fig. 6.9 Images of the three core samples of caprock from East Brae

The flow experiments were run in 4 phases:

- Phase 1—Increasing to in-situ $scCO_2$ conditions
- Phase 2—Maintaining in-situ $scCO_2$ conditions
- Phase 3—Maintaining in-situ gaseous CO_2 conditions
- Phase 4—Re-run of the increasing upstream CO_2 fluid pressure and lock in under supercritical and gaseous conditions to investigate wettability changes after CO_2 exposure along the fracture face

The results of Phase 1, 2 and 3 for samples B-a and B-b are presented as confining pressure (in MPa, MegaPascal), upstream pressure (MPa) and downstream pressure (MPa) on the primary axis and temperature (°C) on the secondary axis all against time (Figs. 6.10 and 6.11). There was no detectable flow of CO_2 (downstream pressure) measured across both fractured samples, even with a pressure differential across the fractured of 43 and 51 MPa, for samples B-a and B-b respectively, which equates to a pressure gradient of 866 and 942 MPa/m. As soon as the fluid pressure, confining pressure and temperature were dropped to gaseous CO_2 conditions of 20 MPa confining pressure, 5 MPa upstream pressure (the same

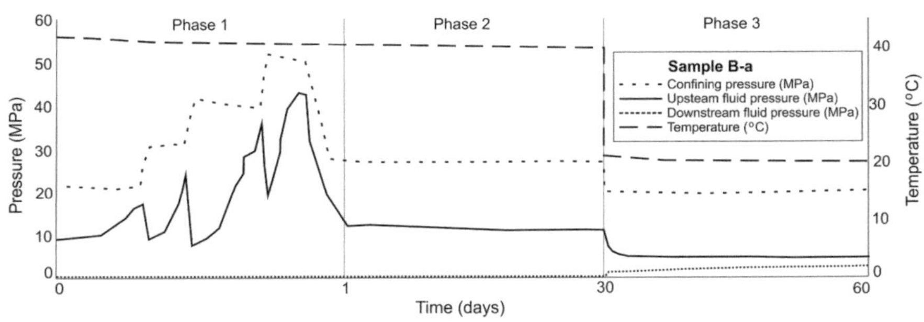

Fig. 6.10 Results of the fractured caprock sample B-a, plotted as confining pressure (MPa), upstream pressure (MPa), downstream pressure (MPa) and temperature (°C),with time

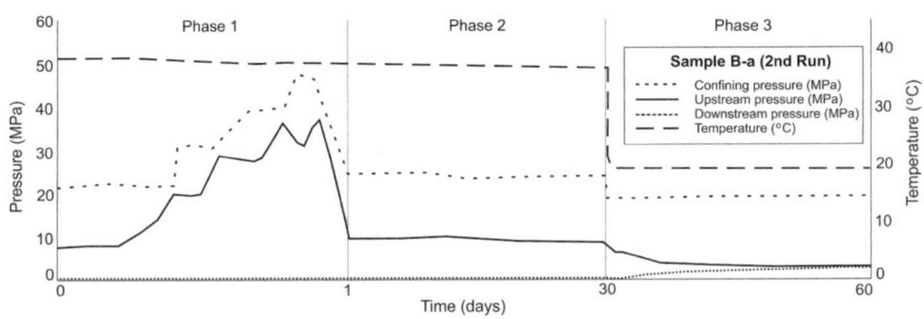

Fig. 6.11 Results of the fractured caprock sample B-b, plotted as confining pressure (MPa), upstream pressure (MPa), downstream pressure (MPa) and temperature (°C), with time

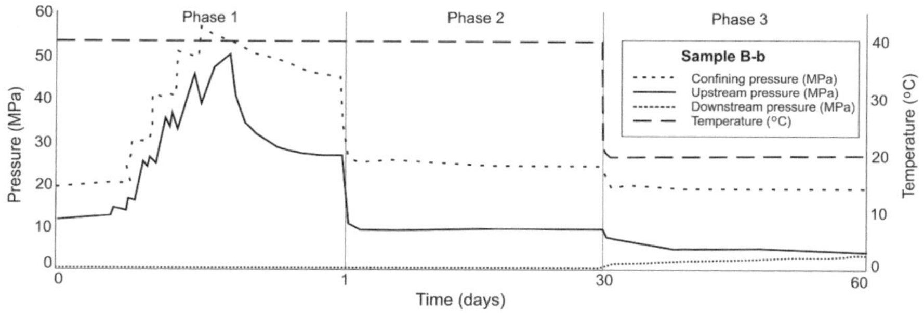

Fig. 6.12 Results of the second flow run on fractured caprock sample B-a, plotted as confining pressure (MPa), upstream pressure (MPa), downstream pressure (MPa) and temperature (°C),with time, to investigate any wettability effects after CO_2 exposure along the fracture face

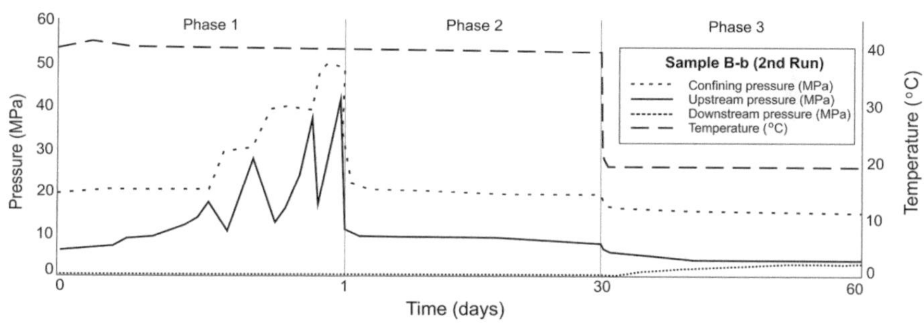

Fig. 6.13 Results of the second flow run on fractured caprock sample B-b, plotted as confining pressure (MPa), upstream pressure (MPa), downstream pressure (MPa) and temperature (°C), with time, to investigate any wettability effects after CO_2 exposure along the fracture face

15 MPa pressure difference as with the supercritical lock in) and 20 °C temperature, the downstream pressure immediately and steadily increased over the 30 days towards that of the upstream pressure. This introduces the possibility that for these particular naturally occurring fractures gaseous CO_2 does flow across the fracture, whereas there was no supercritical CO_2 flow across the fracture, even when the fracture aperture is held under the same differential pressure.

During phase 4, the re-run after gas phase CO_2 exposure to the fracture, the same pattern of pressure (flow) across the fracture was observed during the second run as was seen during the first run, Figs. 6.12 and 6.13. There was no flow measured under increasing $scCO_2$ flow even with pressure differentials of 37 and 42 MPa for samples B-a and B-b respectively. This was also under conditions of fluctuating fracture aperture linked directly to a fluctuating difference in pressure between the upstream fluid and confining pressure. There was little or no detectable flow of $scCO_2$ measured across both fractured samples under $scCO_2$ lock over 30 days. However under gaseous CO_2 lock in (with the fracture aperture held under the same pressure difference between the upstream fluid pressure and confining pressure as it

Table 6.2 Average mineral weight percentages for sample B-a matrix and fracture before and after CO_2 exposure

	wt% Al	wt% Si	wt% Ca	wt% Mg	wt% Fe
Sample B-a matrix before CO_2 exposure	7.45	37.30	0.51	0.54	0.68
Sample B-a matrix after CO_2 exposure	9.51	35.24	0.47	0.61	0.83
Sample B-a fracture before CO_2 exposure	8.21	35.86	0.55	0.51	1.17
Sample B-a fracture after CO_2 exposure	10.40	32.65	1.07	1.33	0.99

was under supercritical conditions), the same pattern of immediate and increasing downstream pressure towards that of the upstream pressure was observed.

This indicates that under these experimental conditions and timescales, the wettability (and in turn capillary entry pressures) of these natural fractures was not influenced by exposure to gaseous CO_2.

Table 6.2 presents the average mineral weight percent of the caprock matrix material and fracture material before and after the CO_2 flow experiments on sample B-a from SEM energy dispersive X-ray elemental analysis. It can be ascertained that although there was a slight reduction in Silicon (2 %) and corresponding increase in Aluminum (2 %) after exposure to CO_2 as cation exchange occurs. However it is not significant enough to cause fracture/matrix pore network opening or closing that would influence the CO_2 flow through the duration of the experiment.

This lack of any chemical reactivity is borne out by comparing the sample dimensions before and after the experiment, the confining pressure data logs also corroborate the lack of any volume changes within the samples at equilibrium during experimentation.

The conclusions we can draw from the experimental work undertaken so far on the sealing capability of naturally fractured shale caprocks to supercritical CO_2 flow are that:

- Supercritical CO_2 did not flow through these particular tight natural caprock fractures under supercritical reservoir conditions.
- When the temperature and fluid pressure were reduced to below the critical point, CO_2 in its gas phase did flow through the tight natural caprock fractures (even with a constant pressure difference between the confining pressure and the upstream fluid pressure of 15 MPa under both supercritical and gaseous conditions).
- This contradicts the expected flow profile of a supercritical fluid having enhanced flow properties.
- The contradictory experimental observations are linked to the complex interplay between the fluid conductivity response of the CO_2 phase to the fracture properties, the influence of stress on the fracture aperture, the chemical interaction between the rock minerals and the CO_2 fluid, the fluid pressure influencing the fracture permeability, the influence of CO_2 phase on the capillary entry pressure and the relationship between CO_2 phase on the wettability, interfacial tension and contact angle.

- To facilitate the observed result of gaseous CO_2 flow and inhibited supercritical CO_2 flow it is likely that the interplay between normal stress and pore pressure controlling fracture aperture and the density of CO_2 being inversely proportional to permeability were the dominant influencing factors.
- Calculations indicate that there is a large decrease in hydraulic conductivity as the aperture size reduces, indicating that it is aperture size that is the dominant controlling factor in fracture flow of CO_2. It is possible that there is a critical threshold of fracture aperture size which controls CO_2 flow along the fracture. Above the critical aperture size $scCO_2$ and gaseous CO_2 will flow, on or near the critical fracture aperture size we see gaseous flow but not $scCO_2$ flow and below the critical aperture size we would see little or no gas or $scCO_2$ flow along the fracture and flow would become matrix dominated.
- This has significant implications for the planning of CO_2 storage projects in the North Sea basin, in that the CO_2 should be stored at pressures and temperatures (depth) comfortably above the CO_2 critical point.

6.4.4.2 Experimental Investigation of the Role of Local Heterogeneities on Reactive Transport

Understanding the fundamental processes governing the fluids-rocks interactions is a crucial stake to model, for various temporal and spatial scales, the transport and the storage of CO_2. One of the major difficulties is the identification and the hierarchical organization of the involved imbricated physical and chemical mechanisms. Consequently, an essential step is to perform experiments in controlled conditions for which results can be analyzed from pore-scale to the scale of the samples.

Several percolation experiments of acid fluid (CO_2-rich brines) through different type of rock samples reproducing in situ conditions ($T < 200$ °C, $P < 20$ MPa) will be present here to enlighten the use of experimental results for characterizing the role of heterogeneities on transport and reactions processes (dissolution, precipitation, redox reactions). All the flow-through experiments presented here have been performed using the *ICARE Lab CO_2 sequestration evaluation flow system* presented in Sect. 6.4.3 and described in more details in Luquot and Gouze (2009).

The different analytical techniques mentioned in this chapter have been used in order to characterize the mineral rock composition, the pore structure and geometry, the connectivity of the core before and after each percolation experiment. Besides the classical analytical methods, SEM (Scanning Electron Microscope), Raman spectrometry, Mossbauer, Ionic probe, TEM (Transmission Electron Microscope), have been used as well as X-ray Microtomography images (XMT). As explained in Sect. 6.2.1, one can extract key parameters such as the total porosity, the effective porosity, the pore size distribution, the pore/rock interface surface, the tortuosity, the effective diffusion coefficient, the permeability and other morphological and geometrical parameters such as capillary pressure and pore surface curvature, by processing the 3D micro-tomographic images.

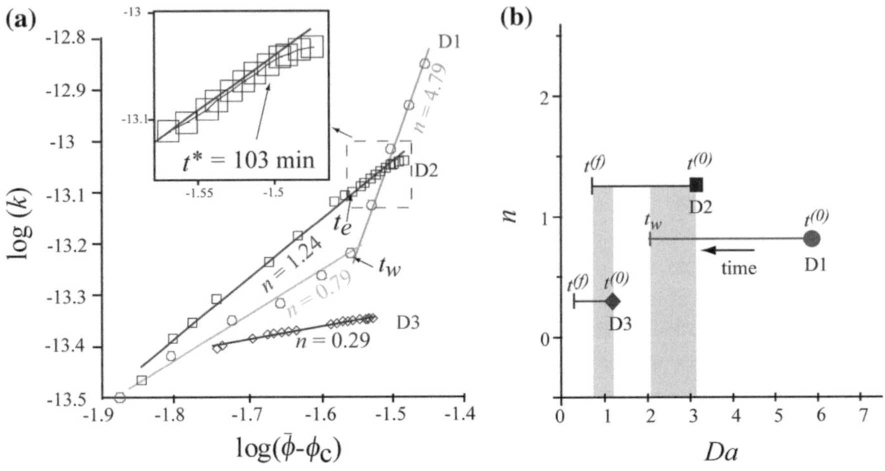

Fig. 6.14 **a** Log-permeability versus log-porosity for the dissolution experiments D1 (*circles*), D2 (*squares*) and D3 (*diamonds*). *Inset* Focus on the experiment D2 data for $t > t_e$. For $t > t^*$, data deviate progressively from the power law model. **b** Exponent n versus $Da(0)$ for experiment D1 (*circles*), D2 (*squares*) and D3 (*diamonds*). The range of $Da(t)$ for each of the experiment is reported (*horizontal lines*) as well as the overlapping between experiments D1–D2 and D2–D3 (from Luquot and Gouze 2009)

The methodology and experimental approach previously mentioned have been first applied to characterize the different dissolution patterns that occurred during CO_2-rich brine injection through limestone core samples (from Paris Basin). Various flow-through experiments have been performed using a same Peclet number but changing the Damköhler number. All the experiments have been done under CO_2 storage conditions ($T = 100$ °C, $P = 12$ MPa, $2.5 < P_{CO_2} < 10$ MPa). Various dissolution patterns have been described during the different dissolution experiments depending of the *Da* number. These dissolution features have been correlated to the flow conditions and initial chemical fluid properties: from localized wormhole formation ($P_{CO_2} = 10$ MPa) to homogeneous dissolution ($P_{CO_2} = 2.5$ MPa). The main results of these set of experiments were published in Gouze and Luquot (2011) and Luquot and Gouze (2009). In particular, the authors demonstrated that porosity, permeability and reactive surface area changes due to dissolution reaction are controlled by initial chemical conditions of the injected fluid and initial parameters (Fig. 6.14). A memory effect has been observed in the variation of all these parameters with time. A phenomenological relationship between porosity and permeability (Fig. 6.15) and reactive surface area have been proposed and later validated by others authors.

Gouze and Luquot (2011) showed that permeability changes during dissolution process are controlled by tortuosity changes in case of heterogeneous dissolution (wormhole localization formation) and to the hydraulic pore diameter in case of homogeneous dissolution (Fig. 6.15).

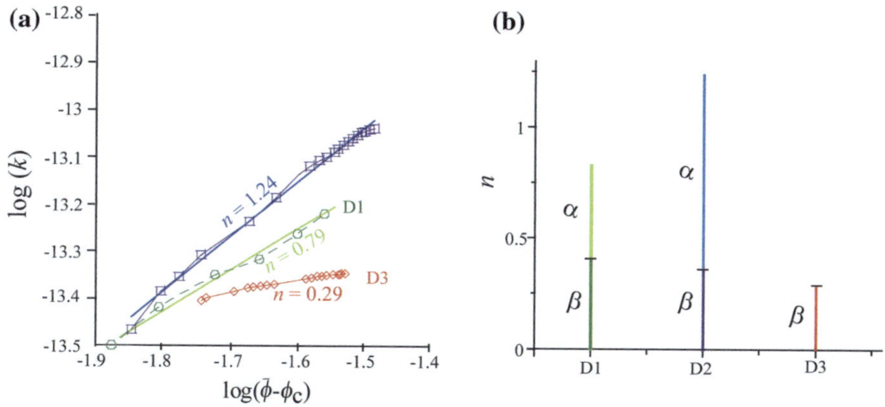

Fig. 6.15 **a** Sample-scale averaged porosity versus permeability for experiments D1, D2 and D3. **b** Values of exponent α (indicator of hydraulic radius increase) and β (indicator of tortuosity increase), with $n = \alpha + \beta$, for experiments D1, D2 and D3 (from Gouze and Luquot 2011)

Others experiments have been performed using different types of limestones confirming that the initial rock structure strongly controls the dissolution pattern and specifically the development of preferential paths and thus controls the permeability-porosity relationship (Fig. 6.14). Furthermore, experiment with lower CO_2 partial pressure using limestone samples characterized by smaller pores have be run in order to try observing homogeneous dissolution processes, but here again the formation of preferential flow path was observed (Fig. 6.16).

From these results on can propose a phenomenological relationship between tortuosity change $\Delta\tau$ and P_{CO_2} and speculated that the porosity–permeability trend is mainly controlled by the change in the effective hydraulic diameter during dissolution with low P_{CO_2} (Fig. 6.17). These results cast doubt on the use of the Peclet and Damköhler numbers to predict dissolution pattern independently on other information on the structural properties and the heterogeneity of the rock (Luquot et al. 2014).

Andreani et al. (2009) and Luquot et al. (2012) studied sandstone core samples coming from potential CO_2 storage reservoir in Australia and dunite samples coming from San Carlos, Arizona, respectively. The interesting point of these two different rocks is the complex mineralogy and the large range of chemical processes that can occur in contact with CO_2-rich brine.

The sandstone rock was coming from de Pretty Hill Formation from Otway Basin and was mainly composed of quartz and feldspar and almost 10 % of Fe-rich chamosite (chlorite) and Laumontite (Ca-rich zeolite). Two interesting processes were observed during CO_2-rich brine percolation through this sandstone: the first one was the feldspar alteration into kaolinite which induces a permeability decrease which was described for the first time. The second key result of this study was the redox reaction characterized between CO_2 and iron-rich chamosite mineral. The results show that a fraction of CO_2 was stored as reduced carbon (graphite) using

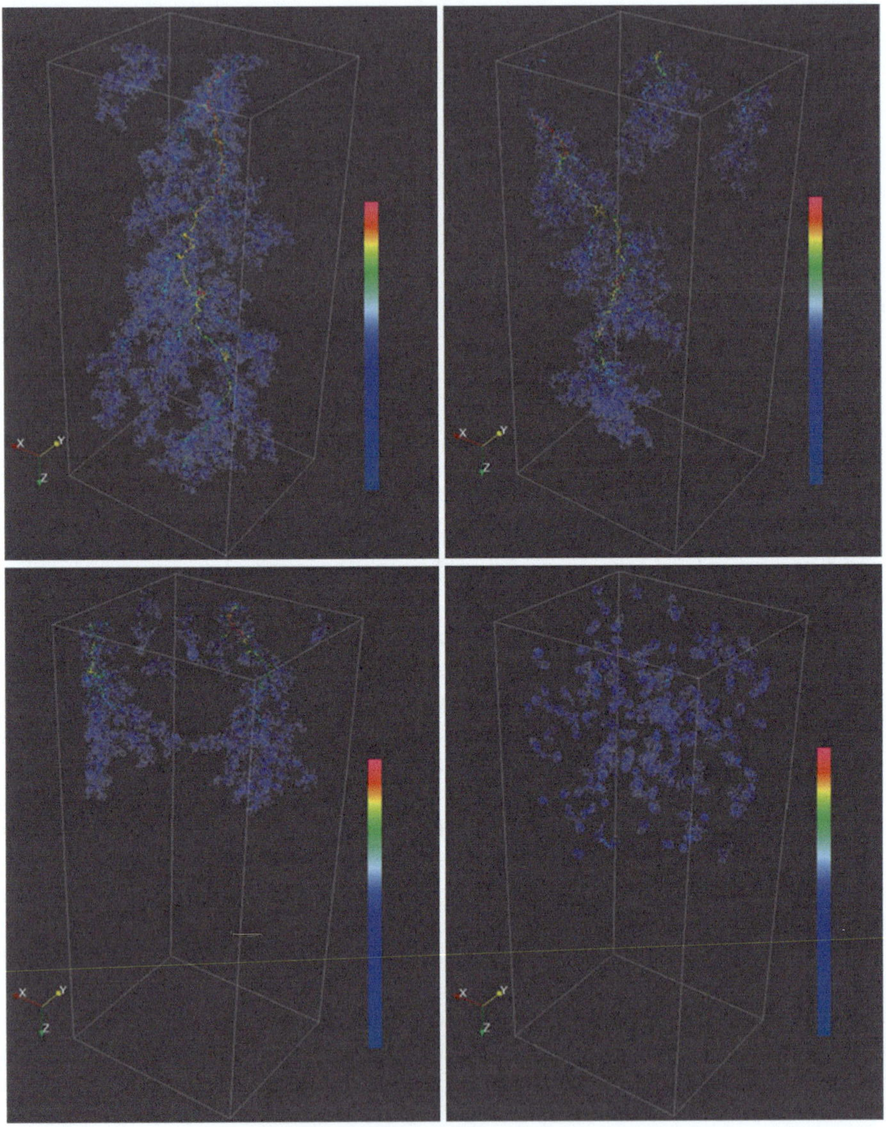

Fig. 6.16 Representation of the nodes of the main components of the skeleton, i.e., parts of the skeleton coinciding with connected components of the void space that were exposed to dissolution at macroscopic scale. The *color map* corresponds to the radius of the largest sphere centered at each node inscribed *inside* the void space (brightest points correspond to largest radii, logarithmic scale in voxel unit), from Luquot et al. (2014)

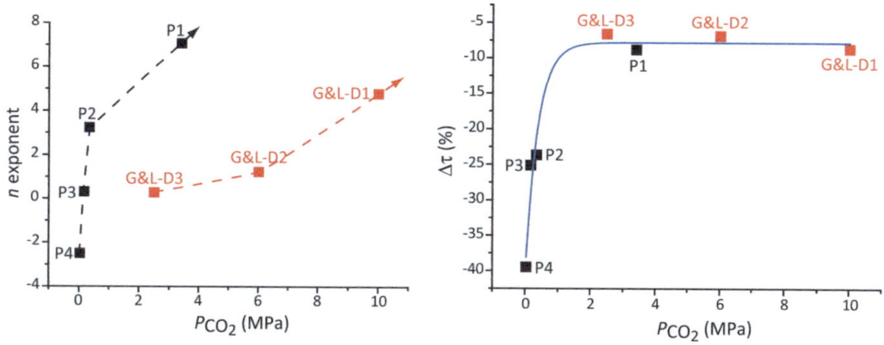

Fig. 6.17 *Left figure* Change of the exponent n of the permeability–porosity relationship versus P_{CO_2}. *Right figure* Tortuosity versus P_{CO_2}. In *black*, the results for experiments P1, P2, P3, and P4 are presented and in red the results for experiments G&L-D1, G&L-D2, and G&LD3 from Luquot and Gouze (2009) and Gouze and Luquot (2011). The *blue curve* represents the equation $\Delta\tau = y_0 + A\ \exp(B \times P_{CO_2})$, with $y_0 = -7.83$, $A = -33.14$ and $B = 0.37$ (from Luquot et al. 2014)

realistic conditions for CO_2 geological storage (Fig. 6.18). This study allows highlighting the role of local mineral heterogeneity on chemical reactions. As observed in Fig. 6.18, the redox reaction occurred very locally closed to chamosite dissolution features.

The study of CO_2-rich brine reaction with dunite was performed at very low flow rates and high temperature and pressure conditions. The rock sample was almost composed of olivine so in this case the initial rock was not characterized by an initial mineral heterogeneity. Nevertheless, various reactions can take place after olivine dissolution triggerd by the CO_2-enriched water: mainly serpentine precipitation (olivine hydration mechanism), carbonate precipitation (mainly Mg, Fe, Ca-carbonates), iron oxide precipitations, hydrogen production. In this study, Andreani et al. (2009) observed the localization of the different reactions depending of the local transport conditions. In zones were transport was controlled by advection, i.e. where the reactant renewal was high, olivine dissolution rate, hydration reactions and iron oxide precipitation were observed. Conversely, carbonate precipitation was observed were the reactant transport was controlled by diffusion such as in dead-end structures (Fig. 6.19). These results clearly indicate that local transport conditions controlled the product of the chemical reaction and that the assumption of *full mixing* discussed in Sect. 6.4.1 is hardly valid in such highly reactive system involving high CO_2 concentration fluids.

More recently, Luquot et al. (2016) confirmed the role of local chemical/mineralogical heterogeneity on reactive processes but also the importance of the local transport conditions. They performed CO_2, sulfate-rich brine percolation experiments at different injection flow rates through sandstone core rock samples coming from Heletz formation in Israel (MUSTANG test site). Experiments were performed under in situ Heletz storage conditions (60 °C and 15 MPa). The authors

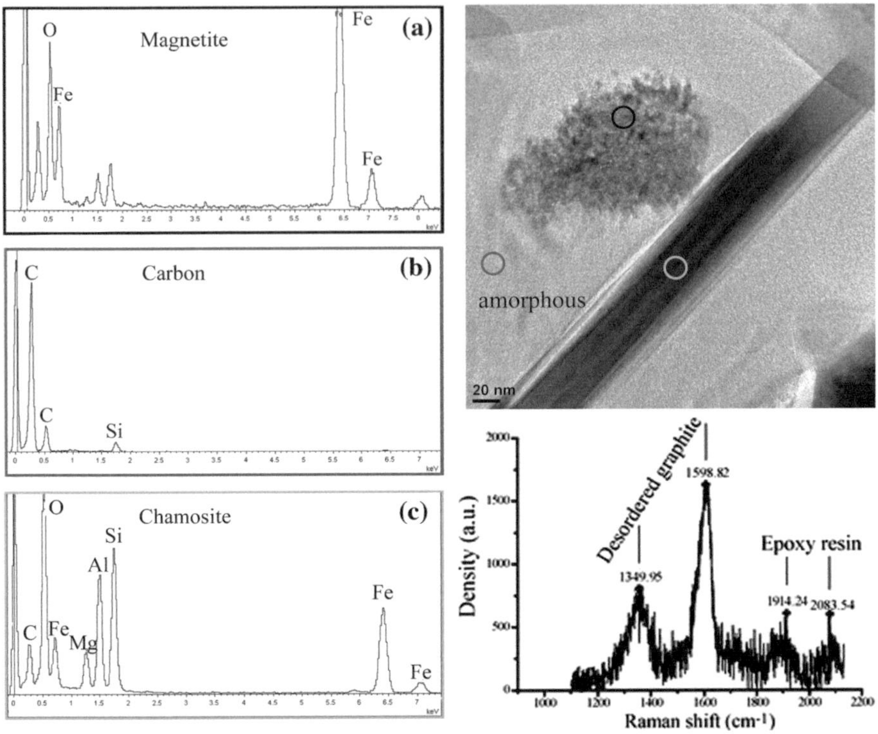

Fig. 6.18 TEM image of the chlorite fibers and precipitated magnesite surrounded by amorphous and poorly crystallized carbon. *On the left* Chemical analysis associated with the TEM image, **a** magnetite chemical analysis, **b** carbon chemical analysis and **c** chamosite chemical analysis. Raman spectra of precipitated carbon near chlorite compared with epoxy-resin spectra and disordered carbon spectra (from Luquot et al. 2012)

observed that different reaction processes took place depending on the transport conditions. First, they concluded that dolomite and ankerite dissolution (the two main minerals displaying fast kinetics) was controlled the hydrodynamic transport of the reactants; second, they observed that dissolution processes were rather homogeneous at high flow rates whereas the dissolution pattern appears to be more localized (wormhole-like) for the lowest flow rate injection experiments; third, permeability increases during all experiments, regardless of flow rate and brine type, but at high flow rates, some particles were apparently dragged through the samples, causing sporadic decreases in permeability due to local pore clogging. Furthermore it was observed that the precipitation of the secondary minerals (kaolinite, muscovite and smectite) as well as the dissolution of K-feldspar was larger at low flow rate injection than at high flow rate. The authors attributed this behavior to dissolution/precipitation localization mechanisms in the immobile zones where local chemical micro-environments may subsist.

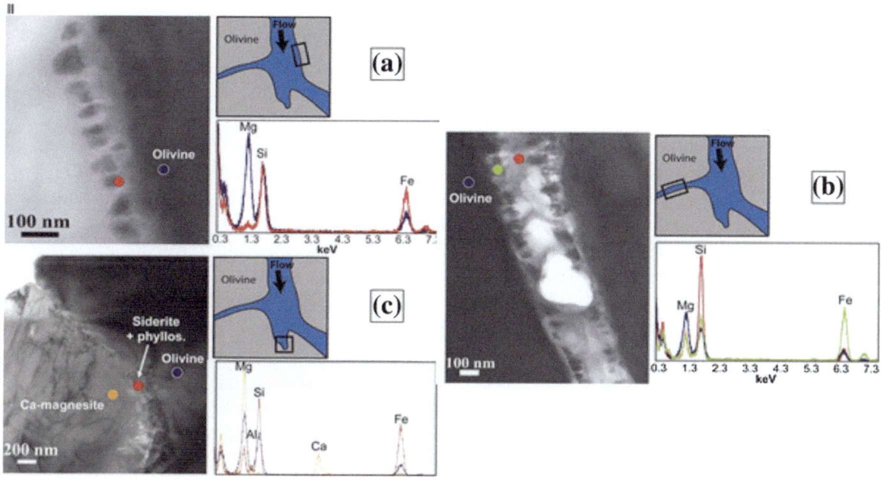

Fig. 6.19 TEM images of the main reaction textures (**a**–**c**) observed in the samples after the experiment end. For each reaction texture, a sketch diagram shows the main type of hydrodynamic zone in which the texture is observed and the corresponding reactants and products identified. This illustrates the variability of transport-reaction processes controlled by the structural heterogeneity of the sample (from Andreani et al. 2009)

The conclusions we can draw from this set of experimental results are:

- For the duration of the experiments, the changes in porosity, permeability and reactive surface area values caused by the dissolution reactions are controlled by the composition of the injected fluid but also by the initial properties of the rock; a memory effect of the initial heterogeneity of the samples has been observed in the variation of all these parameters with time.
- These experiments are quite unique for evaluating the permeability-porosity relationship associated with the dissolution process, while permeability-porosity relationships are generally obtained from cross-plotting permeability and porosity values measured on sets of sampled rock, i.e. not considering the dynamic of the process.
- One can speculate that the porosity–permeability relationship can be parametrized at first order by the effective hydraulic diameter and tortuosity changes triggered by the rock-forming mineral dissolution. Porosity–permeability relationship associated with precipitation mechanisms is presently not settled.
- The type of reactions experimented here cannot be modeled easily by a continuum approach. Local chemical environments triggering strong local concentration gradients at pore scale triggered by the heterogeneity of the velocity field and consequently the spatial variability of the local values of the *Pe* and *Da* numbers.

References

Akin S, Kovscek AR (1999) Imbibition studies of low-permeability porous media. Society of Petroleum Engineers

Al-Bulushi IR, Al-Maamari RS, Wilson OB (2012) Brine versus Klinkenberg corrected gas permeability correlation for Shuaiba carbonate formation. J Pet Sci Eng 92–93:24–29

Andreani M, Luquot L, Gouze P, Godard M, Hoisé E, Gibert B (2009) Experimental alteration of peridotite during injection of CO_2-rich fluids. Environ Sci Technol 43:1226–1231

Andrew M, Bijeljic B, Blunt MJ (2014) Pore-scale imaging of trapped supercritical carbon dioxide in sandstones and carbonates. Int J Greenh Gas Control 22:1–14

Bear J (1972) Dynamics of fluids in porous media. American Elsevier, New York, NY

Berg S, Oedai S, Ott H (2013) Displacement and mass transfer between saturated and unsaturated CO_2–brine systems in sandstone. Int J Greenh Gas Control 12:478–492

Berkowitz B, Scher H, Silliman SE (2000) Anomalous transport in laboratory-scale, heterogeneous porous media. Water Resour Res 36:149–158

Birkholzer JT, Cihan A, Zhou Q (2012) Impact-driven pressure management via targeted brine extraction—conceptual studies of CO_2 storage in saline formations. Int J Greenh Gas Control 7:168–180

Bloomfield JP, Williams AT (1995) An empirical liquid permeability–gas permeability correlation for use in aquifer properties studies. Q J Eng Geol Hydrogeol 28:S143–S150

Brooks RH, Corey AT (1966) Properties of porous media affecting fluid flow. J Irrig Drain Div 92:61–88

Brooks RH, Corey AT (1964) Hydraulic properties of porous media. Colorado State University, Colorado

Brunauer S, Emmett PH, Teller E (1938) Adsorption of gases in multimolecular layers. J Am Chem Soc 60:309–319

Burnside NM, Naylor M (2014) Review and implications of relative permeability of CO_2/brine systems and residual trapping of CO_2. Int J Greenh Gas Control 23:1–11

Carroll S, Hao Y, Smith M, Sholokhova Y, (2013) Development of scaling parameters to describe CO_2–rock interactions within Weyburn–Midale carbonate flow units. Int J Greenh Gas Control, The IEAGHG Weyburn-Midale CO_2 Monitoring and Storage Project 16, suppl. 1, S185–S193

Chang YC, Mohanty KK, Huang DD, Honarpour MM (1997) The impact of wettability and core-scale heterogeneities on relative permeability. J Pet Sci Eng 18:1–19

Chierici GL (1984) Novel relations for drainage and imbibition relative permeabilities. Soc Pet Eng J 24:275–276

Chiquet P, Broseta D, Thibeau S (2007) Wettability alteration of caprock minerals by carbon dioxide. Geofluids 7:112–122

Clennell MB (1997) Tortuosity: a guide through the maze. Geol Soc Lond Spec Publ 122:299–344

Coates GR, Denoo SA (1980) Log derived mechanical properties and rock stress. Presented at the SPWLA 21st annual logging symposium, Society of Petrophysicists and Well-Log Analysts

Corey AT (1954) The interrelation between gas and oil relative permeabilities. Producers Monthly 19:38–41

Cortis A, Berkowitz B (2004) Anomalous transport in “classical” soil and sand columns. Soil Sci Soc Am J 68:1539–1548

Crank J (1975) The mathematics of diffusion. Clarendon Press, Oxford

Cueto-Felgueroso L, Juanes R (2016) A discrete-domain description of multiphase flow in porous media: Rugged energy landscapes and the origin of hysteresis. Geophys Res Lett 43

Cui X, Bustin AMM, Bustin RM (2009) Measurements of gas permeability and diffusivity of tight reservoir rocks: different approaches and their applications. Geofluids 9:208–223

Daccord G, Lenormand R, Liétard O (1993) Chemical dissolution of a porous medium by a reactive fluid—I. Model for the “wormholing” phenomenon. Chem Eng Sci 48:169–178

Dake L (1978) Fundamentals of reservoir engineering. Elsevier, Amsterdam

Dandekar AY (2013) Petroleum reservoir rock and fluid properties, 2nd edn. CRC Press, Taylor & Francis Group, New York

Dentz M, Gouze P, Russian A, Dweik J, Delay F (2012) Diffusion and trapping in heterogeneous media: an inhomogeneous continuous time random walk approach. Adv Water Resour 49:13–22

Duan Z, Sun R, Zhu C, Chou I-M (2006) An improved model for the calculation of CO_2 solubility in aqueous solutions containing Na^+, K^+, Ca^{2+}, Mg^{2+}, Cl^-, and SO_4^{2-}. Mar Chem 98:131–139

Durner W (1994) Hydraulic conductivity estimation for soils with heterogeneous pore structure. Water Resour Res 30:211–223

Ebeltoft E, Lomeland F, Brautaset A, Haugen Å (2014). Parameter based scal-analysing relative permeability for full field application. In: International Symposium of the Society of Core Analysis, Avignon, France, 8–11 Sep 2014

Edlmann K, Somerville JM, Smart BGD, Hamilton SA, Crawford BR (1998) Predicting rock mechanical properties from wireline porosities. Society of Petroleum Engineers, Norway

Edlmann K, Haszeldine S, McDermott C (2013) Experimental investigation into the sealing capability of naturally fractured shale caprocks to supercritical carbon dioxide flow. Environ Earth Sci 70(7): 3393-3409.-17

Fischer G (1992) The determination of permeability and storage capacity: pore pressure oscillation method. In: Wong T, Evans B (eds) Fault mechanics and transport properties of rocks. Academic Press, New York, pp 187–211

Fjar E, Holt RM, Raaen AM, Risnes R, Horsrud P (1992) Petroleum related rock mechanics. Elsevier Science, London

Garrouch AA, Ali L, Qasem F (2001) Using diffusion and electrical measurements to assess tortuosity of porous media. Ind Eng Chem Res 40:4363–4369

Gassman F (1951) Uber die elastizitat poroser mediem. Vier der Natur, Gresellschaft in Zurich 1–23

Georgiadis A, Maitland G, Trusler JPM, Bismarck A (2010) Interfacial tension measurements of the (H_2O + CO_2) system at elevated pressures and temperatures. J Chem Eng Data 55:4168–4175

Gharbi O (2014) Fluid–rock interactions in carbonates; applications to CO_2 storage. Imperial College

Giesche H (2006) Mercury porosimetry: a general (practical) overview. Part Part Syst Charact 23:9–19

Gjetvaj F, Russian A, Gouze P, Dentz M (2015) Dual control of flow field heterogeneity and immobile porosity on non-Fickian transport in Berea sandstone. Water Resour Res 51:8273–8293

Glover P (2009) What is the cementation exponent? A new interpretation. Lead Edge 28:82–85

Golfier F, Zarcone C, Bazin B, Lenormand R, Lasseux D, Quintard M (2002) On the ability of a Darcy-scale model to capture wormhole formation during the dissolution of a porous medium. J Fluid Mech 457:213–254

Gouze P, Le Borgne T, Leprovost R, Lods G, Poidras T, Pezard P (2008) Non-Fickian dispersion in porous media: 1. Multiscale measurements using single-well injection withdrawal tracer tests. Water Resour Res 44:W06426

Gouze P, Leprovost R, Poidras T, Le Borgne T, Lods G, Pezard PA (2009) CoFIS and TELog: new downhole tools for characterizing dispersion processes in aquifers by single-well injection-withdrawal tracer tests. Comptes Rendus Geosci Hydrogeophys 341, 965–975

Gouze P, Luquot L (2011) X-ray microtomography characterization of porosity, permeability and reactive surface changes during dissolution. J Contam Hydrol. Reactive Transport in the Subsurface: Mixing, Spreading and Reaction in Heterogeneous Media 120–121, 45–55

Guarracino L, Rötting T, Carrera J (2014) A fractal model to describe the evolution of multiphase flow properties during mineral dissolution. Adv Water Resour 67:78–86

Haines WB (1930) Studies in the physical properties of soil. V. The hysteresis effect in capillary properties, and the modes of moisture distribution associated therewith. J Agric Sci 20:97–116

Halvorsen C, Hurst A (1990) Principles, practice and applications of laboratory minipermeametry. In: Worthington P (ed) Advances in core evaluation: accuracy and precision in reserves estimation, pp 521–549

Hamouda AA, Karoussi O, Chukwudeme EA (2008) Relative permeability as a function of temperature, initial water saturation and flooding fluid compositions for modified oil-wet chalk. J Pet Sci Eng 63:61–72
Hassanizadeh SM, Gray WG (1993) Thermodynamic basis of capillary pressure in porous media. Water Resour Res 29:3389–3405
Hidalgo JJ, Carrera J (2009) Effect of dispersion on the onset of convection during CO_2 sequestration. J Fluid Mech 640:441
Hilfer R, Øren PE (1996) Dimensional analysis of pore scale and field scale immiscible displacement. Transp Porous Media 22:53–72
Hingerl FF, Yang F, Pini R, Xiao X, Toney MF, Liu Y, Benson SM (2016) Characterization of heterogeneity in the Heletz sandstone from core to pore scale and quantification of its impact on multi-phase flow. Int J Greenh Gas Control
Hoefner ML, Fogler HS (1988) Pore evolution and channel formation during flow and reaction in porous media. AIChE J 34:45–54
Hoek E, Franklin JA (1968) Simple triaxial cell for field or laboratory testing of rock. Trans Inst Min Met 77:A22–A26
Honarpour M, Koederitz L, Harvey AH (1986) Relative permeability of petroleum reservoirs. CRC Press, Boca Raton
Hossain Z (2012) Rock physics modeling of CO_2 bearing reservoir rocks. Society of Petroleum Engineers, Denmark
Huang DD, Honarpour MM (1998) Capillary end effects in coreflood calculations. J Pet Sci Eng 19:103–117
Iglauer S, Paluszny A, Pentland CH, Blunt MJ (2011) Residual CO_2 imaged with X-ray micro-tomography. Geophys Res Lett 38:L21403
Iglauer S, Pentland CH, Busch A (2015) CO_2 wettability of seal and reservoir rocks and the implications for carbon geo-sequestration. Water Resour Res 51:729–774
Jaeger J, Cook NG, Zimmerman R (2007) Fundamentals of rock mechanics, 4th edn. Wiley-Blackwell, Malden, MA
Johnson EF, Bossler DP, Naumann VO (1959) Calculation of relative permeability from displacement experiments 216:370–372
Jones SC, Roszelle WO (1978) Graphical techniques for determining relative permeability from displacement experiments. J Pet Technol 30:807–817
Juanes R, Spiteri EJ, Orr FM, Blunt MJ (2006) Impact of relative permeability hysteresis on geological CO_2 storage. Water Resour Res 42:W12418
Kamath J, de Zabala EF, Boyer RE (1995) Water/Oil Relative permeability endpoints of intermediate-wet, low-permeability rocks. SPE Form Eval 10:4–10
Klinkenberg LJ (1941) The permeability of porous media to liquids and gases. Presented at the Drilling and Production Practice, American Petroleum Institute, pp 200–213
Kosugi K 'ichirou (1996) Lognormal distribution model for unsaturated soil hydraulic properties. Water Resour Res 32:2697–2703
Krevor SCM, Pini R, Zuo L, Benson SM (2012) Relative permeability and trapping of CO_2 and water in sandstone rocks at reservoir conditions. Water Resour Res 48:W02532
Lasaga AC (1998) Kinetic theory in the earth sciences. Princeton University Press, New Jersey, 811p
Lee TC, Kashyap RL, Chu CN (1994) Building skeleton models via 3-D medial surface axis thinning algorithms CVGIP Graph. Models Image Process 56:462–478
Leverett MC (1941) Capillary behavior in porous solids. Trans AIME 142:152–169
Levine JS, Goldberg DS, Lackner KS, Matter JM, Supp MG, Ramakrishnan TS (2014) Relative permeability experiments of carbon dioxide displacing brine and their implications for carbon sequestration. Environ Sci Technol 48:811–818
Levy M, Berkowitz B (2003) Measurement and analysis of non-Fickian dispersion in heterogeneous porous media. J Contam Hydrol 64:203–226
Li K, Horne RN (2006) Comparison of methods to calculate relative permeability from capillary pressure in consolidated water-wet porous media. Water Resour Res 42:W06405

Li K, Horne RN (2002) Experimental verification of methods to calculate relative permeability using capillary pressure data. Society of Petroleum Engineers

Lindquist WB, Venkatarangan A (1999) Investigating 3D geometry of porous media from high resolution images. Phys Chem Earth Part Solid Earth Geod 24:593–599

Lomeland F, Ebeltoft E, Thomas WH (2005) A new versatile relative permeability correlation. In: International Symposium of the Society of Core Analysts, Toronto, Canada, pp 21–25

Lu B, Torquato S (1993) Chord-length and free-path distribution functions for many-body systems. J Chem Phys 98:6472–6482

Luquot L, Andreani M, Gouze P, Camps P (2012) CO_2 percolation experiment through chlorite/zeolite-rich sandstone (Pretty Hill Formation—Otway Basin–Australia). Chem Geol 294–295:75–88

Luquot L, Gouze P (2009) Experimental determination of porosity and permeability changes induced by injection of CO_2 into carbonate rocks. Chem Geol. CO_2 geological storage: Integrating geochemical, hydrodynamical, mechanical and biological processes from the pore to the reservoir scale 265, 148–159

Luquot L, Gouze P, Niemi A, Bensabat J, Carrera J (2016) CO_2-rich brine percolation experiments through Heletz reservoir rock samples (Israel): role of the flow rate and brine composition. Int J Greenh Gas Control

Luquot L, Rodriguez O, Gouze P (2014) Experimental characterization of porosity structure and transport property changes in limestone undergoing different dissolution regimes. Transp Porous Media 101:507–532

Mathias SA, Gluyas JG, de Miguel GJGM, Bryant SL, Wilson D (2013) On relative permeability data uncertainty and CO_2 injectivity estimation for brine aquifers. Int J Greenh Gas Control 12:200–212

McCraw C, Edlmann K, Miocic J, Gilfillan S, Haszeldine RS, McDermott CI (2016) Experimental investigation and hybrid numerical analytical hydraulic mechanical simulation of supercritical CO_2 flowing through a natural fracture in caprock. Int J Greenh Gas Control 48:120–133. http://www.sciencedirect.com/science/article/pii/S1750583616300020

Meijster A, Roerdink JBTM, Hesselink WH (2002) A general algorithm for computing distance transforms in linear time. In: Goutsias J, Vincent L, Bloomberg DS (eds) Mathematical morphology and its applications to image and signal processing, computational imaging and vision. Springer, Berlin, pp 331–340

Metwally YM, Sondergeld CH (2011) Measuring low permeabilities of gas-sands and shales using a pressure transmission technique. Int J Rock Mech Min Sci 48:1135–1144

Moebius F, Or D (2012) Interfacial jumps and pressure bursts during fluid displacement in interacting irregular capillaries. J Colloid Interface Sci 377:406–415

Moghadasi L, Guadagnini A, Inzoli F, Bartosek M (2015) Interpretation of two-phase relative permeability curves through multiple formulations and Model Quality criteria. J Pet Sci Eng 135:738–749

Morita N, Whitfill DL, Massie I, Knudsen TW (1987) Realistic sand production prediction. In: SPE 16989 62nd Annual Technical Conference and Exhibition. Presented at the The SPE Dallas 1987

Morrow NR (1970) Physics and thermodynamics of capillary action in porous media. Ind Eng Chem 62:32–56

Navarre-Sitchler A, Steefel CI, Yang L, Tomutsa L, Brantley SL (2009) Evolution of porosity and diffusivity associated with chemical weathering of a basalt clast. J Geophys Res Earth Surf 114

Nikolaidis N, Pitas I (2001) 3-D image processing algorithms. Wiley-Interscience, New York

Noiriel C, Gouze P, Bernard D (2004) Investigation of porosity and permeability effects from microstructure changes during limestone dissolution. Geophys Res Lett 31:1–4

Noiriel C, Luquot L, Madé B, Raimbault L, Gouze P, van der Lee J (2009) Changes in reactive surface area during limestone dissolution: an experimental and modelling study. Chem Geol 265:160–170

Oikawa Y, Takehara T, Tosha T (2008) Effect of CO_2 injection on mechanical properties of berea sandstone. In: Presented at the The 42nd U.S. Rock Mechanics Symposium (USRMS), American Rock Mechanics Association

Pan C, Hilpert M, Miller CT (2001) Pore-scale modeling of saturated permeabilities in random sphere packings. Phys Rev E 64:66702

Pentland CH, El-Maghraby R, Iglauer S, Blunt MJ (2011) Measurements of the capillary trapping of super-critical carbon dioxide in Berea sandstone. Geophys Res Lett 38:L06401

Perrin J-C, Benson S (2010) An experimental study on the influence of sub-core scale heterogeneities on CO_2 distribution in reservoir rocks. Transp Porous Media 82:93–109

Pinder GF, Gray WG (2008) Essentials of multiphase flow in porous media, 1st edn. Wiley-Interscience, Hoboken, NJ

Pini R, Benson SM (2013) Simultaneous determination of capillary pressure and relative permeability curves from core-flooding experiments with various fluid pairs. Water Resour Res 49:3516–3530

Pinter A, Bodi T (2012) Determination of capillary pressure and relative permeability curves with a novel ultra rock centrifuge. Geosci Eng 1:75–86

Pokrovsky OS, Golubev SV, Schott J, Castillo A (2009) Calcite, dolomite and magnesite dissolution kinetics in aqueous solutions at acid to circumneutral pH, 25 to 150 °C and 1 to 55 atm P_{CO_2}: new constraints on CO_2 sequestration in sedimentary basins. Chem. Geol., CO_2 geological storage: Integrating geochemical, hydrodynamical, mechanical and biological processes from the pore to the reservoir scale 265, 20–32

Purcell WR (1949) Capillary pressures—their measurement using mercury and the calculation of permeability therefrom. J Pet Technol 1:39–48

Rahman T, Lebedev M, Barifcani A, Iglauer S (2016) Residual trapping of supercritical CO_2 in oil-wet sandstone. J Colloid Interface Sci 469:63–68

Ramakrishnan TS, Cappiello A (1991) A new technique to measure static and dynamic properties of a partially saturated porous medium. Chem Eng Sci 46:1157–1163

Rashid F, Glover PWJ, Lorinczi P, Collier R, Lawrence J (2015) Porosity and permeability of tight carbonate reservoir rocks in the north of Iraq. J Pet Sci Eng 133:147–161

Rege SD, Fogler HS (1987) Network model for straining dominated particle entrapment in porous media. Chem Eng Sci 42:1553–1564

Renard F, Gratier J-P, Ortoleva P, Brosse E, Bazin B (1998) Self-organization during reactive fluid flow in a porous medium. Geophys Res Lett 25:385–388

Riaz A, Hesse M, Tchelepi HA, Orr FM (2006) Onset of convection in a gravitationally unstable diffusive boundary layer in porous media. J Fluid Mech 548:87–111

Risnes R, Bratli RK, Horsrud P (1982) Sand stresses around a wellbore. Soc Pet Eng J 22:883–898

Russian A, Dentz M, Gouze P (2016) Time domain random walks for hydrodynamic transport in heterogeneous media. Water Resour Res

Sahimi M (2011) Flow and transport in porous media and fractured rock: from classical methods to modern approaches, 2nd edn. New York, Wiley

Sarda J-P, Kessler N, Wicquart E, Hannaford K, Deflandre J-P (1993) Use of porosity as a strength indicator for sand production evaluation. Society of Petroleum Engineers

Schechter RS, Gidley JL (1969) The change in pore size distribution from surface reactions in porous media. AIChE J 15:339–350

Scheidegger AE (1974) The physics of flow through porous media, 3rd edn. University of Toronto Press, Toronto, Buffalo NY

Schlüter S, Sheppard A, Brown K, Wildenschild D (2014) Image processing of multiphase images obtained via X-ray microtomography: a review. Water Resour Res 50:3615–3639

Seki K (2007) SWRC fit a nonlinear fitting program with a water retention curve for soils having unimodal and bimodal pore structure. Hydrol Earth Syst Sci Discuss 4:407–437

Sen PN (2004) Time-dependent diffusion coefficient as a probe of geometry. Wiley Subscr Serv Inc Wiley Co 23A:1–21

Siddiqi K, Pizer SM (eds) (2008) Medial representations, computational imaging and vision. Springer, Dordrecht

Silin D, Tomutsa L, Benson SM, Patzek TW (2011) Microtomography and pore-scale modeling of two-phase fluid distribution. Transp Porous Media 86:495–515
Smart BDG, Somerville JM, Crawford BR (1999) A rock test cell with true triaxial capability. Geotech Geol Eng 17:157–176
Stein N, Hilchie DW (1972) Estimating the maximum sand free production rates from friable sandstones without using sand control. J Pet Technol
Stein N, Odeh AS, Jones LG (1974) Estimating maximum sand free production rates from friable sandstones for different well completion geometries. J Pet Technol
Succi S (2001) The Lattice Boltzmann equation for fluid dynamics and beyond, numerical mathematics and scientific computation. Oxford University Press, Oxford
Sun R, Dubessy J (2010) Prediction of vapor–liquid equilibrium and PVTx properties of geological fluid system with SAFT-LJ EOS including multi-polar contribution. Part I: Application to H_2O–CO_2 system. Geochim Cosmochim Acta 74:1982–1998
Tenchine S, Gouze P (2005) Density contrast effects on tracer dispersion in variable aperture fractures. Adv Water Resour 28:273–289
Tinker PB (1969) A steady-state method for determining diffusion coefficients in soil. J Soil Sci 20:336–345
Tixier MP, Loveless GW, Anderson RA (1975) Estimation of formation strength from the mechanical-properties log. J Pet Technol 27:283–293
Torquato S (2002) Random heterogeneous materials. Interdisciplinary applied mathematics. Springer, New York
van Brakel J, Heertjes PM (1974) Analysis of diffusion in macroporous media in terms of a porosity, a tortuosity and a constrictivity factor. Int J Heat Mass Transf 17:1093–1103
van Genuchten M (1980) A closed-form equation for predicting the hydraulic conductivity of unsaturated soils. Soil Sci Soc Am J 44:892–898
Van Loon LR, Soler JM, Jakob A, Bradbury MH (2003) Effect of confining pressure on the diffusion of HTO, $36Cl^-$ and $125I^-$ in a layered argillaceous rock (Opalinus Clay): diffusion perpendicular to the fabric. Appl Geochem 18:1653–1662
Virnovsky GA, Skjaeveland SM, Surdal J, Ingsoy P (1995) Steady-state relative permeability measurements corrected for capillary effects. Society of Petroleum Engineers
Wang S, Edwards IM, Clarens AF (2012) Wettability phenomena at the CO_2–brine–mineral interface: implications for geologic carbon sequestration. Environ Sci Technol 47:234–241
Washburn EW (1921) The dynamics of capillary flow. Phys Rev 17:273–283
Weingarten JS, Perkins TK (1992) Prediction of sand production in gas wells: methods and gulf of mexico case studies. Presented at the SPE Annual Technical Conference and Exhibition, Washington
Welge HJ (1952) A simplified method for computing oil recovery by gas or water drive. J Pet Technol 4:91–98
Wojtacki K, Lewandowska J, Gouze P, Lipkowski A (2015) Numerical computations of rock dissolution and geomechanical effects for CO_2 geological storage. Int J Numer Anal Methods Geomech 39:482–506
Wu Y-S, Pruess K, Persoff P (1998) Gas flow in porous media with Klinkenberg effects. Transp Porous Media 32:117–137
Zhang H, He S, Jiao C, Luan G, Mo S, Guo X (2014) Determination of dynamic relative permeability in ultra-low permeability sandstones via X-ray CT technique. J Pet Explor Prod Technol 4:443–455
Zhang Y, Li H, Yang D (2012) Simultaneous estimation of relative permeability and capillary pressure using ensemble-based history matching techniques. Transp Porous Media 94:259–276
Zhu Z, Li M, Lin M, Peng B, Sun L, Chen L (2011) Investigation on variations in wettability of reservoir rock induced by CO_2-brine-rock interactions. Society of Petroleum Engineers, SPE-142979-MS

Chapter 7
Site Characterization

Auli Niemi, Katriona Edlmann, Jesus Carrera, Christopher Juhlin, Alexandru Tatomir, Iulia Ghergut, Martin Sauter, Jacob Bensabat, Fritjof Fagerlund, Francois H. Cornet, Victor Vilarrasa and Christopher Ian McDermott

Abstract A necessary first step in qualifying a specific site for CO_2 storage and for quantifying its relevant properties is a proper site characterization. Site characterization provides the ultimate input data for reservoir modeling and for all the predictions concerning the storage complex and its surroundings. It also provides baseline information for monitoring the behavior of injected CO_2. It also incorporates input from laboratory experiments described in Chap. 6. This chapter gives an overview of site characterization procedures with respect to geological storage of CO_2, by starting

A. Niemi (✉) · C. Juhlin · F. Fagerlund
Department of Earth Sciences, Uppsala University, Villavägen 16, Uppsala, Sweden
e-mail: Auli.Niemi@geo.uu.se

K. Edlmann · C.I. McDermott
School of Geoscience, Edinburgh Collaborative of Subsurface
Science and Engineering (ECOSSE), University of Edinburgh, Edinburgh, UK
e-mail: katriona.edlmann@ed.ac.uk

J. Carrera · V. Vilarrasa
Groundwater Hydrology Group (GHS UPC-CSIC), Institute of Environmental Assessment and Water Research (IDAEA), Spanish National Research Council (CSIC), Barcelona, Spain
e-mail: jesus.carrera.ramirez@gmail.com

V. Vilarrasa
e-mail: victor.vilarrasa@upc.edu

A. Tatomir · I. Ghergut · M. Sauter
Angewandte Geologie, Universität Göttingen, Goldschmidtstr. 3,
37077 Göttingen, Germany
e-mail: alexandru.tatomir@geo.uni-goettingen.de

J. Bensabat
Environmental and Water Resources Engineering Inc., P.O. Box 6770,
31067 Haifa, Israel
e-mail: jbensabat@ewre.com

F.H. Cornet
Institut de Physique du Globe–Strasbourg, Paris, France
e-mail: francois.cornet@unistra.fr

A. Niemi et al. (eds.), *Geological Storage of CO_2 in Deep Saline Formations*,
Theory and Applications of Transport in Porous Media 29,
DOI 10.1007/978-94-024-0996-3_7

from regulatory requirements and guidelines and proceeding to specific methodologies for assessing the sites properties in terms of CO_2 geological storage.

7.1 Background

A necessary first step in qualifying a specific site for CO_2 storage and for quantifying its relevant properties, is a proper site characterization. Site characterization provides the input data for reservoir modeling and for all the predictions concerning the storage complex, as well as baseline information for monitoring the behavior of the injected CO_2. Site characterization models can be divided into static models, describing the existing geological conditions and fluid properties, and dynamic models where the dynamic behavior of fluid flow, stress field and the like are addressed by means of a dynamic reservoir simulations, by using the static model as the input data.

For site evaluation of the large scale structures, i.e. the caprock and the reservoir stratigraphy, their petrophysical and fluid properties must all be characterized. There is a range of relevant data sources that cover measurement scales from seismic data at the scale of kilometers to thin sections at the scale of microns, each capturing specific features and the heterogeneity of the reservoir. Site characterization methodologies are well known from a number of other geological and geo-engineering applications. In addition, geological storage of CO_2 poses some specific requirements in terms of site's suitability.

In the European perspective, detailed instructions on what should be addressed in a CO_2 storage project have been outlined in the EU directive for CCS (EU 2009). There, the procedure is divided in the steps of (i) data collection, (ii) building the 3-dimensional static model and (iii) characterization of the storage dynamic behavior, sensitivity characterization and risk assessment. In terms of data collection it is defined that sufficient data should be collected to construct a static model for the storage site, including the caprock, the surrounding area and the hydraulically connected areas. The data should cover at least geology and geophysics; hydrogeology; reservoir engineering; geochemistry; geomechanics; seismicity; presence and conditions of possible leakage pathways. The 3-dimensional geological model should characterize the storage complex in terms of the EU directive EU (2009) which includes: (a) geological structure of the physical trap; (b) geomechanical, geochemical and flow properties of the reservoir overburden (caprock seals, porous and permeable horizons) and surrounding formations; (c) fracture system characterization and presence of any human-made pathways; (d) areal and vertical extent of the storage complex; (e) pore space volume (including porosity distribution); (f) baseline fluid distribution; (g) other relevant characteristics. The third step, step for characterization of the storage dynamic behavior, sensitivity characterization and risk assessment is defined that it shall be based on dynamic modeling of CO_2 injection into the storage formation, using the above geological model. The directives also state what phenomena needs to be modeled. Dynamic modeling is discussed elsewhere in this book (Chap. 4 in particular). Here we will only exemplify the construction of the structural geological model as well as give examples of data that is needed for the modeling. DNV (2010) has given guidelines for site selection and quantification of sites and projects for CO_2 storage, largely referring to the above EU directives. They also point out that the specific data required to address the requirements may depend on the site and that the developer should therefore be given discretion to select techniques necessary to be used to obtain the information and meet the objectives set.

In the US perspective, NETL (2013) in their recommendations for Best Practices for Site screening, Site selection and Initial Characterization for Storage of CO_2 in Deep Geologic Formations divide the steps of characterizing a potential CO_2 storage site as discussed below. The first ones of these steps are not actual site characterization but will nevertheless provide relevant information for site characterization, and are therefore included here.

According to NETL (2013) site screening is a stage where regional geological data is evaluated and analyzed in terms of (i) Injection formation: identifying regional and sub-regional formations that have suitable characteristics for storage; (ii) Adequate depth: it is ensured that formations have regional depth sufficient so that CO_2 remains in supercritical state; (iii) Confining zone; ensuring that there is adequate confining zone with sufficient lateral extent to avoid vertical migration of the CO_2 and (iv) Prospective storage resources: calculating the prospective storage resources to see that the reservoirs have sufficient pore volumes and can tolerate the pressure changes to accommodate the planned injection volumes. Table 7.1 summarizes in shortened form the NETL guidelines for Site Screening in terms of the regional geological data (excluding the regional and social aspects of site screening that are also discussed in the original report). Some of the issues identified are specific to the conditions in the US but can still be used as reference framework for other locations as well.

Next step, the site selection, is a stage to further evaluate the previously selected areas and develop a shorter list of sites that can be taken further to initial site characterization. The analyses related to geological data at this stage are summarized in Table 7.2. Initial characterization in turn is a stage after site selection stage. The subsurface data analysis is expanded to integrate the elements of the baseline data analysis, including geological, geochemical, geomechanical, hydrogeologic and flux data. Table 7.3 summarizes the issues to be addressed at this stage, in terms of the geoscientific data. Site Characterization then builds on the previous studies to develop a more detail characterization of the site. This could include additional drilling and testing of wells, to analyze geochemical and geomechanical properties, including stimulation testing to analyze injectivity, possible additional seismic surveys and brine injection tests to study the interference and pressure responses. At this stage continuity of the injection zones and confining zones needs to be established and potential leakage issues identified. NETL (2013) does not go into the details of this actual site characterization phase.

In the following sections we will describe some important site characterizations methods, in terms of geological storage of CO_2. We do not attempt to give a full account of all the site characterization techniques, partly as many of these methods are known from a number of different geoscientific and geoengineering applications. We also want to refer to a book specifically focusing on site characterization in the context of geological storage of CO_2 by Surdam (2013). The methods

Table 7.1 NETL guidelines for site screening in terms of geological factors (shortened from NETL 2013)

Element	Guidelines for site screening—geoscientific aspects
Injection formation	Identify formation types for potential injection. Utilize readily accessible data from public sources or from private firms. Data should include regional lithology maps, injection zone data (thickness, porosity, permeability), structural maps, information about structure closure and features that might compartmentalize the reservoir such as stratigraphic pinch outs, regional type logs, offset logs, petrophysical data, and regional seismicity maps
Adequate depth	Assessment of minimum depth of the injection zone to protect drinking water resources. In addition depths of at least 800 m generally indicate CO_2 will be in a supercritical state and may be more cost-effectively stored. Shallow depths (<800 m) may add to the risk profile because CO_2 could be in gas phase and the injection zone may be closer to drinking water sources
Confining system	Injection zones should be overlain by confining system comprised of one or more thick and impermeable zones of sufficient lateral extent. Confining systems can be identified from the same types of information used to identify injection formations. Wells that penetrate potential confining systems should be identified (sources e.g. oil and gas regulatory agencies). Faulting and folding information that may impact confinement integrity should be mapped along with potential communication pathways. Confining system integrity may be validated by presence of nearby hydrocarbon accumulations
Prospective storage resources	Candidate CO_2 formations should contain enough prospective storage resources beneath a robust confining system. Prospective storage resources (and injectivity if permeability data is available) should be estimated at the sub-regional scale utilizing existing data (e.g. state geological surveys) to populate basic numerical models.

Table 7.2 NETL guidelines for site selection in terms of geoscientific aspects (shortened from NETL 2013)

Element	Guidelines for site selection—geoscientific aspects
Injection zone (reservoir)	Define injection zones based on public and acquired regional well data. Analysis should include at minimum the development of a regional stratigraphic column identifying potential storage types and injection and confining zone(s), potential USDWs (underground sources of drinking water); structure and isopach maps of injection and confining zone(s); regional cross-sections; regional tectonic maps, reservoir dip, and analog well data such as lithology, porosity, permeability, pressure, temperature, and dynamic formation evaluation data (DST, well test, production/injection data)
Confining system	Establish the areal extent, thickness, lithology, porosity, permeability, capillary pressure data, and other factors that might affect integrity of the confining system with confining zone(s). Perform a faulting and folding analysis based on tectonic history and analogs. Utilize existing well bore, core, outcrop and regional analog data to identify and map confining zone(s) tops, bases and thicknesses

(continued)

Table 7.2 (continued)

Element	Guidelines for site selection—geoscientific aspects
Trapping mechanisms	There are several mechanisms that effectively "trap" injected CO_2, including physical barriers, as well as physical and geochemical processes. Evaluation of trapping mechanism should be based on the local well, outcrop and any available regional reservoir analyses including analogs in similar formations
Potential injectivity	Utilize collected data and analyses to estimate potential permeability-thickness of target injection zone and identify boundary conditions that will affect injection estimates; assess well stimulation and completion scenarios to achieve target injection rates
Evaluate existing seismic	Existing regional seismic data could be used to validate the regional stratigraphic and structural framework. All available seismic attribute data should be integrated with the injection zone, structure, confining system and capacity evaluations. If existing seismic data is not available, it is recommended that a project developer wait to acquire data during the initial characterization stage—unless regional geology warrants information earlier in process
Prospective storage resources	Prospective storage volumes should be calculated utilizing acquired data, reporting resource volume ranges (low/medium/high) with identification of uncertainties in calculations. The reservoir evaluation should be used in calculation of prospective storage with all parameters and sources defined, such as "efficiency" calculations. Calculations should be reported assuming a maximum storage pressure and either an open or a closed system for brine displacement as endpoints. For more details for methods of calculation see NETL (2013)

Table 7.3 NETL guidelines for initial site characterization in terms of geoscientific aspects (after NETL 2013)

Element	Guidelines for initial characterization—geoscientific aspects
Geological	Develop site specific geologic baseline of qualified site(s) including type log/stratigraphic column; detailed correlation of reservoir architecture including injection intervals within the injection zone and potential confining zones within confining system; detailed structural maps; interpreted depositional model and facies distribution; porosity maps for potential injection intervals and zones; and porosity/permeability log transforms. This evaluation should be updated as additional information is acquired (seismic and well data). During initial characterization any additional data from a new well tests should also be integrated into previous analyses
Geochemical	Develop baseline of groundwater in all overlying aquifers using fluid and fluid level data collected in shallow aquifer formations in offset wells. If available, collect rock and fluid property data (composition, geochemistry, pH, conductivity, mineralogy) from the injection zone to model formation fluid-CO_2- rock reactions in the injection zone and at confining zone interfaces

(continued)

Table 7.3 (continued)

Element	Guidelines for initial characterization—geoscientific aspects
Geomechanical	Develop baselines for injection rates and pressures utilizing drilling data on formation strength and modeling. Analyze advanced logging suites from offset wells and characterization wells (if any exist) to identify faults and fractures. Analyze new or existing core to determine the existing stress state and assess the impact of changes in pore pressure on stress
Hydrogeological	Determine fluid compositions and injection zone flow units from new or offset well data, fluid samples, and hydrologic and other tests; integrate into dynamic injection zone models and compare to the existing hydrological model. Conduct multi-well tests where possible. Injection zone fluids and hydraulic tests should be further investigated during the site characterization phase and fluid samples should be collected if a new well is drilled or an existing well(s) is further tested
Flux baselines	Plan a monitoring system to establish baseline readings of near surface, ground level, and shallow subsurface fluxes. Baseline monitoring should be conducted during initial characterization and conducted for at least a year to account for changes in flux reading due to seasonal changes. Nearby urban, industrial or agricultural expansions and developments may require re-establishing a baseline prior to injection

discussed is Sects. 7.2–7.5 also have a special emphasis on some of the methods developed within the EU FP7 MUSTANG (www.co2mustang.eu) project that has been the starting point of this book. There is also focus on what can be defined as characterizing the dynamic properties of the site in terms of CO_2 transport and trapping (Sects. 7.3–7.5). In terms of building up the geological models and assigning their properties we refer to the number of example sites studied in the previously mentioned MUSTANG project and some of the related Deliverables (Erlström et al. 2010, 2011). A comprehensive site characterization for one of these sites, a research scale CO_2 injection site at Heletz, Israel, is summarized in Niemi et al. (2016), including demonstrating the use of old data from oil investigations and new data specifically collected for the purpose of CO_2 injection studies. Section 8.6 of this book that describes the Frio, USA CO_2 injection experiment also gives a good example of the relevant site characterization in the injection experiment scale. Other examples of thorough site characterization work can be seen in e.g. the US Regional Partnership Studies on CO_2 storage.

7.2 Geological Characterization

Katriona Edlmann, Christopher Ian McDermott and Christopher Juhlin

Geological characterization or construction of the static model of a CO_2 storage site uses data sets, such as cores, wireline logs, seismic and production data along with insight from outcrop analogues to characterise the geological architecture of the storage complex. The input data sources, in decreasing scale are:

1. Seismic data to determine the large scale 3D geometry of the site, combined with an understanding of the geological depositional environment, tectonic history and stratigraphy.
2. Geomechanical facies analysis where different tectonic settings exert different depositional process controls within the tectonic basin, which in turn influence CO_2 storage site suitability in terms of basin architecture, caprock architecture, reservoir quality, stress state, mechanical characteristics, fractures, burial depth, geothermal gradient, risk of orogenic modification, structural stability and preservation potential.
3. Wireline log data to provide information on the geology and stratigraphy, including reservoir and caprock thickness and distribution along with reservoir scale measurements of the petrophysical properties.
4. Production data to provide information on initial reservoir fluids and pressure, as well as possible flow barriers.
5. Core data to provide more detailed petrophysical information.

7.2.1 *Large Scale Geology*

It is necessary to characterise the geology of the storage complex at both local and regional scales. The local scale geological characterisation concentrates on geometry of the storage formation along with the caprock and overburden extent and structural compartmentalisation. The regional scale geology will identify basin wide CO_2 migration patterns and enable the bulk residual and solubility storage potential to be calculated.

7.2.1.1 Seismic Data

The primary data source for large scale geological characterization is seismic data. Acquiring such data, in its simplest form, involves a surface source generating seismic waves which propagate downwards and are reflected in the subsurface at geological boundaries. The reflections propagate upwards and are recorded on surface sensors as a function of time. Large amounts of data are normally acquired and processed. The processed data provide structural images of the subsurface where different reflections generally correspond to different subsurface lithologies. These processed images can be displayed in 2D or 3D, depending upon the acquisition geometries of the sources and receivers, and as a function of time or depth. If displayed as a function of depth, then the seismic velocity of the media needs to be estimated. This can be done by analyzing the data itself to provide a general estimate of the velocity function for the data set. However, it is also useful to have borehole logs and borehole seismic data to increase the accuracy of the estimated velocity function for the surface seismic data. Ideally, a 3D depth converted image should be used for interpretation. In addition to the image, seismic attributes can be determined from the data which can aid in the interpretation. Examples of such attributes are reflection strength, changes in amplitude with offset (AVO) and instantaneous frequency (for further details see Sect. 8.2.2).

The amplitude of a reflected wave depends on the contrast in velocity and density at the boundary of the two media. Normally if the velocity increases the density will also increase, but there are also exceptions. It is the product of the velocity (v) and density (ρ) which controls the reflection amplitude and this product is known as the impedance (Z). For waves that impinge on a boundary between two media at right angles the reflection coefficient is usually defined as:

$$R = \frac{(Z_2 - Z_1)}{(Z_2 + Z_1)} \tag{7.2.1}$$

where Z_2 is impedance in the medium below the boundary and Z_1 is the impedance in the medium above the boundary. With this definition, reflections from interfaces in which the impedances increase below it will have a positive reflection coefficient and reflections for which the impedance decreases below it will have a negative one. This can be used to identify material boundaries. Hard limestones and tight sandstones, for example, generally have high impedances while soft shales and highly porous sandstones have low impedance. Coal layers have generally very low impedance.

An important consideration when interpreting seismic data is the data resolution, both in the vertical and horizontal directions. The reflection coefficient defined above assumes the recorded reflection waveform is from a single interface. However, if the layer is thin, the reflection at the base of the layer will interfere with the reflection off the top of the layer (Fig. 7.1). This interference may either decrease or increase the apparent reflection strength of the layer and it becomes difficult or impossible to separate the reflected signal from the top and bottom of the layer. When the two cannot be separated the vertical resolution limit of the data has been reached, this being governed by the velocity of the media and the frequency of the seismic waves propagating through it. This limit is generally defined as one quarter of the wavelength (λ) of the signals in the data. The wavelength is dependent upon the velocity of the media and the frequency (f) of the signal so the vertical resolution limit (λ_R) will be:

$$\lambda_R = \frac{v}{4f}. \tag{7.2.2}$$

The greater the frequency of the generated seismic waves the better the resolution. For a typical seismic survey the velocity of the media may average 3000 m/s and signal frequencies of 50 Hz may be generated. The resolution would then be 15 m. That is, layers thinner than 15 m cannot be resolved. Note that they will be detected by the seismic method, but one will not be able to quantitatively measure their thickness. The horizontal resolution also needs to be considered when

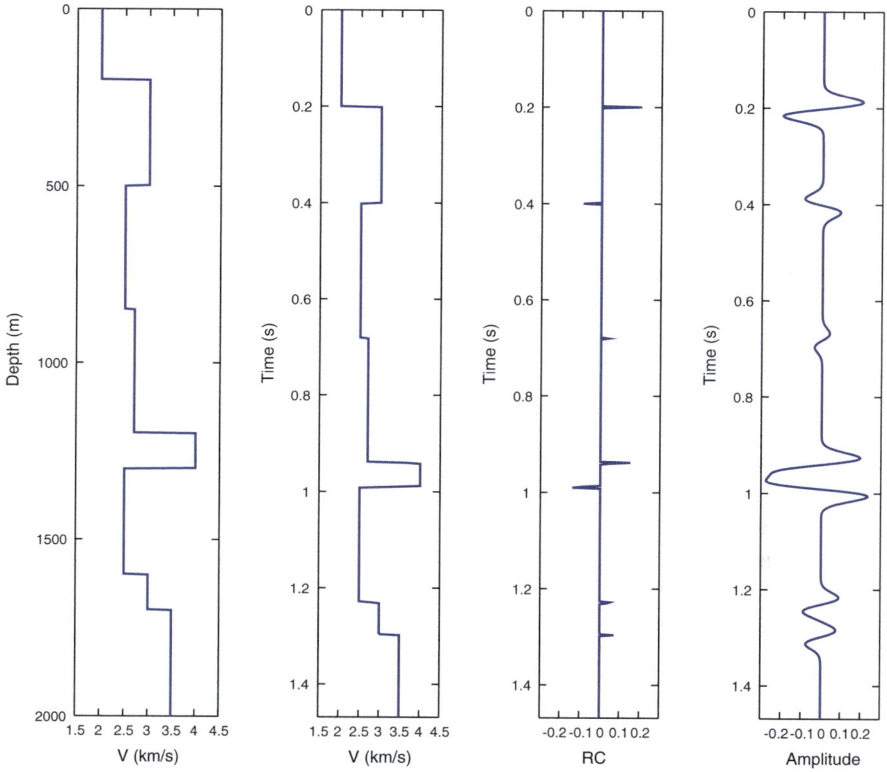

Fig. 7.1 Example of how reflections from interfaces close to one another interfere with one another. *Panel 1* shows the velocity as a function of depth, *panel 2* shows it as a function of time, *panel 3* shows the reflection coefficient as a function of time assuming that the density is constant, *panel 4* shows the resulting seismic response for a wavelet that has a dominant frequency of 30 Hz. Note how the reflections from the *top*, at the *bottom* of the high velocity layer at about 1250 m depth interfere with one another, producing a composite reflection

interpreting seismic data. Horizontal resolution will generally decrease as a function of depth due to spherical spreading of the wave fronts. In general, the horizontal resolution will be less than the vertical resolution.

There are many examples of how seismic data have been used to characterize CO_2 injection sites of different scales. Both depleted gas fields (e.g. Urosevic et al. 2010) and small scale saline aquifer sites (e.g. Alcalde et al. 2013; Doughty et al. 2008; Juhlin et al. 2007) have been characterized. In the Decatur project in Illinois, USA a relatively large 3D seismic survey was performed prior to injection of about 1 million tons of CO_2. For depleted oil and gas fields there will often exist previous seismic data, but of older vintage. Therefore it may be necessary to acquire new data, both for characterization and as baseline for future monitoring. For saline aquifer injection sites there will generally be limited previous seismic data available, especially if there are no overlying producing petroleum accumulations. In most of these cases it will be necessary to acquire new 3D seismic data in order to properly characterize the site.

In spite of the resolution limits, a great deal of information about the subsurface structure can be gained from a seismic survey. The information content can be increased by combining the seismic data with borehole data and information on regional geology. Sedimentary processes in any given depositional environment will have a characteristic distribution of sediment types, termed facies, which can be utilized to provide input data for the large scale geology. These facies can often be identified in the seismic data. Combining the seismic data with an understanding of the depositional environment and regional geology can provide a more detailed picture of the subsurface geology.

7.2.1.2 Depositional Environments and Facies Analysis

Different tectonic settings (Fig. 7.2) exert different depositional process controls within the tectonic basin. These in turn will influence the suitability of the CO_2 storage in terms of basin and caprock architecture, reservoir quality, stress state, mechanical characteristics, fracturing, burial depth, geothermal gradient, risk of orogenic modification, structural stability and preservation potential (Hallam 1981). The different tectonic settings include (i) extensional tectonics resulting in oceanic basins, passive continental margin basins and terrestrial rift basins; (ii) convergent tectonics resulting in trench basins, forearc basins, back arc basins and foreland basins and the (iii) wrench tectonics creating strike slip basins. Geological sedimentary deposits are not randomly formed, but control is exerted by the influence of tectonics over the depositional and structural processes. This leads to the formation of layers of strata which can be grouped together according to their engineering characteristics. By examining the typical characteristics of the geomechanical facies within the different tectonic settings, it is possible to compare and contrast the different tectonic settings to appraise global CO_2 storage opportunities and predict which tectonic settings will be the most suitable for CO_2 storage (Edlmann et al. 2014).

Using the geomechanical facies approach, inputs crucial to the primary CO_2 storage requirements of storage volume and storage security can be evaluated and graded as (i) good, (ii) moderate and (iii) poor, based on an assessment of their net contribution towards providing ideal CO_2 storage conditions. Work by Edlmann et al. (2014) show that foreland basins and passive continental margin basins are more likely to be suitable basins for CO_2 storage than other tectonic settings and should therefore be prioritized for investigation. Strike-slip basins, terrestrial rift basins and back arc basins are also suitable for CO_2 storage with oceanic basins and fore-arc basins being moderately suitable and trench basins unsuitable tectonic settings for CO_2 storage. The geomechanical facies approach is discussed in more detail in Chap. 9.

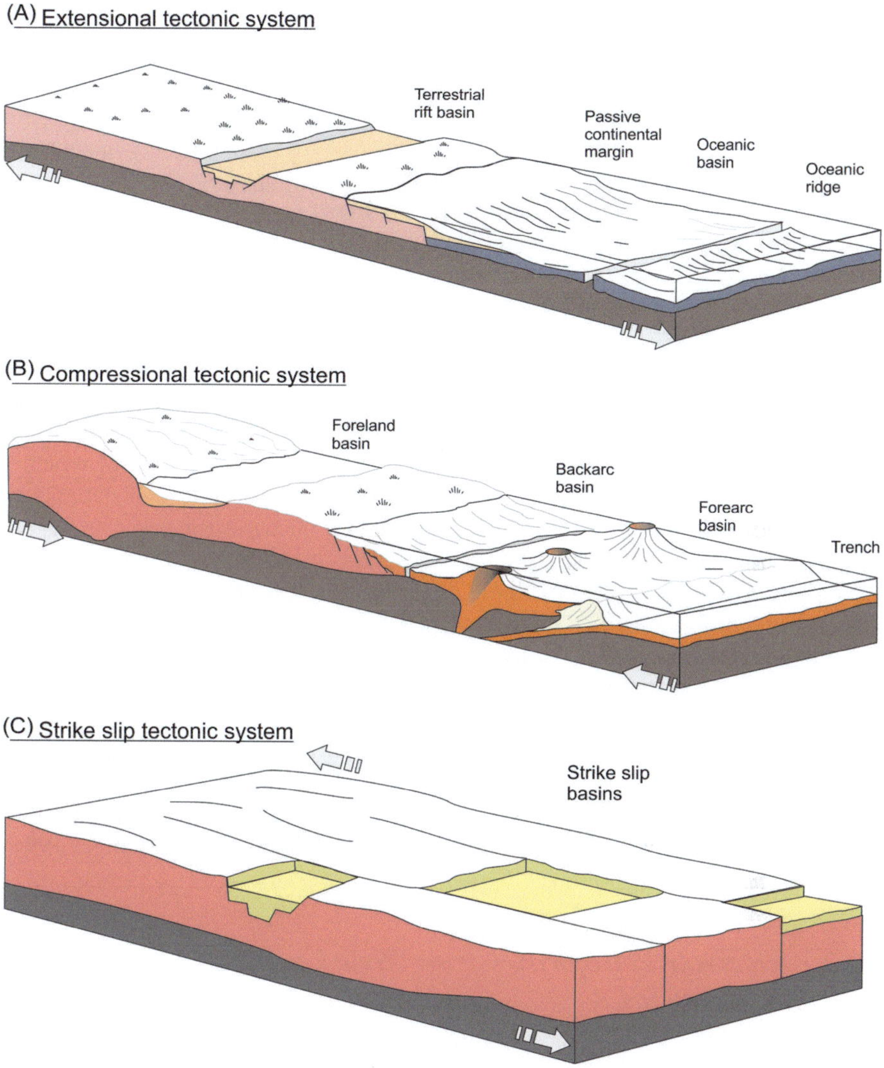

Fig. 7.2 Schematic of the basin types formed under the three primary tectonic settings

7.2.1.3 Wireline Logs

Wireline logs are continuous instrument measurements of the downhole formation where the physical properties of the formation rocks can be inferred from the instrument response. The direct wireline instruments include borehole imaging, electrical resistivity, acoustic response, natural radioactivity, radioactivity response, electrical potential, neutron magnetic resonance and calliper response. These allow the inference of lithology, bulk density, porosity, permeability, fluids type and saturation, stress state, fracture network and mechanical strength. Wireline logs provide a continuous recording of parameters with depth and as such they are very useful for geological characterisation. They provide stratigraphic information through the reservoir and facilitate well to well correlations that can cover the whole field, where intervals of different wells are matched for similarity or characteristic log responses to lithological markers. The wireline logs are also used to provide a geological check to the seismic interpretation. Below some examples are given of what information can be obtained from wireline logs in terms of geological characterization.

To differentiate between the porous and permeable reservoir rocks and the non-permeable clays and shales and to indicate the shaliness of the rocks, the Spontaneous Potential (SP) log and the natural Gamma Ray (GR) are used. The Spontaneous Potential (SP) log shows the electrical potential (voltage) produced by the formation and drilling fluids along with the shales. The natural Gamma Ray (GR) in turn indicates the natural radioactivity of the formation, most of which is within shales as radioactive elements tend to concentrate in clays and shales.

The LithoDensity log measures the density and the photoelectric absorption index of a formation. The density log primarily responds to porosity and pore fluid and secondarily to the rock matrix (lithology) properties, the photoelectric absorption primarily responds to the rock matrix properties and secondarily to the porosity and pore fluid. From these a clear indication of the likely lithology can be obtained.

The neutron, density and sonic logs can be used to determine lithology through the use of crossplots, where various combination of logs respond to lithology. The neutron-density crossplot allows the identification of sandstone, limestone and dolomite. The sonic-density crossplot is particularly useful to identify evaporates. The sonic-neutron and the density-photoelectric crossplots are also useful at facilitating the identification of sandstone, limestone and dolomite (Schlumberger 1991).

7.2.2 Reservoir Petrophysics from Well Log-Scale Observations

Commonly determined petrophysical parameters from logging tools include:

Shale Content: There are primarily two logs that are used in shale identification:

1. The Spontaneous potential (SP) log which measures the difference between electrical potential of a movable electrode in the borehole and the electrical potential of a fixed surface electrode indicates the presence of shales, where shales usually show a straight line on the SP log trace and porous formations will deviated from this shale baseline.
2. The gamma ray log (GR) which records the natural radioactivity of the formation which depends on the concentration of potassium, thorium and uranium, which tend to concentrate in clays and shales.

Rock Porosity: Rock porosity can be obtained from the sonic, density and neutron logs.

1. The sonic log is a recording versus depth of the time taken for a sound wave to traverse 1 ft of formation known as the interval transit time which is a function of lithology and porosity.
2. The density log response is determined by the electron density of the formation which is related to true bulk density which depends on the density of the rock matrix, porosity and fluid density.
3. The neutron log reflects primarily the amount of hydrogen in the formation which is due to the fact that neutrons loose most energy when they collide with a hydrogen nucleus.

Lithology Determination: the measurements of the neutron, density and sonic logs depend not only on porosity but also on the formation lithology and crossplots can be used to demonstrate how various combinations of logs respond to lithology.

Stratigraphic Determination: All logs will reflect sedimentary features; however their vertical resolution is not sufficient to detect thin beds. The dip meter tool is designed to provide detailed downhole information and enhances resolution of strata based on micro resistivity measurements.

Fracture Determination: The array sonic tool can be used to identify fracture zones, where high signal strength at the receiver suggests a competent rock. The calliper tool which measures the borehole shape can also identify breakout or fractured areas within the formation.

Elastic Constants: Dynamic elastic constants can be determined from the measurement of elastic wave velocities from the sonic logging tools.

7.2.3 Reservoir Petrophysics—Core-to-Pore Scale Observations

The physics of deposition and the wide range of geological environments lead to a vast range in geological heterogeneities at different scales of investigation. Within the reservoir analysis there are three primary scales to consider:

- The pore scale, which considers the grain and pore size distribution and shape and is at the micron to millimetre scale, where the laminae scale geological structures are dominant.
- The core scale, where the geological structure of the rock becomes apparent. This is the centimetre scale and depositional bedding structures are dominant and the heterogeneity of the system has increased.
- The reservoir/log scale, where heterogeneities are on the meter to kilometre scale.

These different scales of measurement can, to a certain extent, be addressed by tying scales of data to each other as a calibration. For example, high resolution core data can be used to refine wireline data.

There is a wide range of core analysis that provide petrophysical input data. Typically, a rotary coring bit is used to recover the core, which is hollow in the centre, called the core barrel, where the core is stored and retrieved. The process of coring is expensive and normally only performed through the reservoir interval. Side wall cores of around 1 in. (note; 1 in. = 2.54 cm) in diameter can also be obtained directly from the reservoir.

Any core plug samples are likely to be fully saturated with the drilling fluids. The coring process will also damage the rock, which is not always discernible and there will be a stress release during retrieval which in extreme circumstance may mean the rock is no longer representative of the in situ rock properties. Damage will occur through stress, pore pressure and temperature release and exposure to non-native fluids. This will have an impact on the petrophysical measurements to be performed on the cores. Ideally experiments should recreate the in situ stress conditions, but this is not practically possible and must be considered during analysis, paying attention to the importance of a particular damage mechanism to the rock property being measured, for example a mechanism that changes the structure of the pore fill material will have a great impact on permeability and less so on the mechanical property of the rock.

Routine core analysis procedure involves initial core gamma logging to provide a tie between core depth and logging depth. Then the core plugs are taken and porosity, permeability and saturation are measured. The core is then sliced and core photographs are taken in both white and ultra violet light for both documentation and core description. For more complex reservoir measurement, Special Core Analysis can be undertaken which provides capillary pressure, relative permeability, electrical properties, wettability and nuclear magnetic resonance. In addition rock mechanical testing, miscible gas and chemical flooding and further detailed thin section petrography can be conducted to provide as much information as possible about the rocks.

7.3 Dynamic Characterization—Hydraulic, Tracer and Thermal Properties

Jesus Carrera, Alexandru Tatomir, Iulia Ghergut, Martin Sauter and Jacob Bensabat

This section deals with field tests performed to characterize the hydrodynamic properties of the aquifer prior to actual injection. The most important parameters for the purpose of CO_2 storage are permeability, porosity and geomechanical properties. Permeability of the aquifer controls the injection pressure, the rate at which CO_2 will dissolve into the native brine, and the regional pressure buildup. Permeability of the caprock, together with entry pressure, controls the efficacy of the caprock in containing the injected CO_2. It also controls the migration of brine to overlying water bodies. Porosity determines the storage capacity of the aquifer.

Geomechanical properties control the maximum sustainable injection pressure (rock strength) and the associated deformation (Young's modulus and Poisson ratio). In addition to these parameters, it is also necessary to characterize retention properties, which control capillary trapping and, reactive surface, which controls the rate of chemical reactions.

The evaluation of these properties at the core scale has been described in Chap. 6. However, as discussed in Chap. 5, core scale parameters are rarely the representative ones at the field scale. Heterogeneity, notably fractures or high permeability channels cause small scale (laboratory) parameters to be different from those effective at the field scale. Evaluating these requires field tests. The types of field tests that can and are being used for characterizing CO_2 storage sites include:

1. Hydraulic tests
2. Tracer tests, using both conservative and reactive tracers
3. Thermal tests
4. High pressure injection tests
5. CO_2 injection tests.

The first three of these are described in the following sections while the fourth and fifth type of test are discussed in Sects. 7.4 and 7.5.

7.3.1 *Hydraulic Tests*

Jesus Carrera and Alexandru Tatomir

For almost a century, hydraulic tests have successfully been relied upon for diagnosing aquifer characteristics based on the evaluation of aquifer response data. These tests are commonly performed by pumping or injecting water out of a well while measuring the changes in water level (drawdown) in this well or, if present, in nearby observation wells. The drawdown can be analyzed using various models to obtain estimates of the aquifer parameters, which characterize the transmissive and storage characteristics of the aquifer and the flow system boundaries (Fig. 7.3).

7.3.1.1 Conventional Pumping Tests

The conventional pumping test is performed by pumping an ideally constant flow rate from a fully penetrating well. Hydraulic parameters are determined by matching the measured time-drawdown curves with known type curves of appropriate models or solutions to the groundwater flow equation. The solutions for the type curves require simplifying assumptions such as homogeneity and infinite areal extent. Water is assumed to be released instantaneously from storage with decline of hydraulic head.

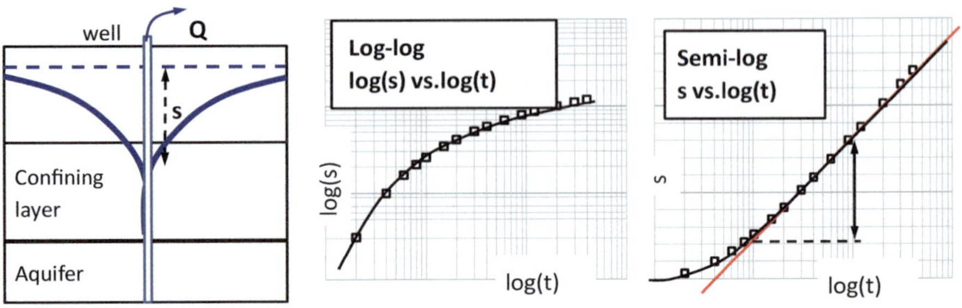

Fig. 7.3 Hydraulic tests are performed by pumping (or injecting) a flow rate Q from a well, and measuring the resulting drawdowns. (s) Parameters are obtained by fitting a model to measurements, which is frequently done graphically using log–log or semi-log graphs

The simplest and most widely used analytical solution to well hydraulics is the one of Theis (1935), which yields the aquifer response to constant pumping from a fully penetrating well in a confined aquifer (Fig. 7.3). The influence of pumping extends radially from the well with time, according to:

$$s = \frac{Q}{4\pi T}\int_{u}^{\infty}\frac{e^{-y}}{y}dy = \frac{Q}{4\pi T}W(u) \tag{7.3.1}$$

where s is drawdown, Q is pumping rate, S is storage coefficient, T is transmissivity, $u = Sr^2/4Tt$, and W(u), implicitly defined by Eq. (7.3.1) is called "well function".

A relevant and enlightening approximation to the well function was proposed by Jacob and Cooper (1946):

$$s = \frac{2.3Q}{4\pi T}\log\left(\frac{2.25Tt}{Sr^2}\right) = \frac{2.3Q}{2\pi T}\log\left(\frac{R}{r}\right) = \frac{2.3Q}{4\pi T}\log\left(\frac{t}{t_0}\right) \tag{7.3.2}$$

where R is the radius of influence $R = \sqrt{2.25Tt/S}$ and $t_0 = Sr^2/2.25T$ is the response time.

Equation 7.3.2 shows two essential properties of the aquifer response to pumping (or injection) (Fig. 7.4). First, when drawdown is plotted versus log t, it tends to display a straight line whose slope m is inversely proportional to transmissivity and whose intercept is t_0. From these, transmissivity and storativity can be obtained as:

$$T = \frac{2.3Q}{4\pi m} = 0.183\frac{Q}{m} \tag{7.3.3}$$

$$S = \frac{2.25Tt_0}{r^2} \tag{7.3.4}$$

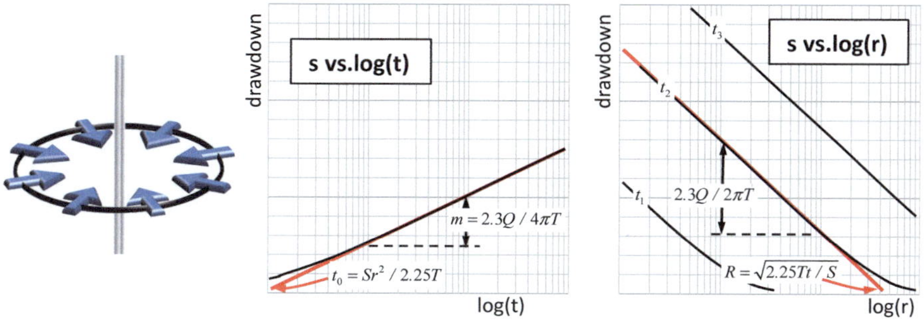

Fig. 7.4 Under radial flow conditions, drawdowns tend to plot as a straight line when plotted either versus log (t), which allows defining the response time t_0, or versus log(r), which allows defining the radius of influence, R. As time grows, so does the cone of depression, but its shape does not change

The second interesting property of Eq. (7.3.2) is that drawdown will also plot as a straight line, with slope equal to 2.3Q/2πT, when plotted versus log r. That is, pumping causes a cone of depression that is displaced downwards (i.e. its pressure drops) as time evolves, but whose shape does not change with time. The cone is always centered at the well and its external radius evolves with R. As we shall see, this property (unchanged shape) remains true even in heterogeneous aquifers.

Data can be represented in several ways to obtain the maximum information from the drawdown (pressure buildup, in the case of injection) data. The traditional method consists of plotting drawdown versus time, both in logarithmic scale (Log-log plot, Fig. 7.3). Ideally, data should be identical to Theis solution; allowing superposition and estimation of T and S from the shift in the vertical and horizontal axes, respectively. Perhaps the simplest representation consists of plotting drawdown versus logarithm of time (semi-log plot). In this case, data should tend to form a straight line for large times if flow is radial (Cooper–Jacob approximation). Transmissivity can be obtained from the slope of this line, whilst the storage coefficient results from its intersection with the log t axis.

The fact that drawdown versus log(t) tends to a straight line under radial flow conditions, prompted Chow (1952) to propose using the derivative of drawdown with respect to log time $\partial s/\partial \ln t = t\partial s/\partial t$ as an interpretation method. The concept was generalized by Bourdet and his colleagues (Bourdet et al. 1983, 1989). They analyzed the behaviour of the log-derivative for a large number of classical models of flow around a pumping well. Doing so, they showed that the joint use of the drawdown and its log-derivative can be highly informative and developed the concept of diagnostic plots. A diagnostic plot (Bourdet et al. 1983) is a simultaneous plot of the drawdown and the logarithmic derivative of the drawdown as a function of time in log–log or semi-log scales (Fig. 7.5). This plot is used to facilitate the identification of the conceptual model best suited to interpret the data. Detailed explanations on the method are provided by Bourdet et al. (1983) and Renard et al. (2009). We will discuss them in some detail later. Prior to that, it is convenient to discuss recovery tests.

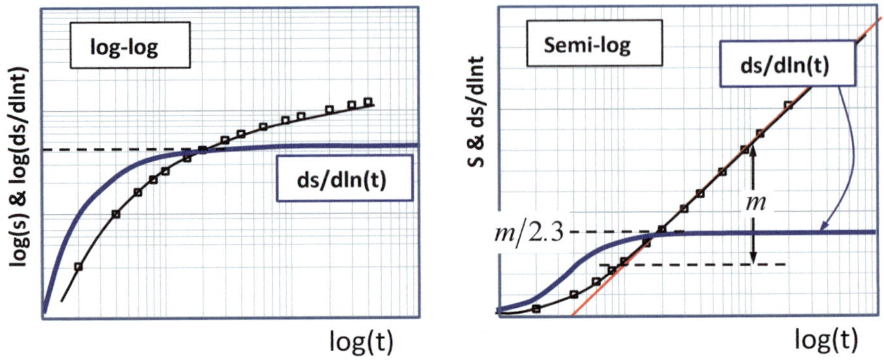

Fig. 7.5 Diagnostic plots include drawdown and derivative graphs. Both log–log (*left*) and semi log (*right*) graphs are useful. In the pure radial flow case, the derivative tends to a constant (*m*/2.3). The 2.3 factor reflects that derivatives are taken with respect to ln(t), while the slope *m* is obtained from the semi-log graph, where the logarithm is decimal

7.3.1.2 Recovery Tests: Theis and Agarwal Methods

Recovery refers to the return of pressure to its natural state. Strictly speaking, it is a "passive" test. However, data collected during recovery may be more informative about the properties of the aquifer than those collected during pumping (or injection). Most aquifer test theory is based on the assumption that the pumping (or injection) rate is constant, which is rarely true. Pumping rate is variable, especially during the early portion of the test, which is often the most informative about aquifer properties. As such early time may not be very reliable. Instead, recovery only requires turning the pump off, which makes these data more reliable.

Solutions to recovery can be obtained in several ways. They generally rely on superposition. The traditional Theis recovery method (Horner method in the petroleum literature) consists of superimposing the head response to pumping and to an injection that starts at the end of pumping (Fig. 7.6). A simple illustration of the concept can be obtained using Jacob approximation (Eq. 7.3.2), which leads to

$$s_R = s_P(t) - s_P(t - t_P) = \frac{2.3Q}{4\pi T}\left(\log\left(\frac{t}{t_0}\right) - \log\left(\frac{t - t_P}{t_0}\right)\right) = \frac{2.3Q}{4\pi T}\log\left(\frac{t}{t - t_P}\right) \tag{7.3.5}$$

where s_R and s_P are the recovery and pumping drawdowns, respectively. This equation, points that recovery data will also tend to a straight line when plotted versus $\log(t/t - t_P)$. This leads to Theis recovery method that consists of deriving transmissivity from such slope using Eq. (7.3.3). This estimate of transmissivity is quite robust (Willmann et al. 2007). Notice that this approach ignores the estimation of storativity, which indeed reflects that late time recovery is not sensitive to storativity.

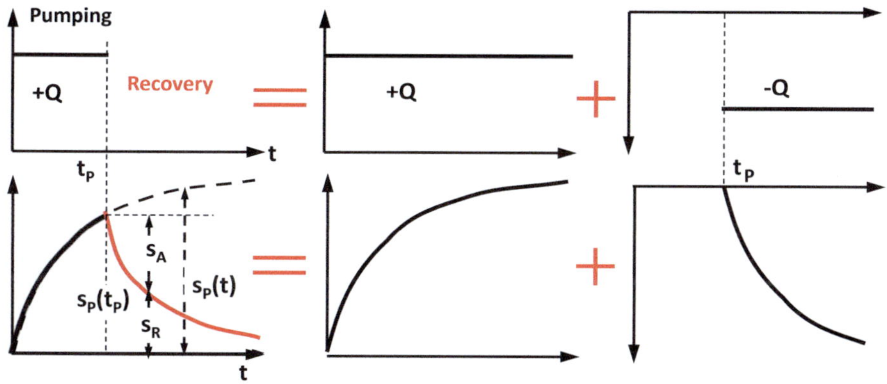

Fig. 7.6 Superposition to obtain recovery plots. Theis recovery method consists of plotting s_R versus $t_p(t - t_P)$, which results from Eq. (7.3.5). Agarwal method consists of plotting $s_P(t_P) - S_R$ versus $t_{P3}(t - t_P)/t$, which should yield a graph very similar to that obtained with a constant pumping rate, see Eq. (7.3.6) and makes it a very attractive method

This limitation can be overcome using Agarwal method, which is a somewhat more sophisticated version of recovery that consists of computing head recovery (i.e., drawdown at the end of pumping minus drawdown during recovery, often termed Agarwal's drawdown) from the fact that drawdown during recovery is directly $s_P(t)$ minus $s_P(t - t_P)$. Using Eq. (7.3.2) to approximate all terms involved, yields:

$$s_A = s_P(t_P) - s_R(t) = s_P(t_P) - (s_P(t) - s_P(t - t_P)) = \frac{2.3Q}{4\pi T} \log\left(\frac{t_P(t - t_P)}{t_0 \cdot t}\right) \tag{7.3.6}$$

Notice that this solution is identical to the one of Eq. (7.3.2), except that time is changed by Agarwal's time $t_A = t_P(t - t_P)/t$. Therefore, Agarwal's drawdown should be very similar to the one caused by a constant pumping rate, provided that t_A is used instead of t. As it turns out, the solution is approximately valid well beyond the Cooper–Jacob solution used here. The only limitations are that pumping should be constant for some time prior to the stop in pumping and that, obviously, Agarwal's time is bound by pumping duration. In fact, the solution is not valid for times close to t_P. It is clear that Agarwal's should be the method of choice, as it only requires careful monitoring of recovery.

7.3.1.3 Effect of Heterogeneity

Heterogeneity is arguably the most ubiquitous feature of natural media. As such, heterogeneity must be acknowledged when dealing with permeable media (recall extensive discussion in Chap. 5). In the hydraulic testing context, two extreme

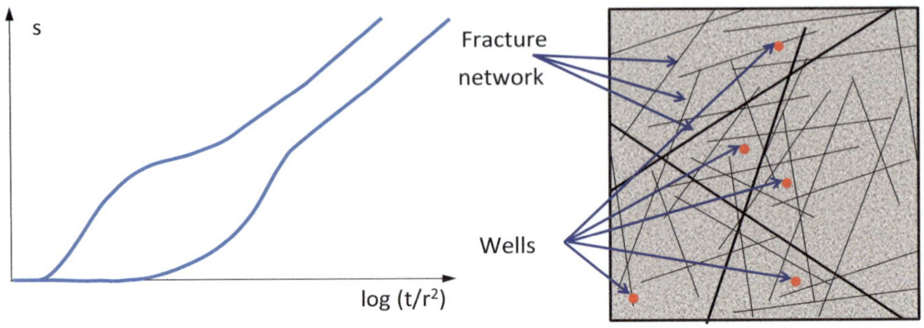

Fig. 7.7 Large scale transmissivity is controlled by preferential flow paths (long, permeable, well connected fractures). The probability of intersecting these fractures is low. So that transmissivity derived from injectivity or short term tests tends to be much smaller. Large scale transmissivity may be derived from the slope of long term tests, which is the same regardless of whether the pumping or observation wells are located in high or low permeability areas. Local features are reflected in the early portion of the drawdown curve

attitudes can be adopted: trying to identify heterogeneity (see, e.g., Yeh et al. 2000; Alcolea et al. 2006) or seeking effective parameters.

The former requires many observation points and several pumping wells, which can be realistic in CO_2 storage project. The latter has been the focus of much research and numerous methods have been developed. All these methods yield similar results (Sánchez-Vila et al. 1995), but no one is generally accepted as best.

An important result from this type of work is the finding that the equivalent transmissivity can be larger than the geometric average of point T values, whenever high T values are well connected (Sánchez-Vila et al. 1996). This is expected to be the rule, rather than the exception (think of fractured media, sedimentary materials, with coarse sediments paleochannels, etc.). An explanation for this "scale effect" is given in Fig. 7.7. Most wells will not intersect the most conductive fractures this yielding relatively small transmissivities. If many wells were available and computed their geometric average, TG (the traditional effective transmissivity for 2D flow), it would be concluded that the aquifer transmissivity is small. In reality, fluids would flow through the fractures. Therefore, one should expect T_{eq} to be larger than T_G. In fact, Knudby and Carrera (2005) use the ratio T_{eq}/T_G to define connectivity. The question is whether one can estimate equivalent transmissivity from field data, which is addressed below.

The value of transmissivity one obtains from a pump test depends on whether the test is interpreted using Theis, Jacob or Thiem method. That is, the resulting transmissivity is as much a property of the real medium as of the adopted model. The question is whether one can draw any general conclusion from such interpretation. This question was addressed by Meier et al. (1998), who simulated pumping tests in heterogeneous media by imposing a constant flow rate at one mode and observing drawdown at all other nodes (Fig. 7.8). They then interpreted the drawdown curves at all nodes using Jacob's methods. They found:

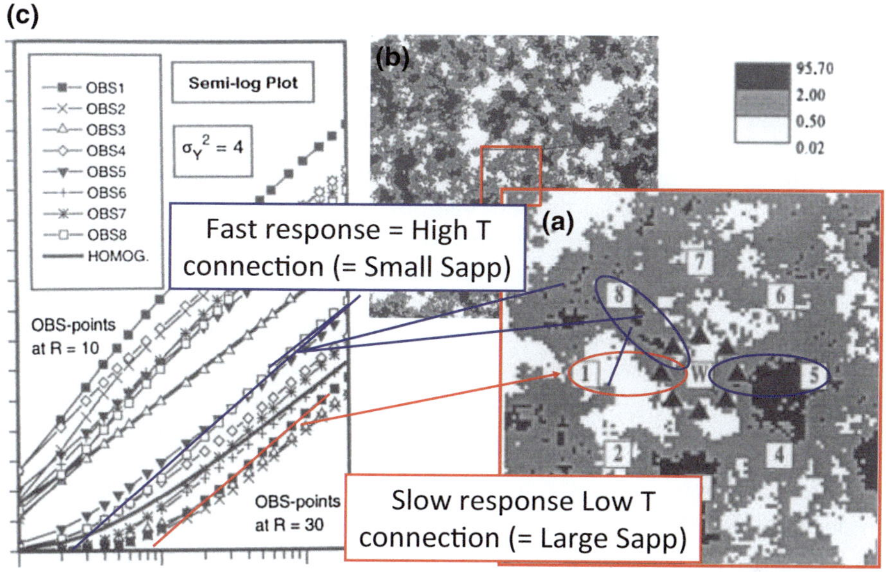

Fig. 7.8 Simulations of Meier et al. (1998) to show the effect of heterogeneity on pumping tests interpretation. **a** Transmissivity field; **b** local region around the pumping well with a few observation points; **c** semi-log drawdown plots. Note that all curves display the same late time slope (i.e., yield the same T), but response times (t_0) are highly variable (i.e., yield equally variable S)

1. The resulting estimated transmissivities were identical at every mode. That is, T estimated with Jacob's method is independent of the location of the observation point.
2. The estimated transmissivity is identical to the equivalent transmissivity for parallel or radial flow.
3. Estimated storage coefficient reflects not only the actual storage coefficient, which was assumed constant, but also the degree of connection (high T) between the pumping well and the observation point.

In summary, conventional tests interpreted using Jacob method will yield highly reliable values of effective transmissivity for radial distances given by R in Eq. (7.3.2) or Fig. 7.4. Therefore, returning to Fig. 7.7, the important point is to make the test long enough to make sure that the fracture network is effectively tested. Unfortunately, the value obtained for storage coefficient is much less reliable, as it reflects the connectivity between pumping and observation well.

7.3.1.4 Perturbations to the Conventional Case

Up to here, we have analyzed the basic Theis case. In practice, a number of perturbations affect this basic theory: wellbore storage may not be negligible; flow may not be radial (e.g., because of fractures or leakage from the caprock), storage may not be released instantaneously; test response may be affected by the presence of boundaries, etc. In the following, and for the purpose of illustration, we discuss these effects.

Wellbore Storage: Theis (1935) neglected the volume of water stored in the well. This assumption may not be valid if the diameter of the well is large (i.e., if the volume of water stored in the well is large compared to the volume of water pumped during the observation interval). Under these conditions, at the beginning of the test, pumped water comes mostly from the well, rather than from the formation. Therefore, drawdown at the well equals $S_w = Qt/A_w$, where A_w is the open area of the well. That is, the well acts as a deposit, which implies that drawdowns will plot as a straight line in arithmetic scale, and both drawdowns and their logarithmic derivative will also plot as a straight line with slope equal to 1 in the log-log plot (Fig. 7.9). Eventually, the drawdown becomes large enough to drive water towards the pumping well, so that Theis solution becomes valid again.

Boundary Effects: Theis (1935) assumed the aquifer to be infinite. Real aquifers are not. The drawdown cone will eventually reach some boundary. This is especially relevant for deep formations which may be compartmentalized (i.e., separated by no-flow boundaries) or intersected by permeable faults connecting the aquifer to another more permeable water body (effectively becoming a prescribed head boundary). The effect of these boundaries can be easily approximated using image wells. An image well is a virtual well located symmetrically to the pumping well with respect to the boundary. If water is injected at the image well with the same flow rate as the pumping well, drawdowns along the boundary caused by the pumping well will be compensated by head buildups caused by the image well. As a result, the boundary will indeed act as a zero drawdown boundary. Elsewhere within the aquifer, observation wells will first notice the effect of the pumping well, with a response analogous to that displayed in Fig. 7.3. Eventually, they will also notice the effect of injection at

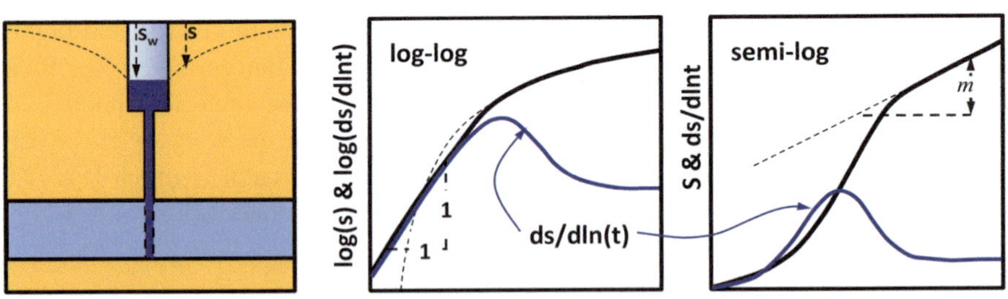

Fig. 7.9 Illustration of well bore storage effects. Well bore storage affects the shape of the response curves while Q_t is comparable to A_wS_w—during this period, both drawdown and derivative plots display a straight line with slope equal to 1, in log–log scale. In semi-log scale, the drawdown *curve* displays a concave shape, while the derivative *curve* displays a maximum

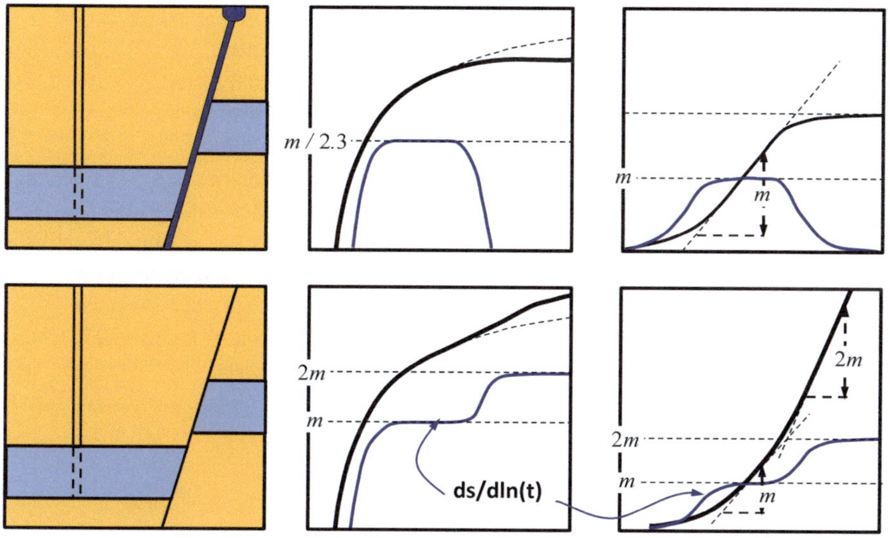

Fig. 7.10 Illustration of boundary effects. A linear prescribed head boundary (e.g., a high transmissivity fault connected to a constant head water body) can be identified by a zero derivative (drawdown becomes constant). If that fault does not allow flow-through, the derivative is multiplied by a factor of 2

the image well, which will cause the drawdown to become constant. Therefore, drawdown plots (Fig. 7.10, top) are characterized by a constant drawdown (the logarithmic derivative goes down as a straight line in the log–log plot).

If the image well is pumped with the same flow rate as the pumping well, flux across the boundary will be compensated by superposition. As a result the boundary effectively acts as a no-flow boundary. In this case, observation wells will also first notice the effect of the pumping well, but the image well, which is also pumping, will cause the drawdown to increase its rate. In fact, by superimposing the solutions to pumping and image wells, it is easy to demonstrate that drawdown plots (Fig. 7.10, bottom) are characterized by a sudden duplication in the slope of the drawdown curve in the semi-log plot or by a step increase the logarithmic derivative. In fact, distance to the image well can be derived from the time at which the effect of the boundary becomes noticeable, using it in the equation for R in Eq. 7.3.2.

More complex configurations of boundaries can be reproduced with several image wells. A case of special importance for CO_2 storage is the one of a close aquifer (i.e., full compartmentalization). In this case, by the time the effect of pumping has reached the whole aquifer, the compartment starts acting as a deposit. That is, drawdowns will tend to grow linearly with time, similar to what was described in the wellbore storage case. The slope will now be equal to Q/(A·S), where A is now the area of the compartment.

Leaky Aquifers: Theis (1935) assumed the aquifer to be confined by perfectly impervious confining layers. In reality, Confining layers over- and underlying an aquifer are never completely impermeable, they are somewhat “leaky”. When a well in a leaky aquifer is pumped, water is withdrawn not only from the aquifer, but also from the over- and underlying layers, thereby creating a hydraulic gradient also in the aquitard. The flow is usually assumed to be vertical in the aquitard and horizontal in the aquifer (Kruseman and De Ridder 1994).

Leaky aquifers are especially relevant for CO_2 storage, because it is leakage across the caprock what will cause pressure buildup in the aquifer to be bounded. The effect of leakage is a slowdown in the rate of head drop, so that the derivative is also reduced (similar to the fixed head boundary, Fig. 7.10, but depending on the aquifer beyond the confining layer, drawdowns may eventually increase). The traditional solution for the leaky aquifer is that of Hantush and Jacob (1955), who neglected aquitard storage. The full solution was derived by Neuman and Witherspoon (1969).

7.3.1.5 Flow Dimensions

Everything discussed up to here, except the leaky aquifer case, was based on the assumption that flow is radial towards the pumping well (or at least that flow is two dimensional). In reality flow needs not be 2D (see Fig. 7.11). Flow will be 1D along a vertical fracture intersecting the well or 3D when pumping from a short interval in a very thick aquifer. Since the extent of the drawdown cone grows as $t^{1/2}$, its volume will grow as $t^{d/2}$. Therefore, the logarithmic derivative will tend to become straight with a slope equal to $1 - d/2$ in log–log plot. That is, if flow is 1D, the logarithmic derivative will tend to display a 0.5 slope. If flow is 3D, this slope will be −0.5. Alternatively the dimension can be derived from this slope m, as $d = 2 \cdot (1 - m)$ (recall that this is the slope of the logarithmic derivative in log–log scale, not to confuse with the drawdown slope in semi-log scale).

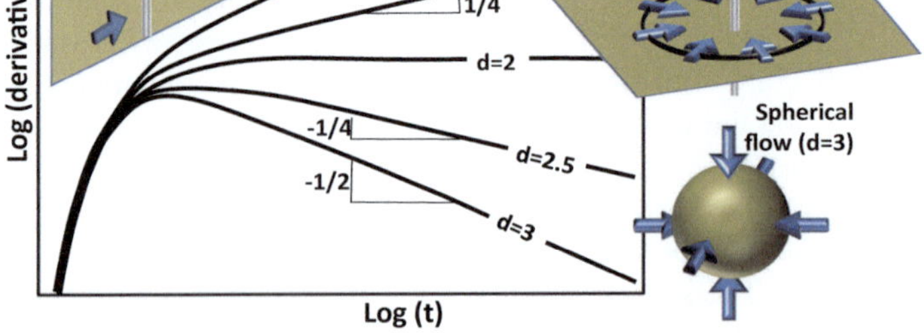

Fig. 7.11 Flow dimension effects

The surprising observation is that the resulting dimension is often non-integer. A body of literature has been developed to address this issue. The fact that the volume of the drawdown cone increases with a non-integer dimension may reflect a fractal connectivity pattern, which is not surprising in fractured media.

7.3.2 Tracer Methods for Characterization of the CO_2 Storage Sites

Alexandru Tatomir, Iulia Ghergut and Martin Sauter

Tracer testing can be defined as the injection into the hydro-geological system of one or more tracers, which usually are chemical compounds, but which can also be energy/temperature, viruses, DNA, etc. Tracer methods are commonly used to study the flow, transport and reactions of fluids and components in regions which are difficult to access and may extend over a wide range of length scales, like in the case of CO_2 storage reservoirs. Tracer methods are suited both for site characterization before CO_2 injection and for monitoring and verification purposes during and after CO_2 injection. The theoretical background of the tracer methods is well developed from the fields of hydrogeology and petroleum engineering. However, for geological storage of CO_2, new challenges arise due to the complex phase behaviour and the range of trapping mechanisms (i.e. structural, residual, solubility and mineral trapping).

The CO_2 storage reservoirs, as well as other technology-relevant geo-reservoirs in the realm of energy production (such as geological formations potentially suitable for spent-radionuclide disposal, gas storage or geothermal energy utilization) contain several fluid and solid phases, as well as mobile and immobile-fluid regions. The nature of utilization of the reservoir, the volumes and interfacial areas of the fluid phases and/or fluid regions, and the processes occurring within the phases are the factors which determine the lifetime of the reservoir. The lifetime of a geo-reservoir subject to a particular utilization can be subdivided and categorized into a hydraulic lifetime, a geomechanical, a hydrogeochemical, and a thermal lifetime. Figure 7.12 schematically illustrates the potential uses of tracer tests to answer fundamental questions related to each lifetime category and the complementarity that exists between the tracer test types. The tracer tests can be performed as single-well injection withdrawal (SWIW), or simply push–pull, tests, or in inter-well configurations. SWIW involves the injection and abstraction of tracer in one well, whereas inter-well testing requires two or multiple wells.

The concept of reservoir lifetime, transferred from the field of geothermal reservoirs, refers to an integral description of the reservoir properties and their change with time due to natural or anthropogenic factors, such as management by wells or CO_2 injection. The key parameters controlling the storage performance are the fluid flow pathways, the reservoir boundaries and hydrogeomechanical integrity, the storage capacity of the reservoir, the heat exchange areas (if existing) and fluid-rock and fluid-fluid interface areas. In designing and dimensioning tracer tests,

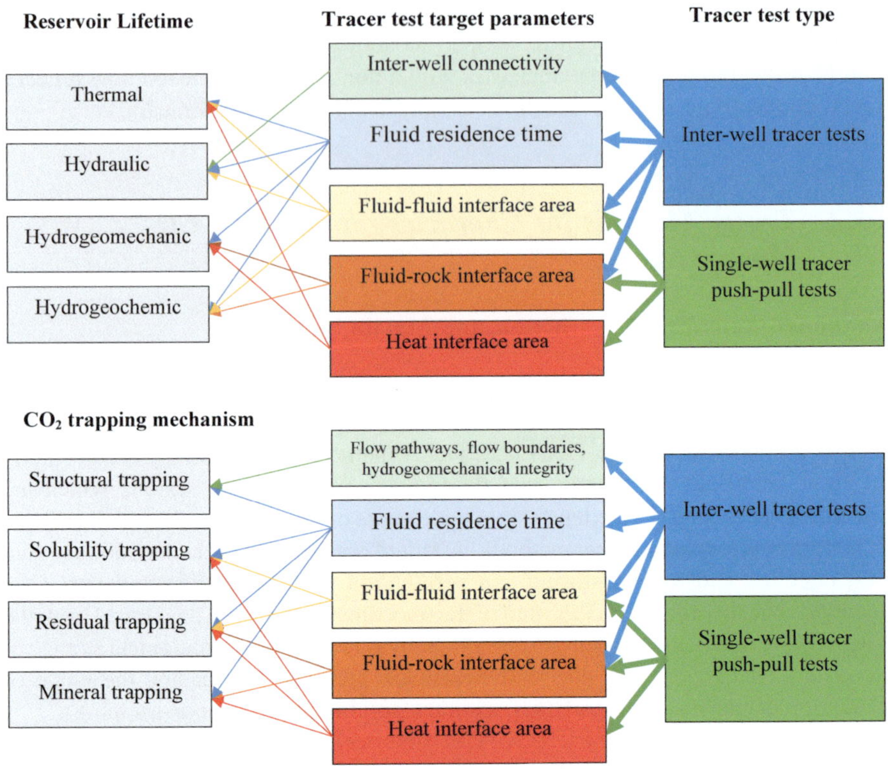

Fig. 7.12 Overview of reservoir lifetime categories (*upper left*) and CO_2 trapping mechanism (*lower left*), the target parameters which can be determined by the tracer tests (*middle*), and the respective tracer methods suitable for measuring them (*right*)

one has to consider that not every test design and any tracer species can be equally sensitive with regard to every fluid transport parameter. Therefore, when referring to tracer sensitivity and applicability range a certain complementarity exists between (Fig. 7.12):

1. Single-well and inter-well methods,
2. Equilibrium and non-equilibrium (kinetic exchange processes),
3. Volume (e.g. fluid phase saturation) and area parameters (e.g. fluid-fluid interfaces, fluid-rock interfaces).

7.3.2.1 Chemical Tracer Types

There is a wide spectrum of tracer applications, capturing various underlying processes and physico-chemical and biological properties, as illustrated in Table 7.4. Depending on their physicochemical and chemical behaviour, tracers can

Table 7.4 Compilation of tracers used for the characterization of geological systems (Schaffer 2013)

Tracer type	Injection phase	Application|target parameters	Underlying processes	Exemplary compound classes	Most relevant tracer properties
Conservative					
Conservative tracers					
Ionic tracers	aq.	Hydromechanical properties (porosity, dispersion, matrix diffusion, arrival times)	None	Organic and inorganic anions (salts, fluorescent dyes)	Anionic, highly water-soluble, non-sorptive, low log D, low pK_a
Tritium tracers			None	Isotopically labeled water	Molecular weight >18 g mol^{-1}
$scCO_2$ plume marker	Non-aq.	Plume migration, arrival times	None	Perfluorocarbons (McCallum et al. 2005 [56])	Inert, non-polar (no partitioning into water)
Temperature (heating/cooling)	aq. (non-aq.)	Groundwater flow, thermal conductivity	Thermal conduction, advection/convection	None	
Pressure	aq. (non-aq.) (gas.)	Permeability, hydraulic conductivity, leakage, arrival times	Pressure stimulation/response	None	
Colloidal tracers (particle tracers)					
Microorganisms	aq.	Hydraulic connection, groundwater drift, mixing, transport of microorganisms	None, (adsorption, coagulation size exclusion)	Bacteria, bacteriophages, yeasts, animal viruses, colored or fluorescently labeled spores (*Lycopodium clavatum*) (Keswick et al. 1982 [41], Becker et al. 2003 [7])	

(continued)

Table 7.4 (continued)

Tracer type	Injection phase	Application|target parameters	Underlying processes	Exemplary compound classes	Most relevant tracer properties
Microspheres/anoparticles)	aq.	Hydraulic connection, groundwater drift, mixing	None, (adsorption, coagulation, size exclusion)	Fluorescently labeled polystyrene or latex microspheres (Becker et al. 1999 [6], 2003 [7]), artificial DNA labeled microspheres (Sharma et al. 2012 [80]), fluorescent silica nanobeads (Agenet et al. 2011 [1])	
Retardation					
Partitioning tracers (equilibrium tracers)					
Sorptive tracers	aq.	Organic carbon content, retardation, (degradation)	Non-ionic sorption processes, (degradation)	Organic, hydrophobic compounds	Neutral molecules, moderate log K_{ow}
Ion exchange tracers	aq.	Surface charge, ionic strength, exchange area, retardation, (degradation)	Cation exchange, (anion exchange)	Inorganic (Dean et al. 2012 [17]) and organic cations (bases) (Leecaster et al. 2012 [46]; Reimus et al. 2012 [72])	Ionic molecules, low log D, high pK_a
Lnterfacial tracers	aq.	Interfacial area between immiscible phases	Adsorption on interface	Liquid/liquid interface (anionic) surfactants (Saripalli et al. 1997 [74], 1998 [73]), liquid/gas interface: high molecular weight alcohols (Kim et al. 1998 [42]; Rao et al. 2000 [70])	Amphiphil, ions with long alkyl chain or other non-polar substructures
Volume-sensitve tracers	aq. non-aq., gas.	Residual saturation of two phases	Partitioning between two phase volumes	Alcohols (Annable et al. 1998 [3]), phenols (Bennett and Larter 1997 [10]), (noble) gases (Vulava et al. 2002 [91]), fluorocarbons (Deeds et al. 1999 [18]), radioisotopes (Hunkeler et al. 1997 [34]; Semprini et al. 2000 [79]; Davis et al. 2002 [16])	Moderate phase partitioning coefficients, low molecular weight
Reaction					

(continued)

Table 7.4 (continued)

Tracer type	Injection phase	Application|target parameters	Underlying processes	Exemplary compound classes	Most relevant tracer properties
Tracer with chemical reaction (non-equilibrium tracers)					
Kinetic interface-sensitive tracers (KIS tracer)	Non-aq.	Interfacial area and development with time	Hydrolysis at interface	Non-polar, hydrophobic esters (Schaffer et al. 2013 [76])	High log *Kow*, reaction products anionic, suitable reaction rates
Thermo-sensitve tracers	aq.	Reservoir temperature, temperature gradients	Temperature-dependent hydrolysis	Polar, anionic esters (Nottebohm et al. 2012 [63])	Low log D, low pK_a, suitable reaction rates
Reactive esters for partitioning	aq.	Residual saturation of immiscible phases	In-situ generation of volume- sensitive partitioning tracers due to hydrolysis	Esters (reaction to alcoholic esters and alcohols) (Tomich et al. 1973 [86]; Myers et al. 2012 [57])	Water soluble, suitable hydrolysis rates of esters and partitioning coefficients of the reaction products
Biogeochemical reactive tracers	aq.	Transformation rates for certain domian, in-situ microbiological activity	Biological and/or chemical decay	Inorganic electron acceptors or donors (O_2, NO_3^-, SO_4^{2-}, H_2) (Istok et al. 1997 [35]), low weight alcohols, benzoate, sugars (Rao et al. 2000 [70]), caffeine (Hillebrand et al. 2012 [33])	Labile (easily biodegradable) or reactive

aq. aqueous phase, *non-aq.* non-aqueous phase, *gas.* gaseous phase, *log Kow* n-octanol/water distribution coefficient, *log D* pH-dependent *n*-octanol/water distribution coefficient of all species of ionizable organics, pK_a logarithmic acidity constant principally be differentiated in conservative and non-conservative tracers. Conservative tracers that are physically and chemically inert and are classically applied for tracking connectivity and pathways for flow, for analyzing travel times, groundwater drift, and flow velocities, for determining recharge and discharge as well as for estimating hydromechanical reservoir properties such as dispersivity and porosity. In contrast, non-conservative tracers experience physico-chemical processes or chemical reactions during their transport. Therefore, they can provide information on physicochemical reservoir properties and the water chemistry (Schaffer 2013; Ptak et al. 2004; Divine and McDonnell 2005).

Table 7.4 gives an overview of the entire range of tracer types, their application and target parameters (e.g., porosity, dispersivity, interfacial area, residual saturation, etc.), the underlying process (e.g., adsorption on interface, hydrolysis), while providing examples of tracer compounds (e.g., perfluorocarbons, esters, etc.) and their most relevant properties (e.g., neutral molecules, water soluble, etc.).

7.3.2.2 Mathematical Models for Characterization of CO_2 Reservoirs

The mathematical models describing the fate of tracers and CO_2 injected into a geological formation are constructed on the equations of reactive single-phase and multi-phase flow and transport in porous and fractured media (Bear 1988). The single-phase formulations are mainly relevant for the site characterization stage while the multi-phase formulations apply more for the monitoring stage.

The mass balance equation for phase α in a multiphase flow porous media system can be expressed as (see also Chap. 3 for more elaborate development and definition of terms):

$$\frac{\partial(\phi\rho_\alpha S_\alpha)}{\partial t} + \nabla \cdot (\rho_\alpha v_\alpha) - \rho_\alpha q_\alpha = 0 \tag{7.3.7}$$

For accounting for dissolved component transport the mass balance equation is written:

$$\frac{\partial(\sum_\alpha \phi\rho_\alpha S_\alpha C_\alpha^\kappa)}{\partial t} - \sum_\alpha \nabla \cdot \{\rho_\alpha v_\alpha C_\alpha^\kappa + \phi\rho_\alpha D_\alpha^\kappa \nabla C_\alpha^\kappa\} - \sum_\alpha \rho_\alpha q_\alpha^\kappa = 0 \tag{7.3.8}$$

where C_α^κ is the concentration of the dissolved component k in the fluid-phase α. The mathematical models in Eqs. 7.3.7 and 7.3.8 can be reduced to a single-phase flow system by considering S_α equal to 1.

The source term q_α^κ contains contributions from intra-phase α (where α can be wetting, w, or non-wetting, n) reactions and from partitioning between different phases. Generally, intra-phase reaction rates need to be multiplied by porosity and saturation to get the correct balance terms. Partitioning between phases can be at equilibrium or kinetic.

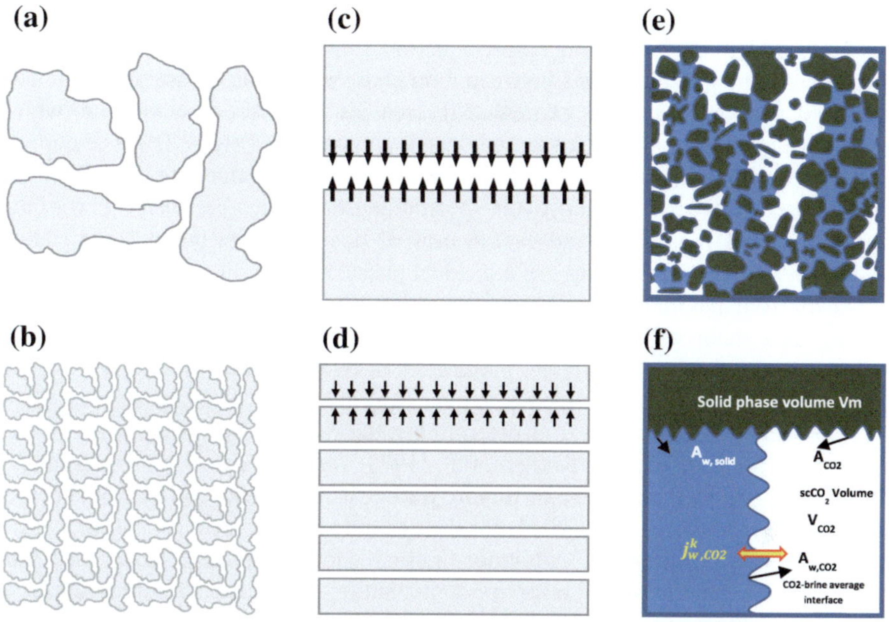

Fig. 7.13 Symbolic representations of fluid-rock (**a**–**d**) and fluid–fluid (**e**–**f**) interface-area density in porous media **a** a low specific contact surface area with the same bulk porosity as **b** high specific contact surface area; **c** fractured porous media: fracture-matrix fluid-rock interface area with low specific contact surface area and **d** high specific contact surface area, and the same bulk porosity; **e** wetting (*blue*)–non-wetting interface-area, **f** upscaled REV phase saturations, interface-area and exchange between phases (after Tatomir et al. 2015)

The Darcy velocity expressing the momentum conservation for fluid phase α is written:

$$v_\alpha = -\frac{k_\alpha}{\mu_\alpha}(\nabla p_\alpha - \rho_\alpha g) \qquad (7.3.9)$$

The key parameter porosity ϕ is essentially a scaling factor in the storage term (time variable), while the size of fluid phase interface densities, $a_{\alpha\beta}$ (e.g., between CO_2 and brine) and fluid-solid interface densities, $a_{\alpha S}$ (Fig. 7.13) are coefficient factors in several of their reaction and solute exchange flux terms.

7.3.2.3 Target Parameters of Tracer Tests

The first of these parameters, ϕ is vital to any CCS project, in that it determines the storage capacity of the target formation and CO_2 plume spreading velocity under given injection rates. The importance of the second parameter $a_{\alpha\beta}$, the specific interfacial areas between fluid phases or between CO_2 and rock, may vary depending on geological, hydrogeochemical and geotechnical details of the intended CCS site operation and fluid conditioning before injection. Fluid-rock interfacial area (density) $a_{\alpha S}$ (the specific interfacial area between fluid phase α and solid phase S) is a lumped averaged parameter, usually expressed in area per bulk reservoir volume, which relates to the effects of fracture spacing, length and aperture (Fig. 7.13). The mineral surface reactive area ($a_{\alpha S}$) is a key parameter for the mineral trapping (e.g., Xu et al. 2005; Luo et al. 2012). Term $a_{\alpha\beta}$ (the specific interfacial area between the two fluid phases α and β, e.g., water and gas) in turn, as determined in the characterization phase, can indicate the presence of a residual phase, such as oil or gas.

Apart from parameters ϕ, $a_{\alpha\beta}$ and $a_{\alpha S}$ the governing equations also contain, among their main variables, fluid phase saturations S_α as well as temperature T. These can also be treated as target parameters if, over the scale of a particular field experiment, they can be assumed to remain approximately constant.

Prominent examples aimed at determining the residual non-wetting phase saturation are the described in Tomich et al. (1973) for the residual oil saturation by means of single-well reactive-partitioning-tracer push–pull tests, and in Zhang et al. (2011) and Rasmusson et al. (2014) for the determination of CO_2 saturation using a combination of single-well hydraulic, thermal and partitioning-tracer push–pull tests. The determination of reservoir temperature by means of thermosensitive tracers is discussed in e.g. Nottebohm et al. (2012).

7.3.2.4 Tracer Tests for CO_2 Site Characterization

Historically, the tracers applied for site characterization have been well described in the field of hydrogeology (e.g., Leibundgut et al. 2009; Divine and McDonnell 2005). The use of liquid-phase conservative tracers, and the use of tracer diffusion, sorption or equilibrium-partitioning is based on well-established principles, models and application methods (Maloszewski and Zuber 1985, 1993; McCallum et al. 2005; Zuber and Motyka 1994; Carrera et al. 1998; Haggerty et al. 2001; Vulava et al. 2002), and shall not be further detailed here. Examples of tracer uses for site characterization is given in Table 7.5. Similarly, a synoptic view on the class of tracers termed "thermal tracers" used for reservoir thermal characterization is given in Table 7.6.

Generally, ***inter-well tracer tests*** are used to determine fluid residence time distributions (RTD). The statistical moments of RTDs provide important information about the reservoir (Ghergut et al. 2011), as follows:

1. The 0th-order RTD moment can tell something about reservoir boundaries.
2. The 1st-order RTD moment, or mean residence time (MRT) represents a measure of reservoir size (the reservoir volume that can be used for fluid storage).
3. The higher-order RTD moments provide information about reservoir heterogeneity. Traditionally, the 2nd-order moment is associated with flow-path dispersion (from hydrodynamic up to reservoir scale). From RTD analysis also a

Table 7.5 Examples of tracer uses for site characterization

Target parameter	Suitable method	Suitable tracer species	Limitations	Application examples
Transport-effective porosity	Inter-well tests; single-well inter-layer circulation	(Liquid-phase only) physico-chemically conservative, non-sorptive	Inter-well tracings cannot be conducted on a large scale at CCS candidate sites; tests remain limited to small-scale, pilot research projects; heterogeneity cannot be captured at reservoir scale	Deep saline aquifers in the N-German Sedimentary Basin (Horstberg, Gr. Schönebeck); geothermal: Kocabas (2005); oilfield: Kocabas and Maier (2013)
Fluid-rock interface area, fissure density	Inter-well tests; single-well tests	(Liquid-phase only) physico-chemically conservative, with contrast in sorption and/or diffusion; heat as a conservative fluid tracer	Parameter interplay; geological heterogeneity; geological uncertainty; parameter inversion requires the accurate knowledge of partitioning/sorption/diffusion coefficients independently of the field test	KTB pilot hole (heat and solute tracers); Bruchsal (attempted with heat as a tracer) (Behrens et al. 2014; Ghergut et al. 2013)
Relative saturation, fluid-fluid interface area (where applicable, in multi-phase systems)	Inter-well tests; single-well tests	Partitioning tracers		NAPL detection in the subsurface: Istok et al. (2002), Fagerlund (2007), LaForce et al. (2014)

Table 7.6 Summary of different meanings for thermal tracers and thermal characterization, single-well tests (SWT), inter-well tests (IWT)

Target information	Suitable method	Suitable tracer species	Limitations	Application examples
Site characterization: in situ measurement of thermal parameters (as an alternative to petrophysical laboratory techniques), for single-continuum descriptions	SWT	Heat as a conservative fluid tracer (fluid temperature signals)	May be disturbed by the presence of fractures/fissures	Kocabas (2005), Freifeld et al. (2008), Oberdorfer (2014)

(continued)

Table 7.6 (continued)

Target information	Suitable method	Suitable tracer species	Limitations	Application examples
Site characterization, cf. Table 7.1 (transport-effective porosity, dispersivity), for single-continuum descriptions	IWT SWT	Heat as a conservative fluid tracer (fluid temperature signals)	Pulse size limitation; not applicable on a large scale	Aquifer characterization: Anderson (2005), Read et al. (2013); theoretical studies of heat transport in porous media, esp. on the incongruence between heat dispersivity and solute dispersivity: Vandenbohede et al. (2009); thermal interference tests: Kocabas (2005), Oberdorfer et al. (2013)
Local characterization of single fractures; fissure aperture, fissure density in fissured media	SWT	Heat as a conservative fluid tracer (fluid temperature signals)	Requires continuous temperature recording downhole (ideally, DTS)	Pruess and Doughty (2010), Jung and Pruess (2012), Maier et al. (2012), Maier and Kocabas (2013)
Tracking thermal fronts in single-continuum media	IWT SWT	"Thermo-sensitive" tracers	pH and/or other non-predictable influences on tracer reaction rates	Nottebohm et al. (2012), Maier et al. (2014), Schaffer et al. (2015)
Tracking thermal fronts in fissured media	IWT	"Thermo-sensitive" tracers	Inversion methods are only developed for single-continuum descriptions; inversion not always unique for highly-heterogeneous, fissured/fractured media	Plummer et al. (2010), Juliusson and Horne (2010)

so-called flow-storage repartition (FSR) can be derived, which is sometimes interpreted as representing reservoir shape (cf. Shook and Forsmann 2005; Shook et al. 2004; Behrens et al. 2010).

Complementarily, ***single-well tracer push–pull tests*** are used to quantify processes other than advection-dispersion. Typically, they are used to quantify exchange of some extensive quantity (mass, energy) between fluid and solid/fluid phases by processes like matrix diffusion or sorption/partitioning, the rate or amount of which depends on phase saturations and/or phase interface densities. Flow-field reversal during the 'pull' phase is supposed to largely compensate the effects of flow-path heterogeneity during the 'push' phase (excepting the hydrodynamic dispersion level), and, under certain conditions, to enhance the effects of tracer exchange processes at phase interfaces. In terms of determining different parameters in Eqs. 7.3.7–7.3.10 from different tracer tests, the following can be summarized:

- Porosity ϕ can only be measured reliably by means of inter-well conservative-tracer tests and it closely relates to tracer residence times.
- Rock-fluid and fluid-fluid interface area densities $a_{\alpha S}$, $a_{\alpha\beta}$ can be determined preferable by inter-well or alternatively by single-well tests using tracer pairs with contrasting properties in terms of: (a) *sorption*: is a process occurring at the fluid-rock interfaces, therefore, tracer testing using tracer components of contrasting sorption properties generate distinct breakthrough curves, implicitly offering a good sensitivity with respect to $a_{\alpha S}$, but rather poor sensitivity to S_α; (b) *rock matrix diffusion* (for fluid-rock interfaces in fractured media): such tracer pair have a good sensitivity to $a_{\alpha S}$, but rather poor sensitivity to S_α; (c) '*intra-particle diffusion*' (for fluid-rock interfaces in unconsolidated porous media): these tracers have a good sensitivity with regard to Sα, but rather poor sensitivity to $a_{\alpha S}$; (d) *liquid-gas equilibrium partitioning* (for liquid-gas interfaces): this pair of tracers have good sensitivity to S_α, but rather poor sensitivity to $a_{\alpha\beta}$; (e) *liquid-gas interface reactivity*, with dichotomic partitioning of reaction products (for liquid-gas interfaces), as proposed by Schaffer et al. (2013): these tracers have a good sensitivity to both $a_{\alpha\beta}$ and S_α.

Use of tracers for CO_2 monitoring is discussed in Chap. 8 and for the specific use of characterizing site's capacity for CO_2 residual and dissolution trapping in Sect. 7.4.

7.3.3 Characterization of the Thermal Properties

Jacob Bensabat

In the context of CO_2 storage, it is important to be able to predict pressure and temperature conditions of the injected CO_2 along the CO_2 injection tubing and at the well bottom. This requires the ability to simulate the injection of the CO_2 in the borehole, from the wellhead to the well bottom. The temperature and pressure of the CO_2 at the well bottom will depend (i) on the pressure and temperature conditions of the CO_2 at the wellhead, which are determined by the operator of the injection, (ii) on the pressure losses along the injection tube and (iii) the heat transfer between the injected CO_2 and its surrounding environment. As the temperature of the injected CO_2 will be different from the temperature of its environment along the

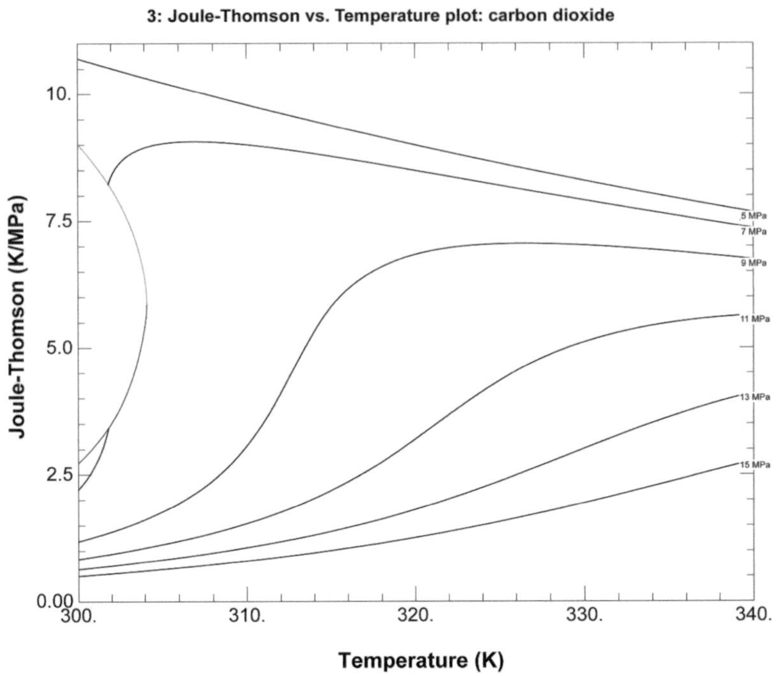

Fig. 7.14 Joule–Thompson coefficient for CO_2 (generated with REFPROP 9; NIST 2010)

borehole, there will be heat exchange, which depends on the configuration of the well and on the thermal properties of the geological materials surrounding the well.

Pressure losses are calculated from the CO_2 properties and from the characteristics of the injection tube. For small scale CO_2 injection experiments, these can be regarded as of secondary importance due to the very low viscosity of the CO_2 and to the generally small injection flow rates (less than 20 tons/h). This process is governed by the Joule–Thompson coefficient, which for CO_2 is presented in Fig. 7.14.

Thermal exchange between the injected CO_2 and its surrounding environment will depend on the well configuration, on the casing and cement properties and on the thermal properties of the surrounding formation and its temperature. This process of heat transfer can be defined as radial heat transfer and the rate of heat flow per unit length of well can be expressed following Hasan and Kabir (2002):

$$Q = -2\pi r_o U_o (T_f - T) \tag{7.3.10}$$

where U_o is the overall heat transfer coefficient, between the CO_2 inside the injection tube and the formation. r_o denotes the radial distance between the injection tube and the formation, T and T_f are the wellbore and the formation temperatures. Simulations of the CO_2 injection in the wellbore indicate that the temperature of the

CO_2 at the well bottom is highly dependent on the heat transfer by radiation between the flowing CO_2 and its environment, described by the heat transfer coefficient.

There are many possibilities of well configuration and as an example we shall consider a well configuration similar to the wells drilled in the frame of the MUSTANG project at Heletz site in Israel (for details see Niemi et al. 2016).

These type of wells are divided in three vertical sections:

1. **The upper section** between ground surface and a depth of ~300 m. This section includes the geological formation, cement, the outer casing ($13^{5/8}$ in. diameter), cement, the middle casing ($9^{5/8}$ in.), cement, inner casing (7 in. diameter), fluid, and the injection tube ($2^{3/8}$ in. diameter).
2. **The middle section** between the depth of 300 m and a depth of 1200 m. This section includes the geological formation, cement, a $9^{5/8}$ in. diameter casing, cement, the 7 in. diameter casing, fluid and the $2^{3/8}$ in. diameter injection tube.
3. **The lower section** between the depth of 1200 m and to the well bottom at 1650 m. This section includes the formation, cement, the 7 in. casing, fluid and the injection tube.

There may be many alternative well configurations than the above one. The overall heat transfer coefficient represents the resistance to heat flow from the tubing to the formation and convective heat flow in the annulus (which may possibly be filled with water in the lower part or air in the upper part). Without loss of generality, we present an expression for this coefficient for lower, and simpler, part of the casing. Developing an expression for the other parts is straightforward.

$$\frac{1}{U_o} = \frac{r_{to}}{r_{ti} h_{to}} + r_{to} \left[\frac{\ln\left(\frac{r_{to}}{r_{ti}}\right)}{\lambda_{tubing}} + \frac{\ln\left(\frac{r_{to}}{r_{ci}}\right)}{\lambda_{annulus}} + \frac{\ln\left(\frac{r_{co}}{r_{ci}}\right)}{\lambda_{ca\sin g}} + \frac{\ln\left(\frac{r_{wb}}{r_{co}}\right)}{\lambda_{cement}} + \frac{\ln\left(\frac{r_{f\infty}}{r_{wb}}\right)}{\lambda_{formation}} \right] \tag{7.3.11}$$

where r_{ti} and r_{to} denote the inner and outer radii of the injection tube, r_{ci} and r_{co} denote the inner and outer diameters of the casing, r_{wb} is the wellbore diameter, $r_{f\infty}$ is a radial distance in the formation at which the field temperature is not affected by the injection and r_{wb} is the wellbore radius. λ_{tubing}, $\lambda_{annulus}$, $\lambda_{ca\sin g}$, λ_{cement} and $\lambda_{formation}$ denote the thermal conductivities of the tubing, the fluid filling the annulus, the casing, the cement and the formation, respectively. h_{to} denotes the rate of heat transfer between the flowing CO_2 and the inside of the injection tubing wall. We assume that the radial heat transfer adjusts instantaneously (quasi-steady state conditions) and we neglect convective processes in the annulus. These assumptions allow us to determine a lumped coefficient, responsible for the heat exchange between the formation and the injection tubing. All of the parameters but two in above formula can be determined. These are thermal conductivity of the formation $\lambda_{formation}$ and the thermal radius of influence of the well, $r_{f\infty}$, or the radial distance from the well at which no thermal effect resulting from the injection of the CO_2 is felt.

The key missing information above is therefore the thermal conductivity of the various formation layers. We present below an experimental sequence aimed at determining these data in situ and suggest the following procedure:

1. If logs are available, so it is possible to have a lithological description of the well section and to construct a discrete division of the well lithological profile with homogeneous geological properties (of homogeneous thermal conductivity for each unit).
2. Measure the temperature along the well, from well head to well bottom. Depending on the well completion, this could be performed in two ways: 1) using distributed temperature sensing (DTS) if that is included in the well completion (see Sect. 8.4) using a temperature probe and attached to a winch for vertical up-and-down scanning of the borehole.
3. Inject cold water, or water with a substantially lower temperature than the one along the well in the formation. Use formation water previously abstracted, in order to prevent any unwanted chemical reaction between formation water and the injected one (such as swelling of clay particles, oxidation and others).
4. Wait until static conditions have been reached, and there is no flow in the tubing.
5. Measure the temperature recovery in the well until the full return to the initial temperature.
6. If there is information available allowing splitting of the well lithology into a series of quasi-homogeneous sections, start from this division and further refine it by analyzing the rate of the thermal recovery along the well. Areas of high thermal conductivity would recover faster than other areas of lower thermal conductivity. If there is no lithological information, use only the temperature time series.
7. Since there is no flow in the well, the temperature recovery in the tubing can be expressed by the following heat balance equation.

$$\frac{\partial T}{\partial t} = \frac{\lambda_w}{\rho_w C_{pw}} \frac{\partial^2 T}{\partial z^2} + U_0(z)\big(T(z) - T_f(z)\big), \quad z \in [z_b, z_t] \tag{7.3.12}$$

$$U(z) \equiv \sum_{i=1}^{N-1} (H_i(z) - H_{i+1}(z))\, U_{oi} \tag{7.3.13}$$

The heat balance equation above has known initial and boundary conditions of temperature, i.e.

$$T(z_t) = T_t \tag{7.3.14}$$

$$T(z_b) = T_b \tag{7.3.15}$$

$$T(t = 0, z) = T_0(z) \tag{7.3.16}$$

where $H_i(z)$ is a Heaviside type function, defined as:

$$H_i(z) = \begin{cases} 0 & if\ z \prec z_i \\ 1 & if\ z \geq z_i \end{cases}. \tag{7.3.17}$$

N is the number of vertical lithological partitions. $T(z)$ is the temperature of the water in the tubing and $T_f(z)$ is the temperature in the formation, which is equal to the initial temperature profile. The heat balance equation (7.3.12) together with the initial and boundary conditions (7.3.14–7.3.16) can be solved analytically or numerically.

In the case of the thermal test the overall heat transfer coefficient is simpler, as there is no flow in the tubing and reduces to:

$$\frac{1}{U_o} = r_{to}\left[\frac{\ln\left(\frac{r_{to}}{r_{ti}}\right)}{\lambda_{tubing}} + \frac{\ln\left(\frac{r_{to}}{r_{ci}}\right)}{\lambda_{annulus}} + \frac{\ln\left(\frac{r_{co}}{r_{ci}}\right)}{\lambda_{ca\sin g}} + \frac{\ln\left(\frac{r_{wb}}{r_{co}}\right)}{\lambda_{cement}} + \frac{\ln\left(\frac{r_{f\infty}}{r_{wb}}\right)}{\lambda_{formation}}\right]. \tag{7.3.18}$$

The determination of the parameter U_o for each vertical partition can be obtained via an inverse procedure, i.e., minimizing the least squares of the errors between measured and simulated temperature over a range of time intervals and vertical sections.

We first divide the time interval that is required for the tubing to return to the initial temperature distribution in N_t equally spaced intervals of size Δt.

The identification procedure can be formally expressed as the following least-squares problem:

$$F(U_o) = \min \sum_{i=1}^{i=N_t} \sum_{p=1}^{p=N} [T(i,p) - T_s(U_o(p), i, p)]^2 \tag{7.3.19}$$

where $T(i,p)$ is the measured temperature, characteristic of the vertical section p at time $t_0 + i\Delta t$. $T_s(U_o(p), i, p)$ is the simulated temperature, which depends on the value of the overall heat transfer coefficient at the same time and vertical section.

7.4 CO_2 Injection Tests as a Characterization Method

Fritjof Fagerlund and Auli Niemi

CO_2 injections can be used to obtain knowledge about formation parameters which are directly related to the behaviour and fate of the injected CO_2. These include parameters which govern (i) the two-phase flow of a CO_2-rich and an aqueous phase, (ii) parameters governing CO_2 trapping in the formation and (iii) interactions between CO_2 and formation fluids and rock. Carefully monitored CO_2 injections have been performed in research projects including sites such as Frio (e.g. Hovorka et al. 2006), Otway (e.g. Paterson et al. 2013a, b, 2014) and Ketzin (e.g. Würdemann et al. 2010) and are underway at sites such as Heletz, Israel (Niemi et al. 2016) and Hontomin. Here the objectives typically have been related to gaining fundamental knowledge about geological CO_2 storage and the fate of the injected CO_2, or the development of field methods. Small-scale CO_2 injections can be useful both at pilot test sites and at sites under consideration for large-scale geological storage, to obtain better knowledge about site-specific properties related to site performance, such as field-scale CO_2 trapping or effective storage capacity.

Trapping of injected CO_2 by processes additional to the accumulation under a structural trap improves the storage security and is essential at many potential storage sites. Due to e.g. heterogeneity at different scales and the resulting preferential flow paths, these processes may differ in both magnitude and timing at the field scale as compared to laboratory investigations on rock cores or theoretical analyses. Field investigations concerning the trapping processes are therefore essential to understand the effective in situ trapping. In following sections, we will present examples field tests designed to study residual phase CO_2 trapping and dissolution trapping in situ. First the field-scale nature of the trapping processes and available measurements will briefly be discussed.

7.4.1 Field-Scale Residual and Dissolution Trapping

Residual phase trapping is a pore-scale process which occurs as small blobs and ganglia of the free-phase CO_2 (the non-wetting phase) are snapped off and become immobilized. However, due to heterogeneity in capillary properties of the medium, the non-wetting phase is also trapped at capillary barriers, thus forming small-scale structural traps.

At the field scale, the total residual phase trapping depends on (i) the amount of residual trapping that occurs per unit volume swept by the free-phase CO_2 (including both the pore-scale trapping and trapping at capillary barriers), and (ii) the sweep efficiency of the free-phase CO_2 plume (Hesse et al. 2009). Much uncertainty remains in quantifying both these parameters in relevant CO_2 storage formations. For poor sweep efficiencies when the CO_2 moves as a thin pancake under the cap rock ceiling, the CO_2 can move very large distances before it becomes immobilized, thereby increasing the risk of reaching a spill point (Hesse et al. 2009).

The pore-scale residual trapping in the aquifer volume swept the free-phase CO_2 depends on the local maximum CO_2 saturation (S_{gmax}) and the local initial gas saturation at the start of imbibition (S_{gi}), and such a dependence can be described by a trapping model (Land 1968). For a comparison of the effect of using different trapping models on CO_2 trapping, the reader is referred to Rasmusson et al. (2016). Thus, locally, the final residual saturation S_{grf} after imbibition of the aqueous phase is a function of the maximum possible residual saturation, Sgrmax, and the initial CO_2 saturation at the start of imbibition, S_{gi}. S_{grmax} is a material property which influences the capacity for residual trapping in a formation. Characterization methods aimed at determining this property have therefore been developed for e.g. the CO_2 injection experiments at Otway (Zhang et al. 2011; Paterson et al. 2013a, b, 2014; LaForce et al. 2014), and the planned experiments at Heletz (Rasmusson et al. 2014).

Heterogeneity is also critical for the residual trapping of CO_2 (e.g. Green et al. 2009; Flett et al. 2007; Hovorka et al. 2004) as it can strongly affect the sweep efficiency and thereby the saturation history. However, the net effect of heterogeneity on residual trapping appears to be dependent both on the nature of the heterogeneities and the general direction of movement of the CO_2 plume, as it can be different for vertical and horizontal migration (see e.g. Tian et al. 2016). Furthermore, it has been suggested that the strategy for CO_2 injection has a large impact on the sweep efficiency and subsequent residual trapping (e.g. Qi et al. 2009, Rasmusson et al. 2016). Qi et al. (2009) suggested that water should be co-injected with the CO_2 at a ratio which produces such mobility ratio between the two fluids that the sweep is maximized.

Dissolution of CO_2 to the aqueous phase can improve storage security as heavier CO_2-rich water sinks and thereby produces a vertical convective mixing which further enhances the dissolution process (e.g. Ennis-King and Paterson 2005; Riaz et al. 2006). Enhanced CO_2 dissolution by convective mixing is, however, a process that has mainly been studied using theoretical analyses and numerical modeling; e.g. Ennis-King and Paterson (2005), Riaz et al. (2006), Pau et al. (2010), and in analogous laboratory experiments (e.g. Kneafsey and Pruess 2010; Neufeld et al. 2010). There is therefore a strong need to measure and demonstrate this process in the field and characterize the key field-scale properties influencing the long-term significance of CO_2 dissolution at given storage sites. While long term experiments are needed particularly to observe convective mixing, much can be learned about the dissolution process and the interaction between the aqueous and CO_2-rich phases including the immobile (residual) water and CO_2 (Paterson et al. 2013a, b; Fagerlund et al. 2013a).

7.4.2 *Field Measurements Related to Two-Phase Flow and Trapping*

Measurements of fluid saturations as well as the parameters affecting two-phase flow and trapping of CO_2 are highly challenging at the depth of typical storage formations (>800 m). Measurements are restricted to inside or in the direct vicinity of boreholes, which are typically scarce due to the difficulties and costs associated with drilling at such depths. Surface geophysical measurements can generally not be used to visualize the two-phase flow and trapping at the typical depths to the required detail. To characterize the related aquifer properties one has to rely on the measurements available in deep wells. A strategy to characterize CO_2 trapping has therefore been to combine the information from several types of measurements which are influenced by the CO_2 trapping and to use inverse modeling to infer the trapping properties (Zhang et al. 2011).

In a characterization experiment involving CO_2 injection, a test sequence can be employed where the same measurements and tests are repeated both without CO_2 in the formation (before CO_2 injection) and with CO_2 in the formation (after CO_2 injection) (e.g. Paterson et al. 2013a, b). Measurements which can be included in such test sequence include:

Pressure or hydraulic test: Water is injected at a given rate (or pressure) and the pressure (or flow rate) in the injection well and other available wells is monitored using pressure sensors in the wells. The pressure signal is affected by the relative permeability to the aqueous phase, which is a function of fluid saturation. When performed with CO_2 at residual saturation the test gives information about the residually trapped saturation. To avoid dissolution and a change in saturation during the test, water saturated with dissolved CO_2 can be injected when creating the residually trapped zone (Zhang et al. 2011).

Thermal test: The concept of a thermal test for CO_2 saturation characterization has been described by Freifeld et al. (2008), and such test was applied at Ketzin (Giese et al. 2009) and Otway (Paterson et al. 2013a, b). The formation is heated from a borehole using a resistance heater and subsequently allowed to cool, while the temperature is measured using a fibre-optic distributed temperature sensor (DTS—more information given in Sect. 8.4 Well instrumentation). The dissipation of heat and thus the temperature signal is influenced by the effective thermal conductivity, which, in turn, is a function of fluid saturation since supercritical CO_2 has significantly smaller thermal diffusivity and specific heat capacity as compared to brine at typical reservoir pressures and temperatures. The sensitivity of this test is related to porosity since for a lower porosity the influence of the fluid in the pore space becomes smaller. A thermal test can penetrate in the order of 1–2 m into the formation depending on the applied heat and the rock properties (Zhang et al. 2011).

Pulsed Neutron Capture Tool (RST): This well-logging technique is sensitive to hydrogen index and thus also to the saturation distribution of CO_2 and brine near the borehole. It was successfully used to measure CO_2 saturation in the direct vicinity of boreholes at Frio (Doughty et al. 2008) and Otway (Paterson et al. 2013a, b). A limitation of the method is however that the penetration depth is only in the order of 0.20 m.

Tracers: Tracers are potentially very useful for the characterization of deep CO_2-brine systems, since the tracers can bring out information from aquifer volumes which are difficult to access with any other technique. However, much testing remains to be done employing tracers and interpreting the tracer response in these systems. Partitioning tracers, which are injected with the aqueous phase and retarded due to partitioning into an immobile fluid phase such as residual CO_2, provide information about the amount of immobile fluid present. This technique has been used to detect non-aqueous phase liquid (NAPL) pollutants (e.g. Istok et al. 2002) and also in single-well push–pull tests to infer residual CO_2 saturation (Paterson et al. 2013a, b). Paterson et al. (2013a, b) used the noble gas tracers Kr and He which were injected with water and produced different breakthrough signals as the water was pulled back to the well due to the different partitioning behaviour (Henry's constants) of the two tracers. The differences in partitioning and comparison to the case of no partitioning (no immobile fluid phase present or a conservative tracer) were used to calculate the residual CO_2 saturation. Reactive partitioning tracers which form daughter products in connection with the partitioning have been used to measure residual oil saturation (Tomich et al. 1973). In the push–pull tests at Otway (Paterson et al. 2013a, b) a similar reactive tracer technique was also employed, but with organic tracers more adapted to the partitioning between brine and supercritical CO_2, as described in more detail by Myers et al. (2012). Furthermore, new reactive tracers—kinetic interface sensitive (KIS) tracers—aimed at quantifying the interface between two immiscible phases are being developed (Schaffer et al. 2013) and can also be used in the characterization of a CO_2-brine system. Tracer techniques can be employed in both single-well push–pull tests and two-well (or multi-well) inter-well tests. Partitioning inter-well tracer tests (PITTs) have been used for measurement of residual oil saturation (e.g. Du and Guan 2005; Tang 2005) with applications in petroleum industry, and have also been used for contamination characterization measuring NAPL saturations (e.g. Nelson et al. 1999; Jin et al. 1995; Mariner et al. 1999). PITTs are also planned at the Heletz site to measure residual CO_2 saturation in a dipole test (Fagerlund et al. 2013a). In these dipole field tests, tracers are also planned to be used to measure the (short-term) dissolution of mobile and immobile (residual) CO_2. As described in more detail by Fagerlund et al. (2013a), the idea is that a tracer with extreme affinity for the CO_2 phase is injected with the CO_2. As the CO_2 dissolves into the formation brine, this tracer (which has negligible aqueous solubility) is enriched in the CO_2 phase. Thus, when the injected CO_2 breaks through to the withdrawal well, it carries information about the amount of mobile supercritical CO_2 which has been dissolved during its migration between the two wells. The dissolution of immobile CO_2 can further be quantified by measuring the CO_2 concentration in the extracted brine. Numerical modeling indicates that tracer techniques potentially can be used to identify the point in time when conditions of residually trapped CO_2 have been established, which can be critical in characterization experiments aimed at quantifying the residual CO_2 trapping. This indicator tracer method has been outlined to more detail by Rasmusson et al. (2014).

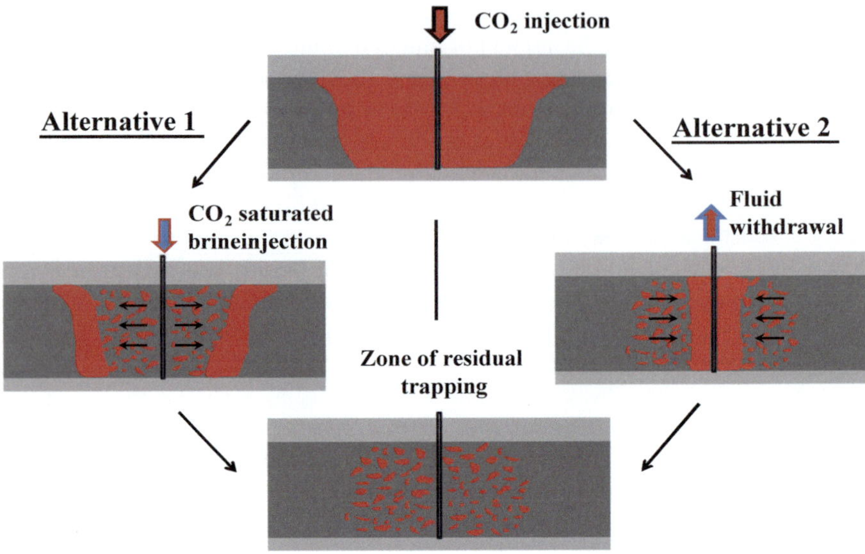

Fig. 7.15 Alternatives for creating a zone of residually trapped CO_2 in a push–pull field test

7.4.3 Single-Well Push–Pull Test for Quantification of Residual Trapping

This section describes the basic concept of a push–pull field test to quantify residual trapping of CO_2 and is mainly based on the test performed at Otway (Paterson et al. 2013a, b; Zhang et al. 2011) and the planned experiments at Heletz (Rasmusson et al. 2014).

Before residual trapping can be quantified, a zone of residually trapped CO_2 must be created in the storage formation. The creation of such zone and verification that the CO_2 phase has become immobile by residual trapping may not be trivial. Two alternatives have been proposed. As illustrated in Fig. 7.15, the first alternative is to inject CO_2 saturated water following the injection of free-phase CO_2. The CO_2 phase is pushed away by the injected brine, but does not dissolve because the brine is already saturated with CO_2. This option was chosen in the field tests at Otway (Paterson et al. 2013a, b). A technical challenge, as experienced in these experiments, can be the mixing of CO_2 into the brine injection and achieving brine just saturated with CO_2 (but not over saturated) at the reservoir pressure and temperature. Following injection of CO_2, a second alternative for creation of the zone of residual trapping (Alternative 2 in Fig. 7.15) is to withdraw formation fluids until the remaining CO_2 phase is immobilized by residual trapping. This method has the drawback that it may be difficult to know when the free-phase CO_2 has become immobile, and thus, to know when to stop withdrawing fluids (Zhang et al. 2011). Too much withdrawal will result in dissolution of the residually trapped CO_2 and

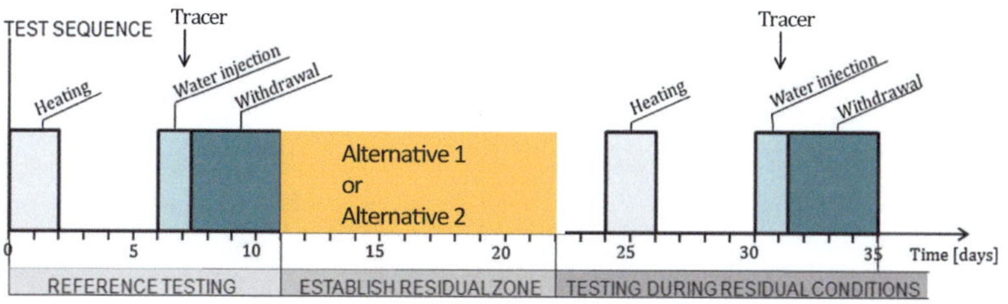

Fig. 7.16 Schematic example of a test sequence for a push–pull field test to characterize residual CO_2 trapping (Rasmusson et al. 2014)

may bias the quantification of residual trapping. A possible solution to this issue has been proposed by Rasmusson et al. (2014) who suggest the use of a conservative indicator tracer which according to modeling results can indicate the time when residual trapping has been achieved.

Both alternatives for creating the zone of residual trapping may be affected by buoyancy flow of the CO_2 phase and geological heterogeneity leading to preferential flow patterns in the formation. Particularly for high permeability, buoyancy flow towards the storage formation ceiling can be significant compared to the pressure driven flow through the well. Thereby a “pancake” of higher saturation free-phase CO_2 will form under the formation ceiling, under the caprock, the saturation distribution becomes uneven in the vertical direction and residual trapping occurs at later times in the top part which is refilled from below, and where the flow along the ceiling is slow.

Geological heterogeneity can influence the flow of both the CO_2 and brine phases. The CO_2 phase may preferentially move in high-permeability, low-entry-pressure channels or formation volumes, thereby affecting the sweep efficiency of the free-phase CO_2 and also the effective amount of residual trapping over the sampled volume. Potential bias in the residual trapping quantification both in terms of effects on sweep efficiency and fluid saturation history must therefore be considered in the test evaluation.

The strategy to quantify the residual trapping in the Otway field test (Paterson et al. 2013a, b; Zhang et al. 2011) was to include reference tests before the creation of the zone of residual saturation so that the same tests and measurements could be performed both with and without residually trapped CO_2 present in the formation. This method reduces some effects of heterogeneity on the test results since the same heterogeneity is seen by the measurements before and after introducing the CO_2. A schematic example of such test sequence is shown in Fig. 7.16, which includes (1) reference tests without CO_2 in the formation, (2) the creation of the zone of residual trapping, and (3) characterization tests with residual CO_2 present. Other tests than the hydraulic, thermal and tracer tests shown here (such as RST logging or cross-hole geophysical measurements) can also be added to the test sequence.

The responses in hydraulic, thermal and tracer tests with and without residual CO_2 in the formation all carry information about the residual trapping.

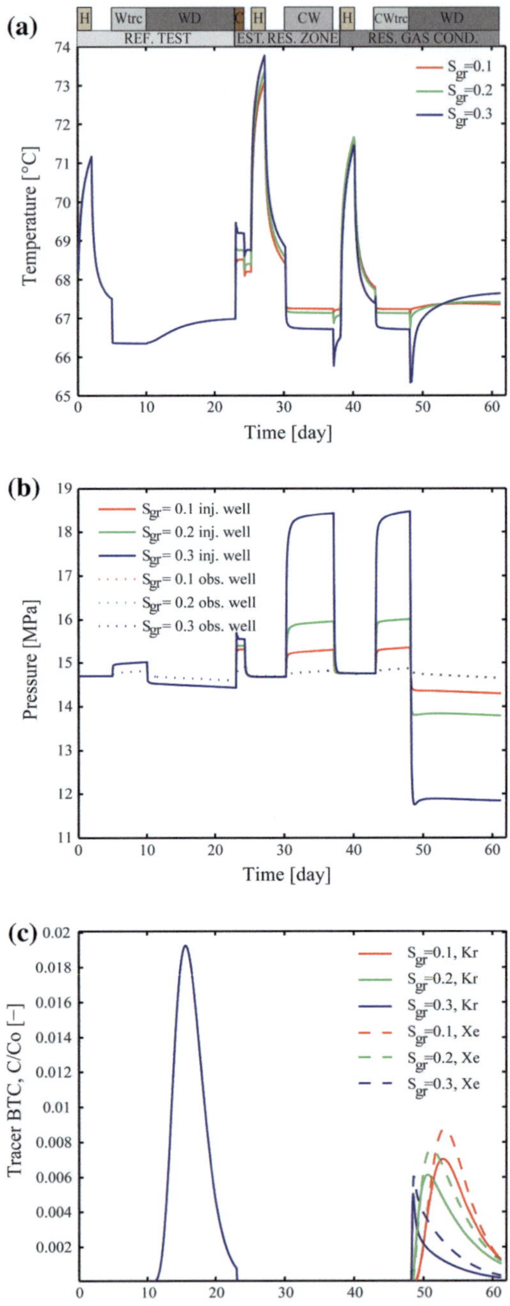

Fig. 7.17 Simulated response for different values of residual trapping (S_{gr}) in **a** temperature, **b** pressure (in both the injection well and an additional observation well) and **c** tracer BTCs for the noble gases Kr and Xe. From Rasmusson et al. (2014)

A process-based model can be used to jointly interpret the information of different measurements and estimate the residual trapping under uncertainty in different model parameters using inverse modeling techniques.

Examples of modeled responses in temperature, pressure and tracer breakthrough curves for different values of residual CO_2 trapping are shown in Fig. 7.17. This example comes from the design modeling for push–pull field tests at the Heletz site (Rasmusson et al. 2014), which is based on the field tests and modeling strategy developed for the Otway site (Zhang et al. 2011), and roughly follows the schematic test sequence presented in Fig. 7.16. As can be seen in Fig. 7.17, the responses in the measurements differ for different values of residual trapping (S_{gr}), however the sensitivities to S_{gr} vary in magnitude and change with time. A clear pressure increase which is sensitive to S_{gr} is for example seen when CO_2 saturated water is injected to create the zone of residual trapping roughly between 31 and 37 days (Fig. 7.17b) as well as in the following hydraulic test (water injection) during the characterization phase.

A prerequisite to successful estimation of the residual trapping is that the sensitivities to S_{gr} of the available measurements are large enough. However, because the measurement responses also depend on other uncertain parameters, the correlations between S_{gr} and these parameters need to be investigated and reduced. A systematic sensitivity and uncertainty analysis is essential with this approach and can also identify additional data needs to constrain the estimates of trapping. As shown by Zhang et al. (2011), combining several data sets and different data types reduces estimation uncertainties and improves the estimates of S_{gr}.

7.4.4 Two-Well Test for Quantification of Residual Trapping and Dissolution

This section describes the concept of a two-well field test to quantify residual trapping of CO_2 and dissolution. It is mainly based on simulation studies by Fagerlund et al. (2013a, b) aimed at exploring two-well field methods planned for the Heletz field site to characterize CO_2 residual trapping and dissolution under influence of geological heterogeneity.

In single-well push–pull experiments, fluids are pushed out and pulled back through the same flow channels, which can reduce the influence of geological heterogeneity. In inter-well tests, the flow and transport is affected by the heterogeneity between the wells and typically the transport also goes through a larger aquifer volume than in a push–pull test. Combining single-well push–pull tests with inter-well tests can therefore provide information about the effect of geological heterogeneity on the flow and trapping processes of injected CO_2. Both passive and actively pumping observation wells can be considered for monitoring of the CO_2 migration from the injection well. In this example we consider an active withdrawal well. This option has the advantages that (i) the flow field to some extent can be controlled by withdrawal of fluids, (ii) a zone of residual trapping between the wells can be established, and (iii) formation fluids and tracers can be measured as they are withdrawn from the observation well. The withdrawal of both formation brine and free-phase CO_2 which has migrated through the formation allows analysis of fluid compositions carrying information about the interphase mass transfer in the formation. Furthermore, tracers in both the aqueous and CO_2-rich phases can be analysed.

This example (based on Fagerlund et al. 2013a, b) will be limited to (i) a hydraulic test aimed at measuring residual trapping of CO_2 in the region between the two wells, and (ii) a tracer in the CO_2-rich phase which together with analyses of the withdrawn fluid compositions can be used for quantification of effective rates of CO_2 dissolution in the formation. Other tests as described above can, of course, be added to the test sequence, adding information and reducing uncertainty in the estimates as described above. The hydraulic test is repeated before and after CO_2 injection and the difference in the pressure response can thus be used to infer the reduction in aqueous phase permeability due to presence of residually trapped CO_2.

The tracer technique involves injection of a tracer with very small aqueous solubility with the CO_2. Thereby, as CO_2 dissolves into formation brine, the tracer is enriched and if tracer dissolution is negligible, the enrichment is directly proportional to the CO_2 dissolution. When free-phase CO_2 arrives at the withdrawal well, the tracer concentration in the CO_2 rich phase carries information about the amount of dissolution of mobile free-phase CO_2 which has occurred during its flow between the wells.

A schematic test sequence for the two-well test is shown in Fig. 7.18. The upper half of the figure shows the injection well activity and the lower part shows the withdrawal (abstraction) well. A reference hydraulic test is included before the injection of CO_2, and the hydraulic test is then repeated at the stage when most of the free-phase (supercritical—sc) CO_2 in the formation has been immobilized by residual trapping. To draw fluids and tracers, and to control the flow field,

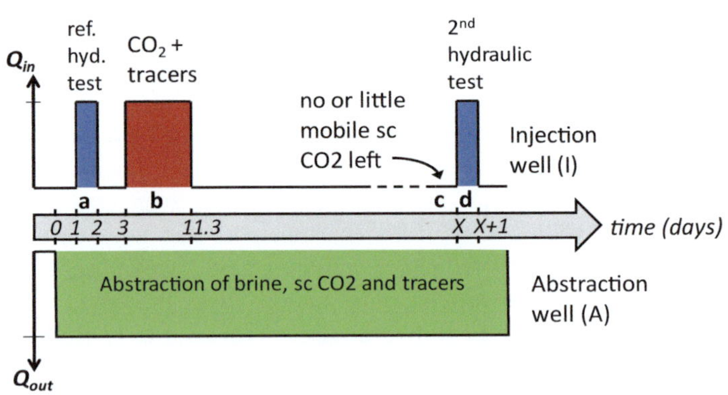

Fig. 7.18 Proposed two-well injection-withdrawal sequence. Time zero is the start of the reference hydraulic test. From Fagerlund et al. (2013a)

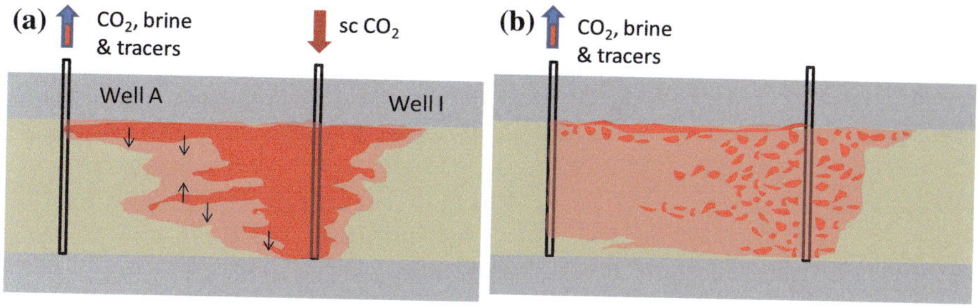

Fig. 7.19 Two-well CO_2 injection experiment at free-phase CO_2 breakthrough to the withdrawal well (**a**) and at residual state of free-phase CO_2 (**b**). From Fagerlund et al. (2013a)

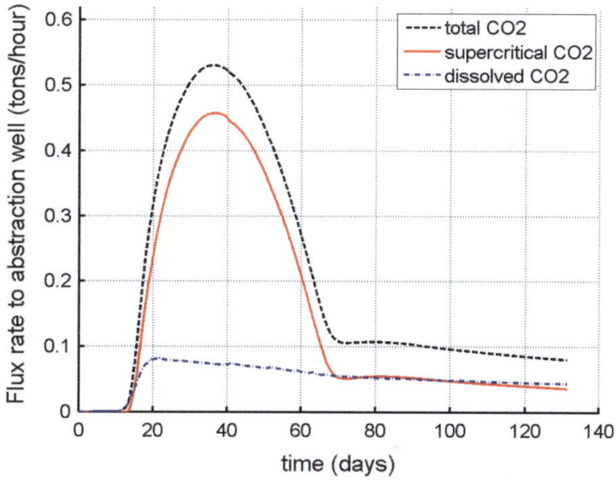

Fig. 7.20 Flux of CO_2 to the withdrawal well. From Fagerlund et al. (2013a)

continuous withdrawal from the second well is maintained throughout the test sequence. Residual trapping and dissolution occurs as the CO_2 migrates through the formation between the two wells, and after a period of fluid withdrawal most of the free-phase CO_2 will be residually trapped. The concept is illustrated schematically in Fig. 7.19. The aim of this two well test is to quantify the residual trapping that occurs under influence of geological heterogeneity and potential preferential flow paths that exist at the field site.

To identify the conditions of residual trapping in the formation, the point in time when very little mobile free-phase CO_2 remains needs to be identified, and this time can of course be different dependent on the formation properties and test configuration. Design simulations for the planned experiments at Heletz indicate that the conditions of residual trapping can be identified by measuring the flux of CO_2 to the withdrawal well. An example from this is shown in Fig. 7.19, where a clear change in the rate of supercritical (and total) CO_2 extraction can be seen at approximately 70 days after start of the test sequence. This change in flux rate corresponds to achieving the state of residual trapping as illustrated schematically in Fig. 7.20.

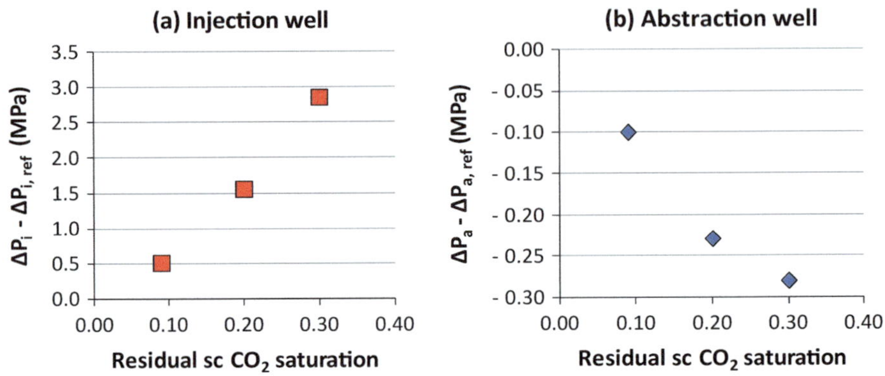

Fig. 7.21 **a** Simulated difference between the maximum pressure during the reference hydraulic test and the hydraulic test with residually trapped CO_2 in the injection well for different amounts of residual trapping. **b** Similarly, difference in minimum pressures in the abstraction well. From Fagerlund et al. (2013a)

The inter-well distance chosen for the test together with the rate of pumping has a large effect on the time required to reach a state of residual trapping (Fagerlund et al. 2013a).

Similar to the push–pull experiments discussed above, simulations of the two-well experiment test sequence have shown that the pressure response in the hydraulic test is highly sensitive to the residual trapping as a result of reduced permeability to the aqueous phase in the presence of trapped CO_2. An example is given in Fig. 7.21 which shows the simulated difference in the maximum pressure during the reference hydraulic test and the hydraulic test at residual trapping conditions in the injection well (Fig. 7.21a), and similarly, the difference between minimum pressures in the withdrawal (abstraction) well (from Fagerlund et al. 2013a). While the total pressure change depends also on the permeability (not shown here), the sensitivity to different amounts of residual trapping is clear both in the injection and withdrawal wells. Both wells can thus contribute to the estimation of the residual trapping between the wells.

Measurement of the concentration of CO_2 in the withdrawn brine allows analysis of the amount of dissolution that occurs in the formation. For the test configuration with active withdrawal of fluids Fagerlund et al. (2013a) found a strong correlation between the rate of CO_2 abstraction and the effective rate of CO_2 dissolution in the formation. As shown in Fig. 7.22, the correlation was linear for a wide range of different simulation scenarios producing different groundwater flow fields and effective total dissolution in the formation. From these modeling studies it was therefore concluded that for the stable flow field maintained using continuous withdrawal of fluids, the total rate of CO_2 dissolution in the formation could be inferred from measurements of abstracted CO_2.

Simulations of the negligible-solubility tracer (NST) test have shown that the enrichment of the tracer in the abstracted free-phase CO_2 is directly correlated to the

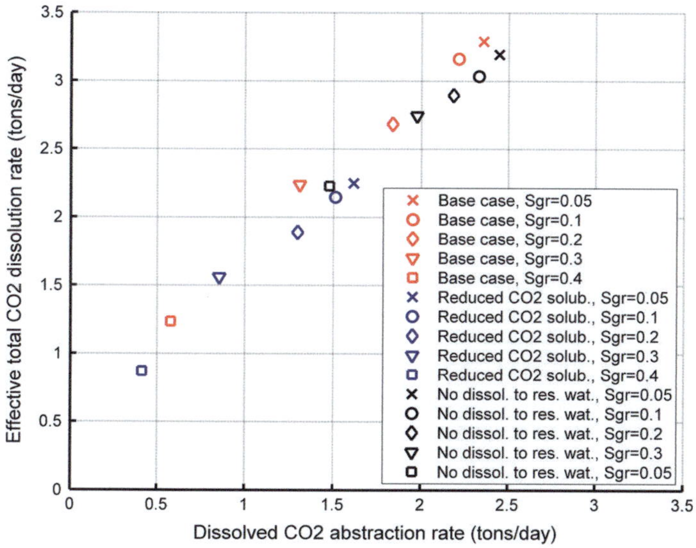

Fig. 7.22 Relationship between rate of CO_2 abstraction and the effective total rate of dissolution in the formation for a range of simulation scenarios producing different flow fields and dissolution rates. From Fagerlund et al. (2013a)

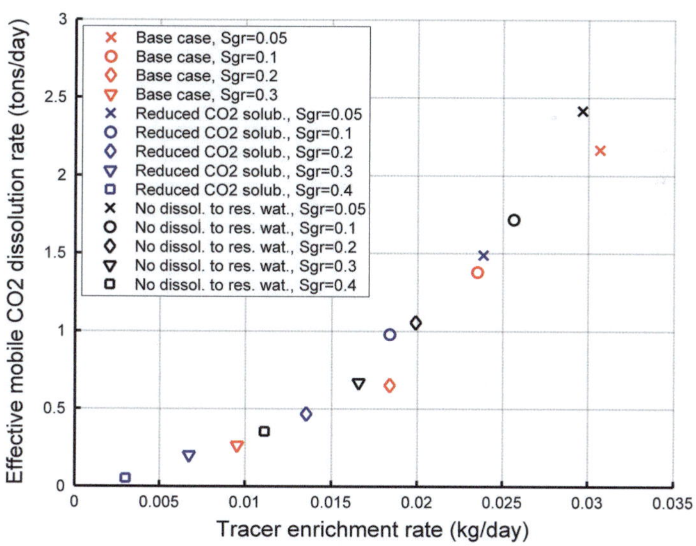

Fig. 7.23 Relationship between dissolution rate of mobile free-phase CO_2 and rate of tracer enrichment in the abstracted free-phase CO_2. From Fagerlund et al. (2013a)

simulated dissolution of the mobile free-phase CO_2 in the formation. This is illustrated in Fig. 7.23 which shows the dissolution of mobile CO_2 as a function of the tracer enrichment rate defined as the amount rate of tracer abstraction above the original injected concentration in the simulation study by Fagerlund et al. (2013a). Combining information about the total dissolution of CO_2 with dissolution of mobile CO_2, also the dissolution of residually trapped CO_2 can be estimated and the dissolution of trapped and mobile CO_2 can be separated.

In the aforementioned study, this tracer method was explored by numerical simulation for a small CO_2 injection (1000 tonnes of CO_2) and a relatively short test period. This test is too short to detect a convective dissolution process. However, with a long enough test period and favourable conditions (high permeability) to get onset of convective mixing, this type of tracer could potentially be used also to quantify the rate of dissolution during convective mixing.

The simulation studies of two-well CO_2 injection experiments show that these tests have the potential to quantify both residual trapping and dissolution of CO_2 at the field scale. Two-well tests will sample a larger reservoir volume and are likely to be more influenced by geological heterogeneity as compared to single-well push–pull tests such those of Paterson et al. (2013a, b).

The experiences from field tests such as Frio (Doughty et al. 2008; see also Sect. 8.6) and Otway (Paterson et al. 2013a, b; Zhang et al. 2011; Myers et al. 2012) have shown that well-instrumented small scale CO_2 injections have the potential to provide valuable information about the CO_2 migration and fate in a storage formation. Given the challenges to access and monitor the flow and transport processes in kilometre-deep reservoirs, the collection of multiple different data set and data types appear to be a vital strategy to understand the fate of the injected CO_2 and constrain estimates of CO_2 migration and trapping properties. A key to the interpretation of the different data sets and data types is a carefully constructed process-based model including the flow and transport processes of interest. With such model the different data can be jointly interpreted, different hypotheses can be tested and inverse modeling techniques can be used to estimate model parameters and related uncertainties.

7.5 Geomechanical Characterization

Francois H. Cornet and Victor Vilarrasa

Geomechanical characterization is an important part of any CO_2 geological storage project as understanding the stress field and the geomechanical strength of the formation rocks is necessary in order to anticipate possible damage to formation rocks and/or induced seismicity.

The drilling of boreholes is an essential part of the characterization program that must be conducted before undertaking any significant large scale operation within a geological formation. Boreholes provide knowledge of the geometry of the various structural elements involved as well as samples for laboratory testing.

Analysis of geomechanical properties from core samples is discussed in Chap. 6. Samples are, however, often perturbed by the sampling process and they typically provide a sample of the intact rock only. Boreholes provide means of comparing results from laboratory tests to in situ field observations. In addition, boreholes provide the opportunity for conducting in situ measurements all along the borehole length of interest and this is very useful for ascertaining the representativity and variability of point observations associated with laboratory tests on samples. We concentrate here on the geomechanical characterization that may be conducted in boreholes, either as stand-alone measurements, or in combination with results from laboratory measurements and other logs.

The section starts with the identification of geomaterials and the characterization of fracture fields and faults. We then discuss the determination of elastic properties and strength characteristics of the various geomaterials. We follow up with a brief review of techniques used for the evaluation of regional stress field from in situ tests in boreholes, with special consideration for the specific needs of underground CO_2 sequestration. Next, we proceed to discuss how in situ stress measurements may help validate hypotheses on the rheological properties of geomaterials. We finally conclude with a discussion of observed scale effects in rock geomechanical properties and introduce a new, still to be tested, procedure that could be used to characterize the geomechanical characteristics of the CO_2 storage formation and the caprock. Please note that the theoretical development of the expressions for the geomechanical relationships is given in Sect. 3.6.

7.5.1 Geomaterials, Fracture Fields and Faults

7.5.1.1 Geomaterials

We define a geomaterial as the continuous material equivalent to the various materials that fill up volumes of rocks over which mechanical properties are considered as being uniform and may be characterized by those of a Representative Elementary Volume (REV) (e.g. Cornet 2015, Chap. 1). Geomaterials are porous and often multiphase, i.e. the pore space is filled by one or more liquids as well as by gas. The solid phase itself may be very heterogeneous at scales smaller than that of the REV. The characterization of geomaterials involves the determination of their mineral composition (including clay content) and various fluids content, their porosity, their density, their electrical conductivity, among other properties. This is the domain of petrophysics and various logging techniques have been developed for determining relevant parameters from in situ observations (see also Sect. 7.2.2). We will hereafter assume that porosity, density, clay content are known for the various geomaterials intersected by boreholes.

7.5.1.2 Fracture Fields and Faults

In addition to geomaterials, great attention must be given to fracture fields and faults. Fractures may be defined as sub-planar structures with one dimension being orders of magnitude smaller than those in the two other directions. It is recognized that some characteristics of fractures (such as roughness, spacing, radius, thickness/aperture) have fractal dimensions and therefore cannot be apprehended with the concept of Representative Elementary Volume (e.g., Brown and Scholtz 1985; Schmittbuhl et al. 1995; Bonnet et al. 2001).

If fractures are large enough to completely intersect a characterization borehole, they are generally identified explicitly both because fractures are weakness planes and exhibit hydraulic properties that strongly differ from those of the surrounding geomaterial. In contrast, smaller fractures are not taken into account explicitly, but only implicitly, through specific characteristics of the equivalent geomaterial. The orientation of fractures (strike and dip, or dip and dip direction) is determined through examination of borehole wall images. These borehole wall images may be obtained either from ultrasonic televiewer logs or from electrical imaging logs. Values for dip and dip-direction are plotted on a Schmitt stereographic projection so as to identify fracture sets. Fracture sets are characterized by their mean direction (two angles) and associated standard deviation, where the dispersion of data may be characterized by a normal (Gaussian) Probability Density Function (e.g., Goodman 1989, Chap. 1; Einstein and Baker 1983; Dershowitz et al. 1998).

It should be pointed out that faults are not big fractures, but are generally made up of a fault core (gouge) surrounded by two highly fractured zones (e.g., Sibson, 1977; Sulem et al. 2004; Micarelli et al. 2006). Faults are commonly longer than 100 m and may reach thousands of kilometers. Faults that are longer than a few kilometers generally involve many branches (e.g., Manighetti et al. 2001; Davis et al. 2005). Faults are ususally assimilated either to a single geomaterial with specific material properties or to a multimaterial system that may be characterized by combining borehole and laboratory observations (Sulem 2007; Lockner et al. 2009). The large scale geometry of faults is often identified by seismic profiles (see Sect. 7.2 and discussion hereafter).

7.5.2 Dynamic Elastic Properties of Geomaterials and Their Spatial Variations

Traditionally, elastic characteristics of geomaterials are measured both in the laboratory by testing specimens prepared from cores and in the field by sonic logs (e.g., Paillet and White 1982), and sometimes by vertical or oblique seismic profiles (VSP).

Simple ***sonic logs*** involve the recording of P and S waves (i.e. the compression and shear waves, respectively) at two different sensors situated about 30–50 cm from each other and at about 2–3 m away from the source. However, specific equipment may involve multiple sources and multiple sensors for sampling different distances between source and receivers. Further, whilst the simplest equipment involves only arrival times detection, more sophisticated equipment delivers the complete recording of signals for time durations long enough to cover the arrivals of Stoneley waves after the P and S waves. Stoneley waves correspond to tube waves; they are very sensitive to fractures and faults and their attenuation is used sometimes as a means for measuring the hydraulic characteristics of fractures (Hardin et al. 1987; Winkler et al. 1989). Sources used in simple sonic logs generate essentially P waves, and the observed S waves are generated by reflections and/or refractions through surfaces of discontinuity. More elaborated sonic tools have been developed and operate two sources in perpendicular directions so as to generate S waves (dipole sources).

Seismic profiles involve vibratory sources located on ground surface and three component sensors that are coupled to the borehole casing, or directly to the rock in open-hole sections, at regular depth intervals along the borehole Sonic logs and seismic profiles involve the measurement of both P wave and S wave velocities and correspond to dynamic measurements that depend on the frequency of the excitation (from ten to twenty kiloHertz for sonic logs, and a few tens of Hertz for seismic profiles). Elastic parameters derived from these observations, which involve strains in the order of 10^{-6} or smaller, are generally referred to as dynamic properties. Sonic logs are much easier to run (and also much less costly) than seismic profiles as well as much more commonly used.

For isotropic materials, the shear modulus G and the Poisson's ratio ν are given by

$$G = \varrho_B V_{S^2} \tag{7.5.1}$$

$$\nu = \left[1 - 2(V_S/V_P)^2\right]/2\left[1 - (V_S/V_P)^2\right] \tag{7.5.2}$$

where V_P and V_S are the P and S wave velocities, respectively, and ϱ_B is the geomaterial mean density.

Given the frequency of sonic logs (1–2 10^4 Hz), the wave lengths of the signals are in the range of a few tens of centimeters, so sonic logs may detect the presence of single fractures. For vertical or oblique seismic profiles, however, the frequency of the signal implies much longer wave lengths (in the range of a few hundred meters). Hence, for seismic profiles, a large number of fractures are included within one wave length and the measured velocity is generally much slower than that measured with sonic logs.

Elastic properties of fractures and microfissures may be characterized by a normal and a shear stiffness (e.g., Cornet 2015, Chap. 10). When these structures are randomly oriented in a material loaded under triaxial stress conditions, the equivalent material becomes stiffer in the maximum principal stress direction than in the minimum principal stress direction. This is also true for microcracks, and extensive work has been done to discuss the effect of stress on microfractured rocks when they are modeled as equivalent anisotropic elastic solids (e.g., Nur 1971; Johnson and Rasolofosaon 1996; Zatsepin and Crampin 1997; Prioul et al. 2007). Randomly microfissured rock under triaxial stress loading conditions usually exhibits an orthorhombic (3 axes of symmetry) anisotropic behavior which implies 9 elastic constants. As a consequence, the velocity of shear waves depends on the relative orientation of their polarization direction with respect to that of the principal stress direction. Hence, the propagation of shear waves in an orthotropic material leads to the splitting into fast and slow shear waves. This phenomenon is known as shear wave splitting, or as the birefringence of shear waves.

Because VSP's involve much longer wave lengths than sonic logs do, they are dependent not only on the density of the microcracks, but also on that of the natural fractures. Experience has shown that the identification of fast and slow shear waves in VSP's may be used for identifying principal stress directions (e.g., Gaucher et al. 1998).

Borehole Seismic profiles are well suited for identifying major faults, not only through the associated local variation in wave velocity but mostly because faults are strong reflectors of both P and S waves (e.g., Place et al. 2011). VSP's have revealed also very useful for identifying limits between the various geomaterials through the reflected waves generated by the corresponding material interfaces and the various arrivals of refracted waves.

7.5.2.1 Strength Characterization

Laboratory measurements of elastic parameters involve the measurement of displacements generated by the application of external loads on specially prepared specimens loaded under uniaxial or triaxial compression conditions. The constant loading rates used for laboratory testing varies commonly between 10^{-4} and 10^{-6} % per second so that laboratory measurements are often referred to as quasi-static measurements. They always yield much lower elastic moduli than do dynamic measurements, unless the minimum principal stress is very high (larger than 100 MPa). This demonstrates that rocks are not linearly elastic and the dependence on time implies some viscosity for the geomaterial behavior. The fact that the discrepancy between quasi-static and dynamic properties decreases as the minimum principal stress applied on the rock becomes larger suggests that the main source of the discrepancy is linked to the mechanical behavior of the discontinuities (micro-fissures and fractures). Micro-fissures and fractures also control the "strength" of these materials, i.e., the stress conditions at which failure develops. Hence, many investigations have been conducted for evaluating whether the difference between dynamic and quasi-static measurements may be combined with the value of the dynamic Young's modulus together with other parameters like the clay content for evaluating the "strength" of the rock. Here the concept of strength refers to the Coulomb failure criterion.

This practice has been developed in the oil industry for the design of drilling operations in sedimentary rocks in order to avoid borehole collapse. An extensive review of these various empirical relationships has been proposed by Chang et al. (2006) for estimating the Unconfined Compressive Strength (UCS) and the intrinsic friction angle implied by the Coulomb failure criterion for sandstones, shales, limestones and dolomite.

However, it should be kept in mind that these empirical relationships are meant to prevent the collapse of boreholes during drilling. They do not provide "strength" data for the safe design of long term CO_2 sequestration schemes. We recommend performing laboratory and field tests to determine the strength site specifically.

7.5.3 Regional Stress Field Evaluation

A crucial part of the design of a CO_2 sequestration project is the identification of the maximum fluid pressure that may be reached during the various injection phases. This is required first so as not to create any hydraulic fracture within the reservoir and second, to avoid leaks through the caprock. Another important issue is the control of induced seismicity. This implies understanding fluid diffusion not only within the aquifer but also in the overlying and underlying geological formations.

These various design considerations require a sound understanding of the regional stress field in the caprock, the aquifer, and the underlying geological formation. An efficient microseismic monitoring, during and after all main injection phases, is required to detect any possible induced seismicity, with particular attention to the growth direction of the microseismic cloud. Here, we only discuss the determination of the regional stress field, i.e., the identification of principal stress directions together with the magnitude of principal stress components for the caprock, the aquifer and the underlying geological formations. These may be evaluated through various techniques that we discuss below.

7.5.3.1 Stress Evaluation from Sonic Logs

The drilling of a borehole in a stressed geological material results in local stress concentrations around the borehole up to distances equal to about 4–5 times the borehole radius (e.g., Cornet 2015). If one of the principal stress components is parallel to the borehole axis (often this is the vertical direction, so that the corresponding principal stress magnitude is called σ_V), then the two other principal stress directions are normal to the borehole axis and are referred to as σ_H and σ_h, with the convention $\sigma_H > \sigma_h$. For such geometry, the tangential stress component, $\sigma_{\theta\theta}$ at the borehole wall is

$$\sigma_{\theta\theta} = (\sigma_H + \sigma_h) - 2(\sigma_H - \sigma_h)\cos\,\theta, \tag{7.5.3}$$

where θ is the angular coordinate defined with respect to the σ_H direction. $\sigma_{\theta\theta}$ varies from $(-\sigma_H + 3\sigma_h)$ in the maximum horizontal principal stress direction to $(-\sigma_h + 3\sigma_H)$ in the minimum principal stress direction. As pointed out above, the elastic constants of a geomaterial depend on the stress field, and the directions of axes of symmetry of the elastic anisotropy correspond to those of the principal stress components, if the rock is isotropic when it is completely unloaded.

This property used in sonic logs conducted with dipole sources for identifying both the directions and the magnitudes of the far-field principal stress components (Lei et al. 2012). Preliminary results from field tests showed good agreement with results from hydraulic tests. However, it should be pointed out that, currently, such measurements are not routinely conducted.

7.5.3.2 Stress Evaluation from Borehole Wall Images (Borehole Breakouts Orientations)

As shown by Eq. (7.5.3), the tangential stress component at the borehole wall varies with the angular coordinate of the point under consideration and takes a particularly simple form in terms of horizontal principal stress magnitudes for vertical boreholes, when the vertical direction is also a principal stress direction. It reaches its maximum value for $\theta = \pi/2$ and $\theta = 3\pi/2$, i.e., in the direction of the minimum horizontal principal stress.

The magnitudes of the far-field principal stress components, i.e., away from the wellbore, increase with depth, and when the tangential stress, at the wellbore wall, reaches the compressive strength of the material, failure occurs.

Hence, when borehole wall images are available (preferentially ultrasonic borehole images, but also electrical borehole images), zones of borehole failure (called borehole breakouts) caused by too large far-field stress magnitudes can be observed at two symmetrical locations. These locations are aligned with the minimum principal stress direction (e.g., Zoback et al. 2003).

Because the compressive strength of shales is much lower than that of limestones and sandstones, borehole breakouts are first detected in these soft formations. Some attempts have been made to extract from the width of borehole breakout the principal stress magnitudes. It should be pointed out, however, that for some materials, like shale or clay, the compressive strength depends on many factors including the composition of the drilling mud. Furthermore, the drill string assembly usually includes, at some distance above the cutting tool, so called reamers that keep the borehole at the required shape so that borehole diameters may not depend solely on far field stress magnitudes. Hence it is considered that borehole breakouts are very reliable markers of the principal stress directions, but not of the principal stress magnitudes.

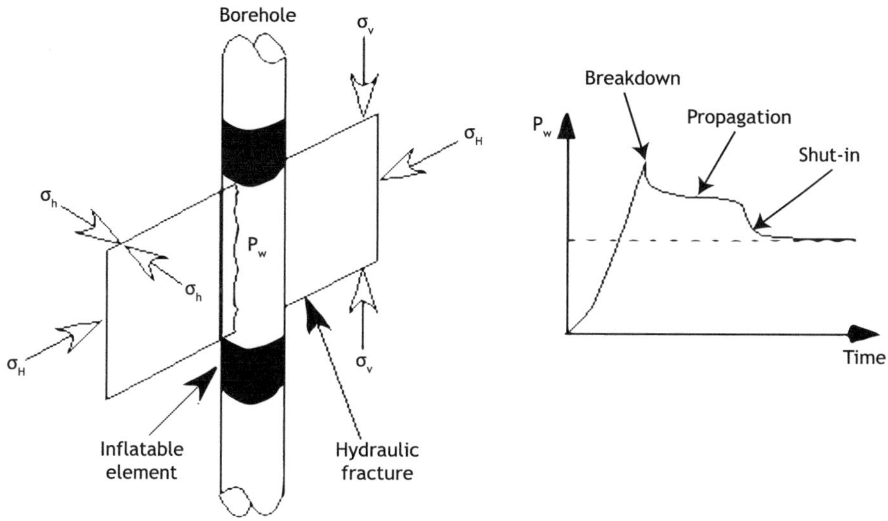

Fig. 7.24 Stress measurement by hydraulic fracturing. *Left* principle of the method; *right* typical pressure-time record

7.5.3.3 Stress Evaluation from Hydraulic Tests in Boreholes (HF and HTPF)

The most robust method for determining in situ stresses from boreholes is by hydraulic tests, with inflatable straddle packers that isolate portions of a borehole (see Fig. 7.24) (Haimson and Cornet 2003). Like borehole breakouts, the method is based on Eq. (7.5.3) that shows that the minimum tangential stress at the borehole wall is in the direction of the far field maximum stress direction, when the borehole is parallel to a principal stress direction. When water pressure P_w is applied to the borehole wall, it generates at all points of the borehole wall a negative tangential stress component equal in magnitude to the applied pressure (compressions are reckoned positive). When the tangential stress reaches the tensile strength of the rock, σ^T, a tensile rupture occurs in the direction of the maximum horizontal principal stress (Fig. 7.24)

$$\sigma_{\theta\theta} = -\sigma_H + 3\sigma_h - P_w = \sigma^T. \tag{7.5.4}$$

On the right panel of the figure it can be seen that, when a fluid is injected at a constant flow rate in between the two inflatable packers, the pressure first rises linearly. Then a peak is reached (breakdown pressure) that is classically considered to correspond to the initiation of hydraulic fracturing. However recent work has shown that the fracture initiates below the packers, for an interval pressure lower than the breakdown pressure. Hence the pressure to be considered for the initiation of fracture is that of the packer when the record of interval pressure versus time gets non-linear.

The fracture initiation pressure and the shut-in pressure may be used to determine both the maximum and the minimum principal stress magnitudes. For HTPF tests (hydraulic testing on pre-existing fractures), the straddle packer is set up on a portion of the borehole where a single pre-existing fracture exists. Then, a slow injection of fluid into the fracture leads to its mechanical opening and the shut-in pressure observed at the end of testing yields a direct measurement of the normal stress applied to the fracture away from the well bore.

Combinations of HF (hydraulic fracturing) and of HTPF yield redundant measurements that help constraining efficiently all the components of the stress field away from the borehole.

7.5.4 Vertical Stress Profiles and the Characterization of the Rheology of Geomaterials

The magnitude of all principal stress components generally increases linearly with depth when stress measurements are conducted in homogeneous formations like granite. But when measurements are conducted in sedimentary formations, which are relevant for geological storage of CO_2, the measured stress magnitudes are found to depend very strongly on the nature of the materials. In particular, it is well recognized that the magnitude of the differential stress ($\sigma_H - \sigma_h$) is much smaller in soft materials, such as shale and evaporites, than in relatively stiff materials like sandstone or limesone (e.g., Cornet and Röckel 2012).

Some attempts have been made at determining the elastic parameters of these formations through numerical modeling (Gunzburger and Magnenet 2014). Such modeling assumes an elastic behavior for the material over millions of years. However, the very fact that dynamic, quasi-static and static elastic measurements may differ markedly, as discussed in Sect. 7.5.2, clearly demonstrates that the hypothesis of a constant value for the elastic parameters independent of time is not valid (Gunzburger 2010).

Various attempts are presently undertaken at evaluating this time dependency (e.g., Sone and Zoback 2014; Cornet 2015), and it is hoped that such approaches will develop further so as to produce reliable values for time scales ranging from thousands to tens of thousands years.

7.5.5 Scale Effects of the Mechanical Properties

As discussed above, mechanical properties of rocks are usually inferred from core samples at the laboratory. These values may not be representative at the field scale because of the existence of fissures, vugs, joints or fractures, which are more deformable than the intact rock that usually constitute core samples. This is

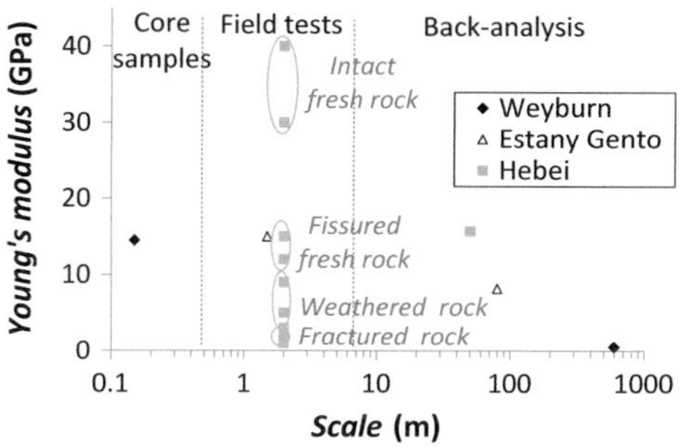

Fig. 7.25 Values of Young's modulus as a function of scale. Three case studies of mechanical properties determination at different scales are included. The rock results softer in back-analysis than in core samples measurements and field tests because of the presence of an increasing number of deformable fractures as the scale increases. The case studies refer to Estany Gento, Spain (gneiss) (Ledesma et al. 1996), Hebei, China (andesite) (Zhang et al. 2006) and Weyburn, Canada (limestone) (Verdon et al. 2011)

illustrated by the dependence of the Young's modulus on the scale of measurement (Fig. 7.25). Figure 7.25 displays three case studies [a cavern excavation in Estany Gento, Spain (Ledesma et al. 1996), a gallery excavation in Hebei, China (Zhang et al. 2006) and CO_2 injection at Weyburn, Canada (Verdon et al. 2011)] in which the Young's modulus is measured at the laboratory from core samples (intact rock), on a rock surface in the field with a flat jack test (meter scale on a variety of rock conditions) and inferred from back-analysis of convergence measurements in underground excavations or microseismicity measurements induced by CO_2 injection (field scale). The results show that field scale values of Young's modulus are much smaller than those obtained from core samples or field tests, typically performed in competent rock. The lower Young's modulus values suggest that rocks become softer for increasing scale. This reflects the fact that discontinuities, which are much more deformable than intact rock, play a major role in the mechanical behavior of the rock mass. Nevertheless, for the case of Hebei, hydraulic jack tests were performed on rock surfaces of varying quality, yielding a wide range of Young's modulus. Therefore, the resulting Young's modulus from the back-analysis of the convergence measurements of a gallery excavation gave an average value representative of the whole rock mass.

The Weyburn case study is a relevant example of scale effects on the mechanical properties in CO_2 storage. Verdon et al. (2011) presented the induced seismicity recorded in the Weyburn CO_2 Storage and Monitoring Project and built up a representative geomechanical model of the CO_2 injection. The model simulates the changes in the effective stress field induced by CO_2 injection and predicts the generation of microseismic events. As microseismic activity is directly linked with induced effective stress changes, it can be used to constrain geomechanical models. The first numerical model they performed used material properties based on laboratory core measurements and did not yield a good match with microseismic

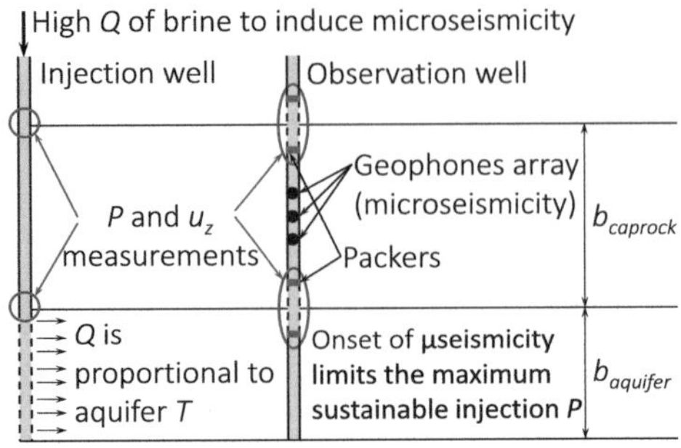

Fig. 7.26 Schematic representation of the hydromechanical characterization test by Vilarrasa et al. (2013). A sufficiently high water flow rate capable of reaching the maximum sustainable injection pressure is injected for several hours. Fluid pressure and displacements or strains should be monitored in the aquifer and caprock in as many places as possible (preferably in both the injection and the observation well(s), but at least in one well)

observations. They found that an alternative model reducing the stiffness by one order of magnitude with respect to the laboratory measurements gives a good correlation with the observed microseismic events.

These examples illustrate that the large-scale geomechanical properties are difficult to quantify (Rutqvist 2012). To address this difficulty, Vilarrasa et al. (2013) have proposed a field test to characterize the macroscopic mechanical properties of the rock layers involved in CO_2 storage in deep geological formations. It should be pointed out that this test has not yet been implemented in the field and is therefore taken here as a conceptual example to characterize the hydromechanical properties rather than an example of how to do a field test. The proposed hydromechanical characterization test consists in injecting water at high pressure, while monitoring fluid pressure, rock deformation and induced microseismicity (Fig. 7.26). The overpressure (several MPa) is proportional to the flow rate, which may become high if the aquifer transmissivity is high. The injected water can be obtained from surface sources, e.g., rivers or lakes. However, aquifer brine must be used if geochemical alteration is not desired. In this case, brine should be pumped and stored at the surface prior to the injection test and leave enough time to recover pre-pumping pressure levels in the aquifer.

The instrumentation for the hydromechanical characterization test consists of equipment to measure fluid pressure, vertical displacement and microseismicity. Fluid pressure and vertical displacement measurements are taken in the injection and observation wells, both in the aquifer and the caprock (Fig. 7.26). Fluid pressure should be measured using pressure transducers located between two packers to isolate an interval of the well that must be in hydraulic contact with the rock. Temperature measurements, which can be made with a fiber-optic (see Sect. 8.4), can be also useful to account for thermal effects if the injected water is not in thermal equilibrium with the storage formation. Vertical displacements in deep boreholes are not easy to measure. Alternatively, strain measurements with fiber-optic can be used. However, it is still difficult to say what deformation is being measured because the measuring equipment will be embedded in the cement between the casing and the rock, which are of different stiffness, and thus deform differently. To detect and locate microseismic events of magnitudes as low as −2, an array of geophones should be placed in the observation well at depth. The distribution of the geophones should be carefully designed to permit distinguishing induced microseismicity from background noise and locating hypocenters with high accuracy (<100 m) (IEAGHG 2013). Additionally, a network of seismographs in surface can complement the microseismicity measurements and help to localize the events..

During the test, overpressure should be progressively increased until the elastic limit is reached and microseismicity is triggered. The initial induced microseismicity still occurs in the elastic domain and can be related to the Kaiser effect (Cornet 2012, 2016). Induced microseismic events may take place both in the aquifer and the caprock. Since microseisms are associated with shear slip, which opens up fractures and enhances their transmissivity, microseismicity will be benificial if it occurs within the aquifer. Microseismicity should, however, be avoided in the caprock as it may jeopardize caprock integrity. Thus, overpressure should be immediately decreased if microseismic events are induced in the caprock to avoid compromizing the caprock sealing capacity. The onset of microseismicity in the caprock can be used as an indicator of the overpressure that must not be exceeded during the operational stage of CO_2 injection.

The characterization of the hydromechanical properties of the aquifer and the caprock is done from pressure and displacement or strain measurements in both rock formations. The hydraulic properties, i.e., aquifer transmissivity and storage coefficient, can be estimated from the interpretation of fluid pressure evolution (Cooper and Jacob 1946). The mechanical properties, i.e., Young's modulus and Poisson ratio, of the aquifer and the caprock can be estimated by introducing field

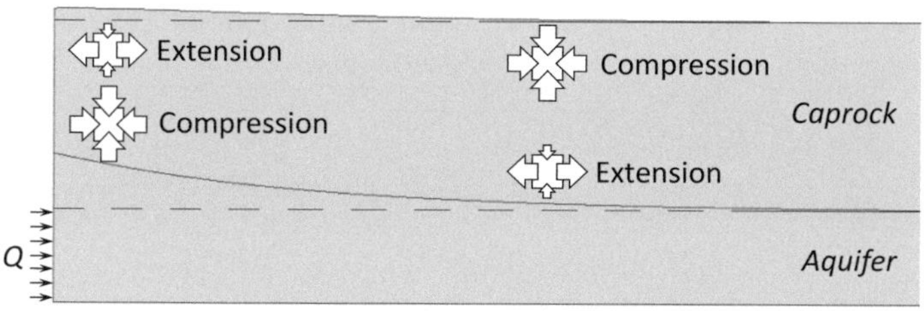

Fig. 7.27 Original (*dashed lines*) and deformed (*continuous lines*) form of the aquifer and the whole caprock when injecting a fluid in the aquifer. The uplift at the *top* of the aquifer generates compression in the *lower part* of the caprock close to the injection well and extension far away from it. However, extension appears in the *upper part* of the caprock close to the well and compression far away from it. While the pore volume decreases where the caprock is compressed, it increases where the caprock extends. Fluid pressure in the caprock is inversely proportional to the volumetric strain change. Thus, fluid pressure increases where the pore volume decreases and decreases where the pore volume increases

measurements in dimensionless plots developed by Vilarrasa et al. (2013). These plots include curves for overpressure and vertical displacement as a function of the volumetric strain term obtained from a dimensional analysis of the hydromechanical equations.

The reason for measuring fluid pressure in the caprock lies in the coupled hydromechanical effect that occurs during injection. When injecting a fluid in an aquifer, fluid pressure increases, changing the effective stress field. This produces an expansion of the aquifer. Vertical displacement presents a shape similar to that of fluid overpressure at the top of the aquifer, which decreases logarithmically with distance. As a result, the caprock is also deformed (Fig. 7.27). The vertical displacement does, however, become smoother at the top of the caprock, because the caprock acts as a spring, dissipating the deformation of the aquifer. But the pressure buildup propagation from the aquifer into the caprock is orders of magnitude slower than that of the aquifer due to the permeability contrast between the two formations. This means that the overpressure caused by injection only affects the first meters of the lower part of the whole caprock. However, fluid pressure changes occur in the whole caprock due to volumetric strain variations caused by caprock deformation (Fig. 7.27).

The uplift at the top of the aquifer generates compression in the lower part of the caprock close to the injection well, so fluid pressure increases. Extension appears at the top of the caprock close to the well, which increases the pore volume and thus fluid pressure decreases. This leads to a reverse-water level fluctuation, which is well-documented in confined aquifers. When fluid is pumped, hydraulic heads in adjacent aquitards rise after pumping starts (Rodrigues 1983; Hsieh 1996; Kim and Parizek 1997). This phenomenon is known as "reverse-water level fluctuation" or "Noordbergum effect", because it was observed for the first time in the village of Noordbergum, the Netherlands (Verruijt 1969). Since in CO_2 sequestration a fluid is injected, an opposite response to that observed in Noordbergum occurs, i.e., fluid pressure drops in the upper part of caprock in the vicinity of the injection well (Vilarrasa et al. 2010). The contrary occurs far away from the injection well, i.e., extension at the lower part of the caprock and compression at its top. These effects are, however, small compared to those close to the well.

References

Agenet N, Moradi-Tehrani N, Tillement O (2011) Fluorescent nanobeads: a new generation of easily detectable water tracers. Procs Int Petrol Technol Conf, IPTC 15312

Aki K, Richards P (1980) Quantitative seismology. University Science Books

Alcalde J, Martí D, Calahorrano A, Marzan I, Ayarza P, Carbonell R, Juhlin C, Pérez-Estaún A (2013) Active seismic characterization experiments of the Hontomín research facility for geological storage of CO_2, Spain. Int J Greenhouse Gas Control 19:785–795

Alcolea A, Carrera J, Medina A (2006) Pilot points method incorporating prior information for solving the groundwater flow inverse problem. Adv Water Resour 29(11):1678–1689

Anderson MP (2005) Heat as a ground water tracer. Ground Water 43(6):951–968

Annable MD, Rao PSC, Hatfield K, Graham WD, Wood AL, Enfield CG (1998) Partitioning tracers for measuring residual NAPL: field-scale test results. J Environ Eng 124:498–503

Bear J (1988) Dynamics of fluids in porous media. Dover, New York

Becker MW, Reimus PW, Vilks P (1999) Transport and attenuation of carboxylate-modified latex microspheres in fractured rock laboratory and field tracer tests. Ground Water 37:387–395

Becker MW, Metge DW, Collins SA, Shapiro AM, Harvey RW (2003) Bacterial transport experiments in fractured crystalline bedrock. Ground Water 41:682–689

Bennett B, Larter SR (1997) Partition behaviour of alkylphenols in crude oil/brine systems under subsurface conditions. Geochim Cosmochim Acta 61:4393–4402

Behrens H, Ghergut I, Sauter M (2010) Tracer properties, and tracer test results—part 3: modification to shook's flow-storage diagram method. Presented at the thirty-fifth workshop on geothermal reservoir engineering, Stanford University, SGP-TR-188

Behrens H, Ghergut J, Bensabat J, Niemi A, Sauter M (2014) Merging single- and inter-well tracer tests into one forced-gradient dipole test, at the Heletz site within the MUSTANG project. Energy Procedia 59:249–255

Bonnet E, Bour O, Odling NE, Davy P, Main I, Cowie P et al (2001) Scaling of fracture systems in geological media. Rev Geophys 39:347–383

Bourdet D, Whittle TM, Douglas AA, Pirard YM (1983) A new set of type curves simplifies well test analysis. World Oil 196:95–106

Bourdet D, Ayoub JA, Pirard YM (1989) Use of pressure derivative in well-test interpretation. SPE Repr Ser 4:293–302

Brown SR, Scholtz CH (1985) Broad bandwidth study of the topography of natural rock surfaces. J Geophys Res 90:12575–12582

Carrera J, Sánchez-Vila X, Benet I, Medina A, Galarza G, Guimerà J (1998) On matrix diffusion: formulations, solution methods and qualitative effects. Hydrogeol J 6:178–190

Chang C, Zoback MD, Khaksar A (2006) Empirical relations between rock strength and physical properties in sedimentary rocks. J Petrol Sci Eng 51:223–237

Chow VT (1952) On the determination of transmissibility and storage coefficients from pumping test data. Trans Am Geophys Union 33:397–404

Cooper HH, Jacob CE (1946) A generalized graphical method for evaluating formation constants and summarizing well field history. Am Geophys Union Trans 27:526–534

Cornet FH (2015) Elements of crustal geomechanics. Cambridge University Press, Cambridge

Cornet FH (2016) Seismic and aseismic motions generated by fluid injections. Geomech Energy Environ 5:42–54

Cornet FH, Röckel T (2012) Vertical stress profiles and the significance of stress decoupling. Tectonophys 581:193–205

Davis BM, Istok JD, Semprini L (2002) Push–pull partitioning tracer tests using radon-222 to quantify non-aqueous phase liquid contamination. J Contam Hydrol 58:129–146

Davis K, Burbank DW, Fisher D, Wallace S, Nobes D (2005) Thrust fault growth and segment linkage in the active Ostler fault zone, New Zealand. J. Struct Geol 27:1528–1546

Dean C, Reimus P, Newell, D (2012) Evaluation of a cation exchanging tracer to interrogate fracture surface area in EGS systems. In: Proceedings, 37th Workshop on Geothermal Reservoir Engineering, Stanford, CA

Deeds, NE, McKinney, Pope, GA, (1999) Vadose zone characterization at a contaminated field site using partitioning interwell tracer technology. Environ Sci Techn 33:2745–2751

Dershowitz W, La Pointe P, Cladouhos T (1998) Derivation of fracture spatial pattern parameters from borehole data. Int J Rock Mech Min Sci 35(4/5):508

Divine CE, McDonnell JJ (2005) The future of applied tracers in hydrogeology. Hydrogeol J 13:255–258

DNV (2010) DNV CO2QUAL: guideline for selection and qualification of sites and projects for geological storage of CO_2. http://www.dnv.com.au/binaries/CO2QUALSTORE_guideline_tcm162-412142.pdf

Doughty C, Freifeld BM, Trautz RC (2008) Site characterization for CO_2 geologic storage and vice versa—the Frio brine pilot, Texas, USA as a case study. Environ Geol 54(8):1635–1656

Du Y, Guan L (2005) Inter-well tracer tests: lesson learned from past field studies. Soc Pet Eng SPE paper 93140

Edlmann K, Edwards MA, Qiao XJ, Haszeldine RS, McDermott CI (2014) Appraisal of global CO_2 storage opportunities using the geomechanical facies approach. Environmental earth sciences. Springer, New York

Einstein HH, Baecher GB (1983) Probabilistic and statistical methods in engineering geology; Rock mech. Rock Eng 84:366–376

Ennis-King J, Paterson L (2005) Role of convective mixing in the long-term storage of carbon dioxide in deep saline formations. Soc Pet Eng SPE paper 84344

Erlström M, Silva O, Vries LM, Shtivelman V, Gendler M, Goldberg I, Scadeanu D, Sperber CM (2010) 3D structures of the test sites. Deliverable D2.2. MUSTANG (a multiple space and time scale approach for the quantification of deep saline formations for CO_2 storage) EU FP7 project report. Nov 2010. www.co2mustang.eu

Erlström M, Rötting T, Sperber CM, Shtivelman V, Scadeanu D, Sperber CM (2011) Report on property values and parameters, related uncertainties. Deliverable D2.3. MUSTANG (a multiple space and time scale approach for the quantification of deep saline formations for CO_2 storage) EU FP7 project report. Deliverable D2.2 July 2011. www.co2mustang.eu

EU (2009) Directive 2009/31/EC of the European parliament and the council of 23 April 2009, on the geological storage of carbon dioxide and amending council directive 85/337/EEC, European parliament council directives 2000/60/EC, 2001/80/EC, 2004/35/EC, 2006/12/EC, 2008/1/EC and regulation no. 101/2006, App 1. Criterial for the site characterization and assessment of the potential storage complex and the surroundings

Fagerlund F (2007) Experimental and modeling studies on the spreading of non-aqueous phase liquids in heterogeneous media. PhD Thesis, Uppsala University

Fagerlund F, Niemi A, Bensabat J, Shtivelman V (2013a) Design of a two-well field test to determine in-situ residual and dissolution trapping of CO_2 applied to the Heletz CO_2 injection site. Int J Greenhouse Gas Control 19:642–651

Fagerlund F, Niemi A, Bensabat J, Shtivelman V (2013b) Inter-well field test to determine in-situ CO_2 trapping in a deep saline aquifer: modelling study of the effects of test design and geological parameters. Energy Procedia 40:554–563

Flett M, Gurton R, Weir G (2007) Heterogeneous saline formations for carbon dioxide disposal: impact of varying heterogeneity on containment and trapping. J Pet Sci Eng 57:106–118

Freifeld BM, Finsterle S, Onstott TC, Toole PLMP (2008) Groundsurface temperature reconstructions: using in situ estimates for thermal conductivity acquired with a fiber-optic distributed thermal perturbation sensor. Geophys Res Lett 35:L14309

Gaucher E, Cornet FH, Brernard P (1998) Induced seismicity analysis for structure identificatioin and stress field determination; SPE paper SPE-47324, pp 545–554 of Euroc-98, Soc Pet Eng

Ghergut I, Behrens H, Maier F, Karmakar S, Sauter M (2011) A note about "heat exchange areas" as a target parameter for SWIW tracer tests. In: Stanford geothermal program workshop report SGP-TR-191, pp 303–312

Ghergut J, Behrens H, Sauter M, Licha T, Nottebohm M (2013) Can Peclet numbers depend on tracer species? Going beyond SW test insensitivity to advective or equilibrium-exchange processes. In: Stanford geothermal workshop proceedings, SGP-TR-198, pp 326–335

Giese R, Henninges J, Lüth S, Morozova D, Schmidt-Hattenberger C, Würdemann H, Zimmer M, Cosma C-G, Juhlin C, CO2SINK Group (2009) Monitoring at the CO2SINK site: a concept integrating geophysics, geochemistry and microbiology. Energy Procedia 1(1):2251–2259

Goodman RE (1989) Introduction to rock mechanics. Wiley, New York

Green C, Ennis-King J, Pruess K (2009) Effect of vertical heterogeneity on long-term migration of CO_2 in saline formations. Energy Proceedia 1:1823–1830

Gunzburger Y (2010) Stress state interpretation in light of pressure-solution creep: numerical modelling of limestone in the Eastern Paris Basin, France. Tectonophysics 483(3):377–389

Gunzburger Y, Magnenet V (2014) Stress inversion and basemen-cover stress transmission across weak layers of the Paris Basin, France. Tectonophys 617:44–57

Haggerty R, Fleming SW, Meigs LC, McKenna SA (2001) Tracer tests in a fractured dolomite: 2. Analysis of mass transfer in single-well injection-withdrawal tests. Water Resour Res 37:1129

Haimson BC, Cornet FH (2003) ISRM suggested methods for rock stress estimation—part 3: hydraulic fracturing (HF) and/or hydraulic testing of preexisting fractures (HTPF). Int J Rock Mech Min Sci 40:1011–1020

Hallam A (1981) Facies Interpretation and the stratigraphic record: freeman, p 291

Hantush MS, Jacob CE (1955) Non-steady radial flow in an infinite leaky aquifer. Trans Am Geophys Union 36:95–100

Hardin EL, Cheng C, Paillet FL, Mendelson JD (1987) Fracture characterization by means of attenuation and generation of tube waves in fractured crystalline rock at Mirror lake, New Hampshire. J Geophys Res 92:7989–8006

Hasan AR, Kabir CS (2002) Fluid flow and heat transfer in wellbores. Society of Petroleum Engineers, Richardson, TX

Hillebrand O, Nödler K, Licha T, Sauter M, Geyer T (2012) Identification of the attenuation potential of a karst aquifer by an artificial dualtracer experiment with caffeine. Water Res 46:5381–5388

Hesse MA, Orr FM Jr, Tchelepi HA (2009) Gravity currents with residual trapping. Energy Proceedia 1:3275–3281

Hovorka S, Doughty C, Benson SM, Pruess K, Knox PR (2004) The impact of geological heterogeneity on CO_2 storage in brine formations: a case study from the Texas Gulf Coast. In: Bains SJ, Worden RH (eds) Geological storage of carbon dioxide, vol 233. Geological Society of London, Special Publications, pp 147–163

Hovorka SD, Benson SM, Doughty C, Freifeld BM, Sakurai S, Daley TM, Kharaka YK, Holtz MH, Trautz RC, Nance HS, Meyer LR, Knauss KG (2006) Measuring permanence of CO_2 storage in saline formations—the Frio experiment. Environ Geosci 13(2):105–121

Hsieh PA (1996) Deformation-induced changes in hydraulic head during ground-water withdrawal. Ground Water 34(6):1082–1089

Hunkeler D, Hoehn E, Höhener P, Zeyer J (1997) 222Rn as a partitioning tracer to detect diesel fuel contamination in aquifers: laboratory study and field observations. Environ sci techn 31 (11):3180–3187

IEAGHG (2013) Induced seismicity and its implications for CO_2 storage risk, 2013/09

Istok JD, Humphrey MD, Schroth MR, Hyman MR, O'Reilly KT (1997) Single-well, push–pull test for in situ determination of microbial activities. Ground Water 35:619–631

Istok JD, Field JA, Schroth MH, Dwarakanath V (2002) Single-well "push–pull" partitioning tracer test for NAPL detection in the subsurface. Environ Sci Technol 36:2708–2716

Jin M, Delshad M, Dwarakanath V, McKinney DC, Pope GA, Sepehrnoori K, Tilburg CE, Jackson RE (1995) Partitioning tracer test for detection, estimation, and remediation performance assessment of subsurface nonaqueous phase liquids. Water Resour Res 31:1201–1211

Johnson PA, Rasolofosaon PNJ (1996) Non-linear elasticity and stress induced anisotropy I rock. J Geophys Res 101:3113–3124

Juhlin Ch, Giese R, Zinck-Jørgensen K, Cosma C, Kazemeini H, Juhojuntti N, Lüth S, Norden B, Förster A (2007) 3D baseline seismics at Ketzin, Germany: the CO2SINK project. Geophysics 72(5):B121–B132

Juliusson E, Horne RN (2010) Study and simulation of tracer and thermal transport in fractured reservoirs. In: Proceedings, 35th workshop on geothermal reservoir engineering, Stanford University, Stanford, CA, SGP-TR-188

Jung Y, Pruess K (2012) A closed-form analytical solution for thermal single-well injection-withdrawal tests. Water Resources Research 48 doi:10.1029/2011WR010979

Kazemeini S, Juhlin C, Fomel S (2010) Monitoring CO_2 response on surface seismic data; a rock physics and seismic modeling feasibility study at the CO_2 sequestration site, Ketzin, Germany. J Appl Geophys 71:109–124

Keswick BH, Wang D-S, Gerba CP (1982) The use of microorganisms as ground-water tracers: a review. Ground Water 20:142–149

Kim J-M, Parizek RR (1997) Numerical simulation of the Noordbergum effect resulting from groundwater pumping in a layered aquifer system. J Hydrol 202:231–243
Kim H, Annable MD, Rao PSC (1998) Influence of air-water interfacial adsorption and gas-phase partitioning on the transport of organic chemicals in unsaturated porous media. Environ Sci Technol 32:1253–1259
Kneafsey T, Pruess K (2010) Laboratory flow experiments for visualizing carbon dioxide-induced, density-driven brine convection. Transp Porous Media 82(1):123–139
Knudby C, Carrera J (2005) On the relationship between indicators of geostatistical, flow and transport connectivity. Adv Water Resour 28(4):405–421
Kocabas I (2005) Geothermal reservoir characterization via thermal injection-backflow and inter-well tracer testing. Geothermics 34:27–46
Kocabas I, Maier F (2013) Analytical and numerical modeling of tracer flow in oil reservoirs containing high permeability streaks. In: SPE middle east oil and gas show and conference, 10–13 Mar 2013, Manama, Bahrain
Kruseman GP, de Ridder NA (1994) Analysis and evaluation of pumping test data, 2nd edn. ILRI publication 47, Wageningen, The Netherlands, p 377
LaForce T, Ennis-King J, Boreham C, Paterson L (2014) Residual CO_2 saturation estimate using noble gas tracers in a single-well field test: the CO2CRC Otway project. Int J Greenhouse Gas Control 26:9–21
Land CS (1968) Calculation of imbibition relative permeability for two- and three phase flow from rock properties. Soc Petrol Eng J Trans Am Inst Min Metall Pet Eng 243:149–156
Lay T, Wallace T (1995) Modern global seismology, vol. 58. Academic Press, New York
Ledesma A, Gens A, Alonso EE (1996) Parameter and variance estimation in geotechnical backanalysis using prior information. Int J Numer Anal Meth Geomech 20:119–141
Leecaster K, Ayling B, Moffitt G, Rose P (2012) Use of safranin T a reactive tracer for geothermal reservoir characterization. In: Proceedings, 37th Workshop on Geothermal Reservoir Engineering, Stanford University, SGP-TR-194
Lei T, Sinha BK, Sanders M (2012) Estimation of horizontal stress magnitudes and stress coefficients of velocities using borehole sonic data. Geophys 77(3), WA181–WA196
Leibundgut C, Maloszewski P, Külls C (2009) Tracers in hydrology, Auflage: 1. Auflage. Wiley, Chichester, UK ; Hoboken, NJ
Lockner DA, Tanaka H, Ito H, Ikeda R, Omura K, Naka H (2009) Geometry of the Nojima Fault at nojima-Hirabayashi, Japan-I. A simple damage structure inferred from borehole core permeability. Pure App Geophys 166:1649–1667
Luo S, Xu R, Jiang P (2012) Effect of reactive surface area of minerals on mineralization trapping of CO_2 in saline aquifers. Pet Sci 9:400–407
Maier F, Kocabas I (2013) Comment on "A closed-form analytical solution for thermal single-well injection-withdrawal tests" by Jung and Pruess. Water Resour Res 49(1):640–643
Maier F, Oberdorfer P, Kocabas I, Ghergut I, Sauter M (2012) Using temperature signals to estimate geometry parameters in fractured geothermal reservoirs. In: Proceedings, COMSOL Conference 2012, Milano, Italy
Maier F, SchafferM, Nur SN, Licha T (2014) Ability of thermo-sensitive tracers for precisely estimating system temperatures in column experiments with thermal gradient. In: Proceedings, 39th workshop on geothermal reservoir engineering, Stanford University, Stanford, CA, SGP-TR-202
Maloszewski P, Zuber A (1985) On the theory of tracer experiments in fissured rocks with a porous matrix. J Hydrol 79:333–358
Maloszewski P, Zuber A (1993) Tracer experiments in fractured rocks: matrix diffusion and the validity of models. Water Resour Res 29:2723
Manighetti I, King GCP, Gaudemer Y, Scholtz CH, Doubre C (2001) Slip accumulation and lateral propagation of active normal faults in Afar. J Geophys Res 106:13667–13696
Mariner PE, Jin MQ, Studer JE, Pope GA (1999) The first vadose zone partitioning inter-well tracer test for nonaqueous phase liquid and water residual. Environ Sci Technol 33(16):2825–2828

McCallum SD, Riestenberg DE, Cole DR, Freifeld BM, Trautz RC, Hovorka SD, Phelps TJ (2005) Monitoring geologically sequestered CO2 during the Frio Brine pilot test using perfluorocarbon tracers. In: Proceedings, Fourth Annual Conference on Carbon Capture and Sequestration DOE/NETL

Meier PM, Carrera J, Sánchez-Vila X (1998) An evaluation of Jacob's method for the interpretation of pumping tests in heterogeneous formations. Water Resour Res 34(5):1011–1025

Micarelli DU, Moretti I, Jaubert M, Moulouel H (2006) Fracture analysis in the south-western Corinth rift (Greece) and implication on fault hydraulic behavior. Tectonophys 426:31–59

Myers M, Stalker L, Ross A, Dyt C, Ho K-B (2012) Method for the determination of residual carbon dioxide saturationusing reactive ester tracers. Appl Geochem 27:2148–2156

Nelson NT, Oostrom M, Wietsma TW, Brusseau ML (1999) Partitioning tracer method for the in situ measurement of DNAPL saturation: influence of heterogeneity and sampling method. Environ Sci Technol 33(22):4046–4053

NETL (2013) Site screening, site selection, and initial characterization for storage of CO_2 in deep geologic formations, 2013 revised edition. DOE/NETL-2013/1605. National Energy Technology Laboratory, USA

Neufeld JA, Hesse M, Riaz A, Hallworth M, Tchelepi H, Huppert HE (2010) Convective dissolution of carbon dioxide in saline aquifers. Geophys Res Lett 37:L22404

Neuman SP, Witherspoon PA (1969) Applicability of current theories of flow in leaky aquifers. Water Resour Res 5(4):817–829

Niemi A, Bensabat J, Shtivelman V, Edlmann K, Gouze P, Luquot L, Hingerl F, Benson SM, Pezard PA, Rasmusson K, Liang T, Fagerlund F, Gendler M, Goldberg I, Tatomir A, Lange T, Sauter M, Freifeld B (2016). Heletz experimental site overview, characterization and data analysis for CO_2 injection and geological storage. Int J Greenhouse Gas Control 48:3–23

NIST (2010) Reference fluid thermodynamic and transport properties database (REFPROP): version 9.1

Nottebohm M, Licha T, Sauter M (2012) Tracer design for tracking thermal fronts in geothermal reservoirs. Geothermics 43:37–44

Nur A (1971) Effects of stress on velocity anisotropy in rocks with cracks. J Geophys Res 76:2022–2034

Oberdorfer P (2014) Heat transport phenomena in shallow geothermal boreholes-development of a numerical model and a novel extension for the thermal response test method by applying oscillating excitations. Ph. D. Thesis, University of Göttingen

Oberdorfer P, Holzbecher E, Hu R, Ptak T, Sauter M (2013) A five spot well cluster for hydraulic and thermal tomography. In: Proceedings, 38th Workshop on Geothermal Reservoir Engineering, Stanford University, Stanford, CA, SGP-TR-198

Paillet FL, White JE (1982) Acoustic modes of propagation in the borehole and their relationship to rock properties. Geophys 47:1215–1228

Paterson L, Boreham C, Bunch M, Ennis-King J, Freifeld B, Haese R, Jenkins C, Raab M, Singh R, Stalker L (2013a) The CO2CRC Otway stage 2B residual saturation and dissolution test: test concept, implementation and data collected. CO2CRC report no: RPT11–3158

Paterson L, Boreham C, Bunch M, Dance T, Ennis-King J, Freifeld B, Haese R, Jenkins C, LaForce T, Raab M, Singh R, Stalker L, Zhang Y (2013b) Overview of the CO2CRC Otway residual saturation and dissolution test. Energy Procedia 37:6140–6148

Paterson L, Boreham C, Bunch M, Dance T, Ennis-King J, Freifeld B, Haese R, Jenkins C, Raab M, Singh R, Stalker L (2014) CO2CRC Otway Stage 2B residual saturation and dissolution test. In: Cook PJ (ed) Geologically storing carbon: learning from the Otway project experience. CSIRO Publishing, Melbourne, Australia, pp 343–380

Pau G, Bell J, Pruess K, Almgren A, Lijewski M, Zhang K (2010) High-resolution simulation and characterization of density-driven flow in CO_2 storage in saline aquifers. Adv Wat Resour 33 (4):443–455

Place J, Sausse J, Marthelot J-M, Diraison M, Géraud Y, Naville C (2011) 3D mapping of permeable structures affecting a deep granite basement using isotropic 3C VSP data. Geophys J Int 186:245–263

Plummer MA, Palmer CD, Mattson ED, Redden GD, Hull LC (2010) Advancing reactive tracer methods for monitoring thermal drawdown in enhanced geothermal reservoirs. Idaho National Laboratory, Preprint INL/CON-10-18881

Poupon A, Loy M, Tixier M (1954) A contribution to electrical log interpretation in shaly sands. J Petrol Technol 6:27–34

Prioul R, Donald A, Koupsell R, El Marzouki Z, Bratoon T (2007) Forward modeling of fracture-induced sonic anisotropy using a combination of borehole images and sonic logs. Geophysics 72:E135–E147

Pruess K, Doughty C (2010) Thermal single-well injection-withdrawal tracer tests for determining fracture-matrix heat transfer area. In: Proceedings, 35th workshop on geothermal reservoir engineering, Stanford University, Stanford, CA, SGP-TR-188

Ptak T, Piepenbrink M, Martac E (2004) Tracer tests for the investigation of heterogeneous porous media and stochastic modelling of flow and transport-a review of some recent developments. J Hydrol 294:122–163

Qi R, LaForce TC, Blunt MJ (2009) Design of carbon dioxide storage in aquifers. Intl J Greenhouse Gas Control 3:195–205

Rao PSC, Annable MD, Kim H (2000) NAPL source zone characterization and remediation technology performance assessment: recent developments and applications of tracer techniques. J Contam Hydrol 45:63–78

Rasmusson K, Rasmusson M, Fagerlund F, Bensabat J, Tsang Y, Niemi A (2014) Analysis of alternative push–pull-test-designs for determining in-situ residual trapping of carbon dioxide. Int J Greenhouse Gas Control 27:155–168

Rasmusson K, Rasmusson M, Tsang Y, Niemi A (2016) A simulation study of the effect of trapping model, geological heterogeneity and injection strategies on CO_2 trapping. Int J Greenhouse Gas Control 52:52–72

Read T, Bour O, Bense V, Le Borgne T, Goderniaux P, Klepikova MV, Hochreutener R, Lavenant N, Boschero V (2013) Characterizing ground water flow and heat transport in fractured rock using fiber-optic distributed temperature sensing. Geophys Res Lett 40:1–5

Reimus P, Williams M, Vermeul V, Rose P, Leecaster K, Ayling B, Sanjuan R, Ames M, Dean C, Benoit D (2012) Use of tracers to interrogate surface area in single-well tracer tests in EGS Systems. In: Proceedings, 37th Workshop on Geothermal Reservoir Engineering, Stanford University, SGP-TR-194

Renard P, Glenz D, Mejias M (2009) Understanding diagnostic plots for well-test interpretation. Hydrogeol J 17(3):589–600

Riaz A, Hesse M, Tchelepi HA, Orr FM Jr (2006) Onset of convection in a gravitationally unstable diffusive boundary layer in porous media. J Fluid Mech 548:87–111

Rodrigues JD (1983) The Noordbergum effect and characterization of aquitards at the Rio Maior Mining Project. Ground Water 21(2):200–207

Rutqvist J (2012) The geomechanics of CO_2 storage in deep sedimentary formations. Geotech Geol Eng 30:525–551

Sánchez-Vila X, Girardi JP, Carrera J (1995) A synthesis of approaches to upscaling of hydraulic conductivities. Water Resour Res 31(4):867–882

Sánchez-Vila X, Carrera J, Girardi JP (1996) Scale effects in transmissivity. J Hydrol 183(1):1–22

Saripalli KP, Kim H, Rao PSC, Annable MD (1997) Measurement of specific fluid-fluid interfacial areas of immiscible fluids in porous media. Environ Sci Technol 31:932–936

Saripalli KP, Rao PSC, Annable MD (1998) Determination of specific NAPL-water interfacial areas of residual NAPLs in porous media using the interfacial tracers technique. J Contam Hydrol 30:375–391

Schaffer M (2013) On the possibility of using organic molecules in the characterization of subsurface processes (PhD Thesis). University of Goettingen, Germany

Schaffer M, Maier F, Licha T, Sauter M (2013) A new generation of tracers for the characterization of interfacial areas during supercritical carbon dioxide injections into deep saline aquifers: kinetic interface-sensitive tracers (KIS tracer). Int J Greenhouse Gas Control 14:200–208

Schaffer M, Niedbala A, Maier F, Idzik KR, Wilke M, Licha T (2015) Recent progress on hydrolyzable compounds as thermo-sensitive tracers for investigating the thermal drawdown of geothermal reservoirs. In: Proceedings, 40th workshop on geothermal reservoir engineering, Stanford University, Stanford, CA, SGP-TR-204

Schlumberger (1991) Log interpretation principles/applications. Schlumberger education services, Texas

Schmittbuhl J, Schmitt F, Scholtz CH (1995) Scaling invariance of crack surfaces. J Geophys Res 100:5953–5973

Semprini L, Hopkins OS, Tasker BR (2000) Laboratory, field and modeling studies of radon-222 as a natural tracer for monitoring NAPL contamination. Transp Porous Media 38:223–240

Sharma AN, Luo D, Walter MT (2012) Hydrological tracers using nanobiotechnology: proof of concept. Environ Sci Techn 46:8928–8936

Shook GM, Forsmann JH (2005) Tracer interpretation using temporal moments on a spreadsheet. Idaho National Laboratory

Shook GM, Ansley SL, Wylie A (2004) Tracers and tracer testing: design, implementation, and interpretation methods. INEEL/EXT-03-01466

Sibson RH (1977) Fault rocks and fault mechanisms. J Geol Soc Lond 133:191–213

Sone H, Zoback MD (2014) Viscous relaxation model for predicting least principal stress magnitudes in sedimentary rocks. J Pet Sci Eng 134:416–431

Sulem J (2007) Stress orientation evaluated from strain localization analysis in Aigion faults. Tectonophys 442:3–13

Sulem J, Vardoulakis I, Ouffroukh H, Boulon M, Hans J (2004) Experimental characterization of the thermos-poro-mechanical properties of the Aigion fault gouge. CR Geosci 336:455–466

Surdam RC (ed) (2013) Geological CO_2 storage characterization—the key to deploying clean fossil energy technology. Springer, New York

Tang JS (2005) Extended Brigham model for residual oil saturation measurement by partitioning tracer tests. Soc Pet Eng SPE paper 84874

Tatomir AB, Schaffer M, Kissinger A, Hommel J, Nuske P, Licha T, Helmig R, Sauter M (2015) Novel approach for modeling kinetic with respect to time-dependent interfacial area change for the optimization of supercritical carbon dioxide injection into deep saline aquifers. Int J Greenhouse Gas Control 33:145–153

Theis CV (1935) The relation between the lowering of the piezometric surface and the rate and duration of discharge of a well using ground water storage. US Department of the Interior, Geological Survey, Water Resources Division, Ground Water Branch

Tian L, Yang Z, Fagerlund F, Niemi A (2016) Effects of permeability heterogeneity on CO_2 injectivity and storage efficiency coefficient. Greenhouse Gas Sci Technol 6:112–124

Tomich JF, Dalton RL Jr, Deans HA, Shallenberger LK (1973) Single-well tracer method to measure residual oil saturation. J Petrol Technol 25(2):211–218

Tong F, Niemi A, Yang Z, Fagerlund F, Licha T, Sauter M (2013) A numerical model of tracer transport in a non-isothermal two-phase flow system for CO_2 geological storage characterization. Transp Porous Med 98:173–192

Urosevic M, Pevzner R, Kepic A, Wisman P, Shulakova V, Sharma S (2010) Time-lapse seismic monitoring of injection into a depleted gas reservoir—Naylor Field, Australia. Lead Edge 29:164–169

Vandenbohede A, Louwyck A, Lebbe L (2009) Conservative solute versus heat transport in porous media during push–pull tests. Transp Porous Med 76:265–287

Vandeweijer V, van der Meer B, Hofstee C, Mulders F, D'Hoore D, Graven H (2011) Monitoring the CO_2 injection site: K12-B. Energy Procedia 4:5471–5478

Verdon JP, Kendall J-M, White DJ, Angus DA (2011) Linking microseismic event observations with geomechanical models to minimise the risks of storing CO_2 in geological formations. Earth Planet Sci Lett 305:143–152

Verruijt A (1969) Elastic storage of aquifers. In: de Wiest RJM (ed) Flow through porous media. Academic Press, New York, pp 331–376
Vilarrasa V, Bolster D, Olivella S, Carrera J (2010) Coupled hydromechanical modeling of CO_2 sequestration in deep saline aquifers. Int J Greenhouse Gas Control 4:910–919
Vilarrasa V, Carrera J, Olivella S (2013) Hydromechanical characterization of CO_2 injection sites. Int J Greenhouse Gas Control 19:665–677
Vulava VM, Perry EB, Romanek CS, Seaman JC (2002) Dissolved gases as partitioning tracers for determination of hydrogeological parameters. Environ Sci Technol 36:254–262
Willmann M, Carrera J, Sánchez-Vila X, Vázquez-Suñé E (2007) On the meaning of transmissivity values obtained from. Hydrogeol J 15(5):833–842
Winkler KW, Liu HL, Johnson DL (1989) Permeability and borehole Stoneley waves: comparison between experiment and theory. Geophys 54:66–75
Wuerdemann H, Moeller F, Kuehn M, Heidug W, Christensen NP, Borm G, Schilling FR, the CO2Sink Group (2010) CO2SINK—from site characterisation and risk assessment to monitoring and verification: one year of operational experience with the field laboratory for CO_2 storage at Ketzin, Germany. Int J Greenhouse Gas Control 4(6) 938–951
Xu T, Apps JA, Pruess K (2005) Mineral sequestration of carbon dioxide in a sandstone–shale system. Chem Geol 217:295–318
Yeh T-CJ, Liu S (2000) Hydraulic tomography: development of a new aquifer test method. Water Resour Res 36(8):2095–2105
Yilmaz Ö (2001) Seismic data analysis, society of exploration geophysicists
Zatsepin SV, Crampin S (1997) Modelling the compliance of crustal rock-I. Response of shear-wave splitting to differential sress. Geophys J Int 129:477–494
Zhang LQ, Yue ZQ, Yang ZF, Qi JX, Liu FC (2006) A displacement-based back-analysis method for rock mass modulus and horizontal in situ stress in tunnelling—illustrated with a case study. Tunn Undergr Space Technol 21:636–649
Zhang Y, Freifeld B, Finsterle S, Leahy M, Ennis-King J, Paterson L, Dance T (2011) Single-well experimental design for studying residual trapping of supercritical carbon dioxide. Int J Greenhouse Gas Control 5:88–98
Zoback MD, Barton CA, Brudy M, Castillo DA, Finkheiner T, Grollimund BR, Moos DB, Peska P, Ward CD, Wiprut DJ (2003) Determination of stress orientation and magnitude in deep wells. Int J Rock Mech Min Sci 40:1049–1076
Zuber A, Motyka J (1994) Matrix porosity as the most important parameter of fissured rocks for solute transport at large scales. J Hydrol 158:19–46

Chapter 8
Field Injection Operations and Monitoring of the Injected CO_2

Auli Niemi, Jacob Bensabat, Peter Bergmann, Christopher Juhlin, Alexandru Tatomir, Iulia Ghergut, Martin Sauter, Barry Freifeld, Larry Myer, Christine Doughty, Axel Liebscher, Stefan Lüth, Sonja Martens, Fabian Möller, Cornelia Schmidt-Hattenberger and Martin Streibel

Abstract Monitoring the fate of the injected CO_2 and possible associated effects, such as hydro-mechanical and chemical effects in the target reservoir and its surroundings, is essential for safe operation of a storage facility. In this chapter, we shall first provide an overview of the technologies available and used for monitoring of CO_2. We shall then proceed to describe specific methods and finally present some important case studies that will demonstrate the use of the discussed monitoring technologies under specific field settings.

A. Niemi (✉) · C. Juhlin
Department of Earth Sciences, Uppsala University, Villavägen 16, 75236 Uppsala, Sweden
e-mail: auli.niemi@geo.uu.se

C. Juhlin
e-mail: christopher.juhlin@geo.uu.se

J. Bensabat
Environmental and Water Resources Engineering Inc., P.O. Box 6770, 31067 Haifa, Israel
e-mail: jbensabat@ewre.com

P. Bergmann · A. Liebscher · S. Lüth · S. Martens · F. Möller · C. Schmidt-Hattenberger · M. Streibel
Helmholtz Centre Potsdam, GFZ German Research Centre for Geosciences, Section Geological Storage, Telegrafenberg, Potsdam 14473, Germany
e-mail: alieb@gfz-potsdam.de

A. Liebscher
e-mail: alieb@gfz-potsdam.de

S. Lüth
e-mail: alieb@gfz-potsdam.de

S. Martens
e-mail: sonja.martens@gfz-potsdam.de

F. Möller
e-mail: fabian.moeller@gfz-potsdam.de

C. Schmidt-Hattenberger
e-mail: cornelia.schmidt-hattenberger@gfz-potsdam.de

A. Niemi et al. (eds.), *Geological Storage of CO_2 in Deep Saline Formations*,
Theory and Applications of Transport in Porous Media 29,
DOI 10.1007/978-94-024-0996-3_8

8.1 Background on Monitoring

Monitoring the fate of the injected CO_2 and possible associated effects, such as hydro-mechanical and chemical effects in the target reservoir and its surroundings, is essential for safe operation of a storage facility. NETL (2012) report on 'Best Practices for Monitoring, Verification and Accounting of CO_2 Stored in Deep Geological Formations' divides monitoring in three sub-groups, according to the domain where the monitoring is taking place and defines them as follows:

Atmospheric monitoring aims at measuring CO_2 density and flux in the atmosphere, to detect any possible leaks. The tools that are used are optical CO_2 sensors, atmospheric tracers and eddy covariance (EC) flux measurements.

Near-surface monitoring measures CO_2 and its effects in the zone ranging from the top of the soil down to the shallow groundwater. Tools include geochemical monitoring (in soil, vadose zone and shallow groundwater), surface displacement monitoring tools and ecosystem stress (e.g. changes in vegetation due to increased CO_2 fluxes) monitoring tools. The latter two are commonly measured by satellite-based remote sensing tools.

Subsurface monitoring where tools are used to detect and quantify the injected CO_2 in the subsurface, as well as the related effects of e.g. seismic activity, as well as to detect faults and fractures. The monitoring tools include well logging, downhole monitoring, fluid sampling including tracer analysis, seismic imaging, high-precision gravity methods and electrical techniques.

NETL (2012) also gives thorough discussions, general as well as case-specific, concerning these methods and their benefits and challenges. Here we will only present the summary tables, giving the description, benefits and challenges for each

A. Tatomir · I. Ghergut · M. Sauter
Angewandte Geologie, Universität Göttingen, Goldschmidtstr. 3, 37077 Göttingen, Germany
e-mail: alexandru.tatomir@geo.uni-goettingen.de

I. Ghergut
e-mail: iulia.ghergut@geo.uni-goettingen.de

M. Sauter
e-mail: Martin.Sauter@geo.uni-goettingen.de

B. Freifeld · L. Myer · C. Doughty
Lawrence Berkeley National Laboratory, 1 Cyclotron Road, Berkeley, CA 94720, USA
e-mail: bmfreifeld@lbl.gov

L. Myer
e-mail: lmyerco2@gmail.com

C. Doughty
e-mail: cadoughty@lbl.gov

Table 8.1 Summary of atmospheric monitoring techniques (adapted from NETL 2012)

Monitoring Technique	Description, benefits, and challenges
Optical CO_2 sensors	Description: sensors for measurement of CO_2 in air
	Benefits: relatively inexpensive and portable
	Challenges: difficult to distinguish release from natural variations and to provide continuous measurements over large areas
Atmospheric tracers	Description: natural and injected chemical compounds monitored in air to detect CO_2 released to the atmosphere
	Benefits: used as a proxy for CO_2, when direct observation of a CO_2 release is not adequate. Also used to track potential CO_2 plumes
	Challenges: In some cases, analytical equipment is not available onsite, and samples need to be analyzed offsite. Background/baseline levels need to be established
Eddy covariance	Description: flux measurement technique to measure atmospheric CO_2 at a specified height above the ground surface
	Benefits: can provide continuous data, averaged over both time and space, over a large area.
	Challenges: specialized equipment and robust data processing required. Natural variability in CO_2 flux may mask the signal

methodology, Table 8.1 is for the atmospheric monitoring, Table 8.2 for the near-surface monitoring and Table 8.3 for the subsurface monitoring. It should be noted that these tables do not include monitoring in off-shore situations where monitoring at seabed is also included, such as in the case of the Sleipner site.

Carbon Sequestration Leadership Forum (CSLF 2013) summarizes the monitoring techniques used in some major storage projects (Table 8.4). Inspection shows that practically all sites monitor well-head pressure and temperature. After that, most used are seismic surveys (2D/3D), downhole pressure and temperature monitoring, fluid sampling and observation wells. Several sites have also seismic downhole (VSP, Crosshole) monitoring, InSAR, soil gas sampling, and atmospheric CO_2 measurements. Microseismic observations and tracers are used in five cases. Other methods are used in four or fewer of the listed projects.

International Energy Agency Greenhouse Gas Control (IEAGHG 2013) reviews a number of test injection projects, summarizes the experiences and based on them, gives suggestions for best practices. Data from altogether 45 small scale projects and 43 large scale projects were compiled. The monitoring techniques used in the small scale projects are summarized in Table 8.5. The method classification somewhat differs from the one used by CSLF above. Inspection of the data in Table 8.5 shows that in these smaller scale, more research oriented projects, reflection seismic, downhole seismic, pressure logging and coring is used in 100 % of the projects. Almost all projects (90 %) use also thermal logging, wireline logging, geological model and reservoir modeling, and 80 % uses some type of geochemical analysis. These are followed by groundwater monitoring (70 %),

Table 8.2 Summary of near-surface monitoring techniques (adapted from NETL 2012)

Monitoring technique	Description, benefits, and challenges
Geochemical monitoring in the soil and vadose zone	Description: sampling of soil gas for CO_2, natural chemical tracers, and introduced tracers. Measurements are made with sensors inserted into the soil and/or with opaque flux accumulation chambers placed on the soil surface
	Benefits: soil-gas measurements detect elevated CO_2 concentrations above background levels and provide indications of releases. Tracers aid in identification of native versus injected CO_2. Opaque flux chambers can quickly and accurately measure local CO_2 fluxes from soil to air
	Challenges: significant effort for null result. Relatively late detection of release. Considerable effort is required to avoid cross-contamination of tracer samples. Flux chambers provide measurements for a limited area
Geochemical monitoring of shallow groundwater	Description: geochemical sampling of shallow groundwater above CO_2 injection zone to demonstrate integrity of freshwater formations. Chemical analyses may include pH, alkalinity, electrical conductivity, carbon, hydrogen, oxygen, and tracers
	Benefits: mature technology, samples collected with shallow monitoring wells. Early detection may be possible
	Challenges: significant effort for null result. Carbon isotopes are difficult to interpret due to complex dynamics of carbonate dissolution in shallow formations
Surface displacement monitoring (includes remote sensing)	Description: monitor surface deformation caused by reservoir pressure changes associated with CO_2 injection. Measurements made with satellite-based radar (SAR/InSAR) and surface- and subsurface-based tiltmeters and GPS instruments. Data allow modeling of injection-induced fracturing and volumetric change in the reservoir
	Benefits: highly precise measurements over a large area (100 km × 100 km) can be used to track pressure changes in the subsurface associated with plume migration. Tiltmeter technology is mature, and has been used successfully for monitoring steam/water injection and hydraulic fracturing in oil and gas fields. GPS measurements complement InSAR and tiltmeter data
	Challenges: InSAR methods work well in locations with level terrain, minimal vegetation, and minimal land use, but must be modified for complex terrain/varied conditions. Tiltmeters and GPS measurements require surface/subsurface access and remote data collection

(continued)

Table 8.2 (continued)

Monitoring technique	Description, benefits, and challenges
Ecosystem stress monitoring (includes remote sensing)	Description: satellite imagery, aerial photography, and spectral imagery are used to measure vegetative stress resulting from elevated CO_2 in soil or air
	Benefits: Imaging techniques can cover large areas. Vegetative stress is proportional to soil CO_2 levels and proximity to CO_2 release
	Challenges: Detection only possible after sustained CO_2 emissions have occurred. Shorter duration release may not be detectable. Natural variations in site conditions make it difficult to establish reliable baseline. Changes not related to CO_2 release can lead to false positives

Table 8.3 Summary of sub-surface monitoring techniques (adapted from NETL 2012)

Monitoring technique	Description, benefits, and challenges
Well logging tools	Description: mature technology used to monitor the wellbore and near-wellbore environment. Logs include porosity, density, acoustic, optical, gamma ray, resistivity imaging, borehole diameter logging, and pulsed neutron capture
	Benefits: easily deployed technology used to detect wellbore release and changes in near-wellbore fluid or formation composition
	Challenges: area of investigation limited to near the wellbore. Sensitivity of tool to fluid change may vary. Some tools are not sensitive to dissolved or mineralized CO_2. Workover fluids may affect log results
Downhole monitoring tools	Description: technology used to monitor CO_2 injection, reservoir conditions, wellbore conditions, CO_2 breakthrough at observation wells; also used to differentiate between CO_2 and brine
	Benefits: indirect and direct measurements of CO_2 transport. Pressure sensors useful for monitoring wellbore mechanical integrity and detecting CO_2 releases. Downhole temperature monitoring data could be used as inputs for history-matching simulation models. Flow meters monitor fluid flow conditions throughout the injection site
	Challenges: sensors need to have little drift over a long time span. Sensors and meters require specific calibrations to conform to regulations
Subsurface fluid sampling and tracer analysis	Description: technology used to monitor changes in the composition of fluids at observation wells and for characterizing CO_2 transport, reactions, dissolution, and subsurface dispersion
	Benefits: Provides information on fluid geochemistry, CO_2 transport properties, and CO_2 saturation to estimate mass balances and distribution of CO_2 in the subsurface
	Challenges: cannot image CO_2 migration and release directly. Only near-well fluids are measured

(continued)

Table 8.3 (continued)

Monitoring technique	Description, benefits, and challenges
Seismic methods	Description: reflection seismic uses acoustic properties of geologic formations and pore fluids to image geologic layers and plume migration in the subsurface. Passive seismic detects microseismic events in the subsurface and can provide information on fluid movement in a formation
	Benefits: reflection seismic is useful for time-lapse monitoring of a CO_2 plume, and possibly for out-of-zone CO_2 migration indicating a release. Borehole seismic (crosswell, VSP) surveys can provide high-resolution imaging of the plume near the wellbore. Passive seismic can be used to detect natural and induced seismicity, to map faults and fractures in the injection zone and adjacent horizons, and to track the migration of the fluid pressure front during and after injection
	Challenges: geologic complexity and a noisy recording environment can degrade or attenuate surface seismic data. Two-dimensional seismic surveys may not detect out-of-plane migration of CO_2. Borehole seismic methods require a wellbore for monitoring, and careful planning is required to integrate these with other surveys. Microseismic monitoring detects fracturing and faulting events that may result from CO_2 injection, but a comprehensive knowledge of reservoir geomechanical properties is needed to properly interpret these events
Gravity	Description: use of gravity to monitor changes in density of fluid resulting from injection of CO_2
	Benefits: fluid density changes due to CO_2 releases or CO_2 dissolution can be detected, unlike seismic methods, which do not identify dissolved CO_2
	Challenges: limited detection and resolution unless gravimeters are located just above reservoir, which significantly increases cost. Noise and gravity variations (tides, drift) need to be eliminated to interpret gravity anomalies due to CO_2

atmospheric monitoring and observation well (60 %). Clearly different from the projects summarized in Table 8.4 that also includes large-scale industrial projects, none of these smaller test injections used InSAR according to this survey. The IEAGHG also gives a full list of monitoring techniques, but does not specify their use in various projects. The full list of individual monitoring techniques can be found on IEAGHG CO_2 Monitoring Technique Data Base.

In the following chapters we will discuss in more detail some essential monitoring techniques (Sects. 8.2, 8.3 and 8.4) as well as case studies, including monitoring experiences from some major large scale industrial projects (Sect. 8.5)

Table 8.4 Monitoring techniques used in some CO_2 injection projects (CSLF 2013)

Monitoring technology	Sleipner	Weyburn	In Salah	Snøhvit	K12-B	Otway	Ketzin	Decatur	Quest	Lacq	Aquistore
Seismic surface (2D/3D)	+	+	+	+		+	+	+	+		+
Seismic surface (3C/9C)		+									
Seismic downhole		+				+	+	+	+		+
Electrical surface	+						+				
Electrical downhole		+					+		+		
Gravity surface/seabed	+	+									+
Tiltmeters			+								+
Satellite interferometry (InSAR)		+	+				+	+	+		+
Downhole P, T		+		+	+	+	+	+	+	+	+
Continuous downhole temperature							+		+	+	
Acoustic seabed imaging	+			+							
Acoustic water column imaging	+										
Geophones								+			
Water column chemistry	+										
Seabed video (ROV/AUV)	+										
Soil gas		+	+			+	+	+		+	+
Surface gas flux		+	+			+		+		+	
Passive CO_2 detectors			+			+					
Ecosystem and biomarkers	+		+	+						+	
Microseismic (passive seismic)		+	+			+	+			+	

(continued)

Table 8.4 (continued)

Monitoring technology	Sleipner	Weyburn	In Salah	Snøhvit	K12-B	Otway	Ketzin	Decatur	Quest	Lacq	Aquistore
Observation wells		+	+		+	+	+	+	+		+
Tracers		+	+		+		+		+		
Microbiology			+				+				
Wireline logs			+		+		+				
Fluid samples		+	+		+	+	+	+	+	+	+
Atmospheric CO_2 mobile/spatial		+	+			+		+	+	+	
Atmospheric CO_2 flux tower		+	+			+				+	
Well head pressure	+	+	+	+	+	+	+	+	+	+	+
Temperature	+	+	+	+	+	+	+	+	+	+	+
Well integrity monitoring (EMIT, PMIT)					+				+		
Downhole camera log					+		+				

Table 8.5 Monitoring technologies used in 18 small-scale projects in saline aquifers as summarized by IEAGHG (adapted from IEAGHG 2013)

Monitoring technology	Percentage of projects using	Monitoring technology	Percentage of projects using
Downhole seismic	100	Thermal logging	90
Groundwater monitoring	70	Wireline logging	90
Soil monitoring	40	Observation well	60
Atmospheric monitoring	60	Geochemical	80
Biological monitoring	20	InSAR	0
Tracer analysis	40	Reservoir modeling	90
Electromagnetic	20	Coring	100
Gravity	0	Reflection seismic	100
Pressure logging	100	Geological model	90

as well as small-scale, more scientifically motivated projects Frio and Ketzin (Sects. 8.6 and 8.7).

8.2 Geophysical Methods

Peter Bergmann and Christopher Juhlin

8.2.1 Overview of Geophysical Methods

Geophysical methods allow for imaging of physical subsurface properties and provide an opportunity for the monitoring of geological CO_2 storage. The objective of any geophysical site monitoring is the development of a baseline model and following changes within it in space and time. Comprehensive site models contain information about the present geometrical structures and composition, such as rock types and pore fluids. Since these models are always simplified representations of reality, they also contain inherent uncertainties.

In order to correctly describe the evolution of these models, they continuously need to be updated with elementary models that are provided by individual survey techniques. This implies that a combination of different geophysical methods are a prerequisite for monitoring the different properties of the models at a (as broad as possible) range of scales. Consequently, monitoring of geological CO_2 storage requires integrated multi-method concepts to allow for comprehensive site descriptions.

A vast number of reported studies underlines the capabilities of geophysical methods for subsurface monitoring. Although most of these studies have been carried out for near-surface hydrogeological purposes or hydrocarbon production, they are of great relevance for CO_2 storage monitoring since many of their methodical and practical aspects are similar. In addition, there are also a number of studies which address CO_2 storage monitoring in particular. The majority of these studies are based on ongoing/completed CO_2 injection projects, such as those located in Norway (Sleipner and Snøhvit), Canada (Weyburn), USA (Frio), Australia (Otway), Japan (Nagaoka), and Algeria (In Salah) and Germany (Ketzin).

These projects are located in diverse environments, concerning factors such as storage depth, reservoir system, reservoir use, pressure and temperature conditions. This variability also results in that different combinations of geophysical methods have been used for monitoring, most of which include seismics and borehole logging, but also electromagnetics and gravity surveying (e.g. Michael et al. 2010). All of these methods provide a certain range of resolution and sensitivity, underlining the importance of using a combination of methods. There are also cases where geophysical methods do not deliver sufficient information or even fail. Therefore, several research initiatives have been initiated (e.g. SACS, CO_2STORE, IEAGHG Monitoring Network, CASTOR, CO_2GeoNet, CO_2ReMoVe, CO_2 Capture Project) in order to compile the gained experiences into best-practice guidelines and to support the definition of regulatory frameworks. Interestingly, these initiatives consistently agree on that monitoring is indeed site-specific, but that is it always needs to be comprised of multi-method geophysical programs.

In this review we focus on two geophysical methods, seismic and geo-electric. Other methods, such as electromagnetic, gravity, passive seismic and InSar, may also be used, but currently it is mainly seismic and geo-electric methods that are being applied at CO_2 storage sites and, therefore, the focus is on these. Even within the fields of applied seismic and geo-electric there is significant research ongoing. The use of sparser arrays to reduce costs, permanent sources and sensors, active seismic interferometry, downhole methods (including fibreoptics) and advanced processing methods are all being tested and their use should eventually provide higher resolution images or allow larger volumes to be investigated without increased cost. Faults and other features can potentially be mapped in greater detail and better geological models produced. The ability to repeat measurements on shorter time scales than what is commonly done with, for example, 3D reflection seismic surveys may also help to better understand CO_2 plume evolution and allow better integration of geophysical results with hydrogeological modeling. However, in the present review we have chosen to focus on the basic principles behind seismic and geo-electric methods that are currently being employed. Furthermore, we refer to the Ketzin site (see Sect. 8.7) to illustrate how changes in physical properties will influence the geophysical response. Note that all CO_2 storage sites will most likely have site specific rock properties and that thorough investigations are required before making predictions on the seismic and geo-electric response at an individual site to CO_2 injection.

8.2.2 *Seismic Methods*

8.2.2.1 Theory

The basis of the reflection seismic method is the controlled activation and measurement of elastic wave fields. Waves which are reflected back to the surface convey information about geologic structures, since the reflections are due to discontinuities in elastic parameters (Fig. 8.1). Wave field properties that are valuable in this context are travel time, amplitude, frequency content, and phase. In the following, the focus will be on the amplitude, because it is the most important property that is monitored in time lapse surveys.

Assume that a compressional wave (P-wave) hits a layer with the wavefront perpendicular to the boundary (normal incidence, $\theta = 0$ in Fig. 8.1), the amplitude coefficients for reflection and transmission are given by (e.g. Kearey et al. 2002)

$$R = \frac{V_2\rho_2 - V_1\rho_1}{V_2\rho_2 + V_1\rho_1} \quad T = \frac{-2\,V_1\rho_1}{V_2\rho_2 + V_1\rho_1} \tag{8.2.1}$$

Here, V_1, V_2 and ρ_1, ρ_2 denote the P-wave velocities and densities in the upper and lower layer, respectively. In this nomenclature, the wave is propagating from within the first layer towards the second layer. The normal incidence assumption implies that a source and a receiver are located on the surface of the first layer at identical position (zero-offset). The receiver will then measure the amplitude of the reflected wave at the zero-offset two-way-traveltime (TWT), which corresponds to the wave propagating forward and backward along the same ray path.

Typically, seismic acquisition is performed at finite offset ($\theta \neq 0$ in Fig. 8.1), which gives rise to two implications: First, forward and backward propagation of a reflected wave will occur along different ray paths. Consequently, the travel time

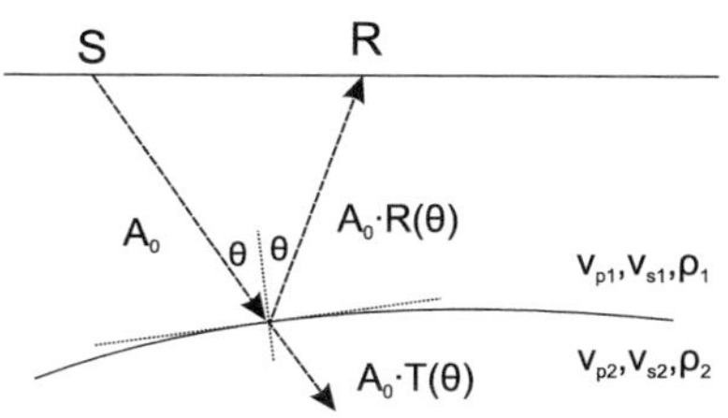

Fig. 8.1 Schematic illustration of a wave propagating from a source location S to a receiver location R after being reflected at an interface. A_0 denotes the amplitude of the wave impinging the interface. R(θ) and T(θ) denote the proportions of A_0 that are reflected and transmitted, respectively

will likely differ from that of a zero-offset ray. Assuming an isotropic medium with horizontal or moderately dipping layers, the onset time of a reflection will be increasing with increasing source-receiver offset. The offset-traveltime relation can then be approximated by hyperbolic functions which define the normal moveout (NMO) of the reflection onsets (Yilmaz 2001). Secondly, acquisition at finite offsets leads to reflections at non-normal incidence, which makes it necessary to consider R for an arbitrary angle of incidence θ. Most often, such a case leads to conversion of P-waves to (vertically polarized) shear waves (S-waves), which implies an equation system that requires knowledge of the P-wave velocities in the upper and lower layers (Vp_1, Vp_2) and the respective S-wave velocities (Vs_1, Vs_2) in order to calculate the reflection and transmission coefficients. The reflection and transmission angles are determined by Snell's law. Respective amplitudes are specified by the Zoeppritz equations (Zoeppritz 1919), which can be derived from the requirement of continuity of displacement and stress at the reflecting interface. Application of the Zoeppritz equations has now become common practice to analyze for so-called amplitude-versus-offset (AVO), or amplitude-versus-angle (AVA), responses to quantitatively assess the elastic properties of the media (Castagna and Backus 1993). Due to the inherent complexity of the Zoeppritz equations, a number of approximations have been introduced to allow for more convenient calculations (e.g. Aki and Richards 1980; Bortfeld 1961; Shuey 1985; Wang 1999). Aki and Richards (1980) presented the following 3-term approximation for layers with small contrasts in elastic properties (see Mavko et al. 2003).

$$R_{pp}(\theta) \approx A + B\sin^2(\theta) + C\tan^2(\theta)\sin^2(\theta) \tag{8.2.2}$$

In the following, only the P-wave reflection from an incident P-wave is discussed, the most common seismic wave that is recorded, and which is indicated by the notation R_{pp}. The angular reflection coefficients A, B and C are (Mavko et al. 2003)

$$A = \frac{1}{2}\left(\frac{\Delta V_p}{\langle V_p\rangle} + \frac{\Delta\rho}{\langle\rho\rangle}\right) \tag{8.2.3}$$

$$B = \frac{1}{2}\frac{\Delta V_p}{\langle V_p\rangle} - 2\left(\frac{V_s}{Vp}\right)^2\left(\frac{\Delta\rho}{\langle\rho\rangle} + \frac{\Delta V_s}{\langle V_s\rangle}\right) \tag{8.2.4}$$

$$C = \frac{1}{2}\frac{\Delta V_p}{\langle V_p\rangle} \tag{8.2.5}$$

with the following contrasts and averages across the interface

$$\Delta V_p = Vp_2 - Vp_1 \quad \langle V_p\rangle = \frac{Vp_1 + Vp_2}{2} \tag{8.2.6}$$

$$\Delta V_s = Vs_2 - Vs_1 \quad \langle V_s \rangle = \frac{Vs_1 + Vs_2}{2} \tag{8.2.7}$$

$$\Delta \rho = \rho_2 - \rho_1 \quad \langle \rho \rangle = \frac{\rho_1 + \rho_2}{2} \tag{8.2.8}$$

A, *B* and *C* can be interpreted in terms of different angle ranges (Castagna and Backus 1993). The term *A* dominates at small angles (near-offsets) and approximates, again assuming small contrasts, the normal-incidence reflection coefficient (Mavko et al. 2003). The terms *B* and *C* dominate at intermediate and large angles (near the critical angle), respectively. In practice, *C* is often neglected, since common acquisition geometries provide reflection data mostly at small and intermediate angles. This leads to a linearized form of the equation, in which *A* is the so-called AVO intercept and *B* the AVO gradient.

Practical AVO analysis is most commonly carried out by crossplots of *A* and *B*, which are used to analyze background trends and search for deviations from them (Ross 2000). For example, the reservoir sandstone where CO_2 is stored at Ketzin shows lower wave velocities and density than the caprock mudstones (Norden et al. 2010), a fact that leads to a negative AVO gradient and a negative AVO intercept. This is also illustrated by the single interface reflection coefficients in Fig. 8.2. However, it is important to recognize that the Ketzin reservoir is of sub-wavelength thickness, which generally poses additional implications on the normal incidence amplitude (e.g. Gochioco 1991; Meissner and Meixner 1969; Widess 1973) and the AVO response (e.g. Bakke and Ursin 1998; Juhlin and Young 1993; Liu and Schmitt 2003). For instance, if the contrasts in elastic properties of reservoir and surrounding rocks increase the main assumption of the AVO equation becomes increasingly invalid. Moreover, the AVO response cannot adequately be approximated by the superposition of the reflections off the top of the layer and off the bottom of the layer only. In such a case interbed multiples and conversions also have to be taken into account (Meissner and Meixner 1969). Based on the Ketzin reservoir model of Kazemeini et al. (2010), Fig. 8.2 illustrates the difference in the AVA response for the reservoir represented by a single boundary and a sub-wavelength layer.

8.2.2.2 Seismic Rock Physics

Seismic wave velocities are governed by the elastic moduli of the rocks they propagate through and their density. The elastic moduli and densities correspond to the whole rock and depend both on the rock matrix properties as well as the properties of the fluids or gases filling the pore space. P-wave (V_p) and S-wave (V_s) velocities are governed by the bulk modulus, *K*, the shear modulus, *G*, and the density.

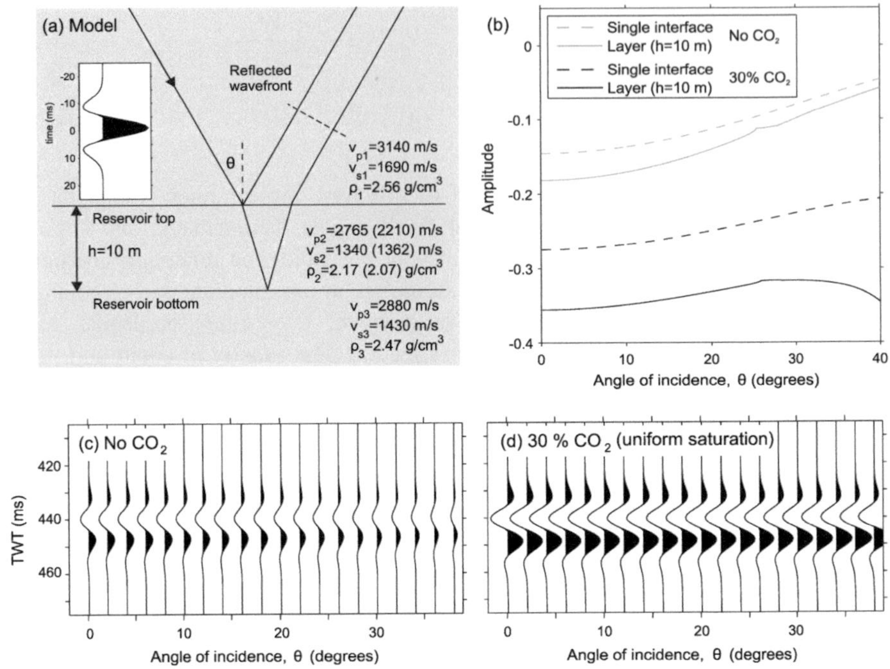

Fig. 8.2 Modeled AVA reflectivity of a thin layer with Ketzin reservoir parameters after Kazemeini et al. (2010) as input model. **a** The input model comprises a single layer representing the reservoir. Modeling was performed with the input model before CO_2 injection (values without brackets) and after CO_2 injection (bracketed values). **b** AVA response of the reservoir top as a single interface and as a layer of 10 m thickness. Thin layer amplitudes were computed with the method of Juhlin and Young (1993) using the 50 Hz Ricker wavelet shown in **a**. Computations include first-order multiples and conversions, and use the Aki-Richards approximation after Guy et al. (2003). **c**, **d** Modeled AVA response of the 10 m layer for the 50 Hz Ricker wavelet. Note that the traces in **a**, **c**, **d** are drawn to the same amplitude scale

$$V_p = \sqrt{\frac{K + \frac{4}{3}G}{\rho}} \tag{8.2.9}$$

$$V_s = \sqrt{\frac{G}{\rho}} \tag{8.2.10}$$

The bulk modulus is defined by the relative volume change caused by an omni-directional confinement pressure. The shear modulus is defined by the relative shear displacement when a shear force is applied (e.g. Lay and Wallace 1995). As there is no restoring force for liquids and gases, their shear modulus is zero.

Injection of CO_2 into porous reservoir rock containing saline fluids will result in the replacement of some of the saline fluid by CO_2. The injection will also increase the pressure in the reservoir. Leakage from a deeper storage formation to shallower levels will result in similar changes in reservoirs at shallower levels. The replacement of saline fluid by CO_2 is referred to as fluid substitution and there are two models for how this replacement affects seismic velocities. These are the uniform saturation model (Gassmann 1951) and the patchy saturation model (Mavko et al. 2003). In both models only the bulk modulus and the density will change due to the replacement of fluid by gas, while the shear modulus is unaffected. This implies that there is very little change in the S-wave velocity when CO_2 is injected into the reservoir, but there will be large changes in the P-wave velocity. However, increased pore pressure in the reservoir caused by the injection will result in a decrease in the effective stress and thereby a decrease in both the bulk and shear modulus of the rock. This difference in behavior between changes in gas saturation and pore pressure can potentially be monitored with seismic methods (e.g. Landrø 2001).

The uniform saturation model gives the following change in bulk modulus (Gassmann 1951):

$$K_{uni} = K_d + \frac{\left(1 - \frac{K_d}{K_m}\right)^2}{\frac{\Phi}{K_f} + \frac{1-\Phi}{K_m} - \frac{K_d}{K_{m^2}}}, \qquad (8.2.11)$$

where K is the bulk modulus of a rock saturated with a frictionless fluid of bulk modulus K_f, K_d is the frame bulk modulus (air-saturated rock), K_m is the matrix bulk modulus of the same rock, and Φ is the porosity.

The bulk modulus K_f of a water/CO_2 mixture after the rock is flooded with CO_2 can be calculated using Wood's equation (Wood 1941),

$$\frac{1}{K_f} = \frac{S_w}{K_w} + \frac{1 - S_w}{K_{CO_2}}, \qquad (8.2.12)$$

where K_w and K_{CO_2} are, respectively, the bulk moduli of brine and CO_2, and S_w is the brine saturation fraction. Wood's equation is based on the uniform stress assumption for fluid mixtures.

On a fine scale, the Gassmann model assumes homogeneous mixing of both phases. However, if mixing is heterogeneous on a coarse scale, a passing wave causes local pore-pressure differences. Assuming that the mixing can be described by geometric patches, which themselves are homogeneously saturated, there will be pressure exchange between nearby patches (Mavko et al. 2003). On a larger scale, wave-induced pore-pressure differences should average to an equilibrated value. At a seismic wave frequency f, these pore pressure heterogeneities will equilibrate for scales smaller than the critical diffusion length L_c (Mavko et al. 2003):

$$L_c \approx \sqrt{\frac{kK_f}{f\eta}} \tag{8.2.13}$$

with k denoting the rock permeability and η the fluid viscosity. If the patches are sufficiently small ($<L_c$), the pore-fluid mixture can be represented by a single effective fluid, which is then considered to be uniformly saturated. If the patches are larger than L_c spatial fluctuations will tend to persist during the passage of seismic waves, a state which is referred to as non-uniform or patchy saturation (Mavko and Mukerji 1998). Patchy saturation can for example be caused by "fingering" of pore-fluids, which might result from spatial variations in wettability, permeability or shaliness (Asveth 2009). Yet, it is possible to describe the individual patches by separate Gassmann models.

The patchy saturation model gives the following change in bulk modulus via the Hill equation (Berryman and Milton 1991; Hill 1963):

$$K_{pat} = \frac{1}{\frac{S_w}{K_0 + \frac{4}{3}G} + \frac{1-S_w}{K_{100} + \frac{4}{3}G}} - \frac{4}{3}G, \tag{8.2.14}$$

where K_0 and K_{100} are the whole rock bulk moduli for 0 % CO_2 saturation and 100 % CO_2 saturation, respectively. In both the uniform and patchy models the density of the saturated rock is given by

$$\rho = \rho_d + \Phi\rho_f, \tag{8.2.15}$$

where ρ and ρ_d are, respectively, the fluid-saturated and dry densities of the rock, and ρ_f is the pore fluid's density. For a mixture composed of water and CO_2 it is determined with an arithmetic average of the separate fluid phases:

$$\rho_f = S_w * \rho_w + S_{CO_2} * \rho_{CO_2}, \tag{8.2.16}$$

where ρ_f is the mixture density, ρ_W and ρ_{CO_2}, and S_W and S_{CO_2} are, respectively, the densities and volume fractions (saturation) of water and CO_2.

Values for the different parameters can either be determined through lab experiments or by theoretical formulas. An online program to calculate fluid properties based on Batzle and Wang (1992) is available at:

www.crewes.org/ResearchLinks/ExplorerPrograms/FlProp/FluidProp.htm

Changes in P-wave and S-wave velocities due to a pore pressure increase may be modeled with second order curves with empirical constants to be determined (Landrø 2001).

$$\Delta V_p = \left(l_\alpha \Delta P + m_\alpha \Delta P^2\right) V_p \tag{8.2.17}$$

$$\Delta V_s = \left(l_\beta \Delta P + m_\beta \Delta P^2\right) V_s \tag{8.2.18}$$

For a hypothetical leak from reservoir depth with accumulation of CO_2 at 300 m depth into a high porosity sandstone the velocities as a function of CO_2 saturation for the uniform and patchy models are plotted in Fig. 8.3 Note the large difference predicted for velocity depending upon which model is assumed. Note also that the S-wave velocity only increases slightly in both models, due to the decrease in density as CO_2 enters the rock. A similar plot for an increase in pore pressure is shown in Fig. 8.4. Note that at 300 m depth pore pressure changes more than 2 MPa are unlikely without fracturing the formations. At greater depth, pore pressure changes due to injection can be significant without fracturing the formations.

8.2.2.3 Time-Lapse Seismics

Reflection seismic based time-lapse methods are the heart of all geophysical monitoring methods in sedimentary environments and, therefore, a very brief outline of reflection seismic processing is given here. Typical processing procedures comprise three main steps: (1) data preprocessing, (2) stacking, and (3) seismic migration (e.g. Yilmaz 2001). (1) The preprocessing aims to extract the

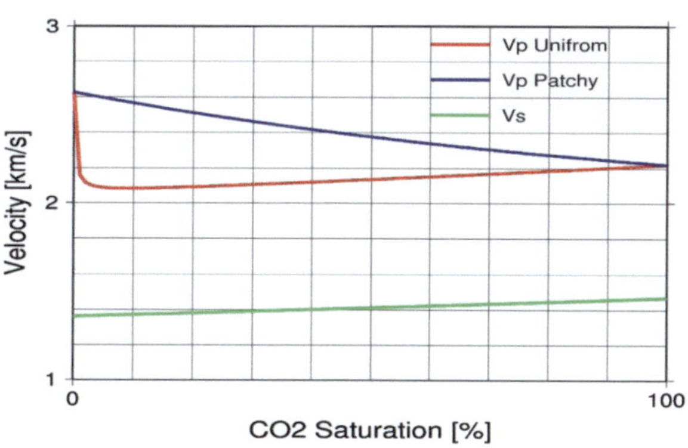

Fig. 8.3 P-wave and S-wave velocities for the uniform and patchy saturation models for a 30 % porosity sandstone at 300 m depth

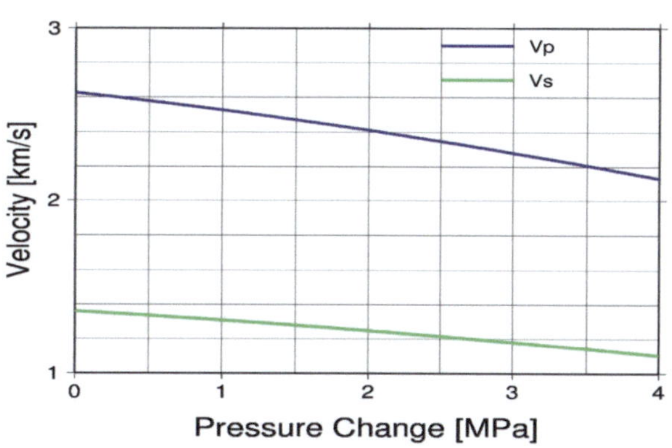

Fig. 8.4 P-wave and S-wave velocities as a function of increased pore pressure

relevant reflections out of the acquired seismograms. Common preprocessing steps are the muting and suppression (filtering) of undesired signals, deconvolution, and amplitude restoration. A further important step is the application of static corrections, which will be explained in more detail below. (2) Seismic stacking comprises resorting of traces into gathers and summation along time-offset trajectories which are defined by velocity model estimations. Most commonly, the traces are resorted into common-midpoint (CMP) gathers. Then, NMO corrections are applied on the basis of velocities that are extracted from velocity analyses in the CMP domain. These velocity analyses are typically carried out in alternation with residual static corrections until the velocity models sufficiently remove the NMO (Yilmaz 2001). Stacking is then completed with the summing of the NMO-corrected traces that belong to the same CMP gathers (Mayne 1962). The number of traces within a CMP gather is termed the fold, which is an indicator of the signal-to-noise improvement that can be obtained in the stacking procedure. Aside from the CMP stack there are also alternative stacking procedures, such as the methodically related common-reflection-element (CRE) stack (Gelchinsky 1988), multifocusing stack (Gelchinsky et al. 1999), or common-reflection-surface (CRS) stack (Jäger et al. 2001). (3) Seismic processing is typically finalized by migration which intends to relocate reflected energy to its true (temporal or spatial) position of origin. Seismic migration generally aims to overcome mis-positioning (e.g. image angle of dipping reflectors) and can be applied in the pre-stack or post-stack domain (see, e.g. Biondi 2006; Yilmaz 2001). In the latter case, migration is typically carried out in conjunction with dip-moveout (DMO) corrections before stacking, which then resembles a pre-stack migration scheme (Deregowski 1986).

The general objective of seismic processing is to modify acquired data into images that can be used for interpretation of subsurface structures. On this basis, time-lapse seismic aims for the detection of changes in the seismic response of the sub-surface by means of repeated data acquisition and processing. There are several metrics which are used to quantify the repeatability of seismic surveys, with the normalized-root-mean-square amplitude difference (NRMS) of (Kragh and Christie 2002) being the most commonly used. The NRMS of two traces a and b is given by

$$NRMS = 100\,\% \frac{RMS(a-b)}{0.5(RMS(a)+RMS(b))} \tag{8.2.19}$$

The NRMS measure ranges from 0 % for identical traces to 141 % for randomly uncorrelated traces, and up to 200 % for 180° out of phase traces (amplitude reversal). It is very sensitive to small changes between the two input traces, whether it is in the amplitude or phase (Domes 2010).

Beyond the impact of noise, there are a number of practical challenges to time-lapse seismic. In the case of onshore surveying, unforeseen acquisition obstacles usually occur. Although the fold reduction caused by these obstacles can be compensated for by relocating source and receiver locations (e.g. acquisition of data that will be binned into the same CMP bin at different offsets), a reduced experimental reproduction inevitably remains. Furthermore, wavelet reproducibility may be limited. This is not only a matter of source technology, but also of source-ground coupling and changes in near-surface velocities (Kashubin et al. 2011). These complications need to be handled by cross-equalization of the frequency and phase characteristics (wavelet matching).

In addition, the seismic response also senses pressure changes (Eberhart-Phillips et al. 1989; Todd and Simmons 1972). It is obvious that time-lapse seismic interpretation for monitoring CO_2 injection must take this into consideration. In this context, Landrø (2001) introduced a method for discriminating the fluid and pressure response in time-lapse seismic data by exploiting the AVO response.

8.2.3 *Geoelectric Methods*

The geoelectric method, here also referred to as Electrical Resistivity Tomography (ERT), uses artificial electrical currents to investigate the distribution of electric resistivity within the subsurface. It serves as a complementary method to seismic methods, and its application to CO_2 storage monitoring is motivated by the expected change in rock resistivity when electrically well conductive brine is substituted by insulating CO_2 (Christensen et al. 2006; Nakatsuka et al. 2010).

8.2.3.1 Theory

Geoelectrics uses diffusive electric fields, as opposed to propagating wave fields as in most seismic applications, which obey Poisson's equation (e.g. Telford et al. 1990)

$$\nabla \cdot \left(\frac{1}{\rho} \nabla \Phi \right) = -I\delta(\vec{r} - \vec{r}_s) \tag{8.2.20}$$

It entails that electric current flow, I, is determined by the spatial arrangement of electrical sources (and sinks) as well as the distribution in electric resistivity ρ. Both factors specify the electric potential ϕ, to which the gradient of the current flow aligns. The right hand side of the equation places an infinitesimal source (represented by Dirac's delta) at r_s, releasing an electric current I. If this source would be located on a perfectly uniform half-space with a resistivity of ρ_0, the potential would be given by

$$\Phi(\vec{r}) = I \frac{\rho_0}{|\vec{r} - \vec{r}_s|} \tag{8.2.21}$$

A combination of current sources can be given by the superposition of their individual potential distributions. Due to the conservation of electric charge, the practical field experiment is typically carried out by a current circuit, which is

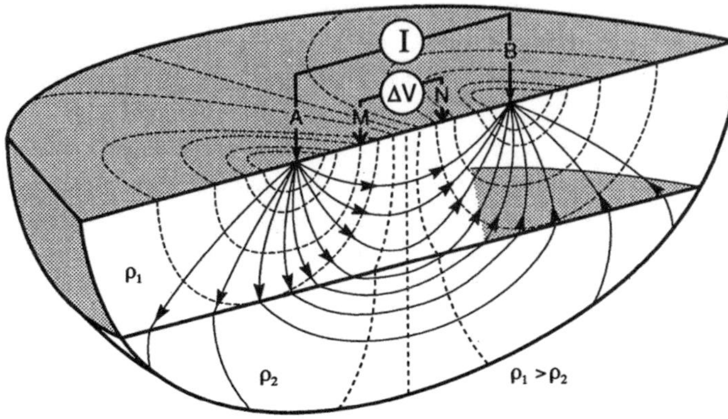

Fig. 8.5 Schematic illustration of a four-point electrode arrangement after Lange (1997). Current flow lines (*solid*) and equipotential lines (*dashed*) are given for a two-layer case with higher resistivity in the first layer

realized through a pair of current electrodes (*A* and *B*). An additional pair of potential electrodes (*M* and *N*) are used to measure spatial differences in ϕ, i.e. the electric voltage *U*. This so-called four-point layout is schematically illustrated in Fig. 8.5.

Geoelectric surveying is commonly performed by using multiple pairs of current electrodes and voltage electrodes with an aim to achieve a dense sampling of the imaging target. From the injected current *I* and the measured voltage, *U*, a resitance, *R*, can be calculated. This resistance has a strong dependence on the geometrical arrangement of the electrodes. Using a uniform half-space again, it is possible to compute geometrical correction factors, *k*, which convert readings of *R* into apparent resistivity values ρ_{app} by

$$k\frac{\Phi(\vec{r}_2) - \Phi(\vec{r}_1)}{I} = k\frac{U}{I} = kR = \rho_{app} \tag{8.2.22}$$

The apparent resistivity represents a weighted mean of the actual resistivity distribution $\rho(r)$. For ERT, the apparent resistivities provide the starting point for assessing the earth's true resistivity by means of inverse procedures. If the electrodes are placed on the surface, the geometric factor *k* is (e.g. Kearey et al. 2002)

$$k = \frac{4\pi}{\frac{1}{AM} - \frac{1}{AN} - \frac{1}{BM} + \frac{1}{BN}} \tag{8.2.23}$$

For current injections below the surface, e.g. electrodes in wells, the positions of the mirrored current electrodes A' and B' also have to be taken into account

$$k = \frac{4\pi}{\frac{1}{AM} + \frac{1}{A'M} - \frac{1}{AN} + \frac{1}{A'N} - \frac{1}{BM} + \frac{1}{B'M} - \frac{1}{BN} + \frac{1}{B'N}} \tag{8.2.24}$$

8.2.3.2 Geoelectric Rock Physics

Electrolytic ion transport is the most efficient conduction mechanism in fluid-filled sedimentary materials, in particular for those which are filled with highly saline brines. The efficiency of the ion transport is determined by the ion concentration in the fluid and the connectivity of the pores (Kirsch 2006). To first-order, porous sediments can be viewed as a composite system comprising the mineral matrix and the pore space. Similarly to the previous discussion, the pore-space may be filled with brine or CO_2 or a mixture of both. Since the electrical resistivity of most matrix-building minerals is high, their contribution to electric current flow is generally neglected. Using this assumption, the empirical Archie equation (Archie 1942) specifies the rock resistivity ρ with regard to the CO_2 saturation S_{CO_2} as

$$\rho = \frac{A\rho_w}{\phi^m (1 - S_{CO_2})^n} \tag{8.2.25}$$

where ϕ now denotes the rock porosity and ρ_w the resistivity of the initially present brine. The porosity exponent *m* reflects the pore geometry, compaction and insulation effects due to possible pore-space cementation. The saturation exponent *n* accounts for the presence of non-conductive fluid in the pore space. The factor *A* reflects the current component being conducted through the matrix. Since *A*, *m*, and *n* are purely empirical parameters, they need to be determined on an experimental or statistical basis. In situations where such a basis is not given, estimates often have to be made from literature values. For instance, the saturation exponent *n* is reported to be in the range of 1.715 for unconsolidated sediments up to 2.1661 for sandstones (Lee 2011). The porosity exponent *m* is reported to vary between 1.8 and 2.1 for sediments (Waxman and Thomas 1974).

Archie-based resistivity models make two crucial assumptions: First, the pore-space mixture of brine and CO_2 is substituted with a virtual equivalent fluid. Electric current flow, however, is affected by complicated geometrical considerations, such as shape and connectivity of the pores and the spatial distribution of these fluids within the pores. For example, Han et al. (2009) reported for fluid saturations <0.2, that the resistivity measured on clay-free sandstone can be notably lower than that predicted by Archie's law. They attributed this observation to liquid films that cover the rock grains and maintain considerable electric current flow even for very low fluid saturations. Secondly, the Archie equation assumes that electric current flow takes place solely within the pore-space. This is a severe simplification, because most sedimentary materials are also composed of conductive minerals, such as clay.

Impact of Clay Content on Rock Resistivity

Electric resistivity in clay-bearing geologic materials has been often studied (e.g. (Butler and Knight 1988; De Witte 1955; Poupon et al. 1954; Waxman and Thomas 1974) and various methods have been proposed to correct for the effect of clay on the formation resistivity (for an overview in the context of shaley sands see Worthington 1985). Frohlich and Parke (1989) extended the Archie equation to a parallel connection of the pore-space resistivity and the clay-related resistivity ρ_s

$$\frac{1}{\rho} = \frac{\phi^m (1 - S_{CO_2})^n}{A} \frac{1}{\rho_w} + \frac{1}{\rho_s} \qquad (8.2.26)$$

In fact, ρ_s is also dependent on the clay content, *cc,* for which Rhoades et al. (1989) presented an empirical calibration (that yields ρ_s in Ohmm)

$$\frac{1}{\rho_s} = \sigma_s = 20.3cc - 00.021 \qquad (8.2.27)$$

The Ketzin site can be used as an example of applying the Archie model as well as the Frohlich and Parke model to make first-order resistivity descriptions of the reservoir. Using an average clay content of about 20 % (Norden et al. 2010) within the reservoir units and assuming a porosity of 30 % a surface resistivity of 22.8 Ω is obtained from the Rhoades equation. In order to improve the choices of the remaining parameters, these can be adjusted by results from laboratory data. Based on two core samples from the Ketzin site, Kiessling et al. (2010) reported an average resistivity ρ_0 of about 0.5 Ω at full brine saturation. Still, *A* and *m* remain unknowns, but the ratio $\phi_m A^{-1}$, the so-called formation factor, can be considered as a single unknown. Thus, given experimental knowledge about ρ_w, ϕ and the rock's resistivity for full brine saturation ρ_0, any set of *A* and *m* can be chosen which satisfies

$$A(m) = \frac{\phi^m}{\rho_w} \left(\frac{1}{\rho_0} - \frac{1}{\rho_s} \right)^{-1} \qquad (8.2.28)$$

Selecting the porosity exponent *m* equal to 2, a value of about 1.24 is obtained for *A*. The respective models are shown in Fig. 8.6 and generally show a rather moderate increase in resistivity for $S_{CO_2} < 0.7$. For larger saturations a more drastic resistivity increase is observed with the respective maxima at full CO_2 saturation. This is a generic behavior of Archie models which has been well discussed regarding its potential for geoelectric monitoring of CO_2 migration (e.g. (Hoversten and Gasperikova 2005).

For CO_2 saturations up to about 70 %, clay has a rather neglible impact (Fig. 8.6). This can be explained by the (highly salinized) pore fluid which strongly exceeds the clay in terms of conductivity at low and intermediate CO_2 saturation. The difference between models is considerable for high CO_2 saturations. In such a

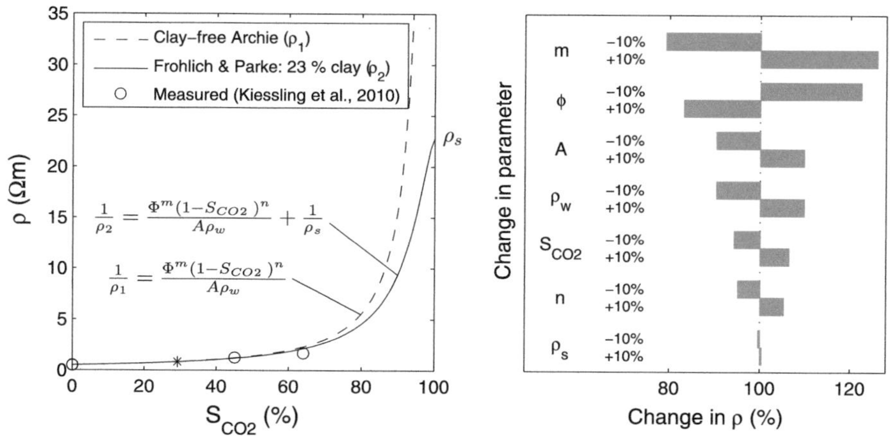

Fig. 8.6 *Left* Electric resistivity models as functions of the CO_2 saturation for the Ketzin reservoir model. A, m, and n were chosen to 1.24, 2 and 1.5, to fit experimental data reported by Kiessling et al. (2010). For further parameters of the reservoir model see text. *Right* Change in model resistivity due to uncertainties in the resistivity model parameters. Analysis had been carried out for the Frohlich and Parke model with a CO_2 saturation of 30 % (see *star symbol* in the *left hand* diagram). Estimation errors in the porosity exponent m and porosity ϕ can be seen to pose the strongest uncertainties on the resistivity predicted by the model

situation, the Archie model does not account for the clay-related conduction, whereas the Frohlich and Parke model allows some conductivity through the clay. Two considerations for geoelectric monitoring of CO_2 storage now arise. Firstly, ERT will be less sensitive at low CO_2 saturations, but will gain sensitivity as the CO_2 saturation increases. Secondly, if quantitative estimation of CO_2 saturations from resistivity measurements is performed in clay-bearing materials at high CO_2 saturations, the utilized resistivity-saturation relations should be based on in situ (laboratory) experiments or adequately calibrated clay models.

Assuming the Archie model to sufficiently describe the resistivity-saturation relation for the Ketzin reservoir within low and intermediate CO_2 saturations, the Archie equation can easily be used in reverse to estimate CO_2 saturations by

$$S_{CO_2} = 1 - RI^{-(1/n)} 1 - \left(\frac{\rho_0}{\rho}\right)^{1/n} \tag{8.2.29}$$

The use of the resistivity index *RI* (Guéguen and Palciauskas 1994) allows the substitution of ρ_w with ρ_0 by

$$RI = \frac{\rho_0}{\rho} = (1 - S_{CO_2})^{-n} \tag{8.2.30}$$

Note that the parameters A and ϕ_m become obsolete when using the ratio of two resistivities at different CO_2 saturations. The saturation exponent n is the only rock parameter required in this equation and with rearrangement of it resulting in

$$S_{CO_2} = 1 - RI^{-(1/n)} = 1 - \left(\frac{\rho_0}{\rho}\right)^{-(1/n)} \quad (8.2.31)$$

and allowing estimation of CO_2 saturation from a measured resistivity ρ and its baseline resistivity ρ_0. However, it is important to note that this assumes the porosity related parameters ϕ_m and A to be constant throughout the fluid-substitution process (no dissolution or cementation).

8.3 Tracer Tests for Monitoring CO_2 Plume Migration

Alexandru Tatomir, Iulia Ghergut and Martin Sauter

The success of CO_2 geological storage projects relies on technologies and capability to efficiently monitor the migration and fate of the injected CO_2 plume. Various types of tracers in the liquid and/or gas phases constitute one such monitoring technology. The overall goal of monitoring tracers is the characterization of processes occurring in the reservoir during and after the CO_2 injection. This involves the determination of the residual and dissolution trapping mechanisms and efficiency, the leakage pathways, the CO_2-brine interface area, extent of the CO_2 plume spreading, etc.

A summary of the various uses of the conservative and reactive tracers for monitoring purposes is given in Table 8.6.

The CO_2 can be present in several states depending on the temperature, pressure and chemical composition of the fluids in the reservoir which makes the assessment of the chemical tracers a challenging issue. CO_2 has been used on regular basis in the oil and gas industry to enhance the oil recovery. Among the most frequently used compounds are (see Table 8.6, Noordman et al. 2000): alcohols (Dwarakanath and Pope 1998), phase partitioning non-condensible gases such as O_2, CO_2, CH_4 (Elodie and Philippe 2012), noble gases (e.g. Kr; Vulava et al. 2002), volatile organic chemical compounds, fluorinated hydrocarbons (McCallum et al. 2005; Wells et al. 2007) and naturally occurring isotopes (e.g., 222Rn; Hunkeler et al. 1997), radioactive isotopes tracers (Johnson et al. 2011a, b).

Within the context of CCS projects, tracer methods can provide understanding over the subsurface movement of the CO_2 plume (Boreham et al. 2011; Freifeld et al. 2005; Underschultz et al. 2011; Vandeweijer et al. 2011), characterize geochemical processes (Assayag et al. 2009; Matter et al. 2007), assess the residual

Table 8.6 Use of tracers for monitoring during storage site operation

Target information	Suitable method	Suitable tracer species	Limitations	Application examples
Tracking or anticipating CO_2 breakthrough	Passive real-time monitoring of gas fluxes	Inert gases; optionally complemented by defined isotopic signatures for the injected CO_2 and accompanying tracers	Some gas species may arrive earlier, some later than the injected CO_2	Ketzin: Nitrogen, Krypton Frio: PFCs, SF_6, Kr Otway: SF_6, Kr, CD_4 (perdeuterated methane)
Tracking thermal fronts	Forced-gradient brine sampling	"Thermo-sensitive" tracers	Discussion on thermo-sensitive tracers in Chap. 6	Table 7.6
Estimating CO_2-brine interfaces	Passive monitoring	partitioning tracers	Requires accurate knowledge of partitioning coefficients, considering various in situ influences on them (temperature, salinity, etc.)	Fagerlund et al. (2013a) Tong et al. (2013) Myers et al. (2013a, b)
Dynamic tracking of CO_2-brine interfaces	Passive monitoring forced-gradient fluid sampling	"KIS" (kinetic interface-sensitive) tracers	Discussion on KIS tracers in Chap. 6	Schaffer et al. (2013) Tatomir et al. (2015)
Brine displacement	Passive monitoring forced-gradient brine sampling	Brine-soluble tracers partitioning tracers	Difficult to detect; extremely high dilution for brine-soluble species; forced-gradient brine sampling is not desirable during CCS site operation	Ketzin (attempted with brine-phase tracers like naphthalene-sulfonic and fluoro-benzoic acids); Behrens et al. (2014)
Cap-rock integrity	Passive monitoring	Geochemical and isotopic tracers, cation dissolution	Requires very sensitive detection (low detection limits); source attribution not always unique	Luquot and Gouze (2009)
Wellbore integrity	Passive monitoring	Geochemical and isotopic tracers, trace elements		Crow et al. (2010)

(continued)

Table 8.6 (continued)

Target information	Suitable method	Suitable tracer species	Limitations	Application examples
Residual trapping of CO_2	Forced-gradient push-pull sequence	Pressure signals—as an alternative to using solute tracers	Geological uncertainty	Zhang et al. (2011)
Residual and dissolution trapping	Inter-well dipole, forced-gradient	Partitioning tracers	Geological heterogeneity	Fagerlund et al. (2013a, b) Myers et al. (2013a, b), LaForce et al. (2014), Rasmusson et al. (2014)

trapping capacity (Myers et al. 2012; Rasmusson et al. 2014; Zhang et al. 2011; Fagerlund et al. 2013a, b). LaForce et al. (2014), determine the containment and leakage rates for monitoring and verification programs (Strazisar et al. 2009; Wells et al. 2010, 2007), measure the rate of CO_2 dissolution (Fagerlund et al. 2013a, b) or provide information about individual trapping mechanisms.

A recent literature review of several relevant CCS tracer applications is done by Myers et al. (2013a). They present case studies from West Pearl Queen, a depleted oil formation in southeastern New Mexico, and from Zero Emission Research Technology Center (ZERT) project in Bozeman, Montana (USA) where several cyclic perfluorocarbon tracers were used for monitoring and verification (Strazisar et al. 2009; Wells et al. 2007, 2010). Furthermore the Pembina Cardium project in Alberta (Canada) is presented, where distinct in situ isotopes were used as tracers for understanding breakthrough and plume migration (Johnson et al. 2011b). At the K12-B gas field in the Dutch sector of the North Sea and Frio Brine I pilot project, perfluorocarbons and inert gas tracers were used for understanding breakthrough and plume migration. At Kezin CO_2 injection site (Germany) the observations obtained from two monitoring wells showed that the two tracers, nitrogen and krypton have arrived prior to the CO_2 breakthrough (Elodie and Philippe 2012; Martens et al. 2011; Zimmer et al. 2011).

In general, when designing the tracer tests for the characterization and monitoring stages of a CO_2 storage site several key aspects must be considered (Myers et al. 2013a, b): chemical stability, environmental safety, cost effectiveness, ease of detection, toxicity, injection/sampling protocols and behavior in the reservoir conditions. Additionally, further restrictions have to be considered in the test design considering single-well and inter-well testing (Ghergut et al. 2013):

1. mobile-fluid volumes (e.g. CO_2 plume) can be measured from inter-well conservative-tracer tests, whereas single-well push-pull tests are generally insensitive to mobile-fluid volumes;
2. immobile-fluid volumes, in single-phase systems, are rather difficult to measure, by either kind of test;
3. fluid-phase saturations can be determined from inter-well tests using partitioning tracers at equilibrium exchange between phases; whereas single-well tracer push-pull tests are rather insensitive with respect to tracer exchange processes at equilibrium;
4. mobile and immobile fluid regions, or fluid-fluid interface areas can, in principle, be determined from single-well tracer push-pull tests relying on kinetic exchange processes between compartments or phases (Schaffer et al. 2013; Tatomir et al. 2013, 2015).

An interesting trade-off between the advective- or equilibrium-dominated parameter sensitivity regimes, and the advection- or equilibrium-insensitive regimes is obtained using in situ tracer creation, or conversion, in a time-dependent manner (from another initially-injected tracer with different phase-partitioning properties), as had originally been proposed by Tomich et al. (1973) for determining residual-oil saturations.

Reactive tracers constitute a promising research area for innovative process characterization, especially the thermo-sensitive tracers for tracking in situ temperature fronts, as proposed by Nottebohm et al. (2012), and the new class of KIS tracers (CO_2-brine interface sensitive), as proposed by Schaffer et al. (2013) and modeled in (Tong et al. 2013) and Tatomir et al. (2015).

Examples of tracer uses in terms of classical site characterization is given in Table 7.5, for characterizing the sites CO_2 trapping properties through CO_2 injection experiments in Sects. 7.4 and 8.6.

8.4 Well Instrumentation

Barry Freifeld

8.4.1 Objective of a Borehole Monitoring Program

The overarching objective of monitoring geologically sequestered CO_2 is to demonstrate the safe and effective long-term storage and integrity in the target reservoir. This is accomplished through a multi-faceted monitoring program by which data is acquired that (1) assures the public and regulators that the reservoir is behaving as intended, (2) validates conceptual models developed for reservoir engineering and storage management, and (3) demonstrates protection of drinking water and the greater environment. Dedicated monitoring wells are expensive and it is most efficient to use them to simultaneously acquire a diverse set of complementary data sets. The concept of integrated well monitoring is to engineer each completion such that tradeoffs necessary to deploy disparate technologies are considered in a holistic way so that the end result is an optimal suite of tools to accomplish monitoring objectives. Examples of integrated monitoring completions for CO_2 storage are available in both demonstration and pre-commercial deployments. These include the Ktzi 200, 201 and 202 boreholes at the Ketzin pilot site, Germany (Prevedel et al. 2008), the CO_2CRC Otway Project Naylor-1 Well (Jenkins et al. 2012), and the Modular Borehole Monitoring system developed at Lawrence Berkeley National Lab for the D9-8 Well (CO_2 Capture Project 2013).

8.4.2 *Monitoring Environmental Challenges*

The geologic reservoirs targeted for geological storage of CO_2 are normally at pressures and temperatures above CO_2's critical point: 31.1 °C at a pressure of 73.8 bar. These temperatures and pressures are typically found at depths greater than ~750 m, and sequestration pilots have often been at 2–3 km depth where multiple sealing layers provide redundant barriers to migration of CO_2 to the surface. There are many engineering challenges associated with the environmental conditions encountered, which include elevated pressures, temperatures, and aggressive groundwater chemistries to name a few.

The depth of the target reservoir and the corresponding hydrostatic pressure provides a significant challenge to the design and survivability of complex monitoring instruments. A 3000 m deep well will develop around 300 bar static pressure at bottom. This can be even greater depending on the salinity of the fluid.

In addition to pressure, elevated temperatures present additional engineering challenges for MVA (monitoring, verification and accounting) tool design. Downhole electronics experience increasing rates of failure at temperatures above 100 °C, with lifetimes of downhole electronic circuitry decreasing nonlinearly with increased temperatures. A study conducted by Quartzdyne, Inc. a major manufacturer and OEM supplier of quartz crystal and electronic circuit boards for permanent pressure/temperature gauges found that surface mounted electronics could be used reliably at up to 150 °C, with lifetimes of 5 years at 125 °C. Hybrid electric circuitry assemblies can last up to two years at 200 °C or five years at 180 °C (Watts 2003). These durations are frequently much shorter than would be expected during a permanent CO_2 monitoring program, and hence some means for removal and replacement of electronic based sensors would be needed for a "life-of-the-well" solution.

Similarly, fiber-optics also suffer from degradation at elevated temperatures. Standard acrylate coated optical fibers are rated for use up to 85 °C, with high temperature acrylate fibers acceptable for extended usage at 150 °C. Polyimide coatings are used at temperatures up to 300 °C, while difficult to manufacture fibers using metallic coatings are available beyond this temperature. Two of the challenges that metallic coated fibers face is in the reliable fabrication of long lengths and the difficulty in recoating after splicing. Optical fibers in general suffer a condition known as hydrogen darkening at elevated temperatures, where hydrogen diffuses into the fiber and degrades the optical characteristics. In high temperature boreholes (>200 °C) with hydrocarbons present, the diffusion of hydrogen into fibers can be severe and seriously degrade the life of a fiber-optic cable in the timespan of several months (Rassenfoss 2012).

Corrosion and chemical resistance of the materials selected for downhole use in the MBM (modular borehole monitoring) system is an important consideration and is related to the temperature issue because of the exponential dependence of reaction rates on temperature. Deep sedimentary aquifers, often rich in dissolved salts, are considered the largest potential targets of CO_2 sequestration. Monitoring in wells used for fluid sampling means exposure to CO_2 rich fluids. CO_2 dissolved in formation waters will form carbonic acid, with the resulting acidity determined by the host formations buffering capability. Acidic waters form a hostile environment to most ferritic materials commonly used in well completion. To mitigate potential high corrosion rates carbon steel is often replaced by high chromium alloys, which in turn increases well costs. Fiberglass is an alternative casing material to consider, but the structural integrity needs to be considered in deep well installations, particularly in designing cementing operations that limit compressive forces.

8.4.3 Monitoring Technologies

Many of the technologies that have been employed for monitoring CO_2 sequestration sites are derived from the oil and gas industry. These include permanent pressure and temperature gauges, fiber-optic temperature, acoustic, and strain, as well as numerous wireline logging technologies. For geophysical logging there has not been a broad adoption of permanent sensing in the oil and gas industry, but there have been examples of in well electrical and seismic sensor arrays. Permanent microseismic sensing has frequently been employed for monitoring unconventional hydraulic fracturing operations. Downhole fluid sampling in the oil and gas industry is typically performed using wireline tools to acquire accurate PVT information during the reservoir appraisal process, as wellhead samples are normally used after a well is put into production. For continuous monitoring of brines for CO_2 sequestration alternative methods have been developed such as U-tube fluid sampling (Freifeld et al. 2005) or Schlumberger's Westbay Multilevel sampling system (Picard et al. 2011).

8.4.3.1 Pressure/Temperature

Subsurface pressure and temperature are fundamental parameters used in all reservoir models. Hydrologic testing requires knowledge of the evolution of a pressure transient during fluid injection or withdrawal in order to assess a reservoirs permeability and storativity (see Chaps. 3 and 8 for definitions). In a CO_2 storage reservoir having pressure gauges deployed both at the bottom and top of a perforated interval permits an estimate of the fluid density, and hence the height of a column of CO_2 in brine.

Permanently deployed discrete pressure/temperature gauges are commercially mature products with dozens of vendors that will supply and install the instruments. Pressure gauges operate using a variety of measurement methods, with the deep well environment sensors dominated by piezoresistive and quartz gauge technology. Resonating quartz cells are considered the most stable and accurate. Data from permanent gauges are typically read out at the surface through single conductor

TEC (tubing encapsulated conductor). Alternatively, memory gauges can be installed in side pocket mandrels and retrieved periodically to download data and replace batteries. The benefit of retrievable gauges is that they can be replaced upon a gauge failure, whereas a permanent gauge with surface readout cannot be replaced if it fails. Because of the high value of real-time data early in the life of a project it is possible to install permanent gauges that may fail in five or ten years, but with either side-pockets or landing subs that would allow easy deployment of retrievable gauges in the future.

8.4.3.2 Fluid Sampling

There are numerous methods for obtaining subsurface fluid samples, including wireline samplers, formation testers, gas lift systems, and U-tube samplers (Freifeld et al. 2005). For fluid samples from two-phase reservoirs, such as exist in mixed brine CO_2 systems, methods that preserve the relative ratio of the separate phases are preferred as they provide information deemed important to understanding the state of the reservoir. Electrical pumps and gas lift significantly distort the composition of the fluid, and hence downhole wireline and U-tube samplers are the preferred techniques for monitoring CO_2 sequestration reservoirs. A comparison of all of these sampling methods was conducted at the Citronelle field site by a team led by Yousif Kharaka, USGS Menlo Park. Unpublished results showed that the wireline and U-tube samples provided the least disturbed dissolved gas chemistry, resulting in more representative samples than submersible pumps and gas lifting fluids.

Additional tools have been developed by major oilfield service provides for sampling fluids through casing. This involves creating a hole, extracting fluid, and repairing the hole. As expected these tools are highly specialized and carry significant costs to mobilize and use. They however can provide one of the few methods by which suspected leakage above zone can be investigated.

If it is known in advance that fluid samples are required to be collected above the reservoir, there are a couple of different experimental methods by which a permanent sampling system can be installed outside of the casing. As part of the PTRC (Petroleum Technology Research Centre) Aquistore Project, a cement diverter has been installed with a U-tube sampling port and fluid sampling lines cemented outside of casing. To date, the performance of the system is unknown as it has not been function tested since installation, which occurred shortly before the writing of this report.

An alternative method is to deploy a U-tube as part of a behind casing perforation system. Behind casing perforation systems have been used to couple discrete pressure/temperature gauges to the formation. This works by installing a hollow perforation charge carrier connected through capillary tube to the pressure sensor. The perforations create a fluid pathway between the formation and the pressure gauge. This type of device has been marketed by several companies including Promore, Houston, TX and Sage Rider, Rosharon TX. Alternatively this same deployment method can be used to couple the formation to a U-tube fluid sampler.

8.4.4 Fiber Optic Technologies

8.4.4.1 State-of Sensor Technology

Fiber optic based sensor systems are either distributed, based upon Raman or Brillouin scatter or discrete or multi-point, based upon Fabry-Perot cavities or Fiber Bragg Gratings (FBGs). Distributed temperature sensing is by far the most widely adopted well monitoring technique, having been first developed in the early 1980s at the Southampton University in England. The technique was commercialized initially by York Sensors Ltd and several other companies including Sensortran, Sensornet, LIOS Technology and APSensing (a spin-off from Agilent Systems) have since developed commercial products. Performance specifications for RAMAN based DTS systems are usually a function of the overall cable length and the integration period for each measurement cycle, with spatial resolutions typically 15 cm to 1 m and temperature resolution as high as 0.01 °C.

Brillouin based temperature monitoring systems typically have lower measurement resolution and accuracy than Raman Systems, but because strain induced variations in optical properties can be decoupled from the temperature measurements, the technique is less susceptible to noise induced by strain on the cables. Because the Brillouin technique uses low loss single-mode fiber it can be operated at ranges as long as 100 km. Brillouin measurements use single mode fiber in comparison to the multimode fiber employed for Raman based temperature measurement. Brillouin sensing is also used for monitoring fiber-strain. Typical sensitivity limits for stain are from 2 $\mu\varepsilon$ to 10 $\mu\varepsilon$ up to as high as 4 % strain depending on the cable material. One difficulty in monitoring strain is the challenge of transferring environmental strain onto the cable in a way that accurately transfers the strain but does not degrade the environmental integrity of the fiber-optic cable encapsulation, which needs to still resist the elevated pressures of the deep subsurface environment. This is still an area of active research. FBG strain sensors are more commonly deployed to monitor strain at discrete locations because of the difficulty of imparting strain onto a continuous fiber. Baker Hughes and Shell jointly developed an FBG based real-time compaction imaging system to monitor sand screen deformation and casing shape which used FBG strain sensors.

A technology that is more recent than DTS, but has rapidly evolved in only a few years is distributed acoustic sensing (DAS). Discrete fiber-optic based geophone sensors have been marketed for many years based on FBG technology. However, there was little commercial uptake of the technology as the advantage over conventional copper wire based geophone sensors was not significant enough to overcome the price for utilizing the fiber-optic technology. DAS uses commercial grade single-mode telecom fibers to monitor with high spatial resolution (up to 1 m) to provide truly distributed sensing over kilometers of cable.

Fiber-optic DTS monitoring specifically for CO_2 sequestration has been deployed at the CO_2SINK site at Ketzin, Germany (Giese et al. 2009), the CO_2CRC Otway Project and the SECARB Cranfield Site, in Mississippi (Daley et al. 2013) and at the Quest project in Alberta, Canada. Both the CO_2SINK and Otway Project sites deployed a variant of passive DTS monitoring, referred to as heat-pulse monitoring (Freifeld et al. 2008) which provides for the creation of a thermal pulse to investigate the thermophysical setting of the near wellbore environment.

Many technologies have been developed for borehole deployment as stand-alone measurements. We will consider these to the extent they could possibly be integrated into a modular deployment. A good example is strain. Current fiber optic technology, typically used for distributed temperature sensing, is being applied to strain measurements. Current measurement sensitivity is sufficient for sensing casing damage.

8.4.5 Instrumentation Deployment Strategies

There are several different methods for installing instrumentation in boreholes, but by far the most common method is run-in-hole on tubing, where the instruments sit in the annular space between tubing and casing. The hardware associated with a tubing deployment has a mature supply chain, and the engineering expertise is readily available. Less common but still considered relatively mature is behind casing installation. In a behind casing installation the instruments sit outside of the casing, allowing the full interior space within the well to be available for temporary deployments. The deployments at the Ketzin pilot site were an example of a hybrid installation, where some instruments sat outside of the casing and others were affixed to tubing (Prevedel et al. 2008). Considered as experimental techniques are coiled tubing installations and wireline/umbilical installation of instruments.

8.4.5.1 Tubing

In many ways tubing instrumentation deployments are operationally similar to ESP (Electrical Submersible Pump) deployments, as the specialized equipment to protect and run-in-hole with instrumentation control lines are identical. Specialized vendors are required to oversee the installation and operation of their particular instruments and a spooling operator coordinates with the rig floor workers for the installation of mandrels, clamps, and bands during the installation. The wellhead will need to accommodate control lines feeding through the tubing hanger and out through the tubing head adapter flange. Tubing deployment of instruments is more common than installation outside of casing, and the variety of vendors and service organizations with familiarity with the process is greater. However tubing deployment lacks the benefit of behind casing sampling for sensors that require close contact to the formation, particularly seismic and electrical sensors.

8.4.5.2 Cemented Outside Casing

As part of standard techniques within the oil and gas industry, methods for instrumenting the outside of a well casing with control lines that are cemented in place have been developed. The installation of DTS cables outside of casing provides a real-time and continuous evaluation of cement operations, allowing the concentration of cement to be assessed by its exothermic curing process. Other instrumentation can be deployed on casing as part of an MVA effort. Many MVA tools such as ERT (Electrical Resistivity Tomography), seismic sensors, samplers, etc., have been installed using casing deployment in demonstration programs such as the Ketzin pilot site and SECARB's Cranfield DAS test in Cranfield, Mississippi. There are several significant benefits to deployments of instruments behind casing, which includes leaving the wellbore available for wireline logging and other temporary tool deployments and better coupling to the formation for seismic or electrical sensors. The entire deployment of instrumentation on casing requires the use of specialized subcontractors that have experience in completion operations that are modified to accommodate the physical presence of the instrumentation.

While casing deployment is similar in many ways to tubing deployment, as spooling units and control line protectors are also used, there are numerous complexities that arise that are not encountered with tubing deployment. The cementing operation of the casing has to take into consideration the damage that could occur during casing movement which is used to improve the cement job. Rotation of the casing is not permitted, however reciprocation can usually still be performed. Perforation needs to be performed in such a way as to mitigate the risk of the perforation charges damaging the instruments. One way to do this is to install behind casing charges which are aimed away from the instruments. This method has most frequently been used for the installation of behind casing pressure/temperature sensors. If the perforation will be performed after cementing than some method for oriented perforating as well as "blast shield" or other protective housings placed over critical instruments are usually employed.

8.4.5.3 Coiled Tubing (CT)

A coiled tubing rig is potentially more economical than a standard workover rig used for conventional tubing deployment. Deployment is more rapid because joints don't have to be made up and there are no control line protectors to be positioned on each joint. However the engineering for instrumented deployments using coiled tubing is far less mature than for convention tubing deployment, and the availability of CT rigs and specialized personal considerably lower leading to large variability in the ability to performed instrumented CT deployments. An example of a service provider offering instrumented CT is Precise Downhole Services Ltd., located in Nisku, Alberta, Canada. To date there has not been a CO_2 monitoring well completed with instrumented coil tubing, although a temporary seismic hydrophone cable was deployed at Weyburn with CT.

8.4.5.4 Wireline/Umbilical

An umbilical system as used in subsea applications that runs from platform to wellhead could bridge the gap between flatpack coiled tubing and standard wireline deployment. CJS Production Technologies, Calgary Alberta, Canada, have been commercializing an umbilical style flat-pack. They have modified a conventional CT rig to use rectangular shaped push blocks that can grip and deploy a rectangular umbilical. More significantly, they have worked on methodologies for performing pressure control, which is one of the significant engineering challenges in an umbilical style deployment. The flat-pack at Citronelle dome is really a hybridization of a conventional tubing deployment with a flat-pack encapsulated instrumentation bundle. Problems that CJS Production Technologies have encountered include leakage between the encapsulant material and the instrumentation lines as well as the need to engineer highly customized wellhead components.

8.4.5.5 Deployment Pressure Control Issues

For both casing and tubing deployment pressure control is critical. Pressure control must be maintained at all times in open hole casing deployment and for tubing deployment in a perforated well. For completed wells this means having the previously mentioned zonal isolation at some depth above the perforations (such as a packer or seal bore) or a well head with a gate valve. All such zonal isolation requires more engineering when monitoring control lines need to be passed through seals. While running in well, often only 'kill-fluid' (high density fluid) is primary well control, with secondary control additional devices such as a hydril, blind ram or shear ram as part of a BOP stack.

8.4.6 Example of an Integrated Monitoring Installation: Heletz H18a

8.4.6.1 Project Background

Heletz is a depleted oil field, filled with brine at its edges. The site is instrumented for scientific CO_2 injection experiments (Niemi et al. 2016). The Heletz H18a is one of two wells drilled in the frame of the EU-FP7 funded MUSTANG project on the characterization of deep saline formations for the storage of CO_2. The two wells were installed into the saline aquifer part of the formation with the objective to develop field scale methods for assessing the capacity and safety of a CO_2 storage reservoir using a combination of both single-well and cross-well experimental tests. The H18a well was drilled from January to May of 2012 to a total depth of 1649 m. The well was perforated through two of three sandstone intervals at depths of 1627–1629 m and 1632–1641 m.

8.4.6.2 H18a Integrated Monitoring Well

The technologies chosen for the H18a injection well include U-tube fluid sampling, permanent quartz pressure/temperature gauges and an integrated fiber-optic bundle to facilitate temperature, seismic, and heat-pulse monitoring. In addition, a chemical injection mandrel and gas lift mandrel facilitate both push-pull injection testing and production of fluids by artificial gas lift. Figure 8.7 provides a schematic layout of the borehole completion package. The primary tubing is 2–7/8″ 6.5 ppf L-80 RTS-8 with an internal coating of Tuboscope TK-805 to improve resistance to exposure to carbonic acid from conventional carbon steel. The 2–7/8″ tubing permits conducting periodic logging campaigns using industry standard 1–11/16″ slim-hole tools.

8.4.6.3 Packer and Overshot Design

In considering zonal isolation for the bottom hole assembly (BHA) both inflatable and hydraulic set packers have been used in the past. Inflatable packers are generally considered not as reliable since any slight leak that develops in the gland or seals can lead to deflation, and the multi-year life required of the completion string requires the highest dependable installation possible. Mechanical set packers require twisting of the string which is not permitted at the packer because of the three control lines that pass through the seal location. For Heletz H18a, a hydraulic set packer coupled with an overshot to connect the tailpiece to the packer was selected for coupling the BHA to the support string based upon recommendations by Denbury Resources and experience they have in long-life installations.

The packer selected was a D&L Hydroset II Packer, which is a hydraulic set, mechanically held dual string packer with asymmetric short and long string connections. The 2–7/8″ long string connection was used for the production tubing while the smaller 1.900 EUE facilitates pass-throughs for the fiber-optic, pressure/temperature gauge, and U-tube sampling lines. Figure 8.8 shows the dual-mandrel packer with an inset picture highlighting the pass-throughs that penetrate the short string coupling. An overshot was used to couple the tailpipe to the packer to avoid twisting the lines running through the packer.

8.4.6.4 H18a Installation

The installation was conducted by running a work string into H18a with a casing scrapper and then circulating 30 m^3 of fluid once on bottom. Starting with the reentry guide, the bottom-hole assembly was assembled and the control lines and pressure/temperature gauges installed on special instrumentation mandrels. Pneumatic spooling units are used to tension the control lines as they were led over a multi-line sheave hung off the derrick board (Fig. 8.9). Total time to install the integrated monitoring completion was two and a half days for well and equipment

SAGE RIDER

Company: Class 6 Solutions Heletz Mon	Date: 27-Jan-14	Temp: 130 °F	Gun Size: 0"	SPF: 0	Phasing: 0°	Casing Size/Wt: 7
Well Name: H18 Injection	Field: 0	Expl. Type: 0	Fluid Wt. 8.5#	Fluid Type 2% KCL	Job Type: Horizontal TCP	
County/Parish: 0	State: 0	Completion Intervals: VI		Firing Device #1: 0	Firing Device #2: 0	

Item	Description	O. D.	I. D.	Length	Depth Feet	Depth Meters
	Run 162 total joints and 1 perf joint					-
24	Tubing to Surface	3.660	2.441	2,929.55	-	-
23	Chemical Injection Sub	3.660	2.441	4.09	2,929.55	892.93
22	10 Joints 2 7/8" Pipe Joint 63-72	3.660	2.441	325.73	2,933.64	894.17
21	Gas Lift Mandrel	3.660	2.441	4.09	3,259.37	993.46
20	58 Joints 2 7/8" Pipe Joints 4-62	3.660	2.441	1,922.02	3,263.46	994.70
19	Marker Sub	3.660	2.441	6.30	5,185.48	1,580.53
18	1 Joint 2 7/8" Tubing #3	3.660	2.441	32.57	5,191.78	1,582.45
17	X-Nipple	3.660	1.875	0.99	5,224.35	1,592.38
16	1 Joint 2 7/8 Tubing #2	3.660	2.441	32.59	5,225.34	1,592.68
15	7" 23# Dual String Packer	6.062	2.441	7.83	5,257.93	1,602.62
14	1 Joint 2 7/8" Tubing #1	3.660	2.441	32.58	5,265.76	1,605.00
13	On/Off Tool w/ X profile	4.500	1.875	1.85	5,298.34	1,614.93
12	X-O Pup	3.660	2.441	6.09	5,300.19	1,615.50
11	Dual Gauge Carrier	4.850	2.441	5.67	5,306.28	1,617.35
10	Sample Filter Carrier	3.990	2.441	3.00	5,311.95	1,619.08
9	X-O 2 7/8" NU Box X 2 3/8" Pin	3.660	2.100	0.40	5,314.95	1,620.00
8	Pump out plug	3.063	1.875	0.68	5,315.35	1,620.12
7	X-O 2 3/8" Box X 2 7/8" Pin and 2' pup	3.660	1.995	2.67	5,316.03	1,620.33
6	1 Joint 2 7/8" Perforated Pipe	3.660	2.441	32.55	5,318.70	1,621.14
5	X-O Pup	3.660	2.441	6.08	5,351.25	1,631.06
4	Dual Gauge Carrier	4.850	2.441	5.67	5,357.33	1,632.91
3	X-O Pup and Fiber Termination Sub	3.660	2.441	5.62	5,363.00	1,634.64
2	Wireline Re-entry Guide and XN Nipple	3.660	2.441	1.38	5,368.62	1,636.36
1	Bottom of W/L Re-entry Guide				5,370.00	1,636.78
						-

Comments: Perforations:	Total Length of BHA =	2440.45	743.85
	Total BHA w/o Pipe=	94.96	28.94

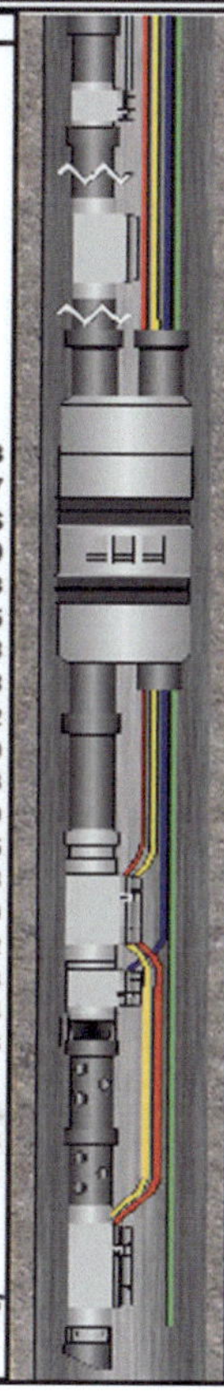

Fig. 8.7 Borehole completion package for the Heletz Site H18a injection well

preparation, one day to assemble the bottom assembly and run-in-hole and a final day to complete the well head and install surface lines and equipment.

8.4.7 Conclusions

A variety of permanent monitoring technologies can be engineered for installation into a single integrated package for comprehensively monitoring a CO_2 storage site. Well designs exist that facilitate simultaneous geophysical monitoring, permanent

Fig. 8.8 D&L dual-mandrel hydraulic set packer with short string fitted with adapter to seal around control lines using compression fittings

discrete instrument gauges, and repeat wireline logging. While some technologies such as permanent pressure/temperature gauges have been available for decades, new and emerging technologies such as distributed fiber-optic acoustic sensing are making rapid strides in becoming accepted technology and have been demonstrated in carbon sequestration pilot tests. Given the requisite long duration for a CO_2 monitoring program only the most robust technologies and carefully selected materials and installation methods will provide life-of-the-field solutions.

Fig. 8.9 Workover operation in progress at H18a showing rig with double stands of tubing and pneumatic spooling units used to tension control lines as they are fastened to the tubing

8.5 Monitoring Results from Selected Large Scale Field Projects

Larry Myer

The following sections summarize the findings from monitoring programs at selected, major, large scale CO_2 storage projects, worldwide, which have made significant technical contributions toward enabling broad, global, geologic storage of CO_2. The projects discussed are: Sleipner, offshore saline formation storage, Europe; In Salah, onshore saline formation storage, Africa; and Weyburn-Midale, onshore EOR/storage, North America.

8.5.1 *Sleipner*

8.5.1.1 Project Overview

The Sleipner CO_2 storage project is the world's longest running geologic storage project. Since 1996, approximately 1 M tons of CO_2 per year have been injected from a single well drilled into the saline water-saturated Utsira Formation (Alnes et al. 2011). The Sleipner storage project is being carried out in conjunction with a commercial natural gas production project operated by Statoil. Located about 240 km off the coast of Norway in the North Sea, natural gas is produced from the Sleipner West field from a reservoir below the Utsira. In order for the natural gas to meet the sales gas specification, its CO_2 content is reduced from about 9 % down to 2.5 % (Nooner et al. 2007).

The regional geometry of the Utsira and overlying units was well defined from interpretation of nearly 14,000 line kilometres of 2D seismic data and over 300 wells (Chadwick et al. 2000). The Utsira sand is a tabular, basin-restricted unit stretching about 450 km from north to south and 40–90 km west to east. It lies at depths of about 800–1100 m below the sea floor with a thickness of about 250 m around the injection site (Arts et al. 2008). Overlying the Utsira sand is the Nordland shale, which, in the Sleipner area is between 200 and 300 m think (Arts et al. 2008). Immediately overlying the sand is a shale drape, which is a tabular, basin-restricted, seal (Chadwick et al. 2000). The Utsira sand is poorly consolidated, highly porous (30–40 %) and very permeable (1–3 Darcy) (Arts et al. 2008). The very high permeability, high porosity, and large reservoir volume has resulted in negligible pressure increases in the reservoir.

8.5.1.2 Seismic Monitoring

At Sleipner, the primary monitoring method has been time-lapse 3-D seismic. It is a very important case history because Sleipner was the first project to clearly demonstrate the potential of seismic surveys for monitoring CO_2 storage. By 2010, nine 3-D surveys had been carried out, with the first, in 1994 providing the pre-injection baseline. The time-lapse seismic results clearly show the steady expansion of the plume over time. The results also show that the expansion is affected by mudstone layers in the reservoir, leading to new understanding of the effects of internal reservoir structure and heterogeneity on plume movement (Fig. 8.10). Well logs revealed the presence of the thin (on the order of one meter thickness), laterally discontinuous mudstone layers, but they were not visible in the pre-injection seismic data (collected in 1994) and their significance not recognized until the first repeat 3D seismic survey carried out in 1999. That survey showed reflections from CO_2 in a stack of layers, which were then correlated with the mudstone layers observed in the well logs. A seismic reflection would be expected from increases in the acoustic impedance contrast between sandstone and a

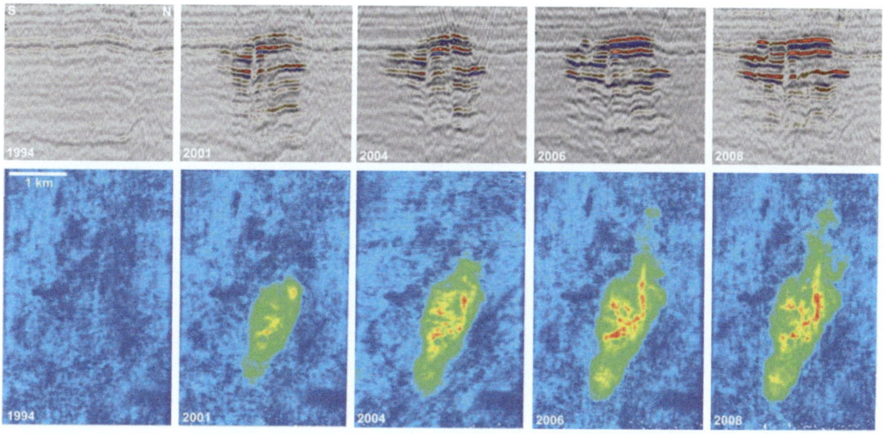

Fig. 8.10 Time-lapse seismic images of the Sleipner CO_2 plume—NS inline through the plume (*top*); plan view of total reflection amplitude in the plume (*bottom*) (Chadwick et al. 2010)

mudstone layer, resulting from high saturations of CO_2 accumulating at the top of the sandstone layer. The mudstone layers baffle the upward migration of the CO_2 within the reservoir, having a significant effect on the storage efficiency of the reservoir.

In addition to the effects of mudstone baffles, seismic data from Sleipner also show that the expansion of the plume is significantly influenced by the topography of the interface between the sand reservoir and the caprock. This interface undulates, creating topographic highs. Under buoyancy drive, the CO_2 fills one high spot before spilling laterally to fill the next. Seismic reflection amplitude maps of the topmost layer show that CO_2 first reached the reservoir top in 1999, as two small separate accumulations within a local topographic dome. It then spilled northwards along a prominent north-trending linear ridge before entering a more vaguely defined northerly topographic high. Lateral migration was particularly rapid along the linear ridge where the CO_2 front advanced northwards at about 1 m per day between 2001 and 2004 (Chadwick and Noy 2010).

Boait et al. (2012) extended previous analyses by detailed mapping of the seismic data acquired between 1999 and 2008. The mapping revealed nine distinct reflective horizons. In each horizon, the area of reflectivity, interpreted as the CO_2 plume, is roughly elliptical with eccentricities ranging between two and four. In the top half of the reservoir, the interpreted plume grows linearly with time. In the bottom half, the interpreted plume initially grows linearly for about eight years and then progressively shrinks. The detailed analysis of Boait et al. (2012) also found a decrease in reflectivity over time in the central portion of several of the horizons. This was interpreted as being caused by flow of CO_2 between layers.

The Sleipner seismic dataset has also been valuable for testing of methods for quantitative interpretation/analysis of plume characteristics. Eiken et al. (2011) reported that the sum of the seismic amplitudes was observed to track linearly with the volume of CO_2 injected. Chadwick et al. (2010) performed prestack and poststack inversion and found that prestack inversion provided improved characterization of the sand unit between the reservoir top and the uppermost intra-reservoir mudstone. They used specialized spectral decomposition algorithms to identify frequency tuning, from which CO_2 layer thicknesses could be derived. They found that AVO analysis to estimate CO_2 layer thickness proved challenging, in part because the CO_2 layers are thin. They also used a technology called extrema classification (Borgos et al. 2003) in order to better detect and map the intra-reservoir mudstones.

Finally, the plume migration shown by the seismic data has also been used as a basis for validation and refinement of numerical reservoir simulators (Bickle et al. 2007; Cavanagh 2013; Chadwick and Noy 2010; Estublier et al. 2013; Fornel and Estublier 2013; Nilsen et al. 2011; Singh et al. 2010). These studies involved conventional simulators based on Darcy flow, as well as invasion percolation simulation, which assumes that gravity and capillary forces dominate flow. Results show that available simulators are able to reproduce the Sleipner plume migration reasonably well, but the layering, which produce thin plumes with large differences in horizontal and vertical dimensions, and the complex topology of the flow paths, create challenges.

8.5.1.3 Other Monitoring at Sleipner

Sleipner is also the first project to employ gravity methods as part of the monitoring program. Gravity measurements have much lower spatial resolution than seismic measurements. However, gravity can provide information in situations where seismic methods do not work as well, and gravity measurements can be used to assess the amount of dissolved CO_2, to which seismic measurements are insensitive.

At Sleipner, precision gravity measurements were carried out using a ROVDOG (Remotely Operated Vehicle deployable Deep Ocean Gravimeter) at 30 seafloor stations above the CO_2 plume in the years 2002, 2005, and 2009 (Alnes et al. 2011; Nooner et al. 2007). About 5.88 million tons of CO_2 had been injected over this time period. Inversion for average density using geometry constraints from seismic gave 675–715 kg/m^3 for the density of the separate phase CO_2 in the reservoir. Combining this with temperature measurements, Alnes et al. (2011) concluded that the rate of dissolution of the CO_2 into the water did not exceed 1.8 % per year.

A Controlled Source Electromagnetic (CSEM) survey was carried out in 2008 (Eiken et al. 2011) using conventional surface-to-surface techniques. Modeling by Park et al. (2013) showed that the expected resistivity anomaly is around 5 % and probably close to the noise level of surface-to-surface CSEM data. Their modeling results also suggest, however, that the surface-to-borehole CSEM survey could provide high sensitivity data, opening a new possibility of applying CSEM to CO_2 reservoir monitoring in the future.

8.5.2 In Salah

8.5.2.1 Project Overview

The In Salah project, located at the Krechba Central Processing Facility in the central Algeria Sahara, is a joint venture of BP, Statoil, and Sonatrach. It is a commercial natural gas production project in which CO_2 is removed from the natural gas in order to meet the gas export specification of 0.3 % CO_2. The CO_2 content of the natural gas is 5–10 % (Ringrose et al. 2009). From 2004 to 2010 more than 3 million tons were stored. An extensive monitoring program was undertaken, both to meet the commercial needs of the project, and to support development of monitoring technologies. Monitoring was provided by the Joint Industry Project (JIP). This project is an interesting case history because of the unique monitoring technologies applied.

The Krechba Carboniferous reservoir is a sandstone rock which on average is 20 m thick with a porosity of about 13 % and a permeability of 10 mD (Ringrose et al. 2009). Structurally, it is a four-way dip (dome-like) closure, in which the hydrocarbons have accumulated at the high part of the dome. Down-dip from the natural gas, the rock was saturated with saline water, and the CO_2 was injected in this portion of the reservoir at a depth of about 1950 m. There were three CO_2 injection wells at Krechba, injecting up to about 2800 metric tons per day of CO_2. The reservoir is overlain by about 900 m of mudstone rock which acts as a seal against vertical migration of both the natural gas and the CO_2.

8.5.2.2 Monitoring at In Salah

The JIP mentioned above was set up in 2005 to monitor the CO_2 storage process using a variety of geochemical, geophysical, and production techniques (see Table 8.7) over a 5-year period. To help select monitoring technologies, the JIP also used a "Boston Square," which allows a comparison of techniques based on two criteria—cost and benefit to the project (Ringrose et al. 2013). Of the 29 monitoring technology options assessed, repeat 3D surface seismic technology had the highest benefit but also the highest cost. The use of surface seismic technology is challenging at Krechba. The Krechba sandstone storage domain is onshore, deep, thin, and has low porosity and permeability compared to other sites, such as Sleipner.

An extensive 3D seismic survey was carried out at Krechba in 1997. This survey defined the overall structure of the reservoir and provided information about its internal architecture and distribution of the sandy portions with the best porosity and permeability, but no significant faults were identified in this survey (Iding and Ringrose 2009). In 2002, when drilling began in the development phase of the project, it became evident that fractures and faults could play a role in production and injection operations. Data from the wells suggested that the injection horizon

Table 8.7 In Salah monitoring and verification technologies (adapted from Mathieson et al. 2011)

Monitoring technology	Application	Comment
Repeat 3D seismic	Plume migration	Initial survey in 1997
	Subsurface characterization	High resolution repeat 3D survey acquired in 2009
Microseismic	Caprock integrity	500 m test well drilled and recording information above KB502
InSAR monitoring	Plume migration	Images captured using X-band (8 days) and C-band (32 days)
	Caprock integrity	
	Pressure development	Used to develop time lapse deformation images
Tiltmeters/GPS	Plume migration	Used to calibrate satellite data
	Caprock integrity	
	Pressure development	
Shallow aquifer wells	Caprock integrity	5 wells drilled to 350 m—one beside each injector, one remote and one between KB5 and KB502
	Potable aquifer contamination	
Wellhead/annulus samples	Wellbore integrity	2 monthly sampling beginning 2005
	Plume migration	
Tracers	Plume migration	Different perflourocarbon tracers into each injector
Surface flux/soil gas	Surface seepage	Initial survey pre-injection
		Two surveys in 2009
Microbiology	Surface seepage	First samples collected in late. 2009/early 2010
Wireline logging/sampling	Sub-surface characterization	Overburden samples and logs in new wells
	Geomechanical and geochemical modeling	

and the immediate overburden are naturally fractured with a preferred NW-SE orientation (Iding and Ringrose 2009). A repeat 3D seismic survey was acquired in 2009 with improved shot spacing and fold to gain better imaging of the storage interval and caprock sequence. Two NW-SE trending linear features in the vicinity of the KB502 and KB503 CO_2 injectors were observed as slight depressions (velocity/amplitude pull-downs) on the 2009 3D seismic. These features are aligned with the dominant fracture orientation as identified in the well data.

When seismic is challenging, other methods take on additional importance in a monitoring program to provide information on the behavior of the plume. At In Salah, these methods included ground surface displacement measurement, annulus and wellhead monitoring, including tracer analysis and pressure monitoring, combined with history matching.

The tracer monitoring approach involved injection of small amounts of perfluorocarbons along with the CO_2, and sampling of well bore fluids in observation wells. Different perfluorocarbons were used to 'tag' the CO_2 injected at each injection well, so that any CO_2 detected can be differentiated from the natural CO_2 in the subsurface and traced back to an individual injection well. The results of the perfluorocarbon tracer measurements confirmed that the CO_2 migrated from the injection well KB502 to well KB5, further demonstrating the impact of the NW-SE preferred fracture direction on plume migration.

Surface displacement measurements are a unique and significant aspect of the monitoring program at In Salah, which is the first application of satellite InSAR technology for monitoring of geologic storage. InSAR, which stands for satellite airborne radar interferometry, detects changes in elevation at the earth's surface. Injection of the CO_2 causes an increase in the pore pressure in the reservoir, and that pressure increase results in small displacements at the ground surface above the reservoir. The amount of surface displacement depends on the magnitude of the pressure, as well as geometry of the pressurized region, depth, and rock properties.

One advantage of InSAR data is the relative low cost and ease of acquisition compared to seismic data. The satellite is collecting data all the time, so the frequency with which data is available for a specific site is related to the orbit of the satellite and how often it passes over the site of interest.

A major challenge in application of this technology is to be able to resolve the very small surface displacements associated with CO_2 injection. At In Salah, the surface uplift due to CO_2 injection was about 3–5 mm/year, compared to approximately 200 mm per day due to earth tides. Methodologies for processing the satellite data to obtain higher resolution displacement measurements continue to evolve. PSInSAR (Permanent Scatterer InSAR), which has been applied at In Salah, gives an accuracy of around 5 mm/year and up to 1 mm/year for a longer term average (Ringrose et al. 2009).

Integration of InSAR data with geomechanical models, along with seismic and fracture data, provided important additional understanding of the impact of fractures/faults on plume migration at In Salah. InSAR data collected in 2006 and 2007 showed that, above the active injection wells, there was surface uplift, which was elongated and extended several km in the direction of the well KB5. The location and orientation of the uplift was found to be well aligned with the NW-SE trending linear features in the 2009 3D seismic data (Fig. 8.11). The uplift above injector KB-502 was in the shape of a double lobe, which Vasco et al. (2010) showed could be caused by opening of a vertical fracture extending above and below the injection zone. Evaluation of the rate and pattern of surface uplift at In Salah and its relationship to fluid pressure changes and fractures in the subsurface, has been the subject of several geomechancial modeling studies including Vasco et al. (2008), Rutqvist et al. (2010), and Gemmer et al. (2012).

Limited microseismic monitoring was also carried out at In Salah. A set of vertical 3-component geophones was deployed in a microseismic pilot well drilled to a depth of 500 m directly above the trajectory path of the KB-502 injection well. P-S arrival times, shear wave polarisation and time series analyses indicated that most of the observed events (over 1000 microseismic events) were related to CO_2 injection (Oye et al. 2013). Event location was very limited because the array was limited to a single pilot well, but analysis of the microseismic waveform data using

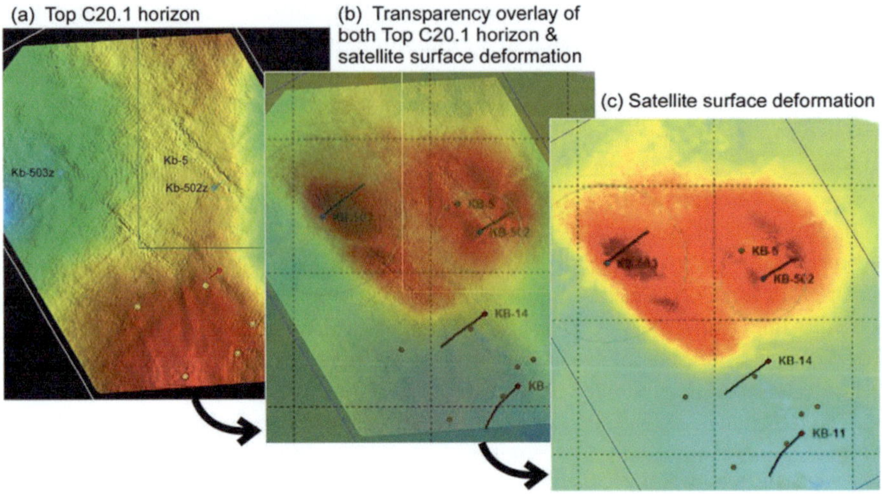

Fig. 8.11 NW-SE linear features seen on 2009 3D seismic data compared with InSAR surface deformation data (Ringrose et al. 2013)

cross-correlation techniques indicated that most events occurred within distinct spatial clusters (Oye et al. 2013).

Assessment of the microseismic data in combination with the other monitoring data discussed above has led to the conclusion (Ringrose et al. 2013), that CO_2 injection at In Salah stimulated natural fractures, and may have introduced new hydraulic fractures, in the vicinity of injection well KB-502. While analyses indicate that these fractures did propagate upwards into the lower caprock, it is considered unlikely that they propagated further through the upper caprock (Ringrose et al. 2013).

8.5.2.3 Other In Salah Monitoring Results

In addition to 3D seismic and InSAR measurements, data acquisition prior to injection included extensive sampling and logging programs (including image logs) in the new development wells, saline aquifer sampling and headspace gas sampling through the overburden, soil gas surveys around each of the new wells, and soil gas sampling from the shallow aquifer water wells.

Shallow soil gas and flux measurements were difficult because of the hard ground, but not impossible. Loose sand and gravel was found where the ground was not hard and these loose materials also presented difficulties because of the potential for contamination of samples due to movement of atmospheric gases through the highly permeable materials. Despite these difficulties, elevated CO_2 soil gas and flux measurements were observed near the KB-5 well, as would be expected given the breakthrough of CO_2 at the well (Jones et al. 2011).

Soil gas and flux measurements also provided some data on background CO_2 levels in a harsh desert environment. In comparison with more vegetated sites from temperate regions, the soil gas values were found to be lower by at least an order of magnitude compared with vegetated sites from temperate regions.

Changes which occur in plants in response to elevated levels of CO_2 in the soil, though not a direct measure of CO_2, are considered to be another indicator of leaking CO_2. Finally, it has been proposed that elevated levels of CO_2 in the soil might also affect microbial communities. At In Salah, vegetative cover is very low, commonly 10 % or less, though somewhat higher in topographic lows, reflecting the desert environment. Some of the plants represented species which might be affected by CO_2, if exposed to elevated soil gas concentrations. Microbial populations were also low, but were present (Jones et al. 2011).

8.5.3 Weyburn-Midale

8.5.3.1 Project Overview

The IEA GHG Weyburn-Midale CO_2 Monitoring and Storage Project began in 2000 in close collaboration with EnCana, which is the operator of the CO_2 EOR project in the Weyburn Field in Saskatchewan, Canada. While CO_2 EOR is considered a commercial technology, this project is unique because of its research focus on storage in conjunction with EOR. The Weyburn CO_2 EOR flood is likely the most intensely studied operation of its kind in the world.

The CO_2 EOR reservoir is the Midale beds of the Charles Formation. The Midale consists of a layer <30 m thick of fractured carbonate rock at a depth of about 1500 m. The reservoir is comprised of vuggy limestone ("Vuggy") and overlying marly dolostone ("Marly"). The reservoir is overlain by a seal of evaporate rocks (anhydritic dolostones and anhydrites). Above these are a series of additional sealing formations, including the Lower Watrous Member, which forms the most extensive primary seal to the Weyburn system (Whittaker 2004).

The Midale reservoir has been under oil production for decades. At the end of primary production in 1964, water flooding was begun to enhance production. Further field development, including application of horizontal wells, began in 1991 (Preston et al. 2005), and CO_2 injection began in 2000. By the end of 2011, a total of 21 M tons of CO_2 had been stored in the Weyburn-Midale field with total field injection rates of approximately 13 k tons per day (White 2013a). The CO_2 (a byproduct of gasification of lignite) is purchased from the Dakota Gasification synthetic fuel plant in Beulah, North Dakota, and transported through a 320 km pipeline to Weyburn.

8.5.3.2 Monitoring Activities in the Weyburn-Midale Project

Monitoring methodologies investigated as part of the project included:

- geochemical fluid sampling,
- surface seismic, augmented by VSP,
- passive seismic,
- shallow well monitoring and sampling,
- soil gas surveys, and
- tracers.

A comprehensive set of papers presenting results of Weyburn-Midale Project monitoring efforts has been published elsewhere (Wildgust et al. 2013). Some of the key findings are discussed below.

Unique among the global storage demonstrations is the geochemical fluid sampling campaign at Weyburn. A baseline geochemical sampling survey was followed by sampling on 16 occasions over the course of two time periods—from 2000 to 2004, and from 2008 to 2010. Wellhead fluid and gas samples from about 50 wells were analyzed for over 40 compositional and isotopic parameters, generating a unique, comprehensive database. The spatial and temporal changes in pH, alkalinity, concentrations of Ca and Mg, and carbon isotopes were found useful in monitoring the movement and fate of the CO_2 in the subsurface and providing indication of incipient CO_2 breakthrough at wells (Emberley et al. 2005; Gunter and Perkins 2004). The results show that geochemical monitoring provides valuable information for identifying the time scales required for solubility and ionic trapping of injected CO_2. Solubility trapping, i.e. the formation of H_2CO_3, was observed within six months of the onset of CO_2 injection and ionic trapping, i.e. the reaction of CO_2 with carbonate minerals, commenced within one year of injection (Shevalier et al. 2013). Results also showed that brine resistivity can be used to indirectly track the movement of injected CO_2 within the reservoir (Shevalier et al. 2013).

Advances in the application of surface seismic technology for monitoring were also made in the Weyburn-Midale project. The Midale reservoir was a challenging surface seismic because of its thinness and rock properties, but, through application of advanced acquisition and processing methods, it proved successful. 3D, three-component, time-lapse seismic data were acquired over a portion of the project area in 1999 (baseline survey), 2001, 2002, 2004, and 2007. Waveform correlation techniques were used during post-stack normalization of the time-lapse seismic data and then subsequently in determining the amplitude and travel time variations (White 2013a, b). The effects of CO_2 injection and oil production were observed in the 3D difference seismic volumes in both the amplitude changes for reflections from the reservoir, and in travel time changes for travel paths through the reservoir. Figure 8.12 is a plan map of the seismic amplitude changes which shows the spatial correlation between the amplitude changes and CO_2 injectors, and the temporal increase in area of these changes as injection increased. Comparison of the time-lapse seismic results to reservoir flow simulations demonstrated a clear

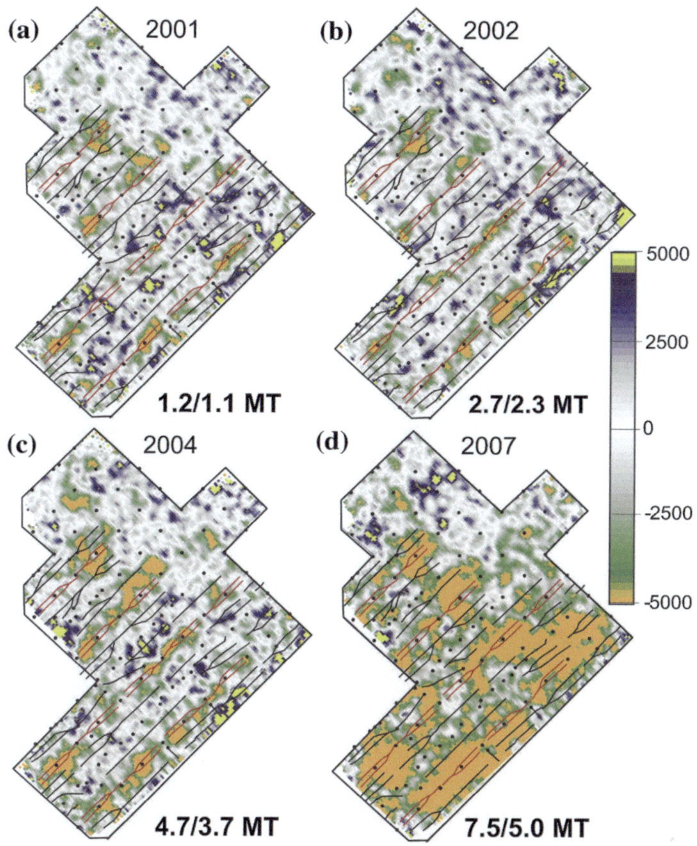

Fig. 8.12 Seismic amplitude difference maps at the Marly reservoir level for time-lapse surveys from 2001 to 2007. Estimates of the quantities of CO_2 injected and CO_2 stored (injected amount minus produced amount) are shown for each time. The amplitudes in each panel are scaled the same and are unitless. Horizontal CO_2 injection wells are shown in *red* whereas horizontal production wells are shown in *black* (White 2013a, b)

correlation between the injection-related reservoir changes and the resultant seismic response (White 2013a, b).

Additional processing (Meadows and Cole 2013), combined with extensive laboratory testing and rock physics modeling (Meadows 2013; Njiekak et al. 2013) was carried out to more quantitatively interpret the time-lapse seismic data at Weyburn. Prestack seismic data was migrated and used in an impedance inversion to obtain P- and S-wave impedance volumes. The resulting time-lapse impedance changes were input, along with the rock physics model, into a direct inversion algorithm to generate volumes of pore pressure and CO_2 saturation changes over time (Meadows and Cole 2013).

The 3D time-lapse seismic data was also analyzed to evaluate caprock integrity and to look for CO_2 which might have migrated vertically from the reservoir (White

2011, 2013a). White (2013a) used log-based fluid substitution modeling for calculating the seismic sensitivity to the presence of CO_2 within the various intervals, accounting for the effects of rock lithology, porosity, pore pressure and temperature. His analyses indicated that the maximum estimated proportion of CO_2 residing in either interval above the regional seal was ≤ 1 %, and the maximum amount of CO_2 potentially residing above the regional seal by 2007 was <56,000 tons.

Though limited in array size, the Weyburn Project was the first large-scale CCS pilot project to include downhole microseismic recording as a monitoring tool. The microseismicity was low in rate and intensity. During the monitoring period of 2003–2011, approximately 200 microseismic events were located (Verdon et al. 2013). The microseismicity predominantly occurred in episodic temporal clusters that were linked to specific operational field activities. Verdon et al. (2011) carried out coupled fluid flow-geomechanical simulations and found that stress changes induced by deformation of the reservoir were transferred into the overburden, leading to an increase in shear stress above the production wells. This stress transfer into the overburden has been interpreted (Verdon et al. 2013) as the likely cause of the events located in the overburden above the producing wells.

8.5.4 Discussion of Field Study Results

The results of these field projects have clearly yielded many advances in geologic storage through validation and demonstration of monitoring technologies. A diverse set of technologies for measurements at the surface and in the subsurface have been field tested. Technologies conventionally used by the oil and gas industry have been validated for application to monitoring of CO_2, and some unique new technologies have been demonstrated. The successful application of seismic techniques for monitoring the movement of CO_2 in the reservoir was clearly demonstrated. Positive results were obtained not only under "ideal" conditions like those at Sleipner, but elsewhere, under more challenging conditions associated with thin, deep, reservoirs. Studies also used seismic measurements to demonstrate that CO_2 has not migrated above confining zones. Though more work is needed, these studies have also provided some insight on the leakage volume detection threshold of surface seismic methods.

Though seismic methods have the highest resolution of the geophysical monitoring methods, it is clear that there are some circumstances where their applicability is limited, and the field studies also showed that other methods can provide complimentary information to improve understanding of plume behavior. Field projects (those discussed above and others) have now demonstrated successful use of satellite-based surface deformation, gravimetry, and electrical techniques, though more work is needed to better determine how broadly applicable they will be. Based on field performance in major pilots and some modeling, Fabriol et al. (2011) have offered the following comparative assessment of seismic, electrical, and gravimetric techniques (Table 8.8).

In addition to geophysical monitoring, the field tests have also demonstrated the value of other types of monitoring measurements, including pressure and temperature, tracers, fluid sampling for geochemical analyses, and well logs of many kinds. Geophysical monitoring is generally considered to be the most expensive type of monitoring, and it is noted that cost effective technologies such as wellhead and annulus monitoring were also proven to be useful.

While the field tests indicate that a portfolio of monitoring technologies is available, they do not yield a single prescriptive list of technologies which are applicable, or necessarily sufficient, for all situations. In fact, the experience to date suggests that monitoring programs will need to be developed to accommodate the unique geology, and risks, associated with each site.

Another observation, not only related to monitoring, but more generally to overall technical management of storage operations, is that injection strategies and monitoring plans need to be adaptable, and should be expected to evolve as experience and monitoring data become available during operation of the project. In none of the reviewed projects was the behavior of the CO_2 in the reservoir exactly as predicted before injection began.

The overall experience represented by the reviewed projects shows that monitoring of geologic storage of CO_2 is technologically feasible in a diverse set of geologic environments. Given the geology-specific nature of the technology, this experience is not sufficient, however, to draw conclusions about all geologic environments. Further work is needed to assess the technical feasibility across the spectrum of depositional environments that might be considered. In addition, very

Table 8.8 Comparison of performance of geophysical monitoring methods (adapted from (Fabriol et al. 2011)

Method	Minimum quantity for verification at reservoir depth (>800 m)	Minimum quantity for leakage detection at reservoir depth	Secondary reservoir detection (at depth ca. 200–300 m)	Minimum quantity in theory detectable in secondary reservoir	Geological limitations specific to CO_2 storage
4D seismic	Hundreds of km	Few km	Yes	Few hundreds of tons	Reservoir: low porosity, thin layers (tuning effect)
Electrical CSEM	1 Mt Few tens of ktons (at Ketzin at 600–700 m deep)	Not yet proved	Yes	Few tens of ktons	Low resistivity, thin layers (either resistive or conductive)
Gravimetry	1 Mt	Not yet proved	Yes	Few tens of ktons	Seasonal surface variations

little data has been developed about the post-injection behavior of CO_2 in the reservoir. The same monitoring tools used during the operational phase of storage are applicable to post-injection phase, but field demonstrations of the processes that lead to plume stabilization and long-term trapping are needed.

8.6 Pilot Scale CO_2 Injection and Monitoring: Frio Site

Christine Doughty

The Frio brine pilot, a research project conducted at the South Liberty oil field operated by Texas American Resources in Dayton, Texas, USA, injected 1600 metric tons of CO_2 over a period of 10 days into a steeply dipping brine-saturated sand layer at a depth of 1500 m (Hovorka et al. 2006). The pilot employed one injection well and one observation well. Pre-injection activities (see Chap. 7) included review of the regional geological setting, development of a detailed local geological model, analysis of wireline logs, laboratory analysis of core samples, collection and chemical analysis of brine samples, pressure-transient analysis of an interference well test, and breakthrough curve analysis for a two-well recirculation tracer test. During CO_2 injection, pressure transients were monitored at both wells and downhole fluid samples were collected frequently at the observation well. Geophysical monitoring of CO_2 movement in the subsurface during and after the injection period provided information on the spatial distribution of CO_2 at several different scales. Frio brine pilot activities are outlined in Table 8.9, and are described in the following sections. Table 8.10 summarizes the material properties and formation conditions inferred from traditional site characterization. This Chapter highlights results from the Frio Brine Pilot monitoring campaign and associated modeling, with an emphasis on lessons learned for future GCS activities. More details on monitoring and modeling of the Frio Brine Pilot may be found in Hovorka et al. (2006), Kharaka et al. (2006), Doughty et al. (2008), Daley et al. (2008, 2011) and Xu et al. (2010).

8.6.1 Geologic Setting and Development of Geologic Model

The Oligocene Frio Formation is an extensive sedimentary formation in the Gulf Coast of the USA, with ample capacity for GCS in thick sandstone layers separated by shale seals, underlying an abundance of CO_2 point sources. The Frio Formation extends over much of the Texas Gulf Coast ($\sim$500 km), at depths ranging from 0 to 3000 m. For depths greater than 800 m (below which CO_2 is supercritical at Frio), porosity ranges from 0.25 to 0.30, net sand thickness is 100–500 m, and permeability is 100 md to 5 d. This setting makes the Frio Formation a logical choice for an early pilot of GCS.

Table 8.9 Activities at the Frio brine pilot

Activity	Monitoring	Information obtained
Review existing data related to historical oil production	3D seismic	Structure of sand and shale layers surrounding salt dome
	Wireline logs in regionally distributed wells	Compartmentalization into fault blocks
Well log analysis	Wireline logs in injection and observation wells	Target sand layer and overlying shale caprock
		Extent, continuity, and variability of layers
		Permeability, porosity, relative permeability parameters (estimated using literature correlations)
Core analysis from newly drilled injection well	Porosity	Calibration for well-log estimates of porosity and permeability
	Permeability	
	Mercury injection	Capillary pressure/saturation relationship
Interference well test	Pressure transients	Inter-well connectivity
		Flow properties of lateral boundaries
		Field-scale permeability
		Estimates of pressure increase during CO_2 injection
Aqueous-phase tracer test	Fluorescein breakthrough curve (BTC)	Single-phase dispersivity
		Porosity-thickness product of sand layer
CO_2 injection and post-injection rest period	CO_2 arrival at observation well	Average CO_2 saturation between wells
	Pressure transients	Two-phase flow properties
	RST (reservoir saturation tool)	CO_2 saturation profiles at injection and observation wells
	Crosswell seismic	CO_2 distribution between injection and observation wells
	VSP (vertical seismic profile)	CO_2 distribution updip of observation well
Two-phase tracer test (concurrent with CO_2 injection)	Two-phase tracer BTC	Two-phase dispersivity
		Evolution of CO_2 saturation distribution with time

In the vicinity of the South Liberty field, the fluvial-deltaic Frio Formation is overlain by the regionally extensive, low permeability Anahuac shale, which acts as a regional upper seal for the Frio Formation sands. Individual sand layers, identified as A, B, C, etc., are separated by more localized shale layers that also serve as barriers to flow. At the South Liberty field, numerous wells drilled for historical oil production at depths around 2400 m provide structural information about the site (Hovorka et al. 2006). The brine-saturated sand layer targeted for CO_2 storage, the C sand, is near the top of the Frio Formation at a depth of 1500 m and is on the flank of a salt dome (Fig. 8.13), where the Frio Formation is laterally compartmentalized by faults (Fig. 8.14). A new injection well was drilled for the Frio brine pilot, sited 32-m down dip from an existing well that served as the observation well. The fault block in which the wells lie is about 800 m across and at least 2500 m

Table 8.10 Material properties and formation conditions obtained from traditional site-characterization

Property	Method	Value	Comments
Injection interval (thickness of high-permeability clean sand), h	Wireline logs	5.5 m	Figure 8.15b
	Match tracer test BTC	8 m	Figure 8.17 Thickness varied, porosity held fixed at 0.34
Porosity, ϕ	Wireline logs calibrated to core analysis	0.34 average over 5.5-m injection interval 0.28 average over 23-m thick C sand	Figure 8.15a
Permeability, k	Wireline logs calibrated to core analysis	2264 md average over 5.5-m injection interval 1001 md average over 23-m thick C sand	Figure 8.15b
	Interference well test	Around wells: kh product consistent with wireline logs, low vertical permeability below clean sand creates leaky aquifer	Figure 8.16 Average k = 1556 md for an 8-m thick layer
Hydrologic properties of small fault	Interference well test	No hydrologic effect	Figure 8.16
Formation compressibility, C	Interference well test	1.28×10^{-9} Pa^{-1}	Figure 8.16 Defined as $C = (1/\phi)\partial\phi/\partial P\|_T$
Capillary pressure parameters	Mercury injection on core samples	van Genuchten (1980) parameters: P_{c0} = 6500 Pa, n = 1.7, S_{lmin} = 0.03 for a sample with k = 837 md	Curve-fit to drainage curves, also used for imbibition curves $P_{c0} \sim k^{-1/2}$
Dispersivity, α	Match tracer-test BTC	0.1 m	Figure 8.17
Pressure, P	Downhole sensor	152 bars	
Temperature, T	Wireline logs, fluid sampling	59 °C	Average of 56–65 °C range obtained for different measurements
Salinity	Fluid sampling	0.093 mg/L	

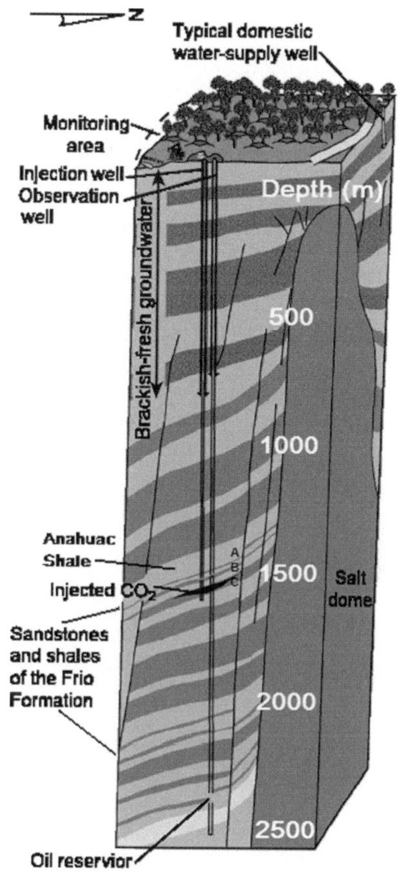

Fig. 8.13 Schematic of the Frio brine pilot site (after Hovorka et al. 2006)

long, and is bounded by mapped faults to the northwest and southeast. Several smaller intra-block faults also exist.

Figure 8.15 shows the porosity and horizontal permeability profiles for the C sand inferred from injection-well logs, with calibration to porosity and permeability measurements made on core samples (Sakurai et al. 2006). Observation-well logs (not shown) contain similar features, suggesting good layer continuity between the two wells. The ratio of vertical to horizontal permeability is assumed to be an increasing function of porosity, and ranges from 0.1 to 1.0. Each well was perforated over approximately 6 m in the upper portion of the 23-m thick C sand, which Fig. 8.15 identifies as a thick interval of clean sand. The lower limit of the injection interval is delineated by a thin marker bed, which is interpreted as low-permeability shale. Capillary pressure as a function of saturation was measured for two core samples, one sandstone and one shale, using mercury injection (Sakurai et al. 2006).

Frio Formation brine samples were collected from both wells at a series of times

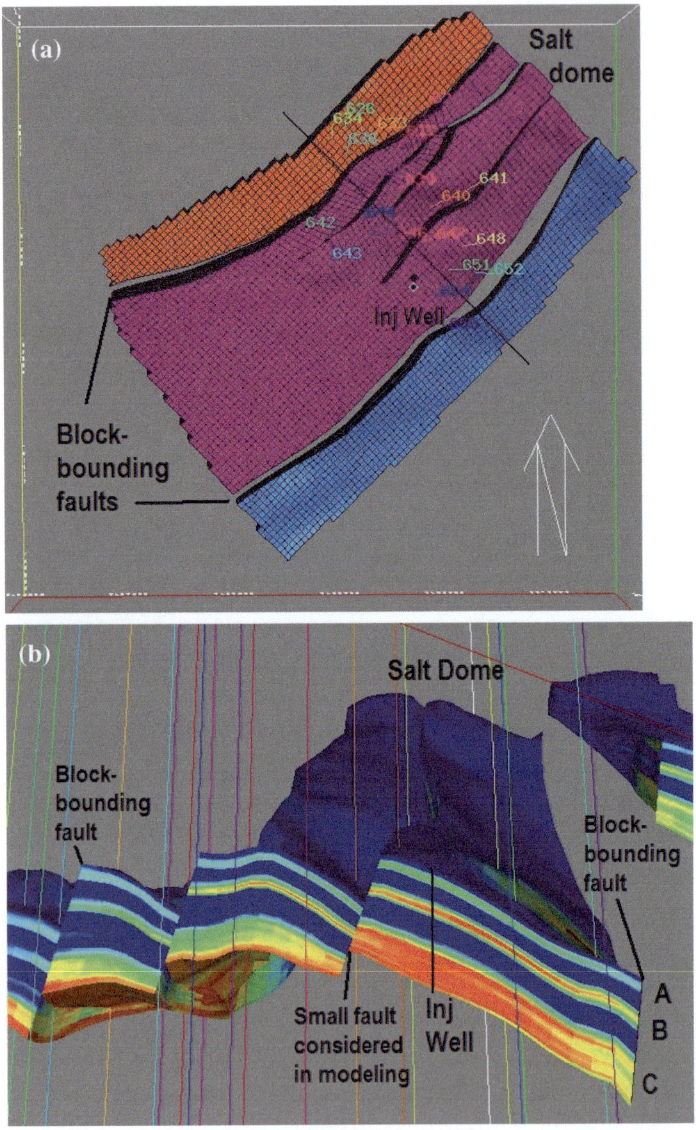

Fig. 8.14 Geological model of the upper Frio Formation at the South Liberty field (courtesy of Joseph Yeh, TBEG). **a** Plan view of the modeled fault block (*pink*) and two adjacent fault blocks. The new injection well is shown as a *black dot with a white border*. The existing observation well is the *black dot* just to the north. **b** Vertical cut through the model along the *black line* shown in the plan view. Vertical exaggeration is approximately a factor of two. *Colored lines* and *numbers* identify wells used to create the geological model

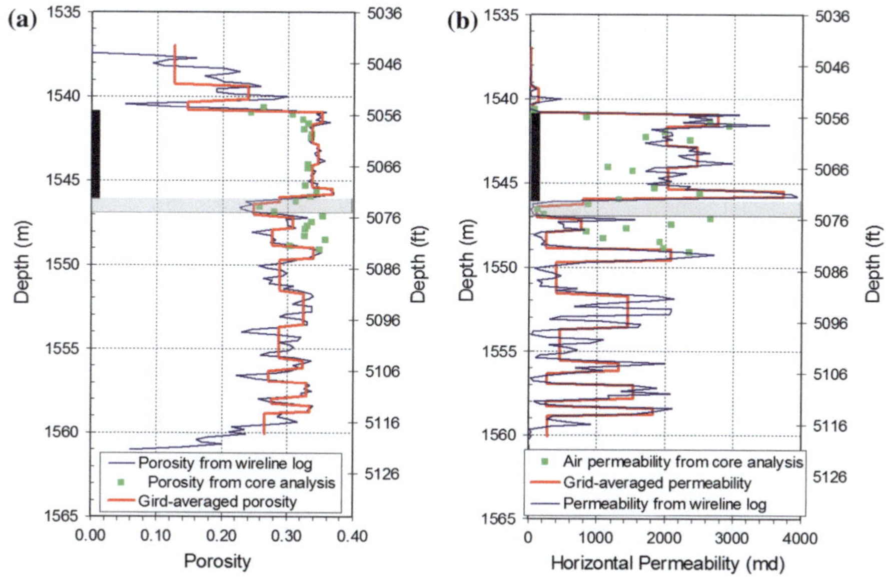

Fig. 8.15 Injection-well property profiles for the C sand (courtesy of Shinichi Sakurai, TBEG), and the grid-averaged values used for the original 3D numerical model: **a** porosity and **b** horizontal permeability

before CO_2 injection. Chemical analysis identified a Na-Ca-Cl-type brine with 93,000 mg/L total dissolved solids (TDS), nearly saturated with methane (CH_4) at formation conditions of about 150 bars and 60 °C (Kharaka et al. 2006). These dissolved salt and methane contents are typical of brine formations found in the vicinity of petroleum resources in the northern Gulf of Mexico basin (Kharaka and Hanor 2003).

8.6.2 Site Characterization

8.6.2.1 Numerical Model for Flow and Transport

Based on the geological model shown in Fig. 8.14, a three-dimensional numerical model employing the TOUGH2 simulator (Pruess et al. 1999) was developed to simulate flow and transport for the Frio brine pilot, and is described in more detail in Doughty et al. (2008). The model represents the C sand and extends over the entire fault block in which the injection and observation wells lie.

8.6.2.2 Interference Test

A 24-h interference well test was conducted by pumping from the observation well and observing pressure changes in both wells. Figure 8.16 shows the pressure transient at the injection well and several modeled pressure transients obtained with different analysis methods. The simplest method is to match the pressure transient to an analytical solution (Theis 1935), which assumes a uniformly thick, homogeneous, flat layer of infinite radial extent that is perfectly sealed above and below. For the early-times behavior a good match to the pressure-transient data could be obtained with the Theis solution (Fig. 8.16), but the linear Theis curve begins to deviate from the field data after about one-half hour of pumping, predicting too large a pressure drawdown.

The pressure drawdown obtained by simulating the interference test with the 3D model becomes too small after only about 100 s ("Original Model" in Fig. 8.16), but has the linear shape of the Theis solution. Figure 8.16 shows that the data do not have the linear character of the Theis solution, but rather show a marked flattening. Such a response is characteristic of a leaky aquifer, in which fluid flows to the pumped well from both the pumped interval and to a lesser extent from above and/or below (Hantush and Jacob 1955), see also Chap. 7. From Fig. 8.15b it can be seen that the permeability just above the pumped interval is very low, but that the permeability of the marker bed below the pumped interval is moderate. By decreasing the vertical permeability of the marker bed from 3 md to 0.25 md, the pressure drawdown labeled "Leaky 5.5-m sand, no small fault" in Fig. 8.16 is obtained, in which the observed flattening beginning at 0.5 h is much better represented.

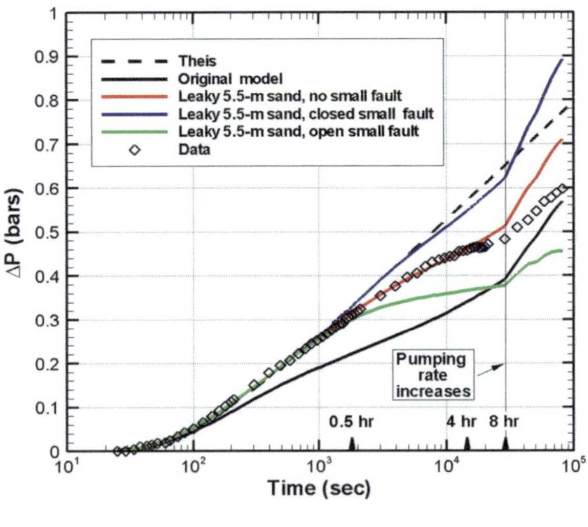

Fig. 8.16 Interference-well-test pressure transient (Doughty et al. 2008) and various model results

To investigate the hydrologic nature of the intersection of the small fault northwest of the observation well and the C sand, three distinct fault conditions are considered with the 3D model: (1) the fault is absent, (2) the fault is a closed boundary, and (3) the fault is a constant-pressure boundary (open small fault). The effect of the small fault is felt about 0.5 h after pumping begins. Both the closed-fault case and the constant-pressure-fault case diverge sharply from the field data, suggesting that the fault does not act as either of these types of boundaries.

There is independent evidence from the tracer test that the thickness of the high-permeability clean sand in which the injection well is perforated is about 8 m, significantly larger than the 5.5 m inferred from the well logs. Because the pressure drawdown response to pumping is primarily sensitive to the permeability-thickness product of a formation, the interference-test data shown in Fig. 8.16 can be equally well matched by a model with a thickness of 8 m, if permeability is decreased correspondingly. The modified model is referred to as the "8-m sand model" and the previous model is referred to as the "5.5-m sand model".

In summary, matching the injection-well (non-pumped well) pressure transient during the interference well test does not provide a single, uniquely-determined hydrologic model of the C sand, but rather two models that bound a range of reasonable models. The clean sand acts as a leaky aquifer, and sensitivity-study results indicate that the small fault within the main fault block should not be considered either a closed or constant-pressure boundary, therefore it was not included in further modeling studies.

8.6.2.3 Tracer Test

After the interference well test had run for 24 h, pumped fluid was reinjected into the injection well, to create a balanced doublet flow field. After another 24 h, when the flow field was steady, a 78-min pulse of fluorescein dye was added at the injection well. The steady flows were maintained for 15 days. Fluorescein arrived at the observation well after 9 days and concentration peaked at 12 days, as shown in Fig. 8.17. The primary parameters to be inferred from the tracer breakthrough curve (BTC) are the porosity-thickness product of the sand layer through which fluid flows and the aqueous-phase dispersion coefficient for the sand, a measure of its heterogeneity. Preliminary attempts to model the tracer test using the 3D numerical model described in the previous section were unsatisfactory, because numerical dispersion smeared out the tracer peak too much for a physical dispersion coefficient to be determined. Therefore, a streamline model (Javandel et al. 1984) is used, with a random walk added to represent dispersion. Figure 8.17 presents modeled breakthrough curves for two different sand-layer thicknesses: the 5.5-m value inferred from the injection well logs (Fig. 8.15), which results in a far too early peak, and an 8-m value, determined by trial and error to best match the observed fluorescein peak time of 12 days. In both cases, porosity is held fixed at 0.34, the average value obtained from wireline logs and core analysis. Matching the width and height of the observed fluorescein peak requires a small single-phase

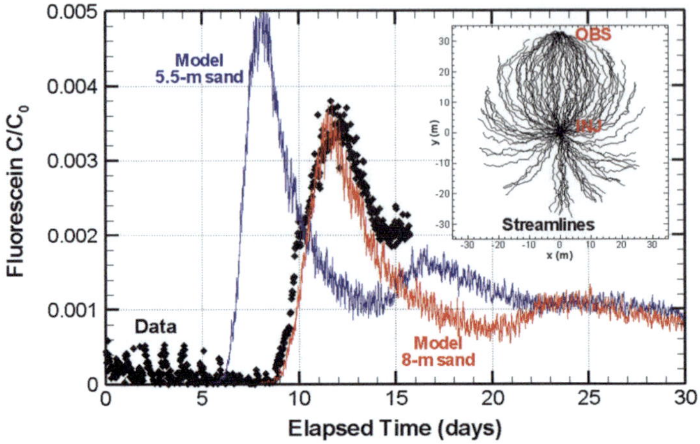

Fig. 8.17 Tracer test data (Doughty et al. 2008) and results of a streamline model

dispersivity (0.1 m), implying that the sand is highly homogeneous, which is considered reasonable for a clean sand. Note that the model breakthrough curves show two peaks, the main peak (8 or 12 days) and a smaller, later recirculation peak (16 or 24 days), which identifies tracer that has arrived at the observation well, been reinjected into the injection well, and traveled to the observation well again. In the field, it does not appear that tracer monitoring continued long enough for a recirculation peak to arrive. The increase in fluorescein concentration observed at 15 days (Fig. 8.17) is considered unlikely to be a recirculation peak, as it would imply a 3-day second trip through the formation, which is too fast.

The 8-m sand thickness inferred from tracer arrival time is certainly possible in terms of the expected variability in sand layer thickness for this geologic setting, and, as described in the previous section, interference test results can be equally well matched assuming either a thickness of 5.5 m or 8 m for the high-permeability zone in which the injection well is perforated. Therefore, the 8-m sand is retained for further modeling studies.

8.6.3 CO_2 Injection and Monitoring

8.6.3.1 Numerical Model for Multi-phase Flow

The well-test and tracer-test described above involve single-phase flow in which gravity does not play a significant role, enabling analysis with an analytical solution or simple single-phase numerical models. However, when CO_2 and brine are both present, multi-phase and gravity effects are significant, requiring a 3D numerical model with two-phase flow capabilities such as TOUGH2 (Pruess et al. 1999).

The simulation results shown here use the capillary pressure curve obtained by fitting to mercury injection data from a C-sand core sample (Sakurai et al. 2006), and hysteretic relative permeability functional forms (Doughty 2007), which are derived from the van Genuchten (1980) formulation (see also Chap. 3). The key parameters of the characteristic curves that need to be identified are the maximum residual gas saturation S_{grmax} and irreducible liquid saturation S_{lr}, below which each phase is immobile, and a parameter describing the interference between the two phases when both are mobile, m, which can range from about 0.4 to 0.9, with lower values of m corresponding to more mobile gas and less mobile liquid, and higher values of m corresponding to less mobile gas and more mobile liquid. During a CO_2 injection period, drainage is the dominant process because the CO_2 plume is growing in all directions. For drainage, the residual gas saturation, S_{gr}^{Δ} , is zero and the parameters controlling plume behavior are S_{lr} and m. After injection ends, the leading edge of the CO_2 plume may still undergo drainage as the plume moves upward and updip by buoyancy forces, but at the trailing edge of the plume imbibition occurs, with S_{gr}^{Δ} depending on saturation history and ranging from zero to S_{grmax}.

Characteristic curves for the base-case simulations of the Frio brine pilot used values of S_{grmax} and S_{lr} taken from the literature: S_{grmax} varies inversely with porosity and averages 0.2 for the C sand (Holtz 2002, 2005); S_{lr} = 0.15 (Bachu and Bennion 2007). The parameter m was chosen to produce a liquid relative permeability curve similar to a Corey (1954) liquid relative permeability curve (m = 0.9). Simulations were also run with values of S_{grmax} that are half the literature values and taking S_{grmax} = 0; with S_{lr} values of 0, 0.30, and 0.45; and with m values of 0.7 and 0.5.

Figure 8.18 shows a time-series of snapshots of the modeled free-phase CO_2 plume during and after the 10-day injection period, using the base-case parameters and the properties shown in Table 8.10. It is clear that buoyancy forces have a large effect on plume evolution. Figure 8.18 also shows that CO_2 distributions for times later than 29 days are very similar to one another, indicating that by that time most gas saturations have decreased to residual values and the CO_2 plume is largely trapped.

8.6.3.2 Downhole Fluid Sampling with U-Tube

Sampling of representative fluids in deep boreholes is challenging because of the need to minimize external contamination and maintain sample integrity during recovery. The U-tube sampling methodology (Freifeld et al. 2005; Freifeld and Trautz 2006) (Sect. 8.4) was developed to collect large volume, multi-phase samples at in situ pressures. At the Frio brine pilot, a U-tube was used to collect a 52-l sample from the observation well every two hours, which was weighed at the surface while being maintained at downhole pressure conditions ($\sim$150 bars). A decrease in sample density from that of formation brine ($\sim$1070 kg/m^3) to that of

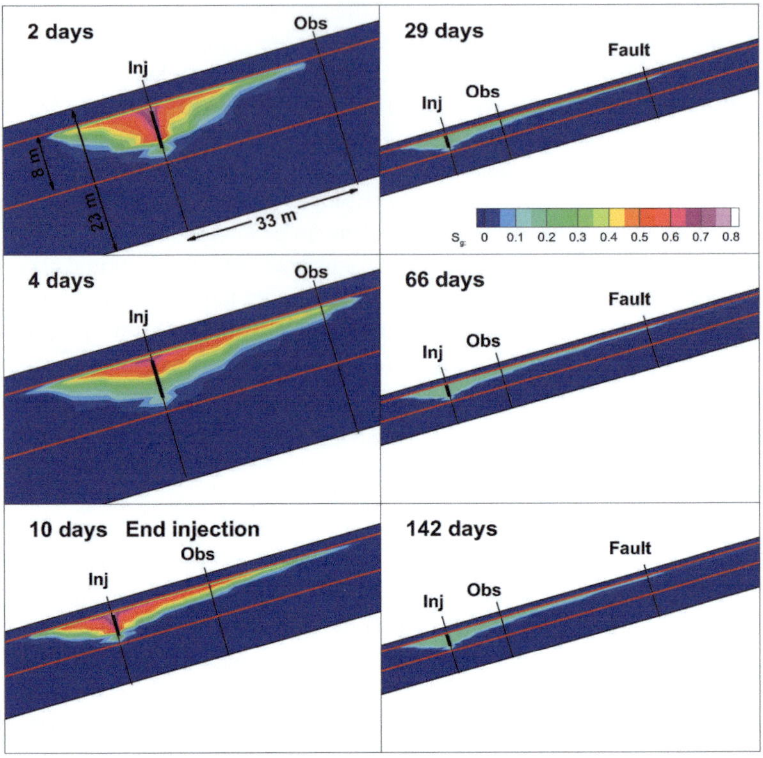

Fig. 8.18 Base-case simulation results showing the evolution of the free-phase CO_2 plume in the vertical cross-section containing the two wells

supercritical CO_2 ($\sim$830 kg/m^3) revealed the transition from single-phase brine to single-phase CO_2 in the wellbore 2.1 days after injection began (Fig. 8.19). Analysis of sample gases (dissolved or separate phase) was performed in the field using a quadrupole mass spectrometer, which also provided unequivocal evidence of the arrival of the CO_2 plume. Additionally, pulses of gas-phase tracers were added to the injection stream at several times during the CO_2 injection period, and their arrival at the observation well provided an indication of changes in CO_2 saturation as injection proceeded.

During the 10-day injection period the CO_2 plume is continually growing, so the formation is undergoing drainage. Therefore, fluid flow (and hence observation-well arrival time) is sensitive to S_{lr} and m, but is not sensitive to S_{grmax}. The CO_2 injection period was modeled using several values of S_{lr} and m, as shown in Fig. 8.19. Note that the U-tube sample density decreases much more than any of the model densities do, because model density represents fluid density in the near-well region, not the density of the wellbore fluid itself, which is what the U-tube samples.

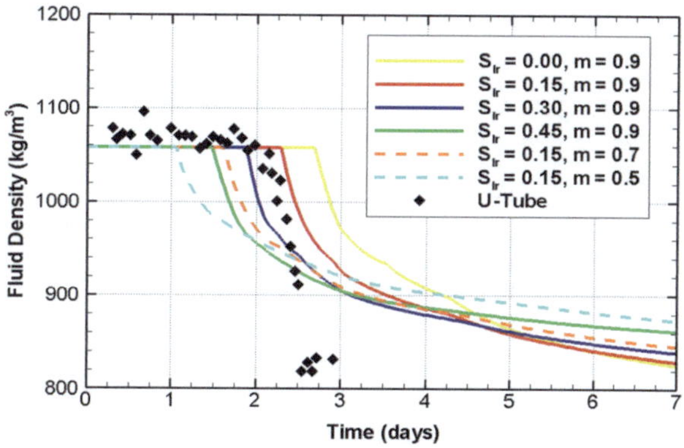

Fig. 8.19 CO_2 arrival at observation well as monitored with U-tube sampling (Doughty et al. 2008) and model results considering different two-phase flow parameters

Figure 8.19 shows that models using $S_{lr} = 0.15$ or 0.30 and $m = 0.9$ give the best match to the field data, and that CO_2 arrival time decreases as S_{lr} increases and as m decreases. Larger values of S_{lr} cause a decrease in CO_2 arrival time two ways. First, with a larger S_{lr}, CO_2 bypasses more immobile brine, so it moves faster through the formation. Second, increasing S_{lr} increases total mobility (the sum of liquid- and gas-phase mobilities), hence it enables more buoyancy flow to occur, resulting in an early arrival of a thin finger of CO_2 shallow in the observation well. Decreasing m also increases total mobility and therefore enhances buoyancy flow.

It is worthwhile to note that the travel time between the two wells is much longer for the aqueous-phase fluorescein tracer (~9 days, Fig. 8.17) than for the two-phase CO_2 plume (~2 days, Fig. 8.19). Various factors contributing to this difference are presented in Table 8.11. Differences in the flow fields imposed by injection and pumping conditions (doublet for the tracer test, single-well for the CO_2 injection, lower injection rate for CO_2) tend to delay the CO_2 arrival, but the delay is more than balanced by the speedup arising from the buoyant, two-phase nature of the CO_2 plume. The modeled distribution of CO_2 (Fig. 8.18) indicates that buoyancy flow and the bypassing of brine within the plume both strongly contribute to the early arrival of CO_2. This finding reiterates the value of a numerical model for interpreting field data. It also illustrates the difficulty of trying to define a simple performance measure such as average CO_2 saturation, which is needed for making capacity assessments of potential CO_2 geologic storage sites (Doughty et al. 2001; Hesse et al. 2006). The average CO_2 saturation within the plume primarily reflects two-phase flow behavior, and indicates the fraction of individual pores that are filled with CO_2. Theoretically it should be consistent with values inferred from laboratory studies. However the average CO_2 saturation over the entire formation is more relevant for capacity assessment, and if buoyancy flow (or heterogeneity) causes the CO_2 plume to avoid large fractions of the formation entirely, the formation-average saturation and the plume-average saturation will be quite different. Neither way of averaging is wrong per se, but care must be taken to use each average in the proper context.

Table 8.11 Comparison between aqueous-phase tracer test and co_2 arrival times

	Tracer test	CO_2 injection	Expected impact on CO_2 arrival time
Arrival at observation well	9 days	2 days	
Flow field	Doublet	Single well	3 times slower
Injection rate	50 gpm	40 gpm	20 % slower
Phase conditions	Single-phase	Two-phase	Faster, bypass pore space containing other phase
Density contrast	None	1.5	Faster, buoyancy flow
Viscosity contrast	None	12	Faster, enhanced buoyancy flow
Density in situ	1060 kg/m^3	~800 kg/m^3	20 % faster

8.6.3.3 Pressure-Transient Analysis

Downhole pressure was measured in both wells throughout the ten-day CO_2 injection period and for about two weeks thereafter (Benson and Doughty 2006; Hovorka et al. 2006). Compared to the 24-h interference test, this longer monitoring period enables the hydrologic properties of more distant features of the fault block to be examined. For example, if the two fault-block boundaries nearest the wells (~600 m to the northwest and ~250 m to the southeast) are considered constant-pressure boundaries, which works for the interference test, modeled pressure increases accompanying CO_2 injection are too small. In contrast, if the more distant salt-dome boundary (~1200 m to the northeast) is considered a constant-pressure boundary, model pressure increases during CO_2 injection are about right, whereas a closed boundary there produces model pressure increases that are too large. Whether this boundary is closed or constant-pressure has no effect on the shorter interference test.

Several stoppages were planned for the injection period, so that pressure-transient analysis could be conducted under two-phase flow conditions. Additional short-term breaks in injection occurred due to operational problems. Pressure-transient responses under two-phase conditions are sensitive to relative permeability parameters, in addition to the intrinsic permeability and formation compressibility, which were inferred from the interference well test. Because only short breaks occur in the injection schedule, drainage is the dominant process occurring throughout the injection period, hence S_{lr} and m are the main parameters to infer. Generally, ΔP decreases as S_{lr} increases and m decreases, consistent with the increase in total mobility that accompanies these parameter changes. The best match to observation-well ΔP is obtained for S_{lr} between 0.15 and 0.30, and m between 0.7 and 0.9, but the sensitivity of the observation-well pressure transient data to S_{lr} and m is rather small, implying that other values of these parameters are also possible.

8.6.3.4 Reservoir Saturation Tool

The wireline reservoir saturation tool (RST), developed by Schlumberger, uses pulsed neutron capture to determine changes in brine saturation as CO_2 displaces brine or vice versa (Hovorka et al. 2006; Sakurai et al. 2006). RST logging was deployed prior to CO_2 injection to determine a baseline and 3 times in the injection well (days 10, 66, 142) and 5 times in the observation well (days 4, 10, 29, 66, 142). Comparing depths at which CO_2 appears in the model to those from the RST logs provides valuable insights into geology, whereas comparing the magnitude of CO_2 saturation S_g provides constraints on two-phase flow properties.

At the injection well, RST logs show that CO_2 extends significantly below the perforated interval. This distribution is best reproduced by the 8-m sand model. This finding supports the hypothesis underlying the 8-m sand model that the thin marker bed located just below the perforations does not have nearly as low a permeability as was inferred from well logs (Fig. 8.15b). At both wells, CO_2 extends almost 1 m shallower than predicted by the model, suggesting that a low-permeability layer identified just above the perforations in both wells may not be continuous, allowing CO_2 to move upward into an overlying sand layer. These findings are consistent with the large sand-layer thickness inferred from the single-phase tracer test, but only the CO_2 injection provides specific information about how this greater thickness may arise.

RST logs collected during the injection reflect a growing CO_2 plume, with drainage occurring throughout the plume, while those obtained during the subsequent rest period reflect the trailing edge of a migrating CO_2 plume, where imbibition occurs. Simulations results for values of S_{lr} from 0 to 0.45, m from 0.5 to 0.9, and S_{grmax} from 0 to 0.2 show the expected dependence on S_{lr}, m, and S_{grmax}. During injection, as S_{lr} increases or m decreases, S_g decreases, as more brine is bypassed rather than being displaced by the invading CO_2, whereas during the subsequent rest period, different values of S_{lr} and m have no impact on the S_g profiles. In contrast, during injection S_{grmax} has no impact on the S_g profiles, whereas during the rest period the amount of CO_2 remaining in the region around the wells decreases dramatically as S_{grmax} is decreased.

The model trends support the use of a small value of S_{lr}, a large value of m, and a large value of S_{grmax} for modeling. However, all model S_g values are significantly smaller than the S_g values obtained from the RST logs. The model results represent the average S_g over a 2-m wide grid block. An RST radius of influence smaller than 1 m could therefore account for some of the discrepancy, especially for the injection well, where conditions can change sharply close to the well.

8.6.3.5 Crosswell Seismic

Crosswell seismic data obtained using source and receiver strings in the observation well and injection well, respectively, were collected shortly before injection of CO_2 and again about six weeks after CO_2 injection ended (Daley et al. 2008). P-wave velocity depends on gas saturation S_g, so a difference tomogram of the seismic velocity before and after CO_2 injection provides an image of the free-phase CO_2 distribution in a vertical plane between the wells, as shown in Fig. 8.20a. A rock physics model for seismic velocity is needed to provide a quantitative relationship between velocity change and S_g. Ideally such a model would be site-specific, derived from core analysis and the relationship between well-log measurements of seismic velocity and well-log measurements of S_g, such as those obtained from the RST. Unfortunately, not all the requisite components for a rock physics model are available for the Frio Formation C sand and a rock physics model calibrated to the Utsira Sand being used for the Sleipner CO_2 storage operation in the North Sea (Carcione et al. 2006; Hoversten et al. 2003) was applied. This modeling suggests that six weeks after CO_2 injection ended, S_g in the vicinity of the injection well was approximately 20 % and S_g near the observation well was approximately 10 % (Daley et al. 2008). RST profiles collected during the CO_2 injection period are shown along the sides of the tomogram, providing a consistency check on the seismic inversion. The crosswell seismic tomogram shows clearly that the inter-well region is heterogeneous, although the resolution of the tomogram (1-5 m) precludes detailed interpretation of specific features.

Plots of the spatial distribution of CO_2 in the vertical plane between the injection and observation wells are shown in Fig. 8.20b for simulations using three values of S_{grmax}. The model plume using the literature values of S_{grmax} ($\sim$0.2) matches the crosswell seismic tomogram best, with $S_g \sim 0.2$ near the injection well, and S_g decreasing and the plume becoming thinner as distance from the well increases. For the case with half S_{grmax}, the shape of the CO_2 plume is similar, but the modeled S_g is too low everywhere, and close examination shows that the maximum S_g does not occur at the injection well, but has migrated away from it. For the case with zero S_{grmax}, the plume migration is extreme and clearly not consistent with the field data. Hence, we infer that the original values of $S_{grmax} \sim 0.2$, which enable significant CO_2 mobility trapping, are optimal.

8.6.3.6 Vertical Seismic Profile

Vertical seismic profile (VSP) data were obtained by creating lines of explosions along the ground surface at three azimuthal directions around the injection well (NW, N, NE), and monitoring P-wave amplitude at the receiver string deployed in the injection well (Daley et al. 2008). VSP data were collected twice: once shortly before CO_2 injection, then again about six weeks after CO_2 injection ended. Figure 8.21 shows the change in P-wave amplitude before and after CO_2 injection, plotted as a function of offset from the injection well, for each direction, along with the corresponding model results for three values of S_{grmax}. The resolution of the VSP data is about 10–30 m, whereas the model resolution varies from 5 m close to the wells to more than 100 m for the largest offsets shown. A quantitative relationship between VSP change in amplitude and CO_2 saturation is not known, so the vertical axes of the plot are adjusted to align these two quantities close to the

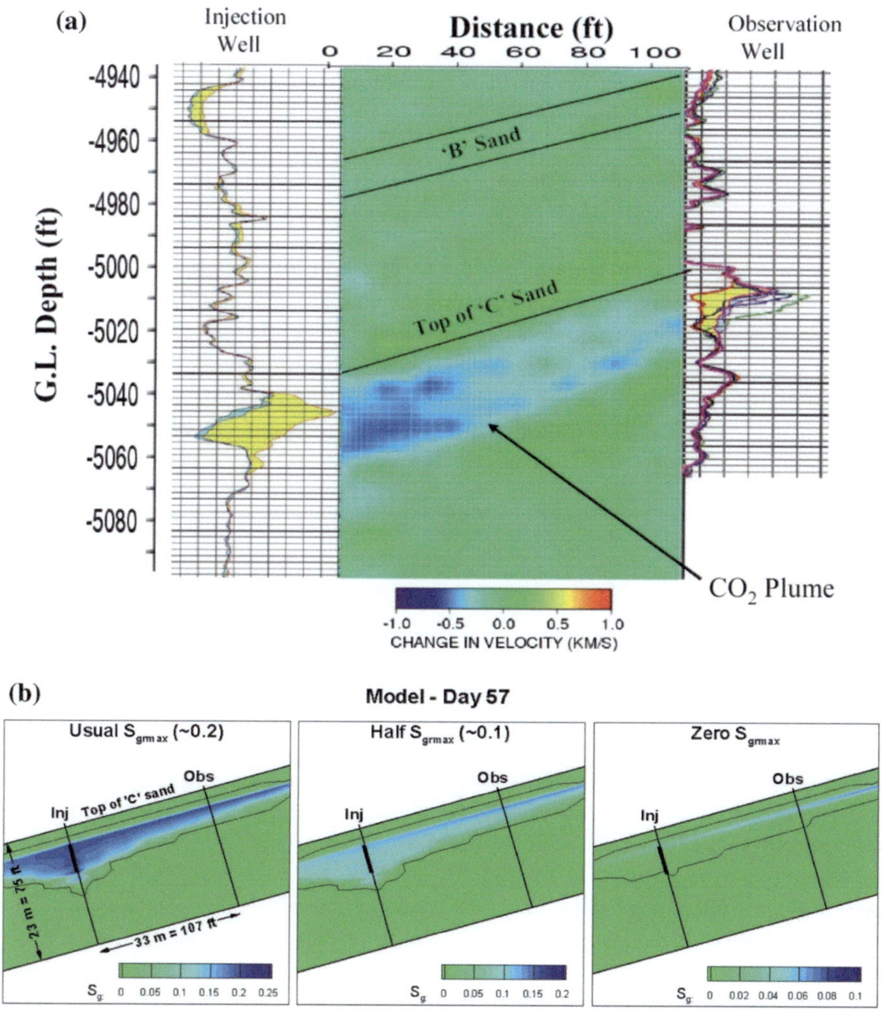

Fig. 8.20 **a** Crosswell seismic tomogram of the difference in P-wave velocity before and after CO_2 injection (Daley et al. 2008). **b** Modeled CO_2 distributions in the plane between the wells considering different values of S_{grmax}. The single black contour line shows $S_g = 0$, an indication of the historical maximum extent of the CO_2 plume

injection well. For the usual S_{grmax} case, Fig. 8.21a shows good agreement between model and VSP in the updip direction (N), but the VSP indicates that the plume has moved farther than the model has predicted to the NE and NW. In fact, the plume has moved as far to the NW as it has to the N, suggesting that either local updip direction is not true north, or there is significant heterogeneity in the permeability distribution beyond the immediate vicinity of the wells, or the planar representation of a warped sand body becomes inaccurate away from the wells. The non-smooth nature of the model profiles far from the injection well indicates the need for a more refined grid.

For the half-S_{grmax} case, the model shows a little too much plume movement in the updip direction (N), while for the zero-S_{grmax} case, there is far too much plume movement updip, so we infer that the original values of $S_{grmax} \sim 0.2$, which allow less CO_2 to move updip, are optimal. Note that the modeled extent of the plume to the NE and NW is independent of S_{grmax}.

Unlike other post-injection monitoring methods, VSP covers such a large spatial extent that it encompasses the entire CO_2 plume, both the trailing edge around the wells, where imbibition is occurring, and the leading edge farther updip, where drainage is still occurring long after injection ceases. Hence, large-offset VSP results are also sensitive to drainage-controlling parameters S_{lr} and m. For the simulation results shown in Fig. 8.21, $S_{lr} = 0.15$ and $m = 0.9$. For a case with $S_{lr} = 0.3$ and $m = 0.9$, the leading edge of the CO_2 plume is about 10 m farther updip than is shown for the N profile in Fig. 8.21a.

8.6.4 Discussion

The fluorescein tracer test indicated that the high-permeability sand through which most fluid travels from the injection well to the observation well is 2.5-m thicker than that inferred from well logs, but did not provide any specific information on how the thickening occurs. The RST profiles for the injection well suggest that the CO_2 plume extends almost 1 m above and several meters below the perforated interval, which was chosen to match the high-permeability zone identified in the well log (Fig. 8.15b). The RST profile for the observation well confirms CO_2 arrival about 1 m shallower than predicted by the well logs, but because of buoyancy, no CO_2 arrives very far below the top of the high-permeability layer. The crosswell seismic tomogram (Fig. 8.20a) shows a CO_2 plume that is about 7.5 m thick at the injection well and thins toward the observation well. Thus, tracer test, RST, and crosswell seismic results all support the choice of the 8-m sand model over the 5.5-m sand model inferred from the well logs. Some evidence for local variability is provided by the crosswell seismic tomogram (Fig. 8.20a), which shows a change in the character of the CO_2 distribution about half-way between the two wells.

The VSP data (Fig. 8.21) allows examining the evolution of the CO_2 plume beyond the immediate vicinity of the wells. Despite matching the updip migration adequately, the model fails to reproduce two other features of the VSP data. First, the CO_2 plume should extend just as far to the NW as it does to the N, suggesting a different local dip direction than currently employed by the model. Second, the lateral extent of the modeled plume is too small, suggesting a short-coming in the conceptualization of geological structure (e.g., the local dip magnitude may be smaller, lateral permeability anisotropy could exist, or the planar representation of a warped sand body becomes inaccurate). Because plume migration is largely a

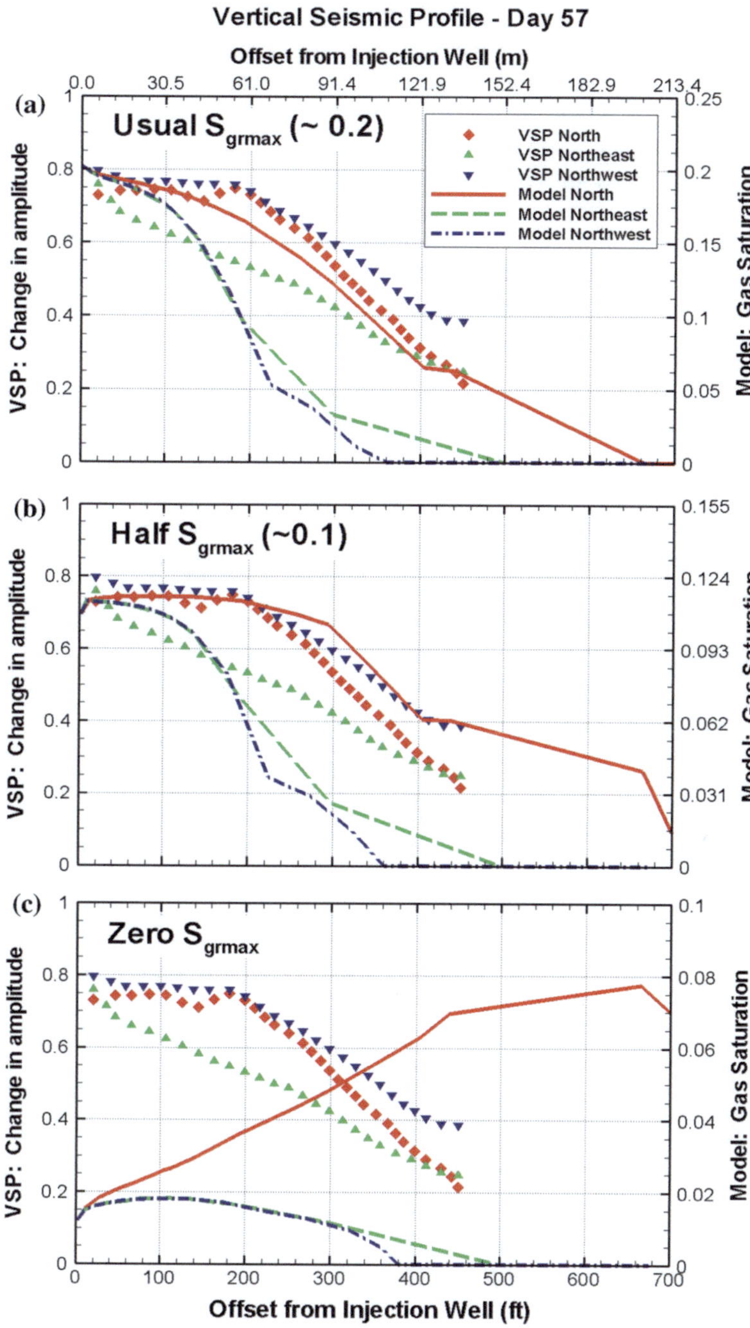

Fig. 8.21 VSP change in P-wave amplitude before and after CO_2 injection (Daley et al. 2008) and model results for far-field CO_2 distribution considering different values of S_{grmax}

consequence of the buoyant flow of CO_2, these features would be difficult to ascertain from traditional site characterization methods.

During CO_2 injection periods, the entire CO_2 plume is undergoing drainage. After injection ends, the leading edge of a migrating CO_2 plume continues to drain, whereas the trailing edge of the plume is undergoing imbibition. Ideally, monitoring should be designed to cover both drainage and imbibition. The drainage process, monitored at the Frio brine pilot via U-tube, early-time pressure transients and RST profiles, and large-offset VSP data, depends on S_{lr} and m, but is not sensitive to S_{grmax}. In contrast, the imbibition process is not very sensitive to S_{lr} and m but depends strongly on S_{grmax}, thus late-time pressure transients and RST profiles, the crosswell seismic tomogram, and small-offset portions of the VSP data provide information on S_{grmax}.

The parameter S_{grmax} is often conceptualized as decreasing with increasing permeability (e.g., Holtz 2002), but these studies strongly support the notion that even for the very high-permeability Frio Formation C sand (>2000 md), S_{grmax} is well above zero, and even provides evidence that the original $S_{grmax} \sim 0.2$ is a better choice than the halved value of $S_{grmax} \sim 0.1$. The late-time RST profiles, the crosswell seismic tomogram, and the VSP data all show consistent results in this regard. Given the importance of a large value of S_{grmax} for trapping free-phase CO_2, this is an important finding for the overall potential for success of CO_2 geologic storage. The results also suggest that S_{lr} is small, consistent with petroleum-literature values (Holtz 2002). The U-tube results for CO_2 arrival at the observation well and pressure-transient analysis supports $S_{lr} = 0.15$, whereas the early-time observation-well RST suggests that S_{lr} is even smaller. A small value of S_{lr} means the CO_2 plume need not bypass so much liquid phase and can form a more compact plume in the subsurface. As one extrapolates from the very high permeabilities of the C sand to moderate permeabilities that may be more typical for CO_2 storage, S_{lr} is expected to increase.

Early in the CO_2 injection period, the simultaneous pressure observations in the injection well, which is surrounded by a two-phase mixture of CO_2 and brine, and the observation well, which is still surrounded by brine, potentially enable deconvolution of multi-phase flow effects, and improved determination of characteristic-curve parameters S_{lr} and m. Unfortunately, the present 3D numerical model is far too coarse (2-m resolution) to adequately resolve near-well effects at the injection well, making such analysis problematic. Studies with a high-resolution model are necessary. For short-time studies, one may be able to increase efficiency by using a 2D axisymmetric model, but for longer times, a 3D model is needed to represent the interplay of heterogeneity, buoyancy, and multi-phase effects.

8.6.5 Concluding Remarks

Well thought-out site characterization is essential for successful geologic storage of CO_2 because of the many physical processes impacting CO_2 plume evolution in the subsurface. At the Frio brine pilot, site characterization techniques such as geological mapping, geophysical imaging, well logging, core analyses, hydraulic well testing, and tracer testing were all valuable and formed the basis of initial site assessment. However, only through the injection and monitoring of CO_2 could the impact of the coupling between buoyancy flow, geologic heterogeneity, and history-dependent multi-phase flow effects truly be appreciated. Thus, the site-characterization process greatly benefited from the addition of CO_2 injection and monitoring. Moreover, development of a numerical model aided in the synthesis of geological, hydrological, and geophysical observations and provided a framework for understanding the coupled flow and transport processes occurring in the CO_2/brine system.

The advantages of using data collected during CO_2 injection to refine reservoir models are numerous. The obvious benefit of CO_2 injection is to provide information on multi-phase flow properties (in particular, the residual gas saturation, below which CO_2 is trapped), which cannot be obtained from traditional site-characterization techniques that examine single-phase conditions. Additionally, the low density and viscosity of CO_2 compared to brine causes the two components to flow through the subsurface differently, potentially revealing distinct features of the geology. Ultimately, to understand stored CO_2 behavior in the subsurface, there is no substitute for studying the movement of CO_2 directly.

It is humbling that even for the small-scale, very well studied, intensively monitored conditions of the research-oriented Frio brine pilot, there are still uncertainties in data interpretation. Even greater uncertainties are projected for full-scale CO_2 geologic storage projects, where economic constraints will limit the availability of data generated through expensive procedures such as tracer tests, fluid sampling, and crosswell seismic. With only one or two monitoring techniques it is generally not difficult to create a model that can reproduce field observations, making it easy to claim a full understanding of the geologic storage system. As different types of observations are added, matching them all becomes much more challenging, which may produce the feeling that understanding of the system has actually decreased. In reality, an appreciation of ones level of ignorance has increased, which is generally a good first step for improving understanding. Despite the commercial pressures attendant to full-scale CO_2 geologic storage, it should be recognized that the coupled flow and transport processes that take place during CO_2 geologic storage can produce subtle and unintuitive effects that will affect the storage efficiency of a reservoir. It is valuable to investigate as many aspects of the system as possible with a range of monitoring techniques to minimize uncertainty.

8.7 Pilot Scale CO_2 Injection at the Ketzin Site: Experiences from the First European On-Shore Storage Site

Axel Liebscher, Stefan Lüth, Sonja Martens, Fabian Möller, Cornelia Schmidt-Hattenberger and Martin Streibel

8.7.1 *Introduction*

Although the implementation of Carbon Capture and Storage (CCS) at industrial scale particularly requires successful demonstration of geological CO_2 storage at scales of >1 Mt CO_2 per year, smaller scale pilot sites on geological storage still play an essential role in progressing the CCS technology. While demonstration projects primarily focus on demonstrating and proving the integration and application of mature technologies, pilot sites are needed to develop, test and progress new storage technologies and to perform field-scale experiments on specific technical and operational aspects that may not be executable at demonstration or industrial scale projects.

The Ketzin pilot site for geological storage of CO_2 in the German Federal State of Brandenburg is the longest operating on-shore CO_2 storage pilot site within Europe and is still the only active CO_2 storage project in Germany. It provides an in situ laboratory for on-shore CO_2 storage in a saline aquifer of the Northeast German Basin. The storage complex is located in an anticline structure above a salt pillow and thereby the Ketzin site shares some fundamental geologic characteristics with future CO_2 storage sites within the European Permian Basin, either on-shore in, e.g., Germany or Poland, or off-shore under the North Sea. The Ketzin site is a pure research and development (R&D) driven project and as a research project limited by national legal regulations to a maximum amount of stored CO_2 of 100,000 t. CO_2 storage at the Ketzin site is accompanied by one of the most comprehensive scientific research and development programs worldwide with key objectives being R&D on CO_2 injection operation, monitoring and modeling. These technological objectives are accompanied by an extensive public outreach and dissemination program. This chapter provides a comprehensive review of the main results and experiences gained at the Ketzin pilot site. For additional information and further reading the reader is referred to Martens et al. (2012, 2013) and Liebscher et al. (2013a, b).

8.7.2 Site Location and Geology

The Ketzin pilot site is located about 25 km west of Berlin and Potsdam (Germany) near the town Ketzin within the Northeast German Basin (Fig. 8.22). The Northeast German Basin is part of the European Permian Basin, which extends from the Polish Trough in the East to present day off-shore areas of the North Sea (Fig. 8.22a) and is made off by sedimentary sequences of Permo-Mesozoic to Cenozoic age. The pilot site itself sits on the southeast flank of the "Roskow-Ketzin" double anticline, which formed above a deep seated, elongated salt pillow (Fig. 8.22b, c). The target reservoir sandstone layers for CO_2 storage belong to the Upper Triassic Stuttgart Formation at about 630–650 m depth and are overlain by more than 165 m of the shaly caprocks of the Upper Triassic Weser and Arnstadt Formations (Fig. 8.22d). The final seal of the multi-barrier system at the Ketzin site is formed by the transgressional Oligocene Rupelian Clay at the base of the Tertiary, which separates the post-Rupelian freshwater horizons from the deep, pre-Rupelian saline formations. The reservoir sandstones are well to moderately-well sorted, immature feldspathic litharenites to lithic arkoses, which are composed of 22–43 wt% quartz, 19–32 wt% plagioclase, 4–13 wt% K-feldspar and subordinate mica, illite, mixed-layer silicates and meta-sedimentary and volcanic rock fragments (Förster et al. 2006). Heterogeneously distributed cement phases make up 5–32 vol.% and include analcime and anhydrite with minor dolomite, barite, and celestine (Förster et al. 2006). Due to the heterogeneity of the reservoir sandstone, reservoir porosity ranges from 12 to 26 vol.% with an averaged permeability of around 100 mD (Zemke et al. 2010; Zettlitzer et al. 2010). The initial reservoir conditions were ~33 °C and 61 bar at 630 m depth (Henninges et al. 2011).

8.7.3 Research Infrastructure at the Ketzin Site

To meet the operational and scientific needs of the Ketzin site a total of five wells have been drilled and an injection facility with a pipeline built (Fig. 8.23). Prior to start of injection in summer 2008, the wells Ktzi 200–202 have been drilled in 2007, each to a depth of 750–800 m (Fig. 8.24). The well Ktzi 201 serves as an injection and observation well, whereas the wells Ktzi 200 and Ktzi 202 are exclusively used for monitoring the injection process and subsurface migration and behaviour of the CO_2. The three wells form the corners of a right-angled triangle (Fig. 8.23). In autumn 2012 a fourth deep well Ktzi 203 was drilled in close vicinity to the injection point (Fig. 8.23) to recover rock cores from the caprock and reservoir sandstones that have been in contact with the injected CO_2 for more than four years. Contrary to the wells Ktzi 200–202, which have steel casings throughout, the well Ktzi 203 has been completed with a glass fibre reinforced

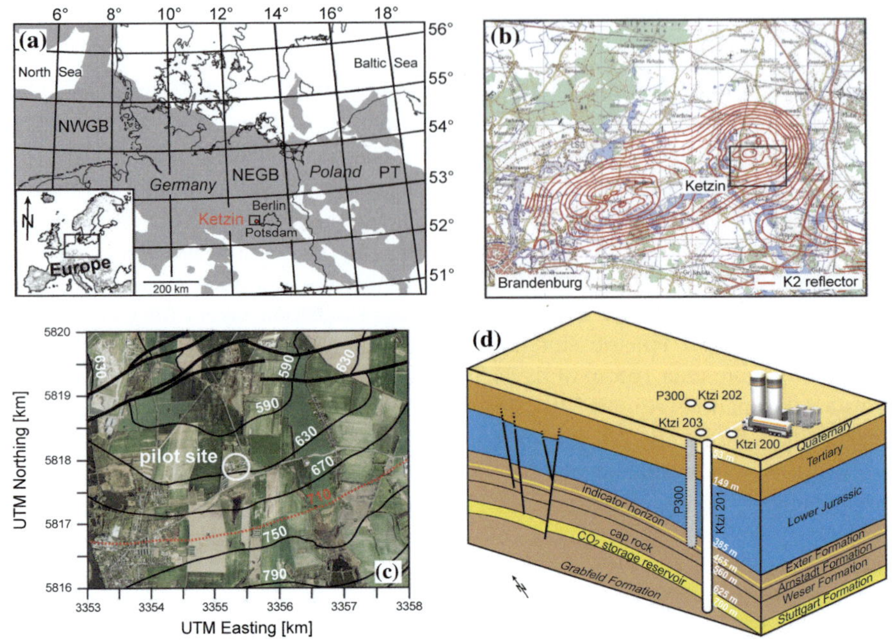

Fig. 8.22 Geographic location and principal geologic environment of the Ketzin pilot site. **a** The Ketzin site is located within the Northeast German Basin (NEGB; NWGB = Northwest German Basin, PT = Polish Trough) (redrawn and modified after Förster et al. 2010) and **b**, **c** sits above the southeast flank of the Roskow-Ketzin double anticline [*red lines* in **b** refer to the depth of the prominent K2 reflector, an anhydrite layer about 80 m above the reservoir; isolines in **c** refer to the depth of top-Stuttgart, *dotted red isoline* at 710 m depth represents the lateral extension of the storage complex. **d** Schematic block diagram showing the general stratigraphic succession at the Ketzin site and the site's infrastructure with four deep wells (Ktzi 200–203) and one shallow observation well (P300) down to the indicator horizon within the Exter Formation (not to scale) [redrawn and modified after Liebscher et al. (2013a, b), Martens et al. (2012)]

plastic casing in the lower-most part to study the applicability of this corrosion resistant material in CO_2 storage operations (Fig. 8.24). To allow for above-zone monitoring within the indicator horizon (Fig. 8.22d) a shallow groundwater observation well P300 was drilled in 2011 to 446 m depth into the lowermost aquifer (Exter Formation) above the caprock.

All four deep wells are completed with a smart casing concept (Prevedel et al. 2009) that allows for quasi-permanent, easily accessible monitoring of the wells and the near-well area (Fig. 8.24). This concept includes behind casing fibre-optic cables for distributed temperature/acoustic sensing (DTS/DAS), behind casing heater cables for heating experiments, permanently installed electrodes (Vertical Electrical Resistivity Array VERA) for electrical resistivity tomography and different pressure and temperature sensors (P-T). The shallow well P300 is equipped with high resolution pressure gauges and a U-tube fluid sampling system to allow for precise pressure and fluid monitoring of the indicator horizon.

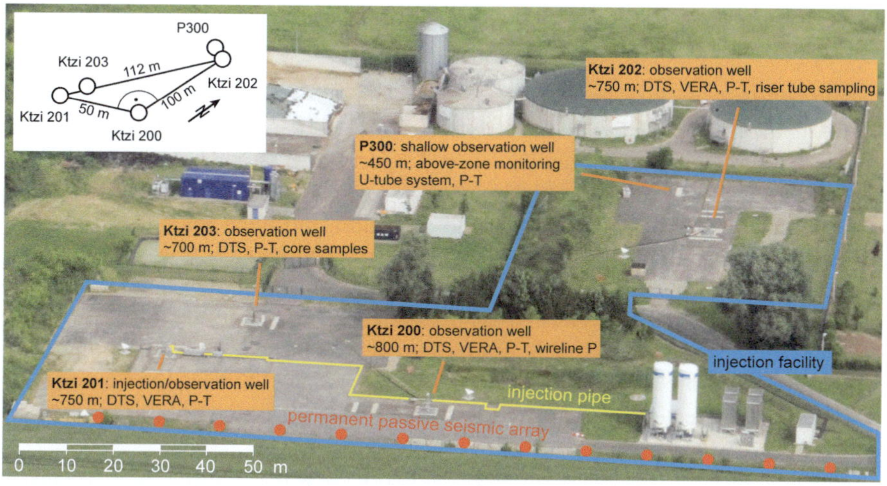

Fig. 8.23 Aerial view of the pilot site Ketzin showing location of the injection and observation wells and the operational and monitoring surface installations

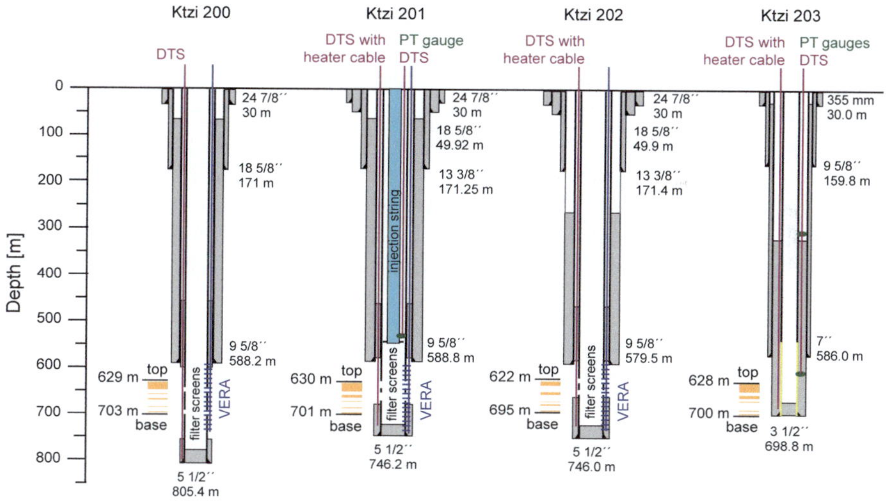

Fig. 8.24 Detailed drawing of the smart casing concept of the four deep wells at the Ketzin site (for location of the four wells see Fig. 8.23). The wells Ktzi 200, 201 and 202 have steel casings throughout whereas Ktzi 203 has a glass fibre reinforced plastic casing in the lower-most part (*yellow*). *Black bars* and *orange boxes left hand side* of the respective wells refer to top and base and main sandstone intervals (=reservoirs) of the Stuttgart Formation (Compiled and re-drawn with additions after Prevedel et al. 2008, 2009; Förster et al. 2010; Liebscher et al. 2013a, b)

8.7.4 Injection Operation and History

Continuous injection of CO_2 at the Ketzin site started on June 30th, 2008 and lasted until August 29th, 2013, when the Ketzin site entered the post-injection phase after more than five years of safe and successful injection of slightly more than 67,000 tons of CO_2. During most time of the injection period, food-grade CO_2 with a purity >99.9 vol.% has been injected at monthly injection rates between 1,000 and 2,300 tons CO_2. From May to June 2011, 1,515 tons of captured CO_2 from the Vattenfall Schwarze Pumpe oxyfuel pilot plant with a purity >99.7 vol.% have been injected. In July and August 2013, a CO_2-N_2 co-injection experiment has been performed to test and demonstrate the technical feasibility of a continuous impure CO_2 injection scenario. A total of 613 tons CO_2 and 32 tons N_2 have been continuously mixed on-site and co-injected resulting in an average CO_2 to N_2 mass ratio of approximately 95:5.

Throughout the entire injection phase, liquid CO_2 was delivered by road tankers and intermediately stored on-site at about −18 °C and 20 bars in two storage tanks with a capacity of 50 t CO_2 each. Prior to injection, the CO_2 was preconditioned in the injection facility to the desired injection conditions. Preconditioning was done in three steps; first up to three plunger pumps for liquid media raised the pressure to the necessary injection pressure, than the CO_2 was heated to the desired injection temperature first by ambient air heaters and afterwards by an electrical heater. The preconditioned CO_2 was transported via the injection pipeline to well Ktzi 201 and injected. Due to the injection process, the reservoir pressure as calculated from the permanently installed pressure sensor increased from initially ~61 bars to about 76–79 bars already after 8 month of injection. After this initial increase the reservoir pressure slightly decreased and stabilized at about 72–75 bars reflecting a stable injection regime (Fig. 8.25; Liebscher et al. 2013b). With stop of injection and entering the post-injection phase, the reservoir pressure started to continuously decrease and evolves back towards initial reservoir conditions (Fig. 8.25).

8.7.5 Monitoring

8.7.5.1 Integrated Monitoring Concept

Research and Development on monitoring is one of the key objectives of the Ketzin pilot site and a comprehensive, integrated state-of-the-art monitoring concept has been established at the pilot site (Fig. 8.26). This concept combines permanent and periodic geophysical, geochemical, operational and remote sensing monitoring techniques. Geophysical monitoring techniques include active seismic (2D, 3D, VSP, MSP), passive seismic and geoelectric (cross-hole, surface-downhole) methods. Geochemical monitoring techniques focus on surface CO_2 soil-flux measurements and fluid sampling either via permanently installed capillary riser

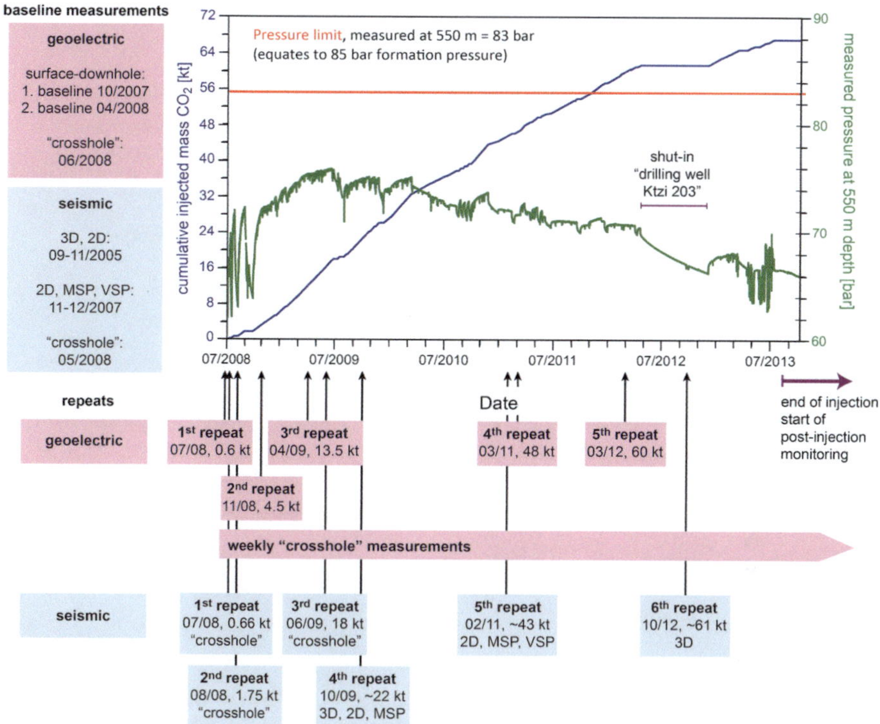

Fig. 8.25 Overall injection history showing cumulative mass of injected CO_2 (*blue*) and measured pressure at 550 m depth in well Ktzi 201 (*green*). The reservoir pressure at 630 m is about 2 bars higher than the measured pressure at 550 m. The *red line* refers to the maximum permitted pressure of 85 bars at reservoir depth. Also shown are the different repeat surveys of the geoelectric and seismic monitoring

tubes or during logging campaigns. These logging campaigns are routinely run and comprise gas-saturation measurements within the reservoir and overburden to test for upward gas migration and magnetic imaging defectoscopy and video logs for casing inspection.

8.7.5.2 Seismic Monitoring

Various active and passive seismic monitoring campaigns were performed during the operational phase at Ketzin (Fig. 8.25). The major part of the active seismic monitoring is the 3D seismic time lapse monitoring with the baseline acquired in 2005 and two repeat surveys in autumn 2009, after the injection of approximately 22,000 tons of CO_2, and in autumn 2012, after the injection of approximately 61,000 tons of CO_2 (Ivanova et al. 2012; Juhlin et al. 2007). The processing and interpretation of the 3D time lapse data delivered amplitude and travel time

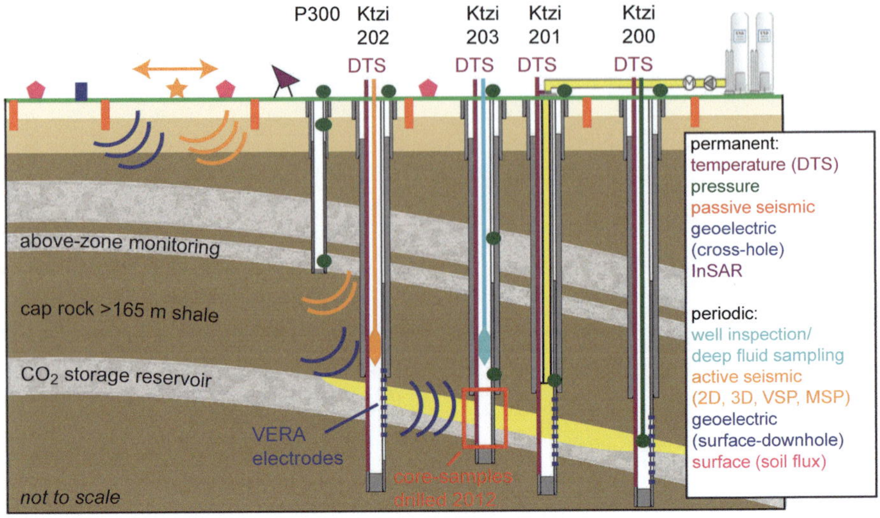

Fig. 8.26 Schematic profile through the Ketzin anticline showing the general geology and the different permanent and periodic monitoring techniques and installations applied at the Ketzin site for an integrated monitoring concept

signatures that could be attributed to the CO_2 stored in the Stuttgart formation and allowed image the lateral spread of the CO_2 plume (Fig. 8.27). The lateral extension of the CO_2 plume as estimated based on the 3D time lapse data was $\sim$0.08 km^2 after approximately 22,000 tons of CO_2 and $\sim$0.15 km^2 after approximately 61,000 tons of CO_2 (Fig. 8.27). The CO_2 plume shows a slightly west-northwest to east-southeast elongated shape with an indicated preferred migration towards west-northwest. The results from the 3D repeat surveys also allowed for an update of the reservoir model with a more realistic description of lateral heterogeneities of physical parameters within the storage formation. Additional experiments include repeat surveys using a focused surface and surface-downhole acquisition pattern in order to achieve an enhanced resolution close to and in the depth range of the injection level (Ivandic et al. 2012).

8.7.5.3 Geoelectric Monitoring

A comprehensive geoelectric surveillance concept has been developed and implemented at the Ketzin pilot site. This concept includes permanent cross-hole measurements and periodic large-scale surface-downhole surveys (Fig. 8.25). Main objectives of the surveillance concept were detection of the early CO_2 migration phase, imaging of the CO_2 plume by electrical resistivity tomography ERT, determination of relative CO_2 and brine saturations as well as correlation between injection operation and geoelectric signal (Bergmann et al. 2012; Kiessling et al.

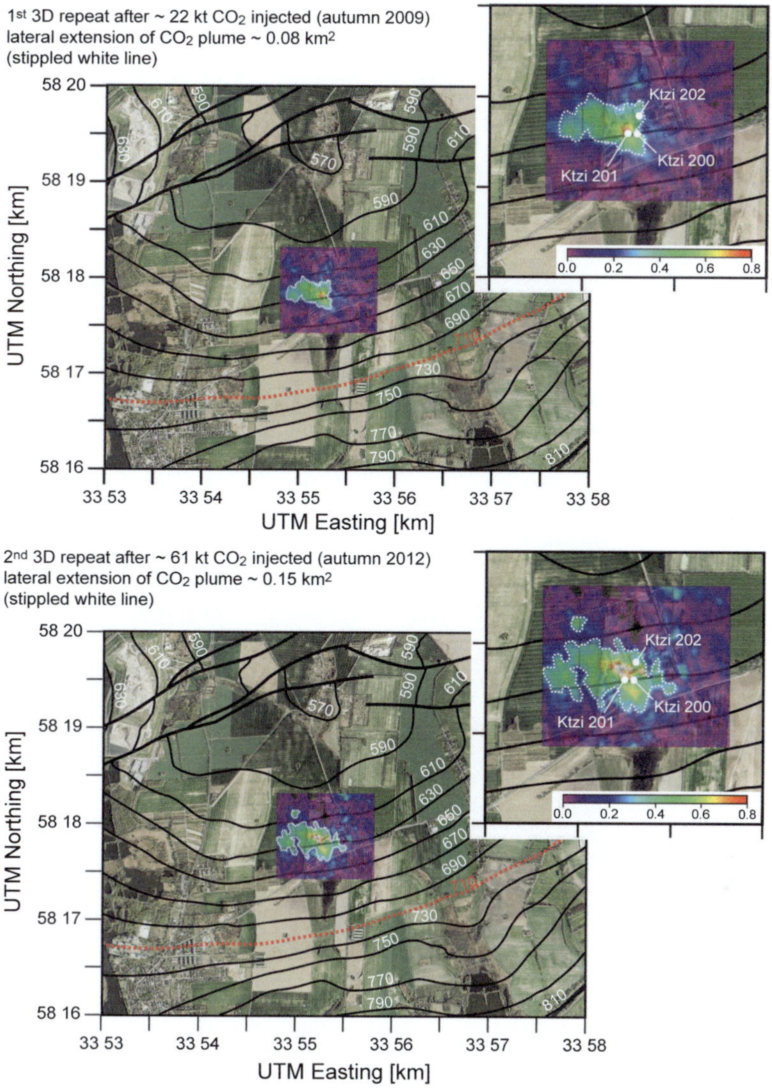

Fig. 8.27 Subsurface extension of the CO_2 plume (*white dotted lines*) as imaged by the 1st and 2nd 3D seismic repeat surveys. *Colour coding* refers to normalized changes in reflection amplitude. *Dotted white lines* are drawn by eye to show changes in normalized reflection amplitude >0.3. (*black lines* refer to top of Stuttgart Formation; *dotted red line* refers to lateral boundary of storage complex as defined by deepest closed top-Stuttgart isobath)

2010; Schmidt-Hattenberger et al. 2011). During the whole CO_2 injection period, a continuous series of time-lapse results based on weekly-measured crosshole data at near-wellbore scale and surface-downhole measurements in 2008, 2009, 2011 and 2012 at a scale of up to 1.5 km was acquired (Fig. 8.25). In the geoelectrical tomographies a noticeable resistivity signature within the target storage zone was observed and could be attributed to the CO_2 plume (Schmidt-Hattenberger et al. 2011) and has been interpreted in terms of the relative CO_2 saturations (Bergmann et al. 2012; Fig. 8.28). The results are very encouraging and show the potential of geoelectric monitoring methods. Especially the relative CO_2 saturations derived for the near-wellbore area correlate well with borehole saturation logs and support the quantitative estimation of injected CO_2 in the near-wellbore area.

8.7.5.4 Surface and Gas Monitoring

To monitor upward migration of CO_2 with potential leakage to the surface, surface soil gas monitoring has been established at the Ketzin pilot site since 2005. The soil gas monitoring network consists of 20 sampling locations for soil gas flux, soil

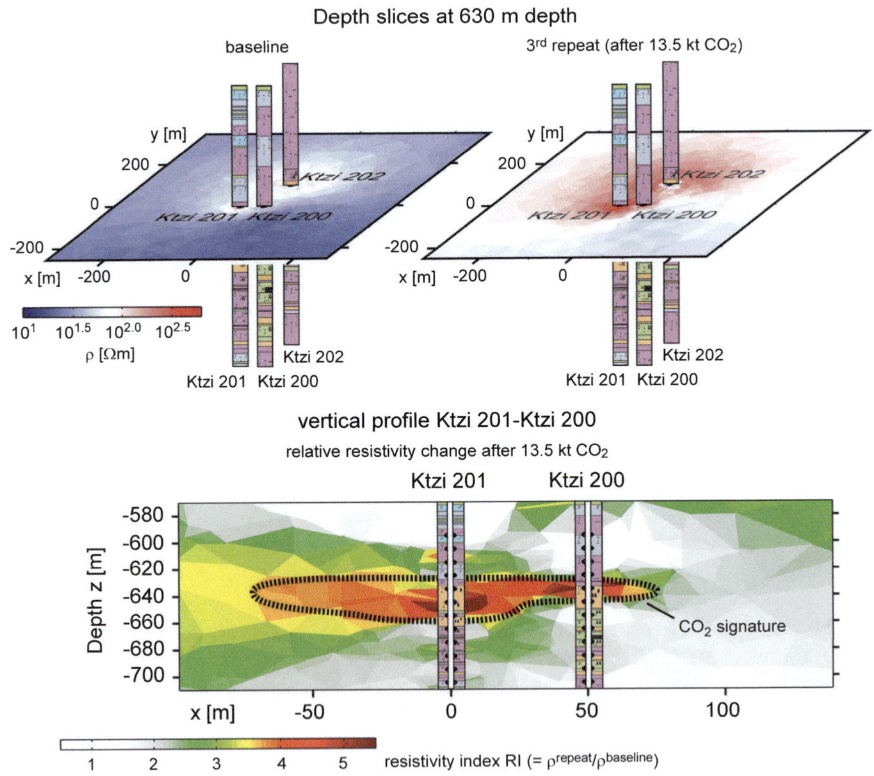

Fig. 8.28 Results from geoelectric surface-downhole monitoring. Resistivity distribution at 630 m depth as estimated for the baseline and repeat survey after 13.5 kt CO_2 injected (*upper part*) and calculated time-lapse signal (*lower part*). Electrical resistivity tomography shows a clear CO_2 signature with up to four to fivefold increase in resistivity within the target reservoir horizon (redrawn and compiled after Bergmann et al. 2012)

moisture and temperature measurements distributed over an approximately 4 km^2 large area around the pilot site. In 2011, the installation of additional eight permanent stations with automated soil gas samplers in the direct vicinity of the injection and monitoring wells expanded the monitoring network. Since the start of injection in June 2008, no change in soil CO_2 gas flux could be detected in comparison to the pre-injection baseline from 2005 to 2007 (Zimmer et al. 2011).

8.7.6 *Public Outreach Activities*

Public acceptance is a key issue for the Ketzin project as it is for any other CO_2 storage project. From the very beginning, the research activities were therefore accompanied by a factual, open and transparent dialogue with the general public, scientists, industry and decision makers. Public outreach activities focus on the local public from the nearby town of Ketzin but also interested people from all over Germany and worldwide. The visitor centre at the Ketzin site is the most important contact point. Here, a computer-based info terminal enables visitors to inform themselves easily and interactively. Weekly visitor tours are offered by GFZ on site. In addition, an annual open house day at the Ketzin pilot site is carried out in close cooperation with the local community of the town of Ketzin and warmly received. Project status and progress are covered and disseminated in brochures and a public website in German and English (www.co2ketzin.de). The Ketzin project is also presented in a broader context in a film entitled "The geological storage of CO_2" made up of six five-minute segments. The research activities at the pilot site are supported by the town of Ketzin, and the regulatory authority has been involved and cooperative since the start of the Ketzin project (Martens et al. 2011). The experience from the Ketzin pilot site shows that an honest, open and factual target group oriented communication and dissemination program is able to overcome critical public perception even for highly debated technologies.

8.7.7 *Conclusions and Outlook*

The Ketzin project demonstrates successful and safe CO_2 storage in a saline aquifer on a pilot scale. The results show that:

- the geological storage of CO_2 at the Ketzin pilot site runs reliably and without any risks for people and environment,
- a meaningful, site related combination of geophysical and geochemical monitoring techniques is able to detect even small amounts of CO_2 and to image its spatial distribution, and
- a targeted communication and dissemination program is able to overcome critical public perception even for highly debated technologies.

Although the CO_2 injection at Ketzin ceased in August 2013, R&D activities on CO_2 storage continues at Ketzin in order to address and close the entire life cycle of a storage site. Well abandonment, post-injection monitoring and transfer of liability are major objectives of the on-going post-injection phase at the Ketzin site. Because the Ketzin site has been permitted under the German Mining Law, transfer of liability will follow the regulations set out by the German Mining Law. The R&D work in the post-closure phase will nevertheless address the three high-level criteria for transfer of liability set out by the EU Directive:

- observed behaviour of the injected CO_2 conforms with the modeled behaviour,
- no detectable leakage, and
- site is evolving towards a situation of long-term stability.

During the post-injection phase, further field-experiments like controlled back-production of CO_2 as reservoir pressure management measure and small-scale brine injection as wellbore leakage mitigation measure are performed. By these experiments, the Ketzin site meets the demands made for pilot sites to perform field-scale experiments on specific technical and operational aspects that may not be executable at demonstration or industrial scale projects.

References

Aki K, Richards PG (1980) Quantitative seismology, 2nd edn

Alnes H, Eiken O, Nooner S, Sasagawa G, Stenvold T, Zumberge M (2011) Results from Sleipner gravity monitoring: updated density and temperature distribution of the CO_2 plume. Energy Procedia. In: 10th international conference on greenhouse gas control technologies, vol 4, pp 5504–5511

Archie GE (1942) The electrical resistivity log as an aid in determining some reservoir characteristics. Trans AIME 146:54–62

Arts R, Chadwick A, Eiken O, Thibeau S, Nooner S (2008) Ten years' experience of monitoring CO_2 injection in the Utsira Sand at Sleipner, offshore Norway. First Break 26:65–72

Assayag N, Matter J, Ader M, Goldberg D, Agrinier P (2009) Water–rock interactions during a CO_2 injection field-test: implications on host rock dissolution and alteration effects. Chem Geol 265(1):227–235

Asveth P (2009) Explorational rock physics: the link between geological processes and geophysical observables. In: Petroleum geoscience. Springer

Bachu S, Bennion B (2007) Effects of in-situ conditions on relative permeability characteristics of CO_2-brine systems. Environ Geol 54:1707–1722

Bakke NE, Ursin B (1998) Thin-bed AVO effects. Geophys Prospect 46:571–587

Batzle ML, Wang Z (1992) Seismic properties of pore fluids. Geophysics 57:1396–1408

Behrens H, Ghergut J, Bensabat J, Niemi A, Sauter M (2014) Merging single- and inter-well tracer tests into one forced-gradient dipole test, at the Heletz site within the MUSTANG project. Energy Procedia 59:249–255

Benson S, Doughty C (2006) Estimation of field-scale relative permeability from pressure transient tests. In: Presented at the CO2SC symposium. Lawrence Berkeley National Laboratory, Berkeley, CA

Bergmann P, Schmidt-Hattenberger C, Kiessling D, Rücker C, Labitzke T, Henninges J, Baumann G, Schütt H (2012) Surface-downhole electrical resistivity tomography applied to monitoring of CO_2 storage at Ketzin, Germany. Geophysics 77:B253–B267

Berryman JG, Milton GW (1991) Exact results for generalized Gassmann's equations in composite porous media with two constituents. Geophysics 56:1950–1960

Bickle M, Chadwick A, Huppert HE, Hallworth M, Lyle S (2007) Modelling carbon dioxide accumulation at Sleipner: Implications for underground carbon storage. Earth Planet Sci Lett 255:164–176

Biondi BL (2006) Three dimensional seismic imaging. Investigation in Geophysics. Society of Exploration Geophysicists

Boait FC, White NJ, Bickle MJ, Chadwick RA, Neufeld JA, Huppert HE (2012) Spatial and temporal evolution of injected CO_2 at the Sleipner Field, North Sea. J Geophys Res Solid Earth 117

Boreham C, Underschultz J, Stalker L, Kirste D, Freifeld B, Jenkins C, Ennis-King J (2011) Monitoring of CO_2 storage in a depleted natural gas reservoir: gas geochemistry from the CO2CRC Otway Project, Australia. Int J Greenh Gas Control 5:1039–1054

Borgos H, Skov T, Randen T, Sonneland L (2003) Automated geometry extraction from 3D seismic data. In: SEG technical program expanded abstracts 2003, SEG technical program expanded abstracts. Society of Exploration Geophysicists, pp 1541–1544

Bortfeld R (1961) Approximations to the reflection and transmission coefficients of plane longitudinal and transverse waves. Geophys Prospect 9:485–502

Butler DB, Knight RJ (1988) Electrical conductivity of steam-flooded, clay-bearing geologic materials. Geophysics 63:1137–1149

Carbon Sequestration Leadership Forum (CSLF) (2013) 2013 Annual report by the CSLF Task force on reviewing best practices and standards for geologic storage and monitoring of CO_2. www.cslforum.org. Oct 2013

Carcione JM, Picotti S, Gei D, Rossi G (2006) Physics and seismic modeling for monitoring CO_2 storage. Pure appl Geophys 163:175–207

Castagna J, Backus M (1993) Offset-dependent reflectivity? Theory and practice of AVO analysis. Investigations in Geophysics. Society of Exploration Geophysicists

Cavanagh A (2013) Benchmark calibration and prediction of the Sleipner CO_2 plume from 2006 to 2012. Energy Procedia. In: GHGT-11 proceedings of the 11th international conference on greenhouse gas control technologies, 18–22 Nov 2012, Kyoto, Japan vol 37, pp 3529–3545

Chadwick RA, Noy DJ (2010) History-matching flow simulations and time-lapse seismic data from the Sleipner CO_2 plume. Geol Soc Lond Pet Geol Conf Ser 7:1171–1182

Chadwick RA, Holloway S, Kirby G, Gregersen U, Johannessen P (2000) The Utsira Sand, Central North Sea—an assessment of its potential for regional CO_2 disposal. In: Presented at the proceedings of 5th international conference on greenhouse gas control technologies. IEA Greenhouse Gas Programme, Cairns, Australia

Chadwick A, Williams G, Delepine N, Clochard V, Labat K, Sturton S, Buddensiek M, Dillen M, Nickel M, Lima A, Arts R, Neele F, Rossi G (2010) Quantitative analysis of time-lapse seismic monitoring data at the Sleipner CO_2 storage operation. Lead Edge 29:170–177

Christensen NB, Sherlock D, Dodds K (2006) Monitoring CO_2 injection with cross-hole electrical resistivity tomography. Explor Geophys 37:44–49

CO_2 Capture Project, CCP CO_2 STORAGE PROGRAM fact sheet, modular borehole monitoring design and field test [WWW Document] (2013) http://www.co2captureproject.org/reports/FACTSHEET_MBM.pdf

Corey AT (1954) The interrelation between gas and oil relative permeabilities. Prod Mon 19:38–41

Crow W, Carey JW, Gasda S, Williams DB, Celia M (2010) Wellbore integrity analysis of a natural CO_2 producer. Int J Greenhouse Gas Control 4(2):186–197

Daley TM, Myer LR, Peterson JE, Majer EL, Hoversten GM (2008) Time-lapse crosswell seismic and VSP monitoring of injected CO_2 in a brine aquifer. Environ Geol 54:1657–1665

Daley TM, Ajo-Franklin JB, Doughty C (2011) Constraining the reservoir model of an injected CO_2 plume with crosswell CASSM at the Frio-II brine pilot. Int J Greenh Gas Control 5:1022–1030

Daley TM, Hendrickson J, Queen JH (2013) Monitoring CO_2 Storage at Cranfield, Mississippi with Time-Lapse Offset VSP—Using Integration and Modeling to Reduce Uncertainty. Energy Procedia 63:4240–4248

De Witte L (1955) A study of electric log interpretation methods in Shaly formations

Deregowski SM (1986) What is DMO? First break 4

Domes F (2010) The influence of overburden on quantitive time-lapse seismic interpretation (Thesis). Heriot-Watt University

Doughty C (2007) Modeling geologic storage of carbon dioxide: comparison of non-hysteretic and hysteretic characteristic curves. Energy Convers Manag. In: Geologic carbon sequestration and methane hydrates research from the TOUGH symposium 2006, vol 48, pp 1768–1781

Doughty C, Pruess K, Benson S, Hovorka SD, Knox PR, Green CT (2001) Capacity investigation of brine-bearing sands of the Frio Formation for geologic sequestration of CO_2. In: Presented at the first national conference on carbon sequestration, Washington, DC

Doughty C, Freifeld BM, Trautz RC (2008) Site characterization for CO_2 geologic storage and vice versa: the Frio brine pilot, Texas, USA as a case study. Environ Geol 54:1635–1656

Dwarakanath V, Pope GA (1998) New approach for estimating alcohol partition coefficients between nonaqueous phase liquids and water. Environ Sci Technol 32:1662–1666

Eberhart-Phillips D, Han DH, Zoback MD (1989) Empirical relationships among seismic velocity, effective pressure, porosity, and clay content in sandstone. Geophysics 54:82–89

Eiken O, Ringrose P, Hermanrud C, Nazarian B, Torp TA, Høier L (2011) Lessons learned from 14 years of CCS operations: Sleipner. In: Salah and Snøhvit. Energy Procedia. In: 10th international conference on greenhouse gas control technologies, vol 4, pp 5541–5548

Elodie J, Philippe S (2012) The relevance of geochemical tools to monitor deep geological CO_2 storage sites. In: Panagiotaras D (ed) Geochemistry—earth's system processes. InTech

Emberley S, Hutcheon I, Shevalier M, Durocher K, Mayer B, Gunter WD, Perkins EH (2005) Monitoring of fluid–rock interaction and CO_2 storage through produced fluid sampling at the Weyburn CO_2-injection enhanced oil recovery site, Saskatchewan, Canada. Appl Geochem 20:1131–1157

Estublier A, Fornel A, Parra T, Deflandre J-P (2013) Sensitivity study of the reactive transport model for CO_2 injection into the Utsira Saline formation using 3D fluid flow model history matched with 4D seismic. Energy Procedia 37:3574–3582

Fabriol H, Bitri A, Bourgeois B, Delatre M, Girard JF, Pajot G, Rohmer J (2011) Geophysical methods for CO_2 plume imaging: comparison of performances. Energy Procedia. In: 10th international conference on greenhouse gas control technologies, vol 4, pp 3604–3611

Fagerlund F, Niemi A, Bensabat J, Shtivelman V (2013a) Design of a two-well field test to determine in situ residual and dissolution trapping of CO_2 applied to the Heletz CO_2 injection site. Int J Greenhouse Gas Control 19:642–651

Fagerlund F, Niemi A, Bensabat J, Shtivelman V (2013b) Interwell field test to determine in situ CO_2 trapping in a deep saline aquifer: Modelling study of the effects of test design and geological parameters. Energy Procedia 40:554–563

Fornel A, Estublier A (2013) To a dynamic update of the Sleipner CO_2 storage geological model using 4d seismic data. Energy Procedia. In: GHGT-11 proceedings of the 11th international conference on greenhouse gas control technologies, 18–22 Nov 2012, Kyoto, Japan 37, pp 4902–4909

Förster A, Norden B, Zinck-Jørgensen K, Frykman P, Kulenkampff J, Spangenberg E, Erzinger J, Zimmer M, Kopp J, Borm G, Juhlin C, Cosma C-G, Hurter S (2006) Baseline characterization of the CO_2SINK geological storage site at Ketzin, Germany. Environ Geosci 13:145–161

Förster A, Schöner R, Förster H-J, Norden B, Blaschke A-W, Luckert J, Beutler G, Gaupp R, Rhede D (2010) Reservoir characterization of a CO_2 storage aquifer: the upper Triassic Stuttgart formation in the Northeast German Basin. Mar Pet Geol 27(10):2156–2172

Freifeld BM, Trautz RC (2006) Real-time quadrupole mass spectrometer analysis of gas in borehole fluid samples acquired using the U-tube sampling methodology. Geofluids 6:217–224

Freifeld BM, Trautz RC, Kharaka YK, Phelps TJ, Myer LR, Hovorka SD, Collins DJ (2005) The U-tube: a novel system for acquiring borehole fluid samples from a deep geologic CO_2 sequestration experiment. J Geophys Res Solid Earth 110

Freifeld BM, Finsterle S, Onstott TC, Toole P, Pratt LM (2008) Ground surface temperature reconstructions: using in situ estimates for thermal conductivity acquired with a fiber-optic distributed thermal perturbation sensor. Geophys Res Lett 35:L14309

Frohlich RK, Parke CD (1989) The electrical resistivity of the vadose zone—field survey. Ground Water 27:524–530

Gassmann F (1951) Über die Elastizität poröser Medien. Inst. für Geophysik an der ETH, Zürich

Gelchinsky B (1988) The common reflecting element (CRE) method (non-uniform asymmetric multifold system). Explor Geophys 19:71–75

Gelchinsky B, Berkovitch A, Keydar S (1999) Multifocusing homeomorphic imaging: part 2. Multifold data set and multifocusing. J Appl Geophys 42:243–260

Gemmer L, Hansen O, Iding M, Leary S, Ringrose P (2012) Geomechanical response to CO_2 injection at Krechba, In Salah, Algeria. First Break 30:79–84

Ghergut J, Bensabat J, Niemi A, Licha T, Ptak T, Sauter M (2013) CCS site characterization by single-well and inter-well tracer tests. Stanford Geothermal Program Workshop Report SGP-TR-198:316–325

Giese R, Henninges J, Lüth S, Morozova D, Schmidt-Hattenberger C, Würdemann H, Zimmer M, Cosma C, Juhlin C (2009) Monitoring at the CO_2 SINK site: a concept integrating geophysics, geochemistry and microbiology. Energy Procedia 1:2251–2259

Gochioco LM (1991) Tuning effect and interference reflections from thin beds and coal seams. Geophysics 56:1288–1295

Guéguen Y, Palciauskas V (1994) Introduction to the physics of rocks. Princeton University Press, New Jersey

Gunter B, Perkins E (2004) Geochemical monitoring and modeling. In: Presented at the 7th international conference on greenhouse gas control technologies, IEA Greenhouse Gas Programme, Vancouver, Canada

Guy E, Radzevicius S, and Conroy J (2003) Computer programs for application of equations describing elastic and electromagnetic wave scattering from planar interfaces, Comput Geosci 29:569–575

Han M, Youssef S, Rosenberg E, Fleury M, Levitz P (2009) Deviation from Archie's law in partially saturated porous media: Wetting film versus disconnectedness of the conducting phase. Phys Rev E 79:31127

Hantush MS, Jacob CE (1955) Non-steady radial flow in an infinite leaky aquifer. EOS Trans Am Geophys Union 36:95–100

Henninges J, Liebscher A, Bannach A, Brandt W, Hurter S, Köhler S, Möller F, CO2SINK Group (2011) P–T–rho and two-phase fluid conditions with inverted density profile in observation wells at the CO_2 storage site at Ketzin (Germany). Energy Procedia 4:6085–6090

Hesse M, Tchelepi HA, Orr FM (2006) Scaling analysis of the migration of CO_2 in saline aquifers. Soc Petrol Eng

Hill R (1963) Elastic properties of reinforced solids: some theoretical principles. J Mech Phys Solids 11:357–372

Holtz MH (2002) Residual gas saturation to aquifer influx: a calculation method for 3-D computer reservoir model construction. Soc Petrol Eng

Holtz MH (2005) Reservoir characterization applying residual gas saturation modeling, example from the Starfak (MSc). University of Texas, Austin

Hoversten GM, Gasperikova E (2005) Chapter 23—non-seismic geophysical approaches to monitoring A2 In: Thomas DC (ed) Carbon dioxide capture for storage in deep geologic formations. Elsevier Science, Amsterdam, pp 1071–1112

Hoversten G, Gritto R, Washbourne J, Daley T (2003) Pressure and fluid saturation prediction in a multicomponent reservoir using combined seismic and electromagnetic imaging. Geophysics 68:1580–1591

Hovorka SD, Benson SM, Doughty C, Freifeld BM, Sakurai S, Daley TM, Kharaka YK, Holtz MH, Trautz RC, Nance HS, Myer LR, Knauss KG (2006) Measuring permanence of CO_2 storage in saline formations: the Frio experiment. Environ Geosci 13:105–121

Hunkeler D, Hoehn E, Höhener P, Zeyer J (1997) 222Rn as a partitioning tracer to detect diesel fuel contamination in aquifers: laboratory study and field observations. Environ Sci Technol 31:3180–3187

Iding M, Ringrose P (2009) Evaluating the impact of fractures on the long-term performance of the In Salah CO_2 storage site. Energy Procedia. In: Proceedings of the 9th international conference on greenhouse gas control technologies 9 (GHGT-9), 16–20 Nov 2008, Washington DC, USA 1, pp 2021–2028

IEAGHG (2013) The process of developing a CO_2 test injection: experience to date and best practice. Report 2013/13. Oct 2013

Ivandic M, Yang C, Lüth S, Cosma C, Juhlin C (2012) Time-lapse analysis of sparse 3D seismic data from the CO_2 storage pilot site at Ketzin, Germany. J Appl Geophys 84:14–28

Ivanova A, Kashubin A, Juhojuntti N, Kummerow J, Henninges J, Juhlin C, Lüth S, Ivandic M (2012) Monitoring and volumetric estimation of injected CO_2 using 4D seismic, petrophysical data, core measurements and well logging: a case study at Ketzin, Germany. Geophys Prospect 60:957–973

Jäger R, Mann J, Höcht G, Hubral P (2001) Common-reflection-surface stack: image and attributes. Geophysics 66:97–109

Javandel I, Doughty C, Tsang C-F (1984) Groundwater transport: handbook of mathematical models. Am Geophys Union

Jenkins CR, Cook PJ, Ennis-King J, Undershultz J, Boreham C, Dance T, de Caritat P, Etheridge DM, Freifeld BM, Hortle A, Kirste D, Paterson L, Pevzner R, Schacht U, Sharma S, Stalker L, Urosevic M (2012) Safe storage and effective monitoring of CO_2 in depleted gas fields. Proc Natl Acad Sci 109:E35–E41

Johnson G, Mayer B, Nightingale M, Shevalier M, Hutcheon I (2011a) Using oxygen isotope ratios to quantitatively assess trapping mechanisms during CO_2 injection into geological reservoirs: the Pembina case study. Chem Geol 283:185–193

Johnson G, Mayer B, Shevalier M, Nightingale M, Hutcheon I (2011b) Tracing the movement of CO_2 injected into a mature oilfield using carbon isotope abundance ratios: the example of the Pembina Cardium CO_2 monitoring project. Int J Greenh Gas Control 5:933–941

Jones DG, Lister TR, Smith DJ, West JM, Coombs P, Gadalia A, Brach M, Annunziatellis A, Lombardi S (2011) In Salah gas CO_2 storage JIP: surface gas and biological monitoring. Energy Procedia. In: 10th international conference on greenhouse gas control technologies, vol 4, pp 3566–3573

Juhlin C, Young R (1993) Implications of thin layers for amplitude variation with offset (AVO) studies. Geophysics 58:1200–1204

Juhlin C, Giese R, Zinck-Jørgensen K, Cosma C, Kazemeini H, Juhojuntti N, Lüth S, Norden B, Förster A (2007) 3D baseline seismics at Ketzin, Germany: the project. Geophysics 72:B121–B132

Kashubin A, Juhlin C, Malehmir A, Lüth S, Ivanova A, Juhojuntti N (2011) A footprint of rainfall on land seismic data repeatability at the CO_2 storage pilot site, Ketzin, Germany. In: Presented at the 2011 SEG annual meeting, Society of Exploration Geophysicists

Kazemeini SH, Juhlin C, Fomel S (2010) Monitoring CO_2 response on surface seismic data; a rock physics and seismic modeling feasibility study at the CO_2 sequestration site, Ketzin, Germany. J Appl Geophys 71:109–124

Kearey P, Brooks M, Hill I (2002) An introduction to geophysical exploration, 3rd edn. Wiley, Hoboken

Kharaka YK, Hanor JS (2003) Deep fluids in the continents: I. Sedimentary basins. Treatise Geochem 5:605

Kharaka YK, Cole DR, Hovorka SD, Gunter WD, Knauss KG, Freifeld BM (2006) Gas-water-rock interactions in Frio Formation following CO_2 injection: Implications for the storage of greenhouse gases in sedimentary basins. Geology 34:577–580

Kiessling D, Schmidt-Hattenberger C, Schuett H, Schilling F, Krueger K, Schoebel B, Danckwardt E, Kummerow J (2010) Geoelectrical methods for monitoring geological CO_2 storage: first results from cross-hole and surface–downhole measurements from the CO_2SINK test site at Ketzin (Germany). Int J Greenh Gas Control 4:816–826

Kirsch DR (2006) Petrophysical properties of permeable and low-permeable rocks. In: Kirsch DR (ed) Groundwater geophysics. Springer, Berlin, pp 1–22

Kragh E, Christie P (2002) Seismic repeatability, normalized rms, and predictability. Lead Edge 21:640–647

LaForce T, Ennis-King J, Boreham C, Paterson L (2014) Residual CO_2 saturation estimate using noble gas tracers in a single-well field test: the CO2CRC Otway project. Int J Greenhouse Gas Control 26:9–21

Landrø M (2001) Discrimination between pressure and fluid saturation changes from time-lapse seismic data. Geophysics 66:836–844

Lange G (1997) Gleichstromgeoelektrik. In: Knödel K, Krummel H, Lange G (eds) Handbuch zur Erkundung des Untergrundes von Deponien und Altlasten. Band 3: Geophysik. Springer, Berlin

Lay T, Wallace TC (1995) Modern global seismology, 1st edn. Academic Press, San Diego

Lee M (2011) Connectivity equation and Shaly-Sand correction for electrical resistivity (USGS scientific investigations report 2011–5005)

Liebscher A, Martens S, Moller F, Kuhn M (2013a) 12-on-shore CO_2 storage at the Ketzin pilot site in Germany. Geological storage of carbon dioxide (CO_2). Woodhead Publishing, Sawston, pp 278–300

Liebscher A, Möller F, Bannach A, Köhler S, Wiebach J, Schmidt-Hattenberger C, Weiner M, Pretschner C, Ebert K, Zemke J (2013b) Injection operation and operational pressure–temperature monitoring at the CO_2 storage pilot site Ketzin, Germany—design, results, recommendations. Int J Greenh Gas Control 15:163–173

Liu Y, Schmitt DR (2003) Amplitude and AVO responses of a single thin bed. Geophysics 68:1161–1168

Luquot L, Gouze P (2009) Experimental determination of porosity and permeability changes induced by massive injection of CO_2 into carbonate reservoirs. Chem Geol 265(1–2):148–159

Martens S, Kempka T, Liebscher A, Lüth S, Möller F, Myrttinen A, Norden B, Schmidt-Hattenberger C, Zimmer M, Kühn M, Group TK (2012) Europe's longest-operating on-shore CO_2 storage site at Ketzin, Germany: a progress report after three years of injection. Environ Earth Sci 67:323–334

Martens S, Liebscher A, Möller F, Würdemann H, Schilling F, Kühn M (2011) Progress report on the first european on-shore CO_2 storage site at ketzin (Germany)—second year of injection. Energy Procedia. In: 10th international conference on greenhouse gas control technologies, vol 4, pp 3246–3253

Martens S, Liebscher A, Möller F, Henninges J, Kempka T, Lüth S, Norden B, Prevedel B, Szizybalski A, Zimmer M, Kühn M, Group K (2013) CO_2 storage at the Ketzin pilot site, Germany: fourth year of injection, monitoring, modelling and verification. Energy Procedia. In: GHGT-11 proceedings of the 11th international conference on greenhouse gas control technologies, 18–22 Nov 2012, Kyoto, Japan 37, pp 6434–6443

Mathieson A, Midgely J, Wright I, Saoula N, Ringrose P (2011) In Salah CO_2 storage JIP: CO_2 sequestration monitoring and verification technologies applied at Krechba, Algeria. Energy Procedia. In: Proceedings of 10th international conference on greenhouse gas control technologies. IEA Greenhouse Gas Programme, Amsterdam, The Netherlands, vol 4, pp 3596–3603

Matter JM, Takahashi T, Goldberg D (2007) Experimental evaluation of in situ CO_2–water–rock reactions during CO_2 injection in basaltic rocks: implications for geological CO_2 sequestration. Geochem Geophys Geosyst 8:Q02001

Mavko G, Mukerji T (1998) Bounds on low-frequency seismic velocities in partially saturated rocks. Geophysics 63:918–924

Mavko G, Mukerji T, Dvorkin J (2003) The rock physics handbook: tools for seismic analysis of porous media. Cambridge University Press, Cambridge

Mayne WH (1962) Common reflection point horizontal data stacking techniques. Geophysics 27:927–938

McCallum SD, Riestenberg DE, Cole DR, Freifeld BM, Trautz RC, Hovorka SD, Phelps TJ (2005) Monitoring geologically sequestered CO_2 during the Frio Brine pilot test using perfluoro-carbon tracers

Meadows MA (2013) 4D rock and fluid properties analysis at the Weyburn Field, Saskatchewan. Int J Greenh Gas Control. The IEAGHG Weyburn-Midale CO_2 monitoring and storage project 16(Suppl 1):S134–S145

Meadows MA, Cole SP (2013) 4D seismic modeling and CO_2 pressure-saturation inversion at the Weyburn Field, Saskatchewan. Int J Greenh Gas Control. The IEAGHG Weyburn-Midale CO_2 monitoring and storage project 16(Suppl 1):S103–S117

Meissner R, Meixner E (1969) Deformation of seismic wavelets by thin layers and layered boundaries. Geophys Prospect 17:1–27

Michael K, Golab A, Shulakova V, Ennis-King J, Allinson G, Sharma S, Aiken T (2010) Geological storage of CO_2 in saline aquifers—a review of the experience from existing storage operations. Int J Greenh Gas Control 4:659–667

Myers M, Stalker L, Ross A, Dyt C, Ho K-B (2012) Method for the determination of residual carbon dioxide saturation using reactive ester tracers. Appl Geochem 27:2148–2156

Myers M, Stalker L, Pejcic B, Ross A (2013a) Tracers—past, present and future applications in CO_2 geosequestration. Appl Geochem 30:125–135

Myers M, White C, Pejcic B, Stalker L, Ross A (2013b) Chemical tracer partition coefficients for CCS (no. EP133018). Chemical Tracer Partition Coefficients for CCS, CSIRO

Nakatsuka Y, Xue Z, Garcia H, Matsuoka T (2010) Experimental study on CO_2 monitoring and quantification of stored CO_2 in saline formations using resistivity measurements. Int J Greenh Gas Control. The ninth international conference on greenhouse gas control technologies 4:209–216

NETL (2012) BEST PRACTICES for monitoring, verification, and accounting of CO_2 stored in deep geologic formations—2012 update. DOE/NETL-2012/1568 second edition. US National Energy Technology Laboratory www.netl.doe.gov

Niemi A, Bensabat J, Shtivelman V, Edlmann K, Gouze P, Luquot L, Hingerl F, Benson SM, Pezard PA, Rasmusson K, Liang T, Fagerlund F, Gendler M, Goldberg I, Tatomir A, Lange T, Sauter M, Freifeld B (2016) Heletz experimental site overview, characterization and data analysis for CO_2 injection and geological storage. Int J of Greenhouse Gas Control 48:3–23

Nilsen MH, Herrera PA, Ashraf M, Ligaarden I, Iding M, Hermanrud C, Lie K-A, Nordbotten JM, Dahle HK, Keilegavlen E (2011) Field-case simulation of CO_2-plume migration using vertical-equilibrium models. Energy Procedia. In: 10th international conference on greenhouse gas control technologies, vol 4, pp 3801–3808

Njiekak G, Schmitt DR, Yam H, Kofman RS (2013) CO_2 rock physics as part of the Weyburn-Midale geological storage project. Int J Greenh Gas Control. The IEAGHG Weyburn-Midale CO_2 monitoring and storage project 16(Suppl 1):S118–S133

Nooner SL, Eiken O, Hermanrud C, Sasagawa GS, Stenvold T, Zumberge MA (2007) Constraints on the in situ density of CO_2 within the Utsira formation from time-lapse seafloor gravity measurements. Int J Greenh Gas Control. In: 8th international conference on greenhouse gas control technologies GHGT-8 1, pp 198–214

Noordman WH, De Boer GJ, Wietzes P, Volkering F, Janssen DB (2000) Assessment of the use of partitioning and interfacial tracers to determine the content and mass removal rates of nonaqueous phase liquids. Environ Sci Technol 34:4301–4306

Norden B, Förster A, Vu-Hoang D, Marcelis F, Springer N, Le Nir I (2010) Lithological and petrophysical core-log interpretation in CO2SINK, the European CO_2 onshore research storage and verification project. SPE Reserv. Eval, Eng 13

Nottebohm M, Licha T, Sauter M (2012) Tracer design for tracking thermal fronts in geothermal reservoirs. Geothermics 43:37–44

Oye V, Aker E, Daley TM, Kühn D, Bohloli B, Korneev V (2013) Microseismic monitoring and interpretation of injection data from the in Salah CO_2 storage site (Krechba), Algeria. Energy Procedia. In: GHGT-11 proceedings of the 11th international conference on greenhouse gas control technologies, 18–22 Nov 2012, Kyoto, Japan 37, pp 4191–4198

Park J, Fawad M, Viken I, Aker E, Bjørnarå TI (2013) CSEM sensitivity study for Sleipner CO_2-injection monitoring. Energy Procedia. In: GHGT-11 proceedings of the 11th international conference on greenhouse gas control technologies, 18–22 Nov 2012, Kyoto, Japan 37, pp 4199–4206

Picard G, Bérard T, Chabora E, Marsteller S, Greenberg S, Finley RJ, Rinck U, Greenaway R, Champagnon C, Davard J (2011) Real-time monitoring of CO_2 storage sites: application to illinois Basin–Decatur project. Energy Procedia. In: 10th international conference on greenhouse gas control technologies, vol 4, pp 5594–5598

Poupon A, Loy ME, Tixier MP (1954) A contribution to electrical log interpretation in Shaly Sands. J Pet Technol 6:27–34

Preston C, Monea M, Jazrawi W, Brown K, Whittaker S, White D, Law D, Chalaturnyk R, Rostron B (2005) IEA GHG Weyburn CO_2 monitoring and storage project. Fuel Process Technol 86:1547–1568

Prevedel B, Wohlgemuth L, Henninges J, Krüger K, Norden B, Förster A, CO2SINK Drilling Group (2008) The CO2SINK boreholes for geological storage testing. Sci Drill

Prevedel B, Wohlgemuth L, Legarth B, Henninges J, Schütt H, Schmidt-Hattenberger C, Norden B, Förster A, Hurter S (2009) The CO_2SINK boreholes for geological CO_2-storage testing. Energy Procedia. In: Proceedings of the 9th international conference on greenhouse gas control technologies 9 (GHGT-9), 16–20 Nov 2008, Washington DC, USA 1, pp 2087–2094

Pruess K, Oldenburg CM, Moridis GJ (1999) TOUGH2 user's guide version 2. Lawrence Berkeley National Laboratory, Berkeley

Rasmusson K, Rasmusson M, Fagerlund F, Bensabat J, Tsang Y, Niemi A (2014) Analysis of alternative push–pull-test-designs for determining in situ residual trapping of carbon dioxide. Int J Greenh Gas Control 27:155–168. doi:10.1016/j.ijggc.2014.05.008

Rassenfoss S (2012) Fiber optic sensing—learning how it really feels downhole. J Pet Technol 64:36–43

Rhoades JD, Manteghi NA, Shouse PJ, Alves WJ (1989) Soil electrical conductivity and soil salinity: new formulations and calibrations. Soil Sci Soc Am J 53:433

Ringrose P, Atbi M, Mason D, Espinassous M, Myhrer Ø, Iding M, Mathieson A, Wright I (2009) Plume development around well KB-502 at the In Salah CO_2 storage site. First Break 27

Ringrose PS, Mathieson AS, Wright IW, Selama F, Hansen O, Bissell R, Saoula N, Midgley J (2013) The in Salah CO_2 storage project: lessons learned and knowledge transfer. Energy Procedia 37:6226–6236

Ross C (2000) Effective AVO crossplot modeling: a tutorial. Geophysics 65:700–711

Rutqvist J, Vasco DW, Myer L (2010) Coupled reservoir-geomechanical analysis of CO_2 injection and ground deformations at In Salah, Algeria. Int J Greenh Gas Control. The ninth international conference on greenhouse gas control technologies, vol 4, pp 225–230

Sakurai S, Ramakrishnan TS, Boyd A, Mueller N, Hovorka S (2006) Monitoring saturation changes for CO_2 sequestration: petrophysical support of the Frio brine pilot experiment. Petrophysics 47

Schaffer M, Maier F, Licha T, Sauter M (2013) A new generation of tracers for the characterization of interfacial areas during supercritical carbon dioxide injections into deep saline aquifers: kinetic interface-sensitive tracers (KIS tracer). Int J Greenh Gas Control 14:200–208

Schmidt-Hattenberger C, Bergmann P, Kießling D, Krüger K, Rücker C, Schütt H, Group K (2011) Application of a vertical electrical resistivity array (VERA) for monitoring CO_2 migration at the Ketzin site: first performance evaluation. Energy Procedia. In: 10th International conference on greenhouse gas control technologies, vol 4, pp 3363–3370

Shevalier M, Nightingale M, Mayer B, Hutcheon I, Durocher K, Perkins E (2013) Brine geochemistry changes induced by CO_2 injection observed over a 10 year period in the Weyburn oil field. Int J Greenh Gas Control. The IEAGHG Weyburn-Midale CO_2 monitoring and storage project 16(Suppl 1):S160–S176

Shuey R (1985) A simplification of the Zoeppritz equations. Geophysics 50:609–614

Singh VP, Cavanagh A, Hansen H, Nazarian B, Iding M, Ringrose PS (2010) Reservoir modeling of CO_2 plume behavior calibrated against monitoring data from Sleipner. Society of Petroleum Engineers, Norway

Strazisar BR, Wells AW, Diehl JR, Hammack RW, Veloski GA (2009) Near-surface monitoring for the ZERT shallow CO_2 injection project. Int J Greenh Gas Control 3:736–744

Tatomir A, Maier F, Schaffer M, Licha T, Sauter M (2013) Modelling of Kinetic Interface Sensitive Tracers for Two-Phase Systems. In: Hou MZ, Xie H, Were P (eds) Clean energy systems in the subsurface: production, storage and conversion. Springer Series in Geomechanics and Geoengineering, Berlin, pp 65–74

Tatomir AB, Schaffer M, Kissinger A, Hommel J, Nuske P, Licha T, Helmig R, Sauter M (2015) Novel approach for modeling kinetic interface-sensitive (KIS) tracers with respect to time-dependent interfacial area change for the optimization of supercritical carbon dioxide injection into deep saline aquifers. Int J Greenhouse Gas Control 33:145–153

Telford WM, Telford WM, Geldart LP, Sheriff RE (1990) Applied geophysics. Cambridge University Press, Cambridge

Theis CV (1935) The relation between the lowering of the Piezometric surface and the rate and duration of discharge of a well using ground-water storage. EOS Trans Am Geophys Union 16:519–524

Todd T, Simmons G (1972) Effect of pore pressure on the velocity of compressional waves in low-porosity rocks. J Geophys Res 77:3731–3743

Tomich JF, Dalton RL, Deans HA, Shallenberger LK (1973) Single-well tracer method to measure residual oil saturation. J Pet Technol 25:211–218. doi:10.2118/3792-PA

Tong F, Niemi A, Yang Z, Fagerlund F, Licha T, Sauter M (2013) A numerical model of tracer transport in a non-isothermal two-phase flow system for CO_2 geological storage characterization. Transp Porous Media 98:173–192

Underschultz J, Boreham C, Dance T, Stalker L, Freifeld B, Kirste D, Ennis-King J (2011) CO_2 storage in a depleted gas field: an overview of the CO_2CRC Otway Project and initial results. Int J Greenh Gas Control 5:922–932

van Genuchten MT (1980) A closed-form equation for predicting the hydraulic conductivity of unsaturated soils 1. Soil Sci Soc Am J 44:892

Vandeweijer V, van der Meer B, Hofstee C, Mulders F, D'Hoore D, Graven H (2011) Monitoring the CO_2 injection site: K12-B. Energy Procedia 4:5471–5478

Vasco D, Ferretti A, Novali F (2008) Reservoir monitoring and characterization using satellite geodetic data: interferometric synthetic aperture radar observations from the Krechba field, Algeria. GEOPHYSICS 73, WA113-WA122

Vasco DW, Rucci A, Ferretti A, Novali F, Bissell RC, Ringrose PS, Mathieson AS, Wright IW (2010) Satellite-based measurements of surface deformation reveal fluid flow associated with the geological storage of carbon dioxide. Geophys Res Lett 37:L03303

Verdon JP, Kendall J-M, White DJ, Angus DA (2011) Linking microseismic event observations with geomechanical models to minimise the risks of storing CO_2 in geological formations. Earth Planet Sci Lett 305:143–152

Verdon JP, Kendall J-M, Stork AL, Chadwick RA, White DJ, Bissell RC (2013) Comparison of geomechanical deformation induced by megatonne-scale CO_2 storage at Sleipner, Weyburn, and In Salah. Proc Natl Acad Sci 110:E2762–E2771

Vulava VM, Perry EB, Romanek CS, Seaman JC (2002) Dissolved gases as partitioning tracers for determination of hydrogeological parameters. Environ Sci Technol 36:254–262

Wang Y (1999) Approximations to the Zoeppritz equations and their use in AVO analysis. Geophysics 64:1920–1927

Watts M (2003) High temperature circuit reliability testing. Quartzdyne, Inc, Salt Lake City

Waxman MH, Thomas EC (1974) Electrical conductivities in Shaly Sands-I. The relation between hydrocarbon saturation and resistivity index; II. The temperature coefficient of electrical conductivity. J Pet Technol 26:213–225
Wells AW, Diehl JR, Bromhal G, Strazisar BR, Wilson TH, White CM (2007) The use of tracers to assess leakage from the sequestration of CO_2 in a depleted oil reservoir, New Mexico, USA. Appl Geochem 22:996–1016
Wells A, Strazisar B, Diehl JR, Veloski G (2010) Atmospheric tracer monitoring and surface plume development at the ZERT pilot test in Bozeman, Montana, USA. Environ Earth Sci 60:299–305
White DJ (2011) Geophysical monitoring of the Weyburn CO_2 flood: results during 10 years of injection. Energy Procedia. In: 10th international conference on greenhouse gas control technologies, vol 4, pp 3628–3635
White D (2013a) Seismic characterization and time-lapse imaging during seven years of CO_2 flood in the Weyburn field, Saskatchewan, Canada. Int J Greenh Gas Control. The IEAGHG Weyburn-Midale CO_2 monitoring and storage project 16(Suppl 1):S78–S94
White DJ (2013b) Toward quantitative CO_2 storage estimates from time-lapse 3D seismic travel times: an example from the IEA GHG Weyburn–Midale CO_2 monitoring and storage project. Int J Greenh Gas Control. The IEAGHG Weyburn-Midale CO_2 monitoring and storage project 16(Suppl 1):S95–S102
Whittaker (2004) Geological framework. In: Presented at the 7th international conference on greenhouse gas control technologies, Vancouver, Canada
Widess MB (1973) How thin is a thin bed? Geophysics 38:1176–1180
Wildgust N, Gilboy C, Tontiwachwuthikul P (2013) Introduction to a decade of research by the IEAGHG Weyburn-Midale CO_2 monitoring and storage project. Int J Greenh Gas Control 16: S1–S4
Wood AB (1941) A textbook of sound: being an account of the physics of vibrations with special reference to recent theoretical and technical developments. Bell, London
Worthington PF (1985) The evolution of Shaly-sand concepts in reservoir evaluation. Log Anal 26
Xu T, Kharaka YK, Doughty C, Freifeld BM, Daley TM (2010) Reactive transport modeling to study changes in water chemistry induced by CO_2 injection at the Frio-I Brine Pilot. Chem Geol 271:153–164
Yilmaz Ö (2001) Seismic data analysis. Investigations in Geophysics. Society of Exploration Geophysicists
Zemke K, Liebscher A, Wandrey M (2010) Petrophysical analysis to investigate the effects of carbon dioxide storage in a subsurface saline aquifer at Ketzin, Germany (CO2SINK). Int J Greenh Gas Control. CO_2 storage at the EGU general assembly 2009 4:990–999
Zettlitzer M, Moeller F, Morozova D, Lokay P, Würdemann H (2010) Re-establishment of the proper injectivity of the CO_2-injection well Ktzi 201 in Ketzin, Germany. Int J Greenh Gas Control. CO_2 storage at the EGU general assembly 2009 4:952–959
Zhang Y, Freifeld B, Finsterle S, Leahy M, Ennis-King J, Paterson L, Dance T (2011) Single-well experimental design for studying residual trapping of supercritical carbon dioxide. Int J Greenh Gas Control 5:88–98
Zimmer M, Erzinger J, Kujawa C (2011) The gas membrane sensor (GMS): a new method for gas measurements in deep boreholes applied at the CO2SINK site. Int J Greenh Gas Control 5:995–1001. doi:10.1016/j.ijggc.2010.11.007
Zoeppritz K (1919) VII b. Über Reflexion und Durchgang seismischer Wellen durch Unstetigkeitsflächen. Nachrichten Von Ges. Wiss Zu Gött Math Phys Kl 1919:66–84

Chapter 9
Natural Analogue Studies

Christopher Ian McDermott, Johannes M. Miocic, Katriona Edlmann and Stuart M.V. Gilfillan

Abstract Lessons learned from sites where CO_2 has naturally been stored for long geologic periods of time provides valuable information for assessing proposed anthropogenic storage sites. This chapter discusses the natural CO_2 storage analogue sites and looks at them worldwide to determine which geological characteristics are preferable for natural CO_2 storage and which are not. Following this, an approach is presented based on geomechanical facies, for a comparative assessment of storage sites, accounting for features observed in the natural analogue sites. Finally, a number of anthropogenic storage sites are classified according to the characterization criteria and a detailed description of a number of natural and anthropogenic storage sites are presented.

9.1 Introduction

In this chapter we define what a natural CO_2 storage analogue is, then we look at an extensive catalogue of analogue storage sites worldwide to determine which geological characteristics are preferable for natural CO_2 storage and which are less preferable. We then apply a holistic approach using geomechanical facies to enable a comparative assessment of storage sites accounting for the various features seen in the natural analogue sites, relating to the individual tectonic and depositional settings. Following this, anthropogenic storage sites are classified according to the

C.I. McDermott (✉) · J.M. Miocic · K. Edlmann · S.M.V. Gilfillan
School of Geoscience, Edinburgh Collaborative of Subsurface Science and Engineering (ECOSSE), University of Edinburgh, Edinburgh, UK
e-mail: cmcdermo@staffmail.ed.ac.uk

J.M. Miocic
e-mail: johannes.miocic@ed.ac.uk

K. Edlmann
e-mail: katriona.edlmann@ed.ac.uk

S.M.V. Gilfillan
e-mail: stuart.gilfillan@ed.ac.uk

A. Niemi et al. (eds.), *Geological Storage of CO_2 in Deep Saline Formations*, Theory and Applications of Transport in Porous Media 29,
DOI 10.1007/978-94-024-0996-3_9

characterization developed from using a geomechanical facies framework, providing a good match indicating that the geomechanical facies approach provides a good first order method of assessment for storage sites. We then provide a detailed conceptual model description and hydro-mechanical parameterization of several natural and anthropogenic storage sites based on the geomechanical facies analysis (Otway, Australia—CO_2 storage project, In Salah, Algeria—CO_2 storage project, Sleipner, Norway—CO_2 storage project, Snøhvit, Norway—CO_2 storage project, Buracica, Brazil—CO_2 Enhanced Oil Recovery, Miller Field, UK North Sea—natural CO_2 reservoir, St. Johns Dome, USA—natural CO_2 reservoir, Fizzy Field, UK Southern North Sea—natural CO_2 reservoir). Finally, in conclusion, we address what we learned from the analogue studies and the application of geomechanical facies approach.

Where CO_2 has been naturally stored over long periods of time in rocks, we can learn lessons from the natural geological conditions regarding what factors are important to retain the CO_2 compared to other rocks where CO_2 is present but not retained. These sites are natural analogues to the proposed engineered storage sites for CO_2. Natural analogue sites provide the possibility to investigate and determine the key factors which ensure the storage of CO_2 and allow scientists to develop selection criteria based on these factors when identifying contemporary engineered storage sites.

CO_2 originating from natural sources such as mantle degassing, volcanism, carbonate rock metamorphism and the degradation of organic matter, is common in sedimentary basins world-wide (Wycherley et al. 1999). Sedimentary basins are formed by layers of strata. A stratum is a layer of sedimentary rock with internally consistent characteristics distinguishing it from other layers. Strata comprise multiple stratum. Typically different types of strata can be found in sedimentary basins, from highly permeable porous rocks such as sandstones, which form good reservoir rocks, to rocks with a very low permeability such as shales, mudstones or evaporites, which can act as effective barriers and thus as seals. Naturally occurring CO_2 is often found in reservoir rocks in which it can reside as either free phase or dissolved within the fluid found within the pores of the rock. Sealing rocks above the reservoir rock can prevent the vertical movement of CO_2 over geological time-scales.

In cases where CO_2 naturally occurs within a reservoir rock-sealing rock sequence located within a structural or lithologic geologic feature that inhibits lateral or vertical movement of the fluids within the reservoir rock, a so-called trap, the site can be identified as a natural analogue for engineered geological CO_2 storage sites. This is because the site has all the features of proposed engineered storage sites such as a reservoir rock—sealing rock pair, a trapping mechanism, and CO_2.

Natural analogues for CO_2 stores can offer unique insights into the long-term behavior and retention of CO_2 in the subsurface (Baines and Worden 2004) and thereby provide truly long-term data on the interaction of CO_2 with the reservoir and caprocks, which are impossible to reproduce in laboratory studies or short-term field experiments. In addition, such sites offer geological evidence of ancient and/or current migration of CO_2 out of the primary reservoir, sometimes all the way to the surface. These can offer insights into the mechanisms by which engineered sites may fail and which properties are optimal for a secure storage site and thus give information relevant for the selection of effective CO_2 storage sites.

Analogue CO_2 storage sites provide experimental evidence of storage performance over long geological time scales, and provide important insights into the key controls of the storage system. The main factors important in enabling CO_2 storage can be shown to be the same for most storage sites. Therefore there are certain common factors which can be compared and contrasted world-wide. Specifically, recognizing that storage sites have holistic characteristics—that is, there is a reservoir rock, a seal rock and an overburden—and by investigating the primary controls on the formation of these systems, including deposition, diagenesis and stress controls, it is possible to identify which sites are likely to provide better storage locations than others.

In order to identify factors that render a site a secure storage site—the opposite of an insecure storage site from which CO_2 is leaking to the surface—over 60 naturally occurring CO_2 reservoirs where CO_2 has been stored over geological time-scales, and in some cases is migrating to the surface, were investigated. This global dataset was then examined for consistent mechanisms leading to the secure retention of CO_2 in the subsurface reservoirs and to identify which processes may lead to the migration of CO_2 out of the reservoir to the surface.

In a second step we compared different sites, including natural analogues and existing CO_2 storage sites, using the framework offered by considering the geomechanical facies present. A geomechanical facies is a conceptual building block for the subsurface. It has specific material characteristics defined by the geology of the rocks and defined by the engineering use to which it will be put. A good analogy is the use of different bricks with the construction of a house. Each brick has certain characteristics, and is used for a certain purpose. The combination of the bricks forms the house, its shape and its individual appearance. We try and identify the characteristics of the "geological" bricks, the geomechanical facies, and then use this concept to compare different sites.

A geological facies is defined as a body of rock with specified characteristics (Reading 1978). A geomechanical facies is described as a series of geological facies grouped together on the basis of engineering parameters that fulfil a specific role within the storage system, e.g. reservoir, caprock, overburden (McDermott et al. 2006). For complete CO_2 storage site assessment, in addition to the basin architecture and sedimentary stratigraphy, the fluid flow characteristics (hydrogeology) and the mechanical characteristics (fractures, rock strength and elastic properties) are particularly important.

Using a geomechanical facies framework, the factors crucial to assessing the CO_2 storage security of a storage basin such as basin architecture, caprock architecture, reservoir quality, stress state, mechanical characteristics, fractures, burial depth, geothermal gradient, risk of orogenic modification, structural stability and preservation potential, can all be taken into account.

Obviously the tectonic setting exerts a principal influence over these components. Different tectonic settings exhibit different depositional process controls.

This directly influences sediment thickness and the distribution of the caprock and reservoir sediments (Hallam 1981). The tectonic setting also determines basin architecture, stress state, mechanical characteristics, fracture properties, burial depths, geothermal gradient, structural stability and preservation potential. An additional factor of importance is the facies distribution within the caprock, and their heterogeneity.

By examining the typical characteristics of the geomechanical facies within the different tectonic settings, it is possible to compare and contrast the different tectonic settings to appraise global CO_2 storage opportunities and predict which tectonic settings will be most suitable for CO_2 storage.

9.2 Natural Analogue Sites and Key Storage Controls

One of the key challenges when studying naturally occurring CO_2 reservoirs as analogues for storage sites, is to correctly determine whether a site is secure, storing CO_2 without any leakage for geological time scales, or insecure, with leakage of CO_2 to the surface occurring. It is crucial that sites are correctly identified if mechanisms that lead to leakage and thus insecure storage site are to be analyzed. Movement of natural CO_2 to the surface can be identified by various surface manifestations often called CO_2 seeps (Roberts et al. 2011). These include

- dry CO_2 degassing via focused vents (a discrete opening that allows gas to pass out of the soil) or diffusely over an area without a discrete vent;
- CO_2-driven mud volcanoes or mofettes (a mofette is a vent from which carbon dioxide and some nitrogen and oxygen issue from the earth in a last stage of volcanic activity); and
- springs with CO_2-rich groundwaters that in some cases are accompanied by travertine deposition (a carbonate rock that precipitates when CO_2-rich waters degas on reaching the surface).

It has to be noted that CO_2 seeps are not necessarily related to a subsurface CO_2 reservoir but may instead represent an open system where CO_2 flow is not constricted and thus CO_2 does not accumulate in large quantities in the subsurface (Roberts et al. 2015). This is important as it shows that not every CO_2 seep on the surface is related to a CO_2 bearing reservoir rock in the subsurface from which CO_2 is migrating. Indeed, many known occurrences of CO_2 seeps are related to volcanic activity and are not linked to an analogue site (reservoir rock—sealing rock pair, a geological trap) in depth.

Natural CO_2 reservoirs at a regional scale have been examined as analogues for saline aquifer carbon storage sites for different regions: the Colorado Plateau (Stevens et al. 2001), Europe (Holloway et al. 2005; Pearce et al. 1996, 2004) and China (Dai et al. 2005), but not on a global scale yet. According to these studies, migration from subsurface CO_2 reservoirs towards the surface occurs mainly along

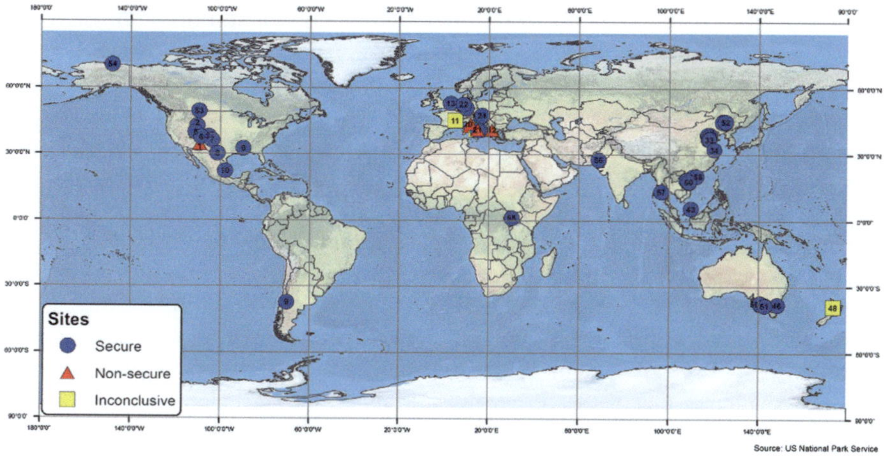

Fig. 9.1 Map showing the locations of natural CO_2 reservoirs included in this study. *Map source* US National Park Service

small and large discontinuities (fractures and faults) within the rocks that are between the reservoir in depth and the surface. Faults (and fractures) can be generally described as planar features and therefore migration of CO_2 along faults is spatially restricted. The fact that faults and fractures are the main migration pathways is not very surprising as fault zones have long been recognized as fluid migration pathways in the subsurface for oil, gas, and groundwater (Faulkner et al. 2010). Considerable research has been completed in the last decades on the hydraulic properties of faults, in particular on the predictability of whether hydrocarbons will or will not flow up or through fault zones (Manzocchi et al. 2010).

We have examined 61 naturally occurring CO_2 reservoirs around the globe in order to better understand the mechanisms that lead to migration of CO_2 out of subsurface reservoirs to the surface and what controls the secure retention of CO_2 within such reservoirs (Miocic et al. 2013; Miocic et al. 2016). The locations of the studied reservoirs are shown in Fig. 9.1. For a reservoir to be classified as insecure, evidence of CO_2 migration to the surface had to be present. This includes all of the above listed types of CO_2 seeps. If such a seep was located within a 10 km surface radius of the subsurface extent of the natural CO_2 reservoir, the reservoir was classified as insecure. The 10 km radius is based on an extensive study of natural CO_2 seeps in Italy by (Roberts 2012) which conclusively found that surface seeps linked to deep naturally occurring CO_2 reservoirs which held CO_2 in a free phase—rather than holding CO_2 dissolved in the pore-fluid—occurred within a 10 km radius of boreholes which encountered the free phase CO_2 in depth.

Of the 61 studied naturally occurring CO_2 reservoirs, six (10 %) show clear evidence of CO_2 migration to the surface and have therefore been classified as insecure. Three (5 %) reservoirs show inconclusive evidence for a successful retention of CO_2 in the subsurface: Montmiral in SE France, which is used as a secure example by Pearce et al. (2004), has many CO_2 rich springs within a 10 km radius of the field which provide evidence for CO_2 migration to the surface. However, it is currently unclear if the CO_2 originates from the reservoir or is sourced from elsewhere. The Monte Taburno reservoir in central Italy is located just 1.6 km from a thermal spring with a small CO_2 content and since there is no further geochemical information about the spring or the CO_2 reservoir, the relationship between the two is unclear (Roberts 2012). The Paritutu reservoir offshore New Plymouth, NZ, is shallow and there is a vent at the surface degassing CO_2 (Lyon et al. 1996). However, the distance between the reservoir penetrating well and the vent is unknown, as are the possible CO_2 migration pathways. Fifty-two reservoirs (85 %) show no evidence of CO_2 migration to the surface above or within a 10 km radius of the subsurface extent of the reservoir, which was concluded to provide sufficient evidence that these reservoirs are successfully sealed. Features of the natural analogues are compared and contrasted below to identify the factors that promote the security of the storage and those which can be associated with leaks.

9.2.1 *Properties of Naturally Occurring CO_2 Reservoirs*

The CO_2 contained in the studied reservoirs is mainly sourced from mantle degassing and igneous processes. This was the case for 62 % of the 35 reservoirs for which stable carbon isotope and noble gas geochemical data is available, with the remainder being sourced from the thermal breakdown of marine carbonates and organic matter. The CO_2 concentrations (vol.% of gas produced) in the reservoirs

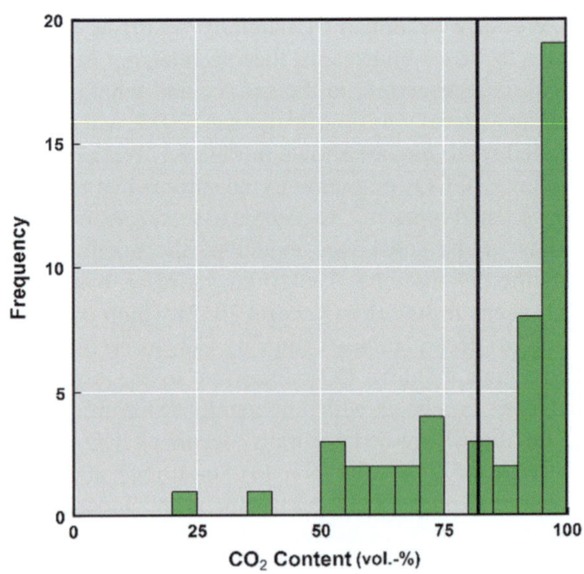

Fig. 9.2 Frequency plot showing the CO_2 concentration in the studied natural reservoirs. The average concentration is 82 %, with the majority of reservoirs holding CO_2 in concentrations of more than 90 %

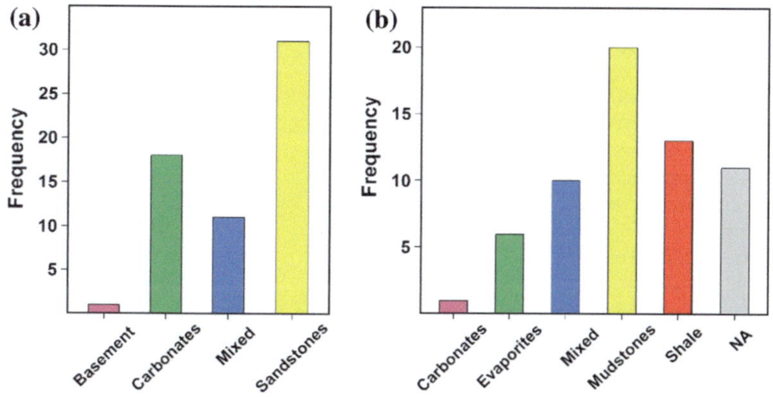

Fig. 9.3 Frequency plots showing **a** the distribution of reservoir rocks and **b** the distribution of sealing rock lithologies of the studied naturally occurring reservoirs

range from 20 to >99 %, with an average concentration of 82 % (Fig. 9.2). This shows that the studied natural reservoirs are good analogues for saline aquifer storage sites where the CO_2 content is to be thought more than 80 %. The natural reservoirs with lower concentrations are analogues for storage sites in depleted oil and gas fields where the CO_2 concentration is naturally lower as residual oil and gas also fills the pores. Other frequently trapped gases include, in order of decreasing abundance; methane, nitrogen, helium and H_2S. It should be noted that there were no notable differences in the CO_2 composition, origin or concentration between the secure and non-secure reservoirs.

The reservoir rocks of naturally occurring CO_2 reservoirs are commonly siliciclastic lithologies (50 %) with sandstones dominating and only minor amounts of siltstones and conglomerates (Fig. 9.3). The other principal reservoir rock lithologies are carbonates (30 %), with limestones and dolomites being equally represented. In some of the analogue sites interlayered carbonate and siliciclastic rocks (17 %) form the reservoir sequence and in one single case basement rocks form the reservoir. While the data on the composition of reservoir rocks is available for all studied naturally occurring CO_2 reservoirs, data on caprock (or sealing rock) is less frequently accessible. This is related to the fact that many of the CO_2 reservoirs are found during exploratory drilling for hydrocarbons. When the operators realize that no hydrocarbons occur within the reservoir they often abandon the well without conducting a detailed analysis of the reservoir-seal interval. For the studied CO_2 reservoirs for which data on the caprock is available, the dominant lithologies are fine grained silicate mudstones and shales (54 %). Interlayered carbonate and mudstone/shales are also common (16 %, Fig. 9.3). Other caprock lithologies are evaporites, with anhydrite and halite both acting as primary seals. There is no relationship between the type of reservoir rock or caprock and the capability of a reservoir to successfully retain CO_2 for geological periods of time. This indicates that both siliciclastic and carbonate rocks can form good reservoir rocks for carbon

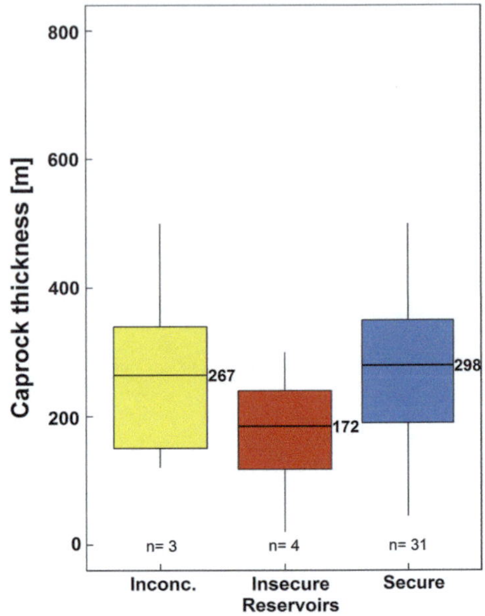

Fig. 9.4 Boxplot of caprock thickness as determined from available geological data against secure and insecure CO_2 reservoirs. The caprock above sealing reservoirs is generally thicker than caprock above insecure reservoirs. The *boxplot* shows the median (*black horizontal line*) and the interquartile range. The *whiskers* (*black vertical line*) depict the 1.5 inter-quartile range

storage sites, and that any caprock, if thick enough (see below), can successfully prevent CO_2 from migrating out of the reservoir.

The thickness of the low permeability and porosity lithology directly above the reservoir seems to have a direct influence on whether a naturally occurring CO_2 reservoir is secure or insecure: Caprocks of secure reservoirs are about twice as thick as caprocks of insecure reservoirs, which have an average thickness of 172 m (Fig. 9.4). Caprocks of the inconclusively secure reservoirs are also on average thicker than the caprocks of insecure reservoirs. Note that here the limitations of the dataset are of importance: There are only three data points for inconclusively secure reservoirs and only four data points for insecure reservoirs while there is data for 31 secure naturally occurring CO_2 reservoirs. Thus there are some uncertainties with this statistical examination and more insecure reservoirs should be added to the dataset in the future. Sites where the caprock directly above the reservoir is not the only low-permeability rock in the rock column above the reservoir but only one of several caprocks, appear to assist the successful retention of CO_2 in the subsurface. Such multi caprock systems or layered compartments occur in at least 30 % of the secure reservoirs, with up to five different reservoir horizons, each corresponding with a caprock. Only one of the insecure reservoirs has layered compartments and these seem to be connected via fracture networks.

For CO_2 storage sites the pressure and temperature within the reservoir rock are important as they govern CO_2 properties such as density and physical state which in turn determine the amount of CO_2 that can be stored within the reservoir. Fluids within sedimentary rocks are under pressure, and this pressure generally increases with depth, the deeper a rock is the higher the pressure of the fluid within the pores is.

There are three pressure gradients within the subsurface which play an important role: (1) in an open system where fluids can move both vertically and laterally through the subsurface and thus dissipate pressure, pore fluids have pressures along the hydrostatic gradient which is at around 10 kPa/m. If movement of fluids is restricted, pore pressures are commonly higher than the hydrostatic gradient (overpressured). (2) If the pore pressure continues to increase, it will overcome the strength of the rock and induce fractures. The pressure at which this occurs is called fracture gradient and is related to (3) the lithostatic gradient. The lithostatic gradient is the pressure caused by the overlying rock material and is a function of depth, rock density. Temperatures in the subsurface generally increase with depth, with deeper rocks having higher temperatures. The gradient that defines the temperature increase differs from sedimentary basin to sedimentary basin, with basins located in areas with a thin lithosphere or strong magmatic activity having a high temperature gradient while basins on cratons having a low temperature gradient. The average global temperature gradient for sedimentary basins is in the order of 30 °C/km.

The studied naturally occurring CO_2 reservoirs around the globe are located in a range of depths below the ground surface (Fig. 9.5). The shallowest reservoir is located in only 300 m depth (Messokampos, Greece), while the deepest reservoir is located at a depth of 4600 m (Jackson Dome, USA). Note that insecure reservoirs are, with one exception, located at depths shallower than 1200 m below surface. Due to the wide range of depths it is not surprising that the reservoir fluid pressures also show a wide range, from 0.5 MPa to more than 60 MPa (Fig. 9.5). Shallow CO_2 reservoirs (<1200 m depth below surface) that are sealing are hydrostatically pressured, whereas insecure reservoirs at these depths exhibit pressures both above and below hydrostatic. Some sealing reservoirs that are deeper than 1200 m below surface show excess pressures 40–50 % above hydrostatic. All insecure and inconclusively insecure reservoirs at these depths exhibit high overpressures. These pressures are close to 85 % of lithostatic pressure, the known fracture pressure of caprocks in the North Sea (Moss et al. 2003), and in other sedimentary basins where the rock fractures (Hillis 2003). The reservoir temperatures of the studied naturally occurring CO_2 reservoirs range from 20 to 200 °C, with insecure reservoirs having either "normal" (30 °C/km) or very high temperature gradients (Fig. 9.5).

Based on the temperature and pressure conditions within the naturally occurring CO_2 reservoirs the CO_2 state and density can be calculated (Fig. 9.6). In the studied sites CO_2 occurs as gaseous phase where pressures are less than the critical pressure of 7.39 MPa, and as supercritical phase where pressures are more than 7.39 MPa and temperatures exceed the critical temperature of 31.1 °C. CO_2 densities in the studied sites range from 100 kg/m^3 to more than 800 kg/m^3. Gaseous CO_2 in the studied analogues has densities of <220 kg/m^3, while the density of supercritical CO_2 ranges from 160 kg/m^3 to more than 800 kg/m^3.

Insecure naturally occurring CO_2 reservoirs tend to be shallow and thus have low reservoir pressures. The CO_2 is in gaseous form and in five out of six insecure reservoirs the CO_2 has a density of <200 kg/m^3. A comparison of reservoirs with gaseous and supercritical conditions shows that reservoirs in which CO_2 is stored in a gaseous state are more prone to leakage than reservoirs with supercritical

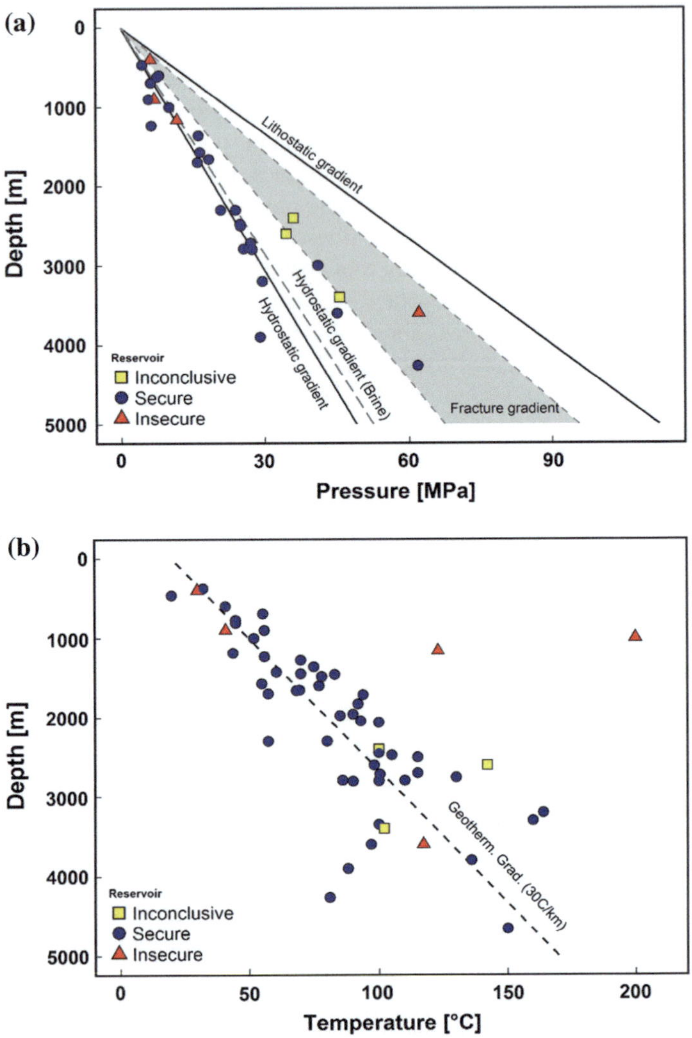

Fig. 9.5 **a** Depth versus pressure plot of natural CO_2 reservoirs with in situ pressure data. Note that non-secure reservoirs are mainly shallow (<1200 m) or within the fracture gradient range. The range of fracture gradients in sedimentary basins is illustrated by the shaded area which ranges from 60 to 90 % of lithostatic stress. The deep, insecure reservoir with reservoir pressure over the fracture gradient is Pieve Santo Stefano, Italy. **b** Depth versus temperature plot of natural CO_2 reservoirs, based on regional temperature gradients as well as in situ data. Note that a high geothermal gradient may lead to migration of CO_2 in shallow reservoirs. Also note that not all reservoirs with temperature data have in situ pressure data and therefore may not be plotted on the pressure graph

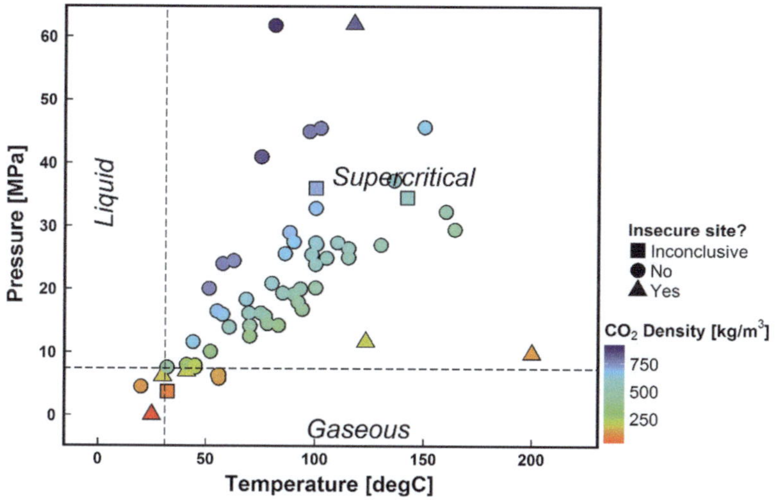

Fig. 9.6 CO_2 state diagram (pressure vs. temperature plot) illustrating the range of reservoir conditions found in naturally occurring CO_2 reservoirs. Reservoirs with gaseous CO_2 and reservoirs with low-density supercritical CO_2 are more likely to be insecure than reservoirs with dense, supercritical CO_2

conditions: 37 % (3 out of 8) of the reservoirs with gaseous CO_2 show evidence for migration of CO_2 out of the reservoir to the surface, while only about 6 % (3 out of 53) of reservoirs with supercritical conditions exhibit such evidence. Here we also relate to recent surprising experimental evidence which suggests that gaseous CO_2 may be more mobile through fractures in the subsurface than supercritical CO_2 (Edlmann et al. 2013).

In some cases CO_2 occurs in several formations of multi-layered reservoirs. Structural geological data indicates that faults are the pathways through which CO_2 migrates from one formation to the next. Faults also play an important role as migration pathways to the surface: for five of the six insecure CO_2 reservoirs, the migrating CO_2 emerges at the surface close to fault tips and traces. Surface manifestations of migrating CO_2 linked to faults are CO_2 rich springs and travertine deposits. While faults are clearly migration pathways that render some of the reservoirs insecure, the mere presence of a fault at a naturally occurring CO_2 reservoir is not equivalent to the reservoir being insecure. More than half (56 %) of the natural reservoirs that securely hold CO_2 over geological timescales are fault bound structural traps. At such traps one (or several) large fault(s) form the boundary of the reservoir and withstand migration of CO_2 through or along the fault. Several more secure reservoirs that are not fault-bound are located in structurally complex and faulted provinces. This is a clear indication that faults often also inhibit CO_2 migration rather than being pathways for leakage. It is noteworthy that the majority of insecure, fault-bound reservoirs are found in tectonically active regions, such as the Apennine thrust belt in Italy or the Florina Basin in Greece.

Therefore active, or close to critically stressed faults, may be more prone to act as migration pathway than faults in tectonically quiet areas. Indeed, the state of stress has a direct influence on the permeability of fault zones (Barton et al. 1995).

9.2.2 *Mechanisms of CO_2 Migration at Naturally Occurring CO_2 Reservoirs*

There are three processes that can lead to the vertical migration of CO_2 from a reservoir through the caprock: Migration through unfractured caprock by capillary flow, migration by fracturing the caprock, and migration through faults (Fig. 9.7). In the following evidence for these processes at the studied naturally occurring CO_2 reservoirs is discussed.

9.2.2.1 Migration of CO_2 Through Unfractured Caprocks

Migration of CO_2 through mudrocks or shales will occur when the pressure in the reservoir exceeds that of the capillary entry pressure of fractures or pores in the seal (Chiquet et al. 2007). The small pore sizes of low permeability rocks require capillary entry pressures of several tens of MPa for this to occur. The density and phase conditions of CO_2 are dependent on pressure and temperature, which is a direct function of the depth of the reservoir. The density contrast between CO_2 and brine in the reservoir decreases with increasing depth, and hence differential buoyancy pressure on the caprock also decreases with increasing depth. For this reason shallow reservoirs (<1000 m depth) are inherently more likely to leak CO_2 through an unfractured caprock. Yet, there are no indications that leakage through the caprock by capillary flow is a leaking mechanism in the studied shallow

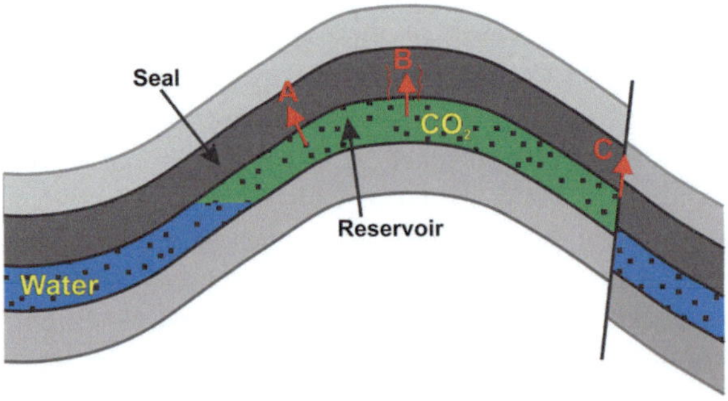

Fig. 9.7 Diagram illustrating three potential migration pathways for CO_2 out of a natural reservoir: (*A*) Migration through unfractured caprock, (*B*) migration by fracturing of the caprock and (*C*) migration along pre-existing faults and fractures. In naturally occurring CO_2 reservoirs only the latter mechanism is found to play a significant role

reservoirs. For leaking CO_2 reservoirs in Italy, Roberts et al. (2015) were able to show that the observed surface CO_2 seep rates greatly exceed the physical possibility of leakage by capillary flow through intact mudrock from the area above the site. Thus CO_2 migration must be via fractures not matrix flow. Lu et al. (2009) showed that even after an estimated 70 Ma, CO_2 only infiltrated 12 m of sealing mudrocks directly over a CO_2 rich oil field in the UK North Sea. Hence, the time-scales required for CO_2 migration through unfractured caprocks are orders of magnitudes longer than those necessary for CO_2 storage to effectively mitigate climate change.

9.2.2.2 Migration by Caprock Fracturing

As burial depth increases, it becomes more likely that fluid pressures will be over pressured, that is, pore pressures deviate from hydrostatic (or "normal") pressures towards lithostatic pressures. If pore pressure in the reservoir exceeds both the pore pressure in the caprock and the tensile strength of the caprock (including any differences in confining stress due to different elastic properties), hydraulic fracturing and/or frictional failure along optimally oriented pre-existing fractures of the caprock occurs (Finkbeiner et al. 2001; McDermott et al. 2013). Both mechanisms can lead to migration of CO_2 from the reservoir through the caprock by fracture flow (Shukla et al. 2010, Fig. 9.7). This effect was induced at the CO_2 injection test site at In Salah (Rinaldi and Rutqvist 2013). Fracturing to form dilatant joints (mode 1 fractures) induced by elevated fluid pressure only occurs when the pressure exceeds the least principal stress of the caprock (Hillis 2003) (this direction of least principal stress is typically horizontal until depths exceed 1.5–2 km (Nara et al. 2011). The pore pressure required to cause such failure is much less than the pore pressure required to overcome the capillary entry pressure of a mudstone caprock (Busch et al. 2010) and so caprocks will transmit CO_2 more readily by fracture flow than by capillary flow. Most sedimentary rocks are fractured during burial and the fracture density depends on the geomechanical properties and thickness of the rock and thus different rock layers fracture differently in response to the same stress (Hanks et al. 1997; Ladeira and Price 1981). Fracture density increases in the vicinity of faults (damage zone) and fractures provide permeability only when open or connected (Faulkner et al. 2003). There is no clear evidence for leakage through dilatant joints in hydraulically fractured caprock in the examined sites.

9.2.2.3 Migration Through Faults

Strong evidence for CO_2 migration through fault induced fractures exists at several insecure reservoirs. CO_2 seeps are frequently located close to active or extinct faults, which may exist prior to CO_2 migration (Roberts et al. 2015; Shipton et al. 2004). Thus fractures and flow through fractures as part of a fault zone play a significant role in permitting CO_2 migration. This indicates that pre-existing faults are important pathways for CO_2 migration, possibly due to increased fracture permeability in the fault damage zone.

9.3 Implications for Engineered CO_2 Storage Sites

Deep (>1500 m) insecure and possibly insecure (inconclusive) reservoirs have high reservoir fluid pressures close to fracture pressure (around 70–85 % of lithostatic pressure) and will therefore readily fail by hydraulic failure (Fig. 9.5). Leak-off data from UK North Sea reservoirs show that the least principal stress in the region is typically within 70–85 % of lithostatic pressure and this has been shown to be the case in other sedimentary basins (Hillis 2003; Moss et al. 2003). Therefore when considering potential sites for engineered CO_2 storage it is critical that the pore pressure, stress history and the present day stress state of the selected storage complex are well understood. It is also imperative that reservoir pressures during CO_2 injection are maintained below the least principal stress of the region. In shallow (<800–1000 m) or hot reservoirs, the low density of gaseous CO_2 compared to the density of reservoir brine leads to a high buoyancy pressure of CO_2 on the caprock. Hence, the differential stress exerted by the state of stored CO_2 needs to be calculated specifically for each individual engineered CO_2 storage sites and especially at potential sites with abnormal temperature gradients. With increasing reservoir pressure (depth) and temperature, CO_2 enters the supercritical state and has a (significantly) higher density, leading to a lower buoyancy pressure. The impact of CO_2 density on migration is highlighted by the fact that two of the three insecure reservoirs where CO_2 is in supercritical conditions have low CO_2 densities (119 and 200 kg/m^3). The larger the difference between the buoyancy pressure and the hydrostatic pressure, the more likely it is that the buoyancy pressure will cause stress change that may cause frictional failure of the fractures. In addition, fault related damage zones and fractures may have an increased permeability when close to critically stressed (Faulkner et al. 2010). Recent experimental investigations of CO_2 flow through naturally fractured caprock indicate that gaseous CO_2 flows more readily through fractures than supercritical CO_2 (Edlmann et al. 2013). From the observations of natural CO_2 reservoirs it can be concluded that migration through faults and fractures is mainly restricted to shallow reservoirs which contain gaseous CO_2 or supercritical CO_2 with a low density.

In recent years studies by international research consortia (including industry, academic and legislative partners) have developed criteria for the selection of industrial storage sites. These criteria, which cover potential risks for CO_2 storage sites from basin-scale (Bachu 2003; Veritas 2010; NETL 2010) to reservoir scale (Chadwick et al. 2008; Delprat-Jannaud et al. 2013; IEAGHG 2009; Smith et al. 2011) are intended to be used to select secure storage sites. Key criteria for all studies include depth, CO_2 state, and the presence of (open) fractures or faults. It is recommended that CO_2 is stored at depths of >800 m (IEA GHG 2009; NETL 2010; Smith et al. 2011) or in depths of >1000 m (Chadwick et al. 2008). Most studies

Table 9.1 Table listing key properties of the six insecure naturally occurring CO_2 reservoirs of this study

Site	Depth (m)	Temperature (°C)	Pressure (MPa)	CO_2 state	Fault
St. Johns Dome	*400–700*	30–49	*6.17*	*Gaseous*	Yes
Farnham Dome	900	41	*7*	*Gaseous*	Yes
Messokampos	*300*	*25*	*0.5*	*Gaseous*	Yes
Latera Caldera	1000	*200*	–	Sc	Yes
Pieve Santo Stefano	3600	117	62	Sc	*Yes*[a]
Frigento Field	1160	*123*	11.7	Sc	Yes

Italic indicates that using the site selection criteria discussed in the text, the reservoir property would render the reservoir unsuitable for CO_2 storage
[a]Seismically active fault, other faults are not known to be active
sc Supercritical

recommend CO_2 to be stored in supercritical state with reservoir temperatures of more than 35 °C at normal (~30 °C/km) temperature gradients, and reservoir pressures at more than 7.5 MPa. The caprocks should be "lateral extensive" (NETL 2010) with "minimal faulting" (Smith et al. 2011), effectively ruling out active faults. The capillary entry pressure of the caprocks should be greater than the pressure increase induced in the reservoir during CO_2 injection (Chadwick et al. 2008).

If these site selection criteria are used to screen the six insecure reservoirs, it becomes clear that all six of them would have been ruled out for failing at least one key selection criteria (Table 9.1). Three of the reservoirs hold CO_2 in gaseous state due to them being at shallow depths (400–700 m at St. Johns Dome; 500 m at Messokampos) or having reservoir pressures of less than 7.5 MPa with the reservoir being located at sufficient depth (7 MPa at 900 m depth, Farnham Dome). Two of the reservoirs are located in suitable depths and hold supercritical CO_2 but have very high temperatures (200 °C at 1000 m, Latera Caldera; 123 °C at 1160 m, Frigento Field). The last of the insecure reservoirs, Pieve Santo Stefano, is located very deep and has supercritical CO_2 at "normal" temperatures and pressures but is located next to a seismically active fault. The fact that all insecure natural reservoirs would have been detected by the selection criteria improves confidence that the internationally accepted selection criteria for engineered storage sites are effective in selecting storage sites which will be able to store CO_2 safely for the timescales required.

9.4 Geomechanical Facies Approach for Characterization

The subsurface is not a random collection of materials. Geological processes have led to structured deposits with distinct material characteristics and geometrical relationships. Processes have formed the geomechanical facies as a conceptual building block for the subsurface. A geomechanical facies has specific material characteristics defined by the geology of the rocks and defined by the engineering use to which it will be put. Using the geomechanical facies framework, the factors

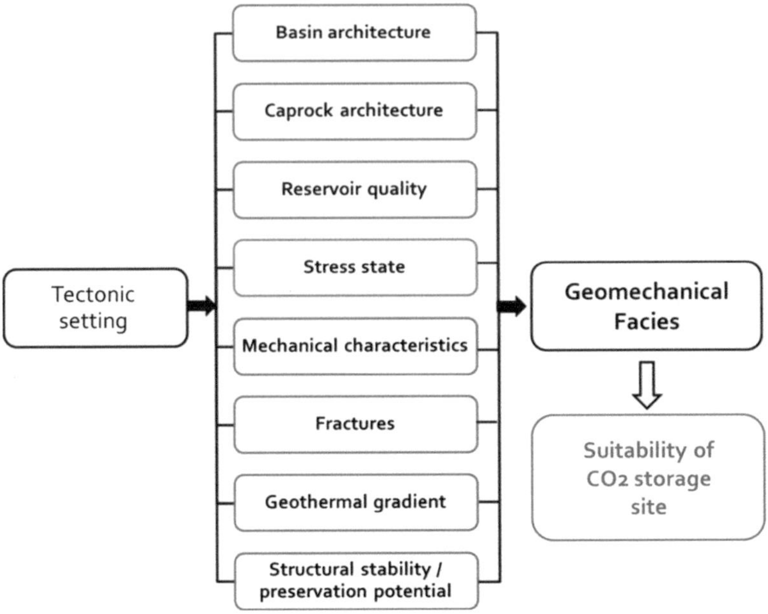

Fig. 9.8 Geomechanical facies parameters as controlled by tectonic setting (after Edlmann et al. 2015)

crucial to assessing the CO_2 storage security of a storage basin such as basin architecture, caprock architecture, reservoir quality, stress state, mechanical characteristics, fractures, burial depth, geothermal gradient, risk of orogenic modification, structural stability and preservation potential (Fig. 9.8) can all be taken into account.

By examining the typical characteristics of the geomechanical facies within the different tectonic settings, it is possible to compare and contrast the different tectonic settings to appraise global CO_2 storage opportunities and predict which tectonic settings will be most suitable for CO_2 storage.

It is important to note that the geomechanical facies approach for CO_2 site selection is the first step in what is a complex and iterative site specific assessment procedure, where uncertainty decreases as the data requirement increases. However this first appraisal step is a crucial stage in the identification and assessment of suitable CO_2 storage sites and underpins the later screening and ranking procedure of sedimentary basins for geological CO_2 storage (e.g. Bachu 2003).

Tables 9.2, 9.3, 9.4 and 9.5 show the results of a detailed analysis of the key parameters of sedimentary basins as determined by Edlmann et al. (2015) using the geomechanical facies approach along with an assessment of their net contribution towards ideal CO_2 storage conditions. These tables were used to apply the geomechanical facies approach to the CO_2 storage projects of Otway, Sleipner, In Salah, Snøhvit, Buracica and the natural CO_2 reservoirs of Miller, Fizzy and St. Johns.

Table 9.2 Summary of the key sedimentary stratigraphy, basin architecture, reservoir quality and caprock architecture as determined using the geomechanical facies approach, along with an assessment of their net contribution towards ideal CO_2 storage conditions

Key sedimentary stratigraphy and facies	Geomechanical facies assessment	
Extensional systems		
a. Oceanic basin		
Sedimentary stratigraphy	Pelagic sediments, fine grained clays and turbidites	
Basin architecture	Long narrow and straight. 10–100 km wide and >2000 km long	Good
Reservoir quality	Thin limited reservoir sands	Poor
Caprock architecture	Caprock muds and silts are extensive and thick	Good
b. Passive continental margin		
Sedimentary stratigraphy	Sedimentation dominated by mud, silt and fine sand laid down in thick sequences	
Basin architecture	Basin has distinct concave upwards base. Straight basins that are a few 10 km's wide and 1000's km long	Moderate
Reservoir quality	Reservoir sands are often extensive and thick	Good
Caprock architecture	Caprock muds and silts are extensive and thick	Good
c. Terrestrial rift basin		
Sedimentary stratigraphy	Sedimentation limited to material derived from neighbouring fault scarps and uplifted blocks. Dominant sediments are alluvial fans, lakes and marine. Fast filling and burial rates	
Basin architecture	Asymmetric geometry based along a boundary fault with a long narrow geometry. 10 km's wide and up to 2000 km long	Moderate
Reservoir quality	Thick rift basin bound or channelised reservoir sands	Moderate
Caprock architecture	Thick caprock muds and silts also rift basin bound or channelised	Moderate
Convergent systems		
a. Trench		
Sedimentary stratigraphy	Variable sedimentation, with trench fans, axial channel sandstones, non channelised sheet flow spreading over and down the trench and starved trench with only hemipelagic mud and turbidites deposited	
Basin architecture	Long and narrow basins that are concave towards the oncoming subducting plate. 10 km's wide and 1000's km long	Moderate
Reservoir quality	Reservoir sands rare and have highly non-predictable geometry	Poor
Caprock architecture	Caprock muds and silts will also have a non-predictable geometry	Poor

(continued)

Table 9.2 (continued)

Key sedimentary stratigraphy and facies	Geomechanical facies assessment	
b. Forearc basin		
Sedimentary stratigraphy	Clastic sedimentation predominates with turbidites and other mass flow deposits, marine sediments commonly grading into deltaic and fluvial sediments	
Basin architecture	Over 100 km wide and >2000 km long	Good
Reservoir quality	Reservoir sands will be thick but non-predictable geometry	Moderate
Caprock architecture	Caprock will also be thick but have non-predictable geometry	Moderate
c. Backarc basin		
Sedimentary stratigraphy	Thick clastic sedimentation predominates, there are pelagic sediments overlying newly formed basin crust; several thousand meters of turbidites in abyssal plains and continental shelves; shallow marine deposition and thick molasses type sediments	
Basin architecture	Small basins, no larger than km's wide and 10's kilometres long. They tend to be linear parallel to the trench	Poor
Reservoir quality	Reservoir sands can be complex and unpredictable	Moderate
Caprock architecture	Caprock muds and silts can be thick and extensive	Good
d. Foreland basin		
Sedimentary stratigraphy	Foreland basinsare filled with sediments that erode from the adjacent mountain belt. The width and depth of the foreland basin is determined by the flexural rigidity of the underlying lithosphere and the characteristics of the mountain belt	
Basin architecture	10's to a few 100 km's wide and 100's to 1000's of km long, varying profile reflecting the geometry of subduction	Good
Reservoir quality	Reservoir sands will be thick and extensive	Good
Caprock architecture	Caprock muds and silts are likely to be thick and extensive	Good
Wrench system		
a. Strikeslip pull apart basin		
Sedimentary stratigraphy	Typically the margins are sites of deposition of coarse facies alluvial fans and fan deltas and these pass laterally over short distance to lacustrine in continental settings or marine deposits	
Basin architecture	Rhomboidal shape elongating with time. km to a few 10's km wide and lengths of km to many 10's km	Moderate
Reservoir quality	Reservoir sands are thick and extensive	Good
Caprock architecture	Caprock muds and silts are thick and extensive	Good

Where an assessment of GOOD means that the input fulfils the ideal attributes of a storage reservoir, an assessment of MODERATE means it fulfils a reasonable number of the ideal attributes and a POOR assessment means it fulfils a low number of ideal attributes of a CO_2 storage reservoir

Table 9.3 Summary of the key stress state, fracture characteristics and geothermal gradient as determined using the geomechanical facies approach, along with an assessment of their net contribution towards ideal CO_2 storage conditions

Key stress state, fracture network and geothermal gradient	Reservoir and caprock quality	
Extensional systems		
a. Oceanic basin		
Stress state	On the stress ellipsoid, the maximum effective stress ($\sigma 1$) is in the vertical direction	
Fracture characteristics	Transform faults. Long near straight parallel fractures perpendicular to the ridge	Moderate
Geothermal gradient	Elevated geothermal gradient	Good
b. Passive continental margin		
Stress state	On the stress ellipsoid, the maximum effective stress ($\sigma 1$) is in the vertical direction	
Fracture characteristics	Normal faults, may flatten with depth	Good
Geothermal gradient	Elevated geothermal gradient	Good
c. Terrestrial rift basin		
Stress state	On the stress ellipsoid, the maximum effective stress ($\sigma 1$) is in the vertical direction	
Fracture characteristics	Normal faulting, commonly half graben with a single boundary fault	Moderate
Geothermal gradient	Average geothermal gradient	Moderate
Convergent systems		
d. Trench		
Stress state	Tectonic stress in a convergent system is characterised by a horizontal oblong stress ellipsoid with $\sigma 1$ in the horizontal direction	
Fracture characteristics	Normal faults, ocean ward of the subduction zone	Moderate
Geothermal gradient	Lower geothermal gradients due to the thrusting of cold, water-filled sediments beneath existing crust	Poor
e. Forearc basin		
Stress state	Tectonic stress in a convergent system is characterised by a horizontal oblong stress ellipsoid with $\sigma 1$ in the horizontal direction	
Fracture characteristics	There are multiple normal fault populations, with off sets of 20 m and dips of 60–70°	Poor
Geothermal gradient	Lower than normal geothermal gradients because of the cooling effect of the relatively cold subducting plate	Poor

(continued)

Table 9.3 (continued)

Key stress state, fracture network and geothermal gradient	Reservoir and caprock quality	
f. Backarc basin		
Stress state	Tectonic stress in a convergent system is characterised by a horizontal oblong stress ellipsoid with σ1 in the horizontal direction	
Fracture characteristics	Extensional faults which form due to the gravitational effects of the subducted crust	Moderate
Geothermal gradient	Lower than normal geothermal gradients because of the cooling effect of the relatively cold subducting plate	Poor
g. Foreland basin		
Stress state	Tectonic stress in a convergent system is characterised by a horizontal oblong stress ellipsoid with σ1 in the horizontal direction	
Fracture characteristics	Predominantly sedimentary wedges and thrust rather than deep faults	Good
Geothermal gradient	Cooler than normal geothermal gradient	Poor
Wrench system		
a. Strike slip pull apart basin		
Stress state	A tectonic stress field in which the maximum and minimum principal stresses σ1 and σ3 are orientated along the horizontal plane and the intermediate principal stress (σ2) is vertical	
Fracture characteristics	Faults range in size from plate boundaries to small scale fractures with only a few hundred meters or even just tens of centimetres of movement. Typical development of flower structure of normal and reverse faults. Rotations of small scalefault blocks	Moderate
Geothermal gradient	Elevated geothermal gradient	Good

Where an assessment of GOOD means that the input fulfils the ideal attributes of a storage reservoir, an assessment of MODERATE means it fulfils a reasonable number of the ideal attributes and a POOR assessment means it fulfils a low number of ideal attributes of a CO_2 storage reservoir

Table 9.4 Summary of the typical risk of overprint (stability), orogenesis modification and preservation potential as determined using the geomechanical facies approach, along with an assessment of their net contribution towards ideal CO_2 storage conditions

Extensional systems	
a. Oceanic basin	
Stability/ risk of overprint	Poor—high risk of overprint or destruction
Risk of major orogenesis modification	Poor—high risk of orogenesis modification
Preservation potential	Poor
b. Passive continental margin	
Stability/risk of overprint	Moderate risk of overprint or destruction
Risk of major orogenesis modification	Moderate risk of orogenesis modification
Preservation potential	Moderate
c. Terrestrial rift basin	
Stability/ risk of overprint	Moderate risk of overprint or destruction
Risk of major orogenesis modification	Good—low risk of orogenesis modification
Preservation potential	Good
Convergent systems	
a. Trench	
Stability/risk of overprint	Poor—high risk of overprint or destruction. Preserved portions will have collided and accreted onshore and are uplifted as mountain regions.
Risk of major orogenesis modification	Poor—high risk of orogenesis modification
Preservation potential	Poor
b. Forearc basin	
Stability/ risk of overprint	Moderate risk of overprint or destruction
Risk of major orogenesis modification	Poor—high risk of orogenesis modification
Preservation potential	Moderate
c. Backarc basin	
Stability/ risk of overprint	Moderate risk of overprint or destruction
Risk of major orogenesis modification	Moderate risk of orogenesis modification
Preservation potential	Moderate
d.Foreland basin	
Stability/ risk of overprint	Moderate risk of overprint or destruction
Risk of major orogenesis modification	Good—low risk of orogenesis modification
Preservation potential	Good

(continued)

Table 9.4 (continued)

Wrench system	
a. Strikeslip pull apart basin	
Stability/ risk of overprint	Poor—high risk of overprint or destruction
Risk of major orogenesis modification	Moderate risk of orogenesis modification
Preservation potential	Moderate

Where an assessment of GOOD means that the input fulfils the ideal attributes of a storage reservoir, an assessment of MODERATE means it fulfils a reasonable number of the ideal attributes and a POOR assessment means it fulfils a low number of ideal attributes of a CO_2 storage reservoir

9.5 Geomechanical Facies Models

Here we demonstrate the application of the geomechanical facies model on several contemporary storage sites, and also provide a generic overview of the sites including material parameters characterization.

9.5.1 Otway, Australia: CO_2 Storage Project

9.5.1.1 Extensional Terrestrial Rift Basin

The CO2CRC Otway Project Pilot Site, the largest demonstration project for geological CO_2 storage and monitoring in Australia, is located in the onshore portion of the Otway Basin, Victoria (Jenkins et al. 2012). Between March 2008 and August 2009 about 65,000 tons of gas, including 58,000 tons of CO_2, have been injected into a depleted, fault bound, natural gas field (Naylor Field). The sandstone reservoir of cretaceous age is in 2050 m depth and has an approximate thickness of 30 m with porosities of up to 30 % and high permeabilities of up to 1–5 Darcy (Dance et al. 2009). The sands are predominantly fluvial channels and tidal fluvial (reworked) sandstones. They are overlain by the 300 m thick Belfast Mudstone which is also the fault bound seal. The bounding faults terminate within the Belfast Mudstone and fluid migration into overlying aquifers is thus unlikely. The maximum horizontal stress orientation is NW–SE and is consistent with the maximum stress orientation in the Otway Basin. Geomechanical analysis shows that pore-pressures could be increased by 1–15.7 MPa, depending on assumptions made about stress magnitude, fault strength, reservoir stress path and Biot's coefficient, before faults would be reactivated (Vidal-Gilbert et al. 2010). Figure 9.9 presents the generic stratigraphy for the Otway CO_2 storage project.

Table 9.5 Summary overview of each of the assessed CO_2 storage component variable as determined by the geomechanical facies approach for each tectonic setting

Tectonic setting	Basin architecture	Reservoir quality	Caprock architecture	Fracture characteristics	Geothermal gradient	Stability/overprint	Orogenesis modification	Preservation potential
Oceanic basins	Good	Poor	Good	Moderate	Good	Poor	Poor	Poor
Passive continental margin	Moderate	Good	Good	Good	Good	Moderate	Moderate	Moderate
Terrestrial rift basins	Moderate	Moderate	Moderate	Moderate	Moderate	Moderate	Good	Good
Trench basins	Moderate	Poor	Poor	Moderate	Poor	Poor	Poor	Poor
Fore-arc basins	Good	Moderate	Moderate	Poor	Poor	Moderate	Poor	Moderate
Back-arc basins	Poor	Good	Moderate	Moderate		Moderate	Moderate	Moderate
Foreland basins	Good	Good	Good	Good Poor	Poor	Moderate	Good	Good
Strike-slip basins	Moderate	Good	Good	Moderate	Good	Poor	Moderate	Moderate

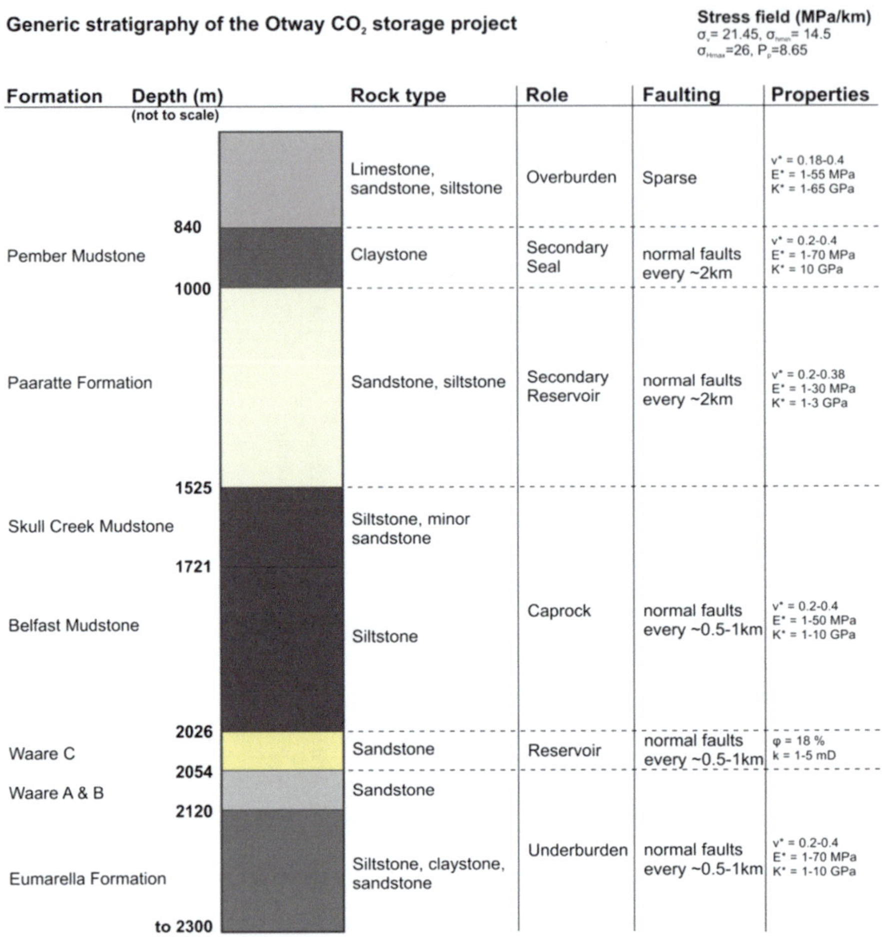

Fig. 9.9 Generic stratigraphy for the Otway CO_2 storage project. *Asterisk* indicates literature values instead of in-situ data. *v* Poisson's ratio, *E* Young's Modulus, *K* Bulk modulus, *φ* porosity, *k* permeability

9.5.1.2 Geomechanical Facies Model

The geomechanical facies model would predict this to be a moderate storage opportunity with long narrow basin architecture, thick but channelized reservoir and caprock architecture and few fractures. The basin will have low risk of overprint, low risk of orogenesis modification and good preservation potential.

The reservoir model and field operation data from the Otway CO_2 storage project exhibits a close first order predictive match with the geomechanical facies model, as summarized in Table 9.6.

Table 9.6 Summary of the match between the geomechanical facies storage potential prediction and the actual Otway field data

Otway	Basin architecture	Sedimentary stratigraphy	Reservoir potential	Caprock extent	Faults	Preservation potential
Geomechanical facies prediction—*terrestrial rift basin*	Long narrow geometry	Alluvial fans, lakes and marine deposits	Thick channelized sands	Thick caprocks	Single boundary normal faults	Good preservation potential
Field data	Elongated basin, reservoirs cut by faults	Fluvial and tidal sandstoneswith marine shales	30 m thick sandstone	300 m shale	Normal bounding faults	Little evidence of tectonic modification
Model/field data match	Good	Good	Moderate	Good	Good	Good

9.5.2 In Salah, Algeria: CO_2 Storage Project

9.5.2.1 Convergent Backarc Basin

The In Salah Gas Project, located in Algeria, is currently the world's largest onshore CO_2 storage site. The CO_2 is separated from natural gas produced from three nearby gas fields. CO_2 is injected into an underground saline aquifer of Carboniferous age through three wells. The storage formation is a tidal-deltaic Carboniferous sandstone overlain by about 900 m of mudstones and siltstones, which form the caprock (Bissell et al. 2011; Ringrose et al. 2011). The reservoir sandstone is approximately 20 m thick and extensively fractured with a predominant open fracture set (NW–SE). This is in close alignment with the present-day stress field, related to tectonic plate convergence between Africa and Eurasia. The storage formation is also segmented by strike-slip faults, indicative of a regional mid-to-late Carboniferous basin inversion (White et al. 2014). Surface deformation has been detected related to the CO_2 injection by DInSAR at In Salah (Onuma and Ohkawa 2009). Figure 9.10 presents the generic stratigraphy for the In Salah CO_2 storage project.

9.5.2.2 Geomechanical Facies Model

The geomechanical facies model would predict this to be a moderate storage opportunity with small linear basins, thick but complex reservoir architecture, thick caprocks with extensional fracturing. The basin will have moderate risk of overprint, moderate risk of orogenesis modification and moderate preservation potential.

The reservoir model and field operation data from the In Salah CO_2 storage project exhibits a close first order predictive match with the geomechanical facies model, as summarized in Table 9.7.

9.5.3 Sleipner, Norway: CO_2 Storage Project

9.5.3.1 Extensional Terrestrial Rift Basin

Sleipner is an offshore gas field in the mid- to eastern edge of the Viking Graben System of the North Sea. The Sleipner project is the first commercial application of storage in a deep saline aquifer in the world and CO_2 is injected into a sand layer called the Utsira formation which is a highly elongated sand reservoir, extending for more than 400 km from north to south and between 50 and 100 km from east to west. The Utsira sand is sparsely faulted and ranges in depth from 550 to 1500 m, the sand thickness is locally about 300 m and the regional top seal is a thick mudstones (Chadwick et al. 2012; Zweigel et al. 2004). It is interpreted as a basin restricted marine deposit. The caprock succession overlying the Utsira reservoir is

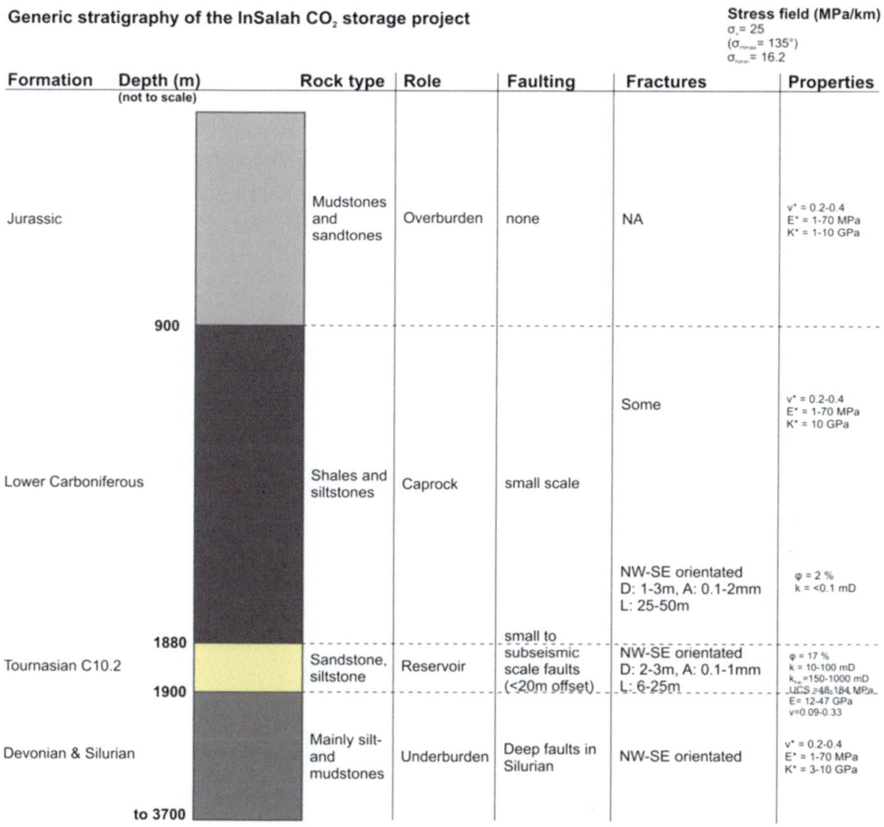

Fig. 9.10 Generic stratigraphy of the In Salah CO_2 storage project. *Asterisk* indicates literature values instead of in-situ data. v Poisson's ratio, E Young's Modulus, K Bulk modulus, φ porosity, k permeability, k_{frac} fracture permeability, *NA* no data available

100 m thick and variable and can be divided into three main units: the lower, the middle and the upper seals (Chadwick et al. 2004; Eiken et al. 2011; Harrington et al. 2009; Torp and Gale 2004). Figure 9.11 presents the generic stratigraphy for the Sleipner CO_2 storage project.

9.5.3.2 Geomechanical Facies Model

The geomechanical facies model would predict this to be a moderate storage opportunity with long narrow basin architecture, thick but channelized reservoir and caprock architecture and few fractures. The basin will have low risk of overprint, low risk of orogenesis modification and good preservation potential.

Table 9.7 Summary of the match between the geomechanical facies storage potential prediction and the actual In Salah field data

In Salah	Basin architecture	Sedimentary stratigraphy	Reservoir potential	Caprock extent	Faults	Preservation potential
Geomechanical facies prediction—convergent backarc basin	Small basins linear parallel to the trench	Alluvial fans, lakes, turbidites, shallow marine and molasses deposits	Thick channelised sands	Thick caprocks	Extensional faults	Moderate preservation potential
Field data	Elongated reservoir	Tidal deltaic	Elongated sands, 20 m reservoir formation	900 m of caprock	Extensively fractured	Subject to regional Carboniferous basin inversion
Model/field data match	Good	Moderate	Good	Good	Good	Good

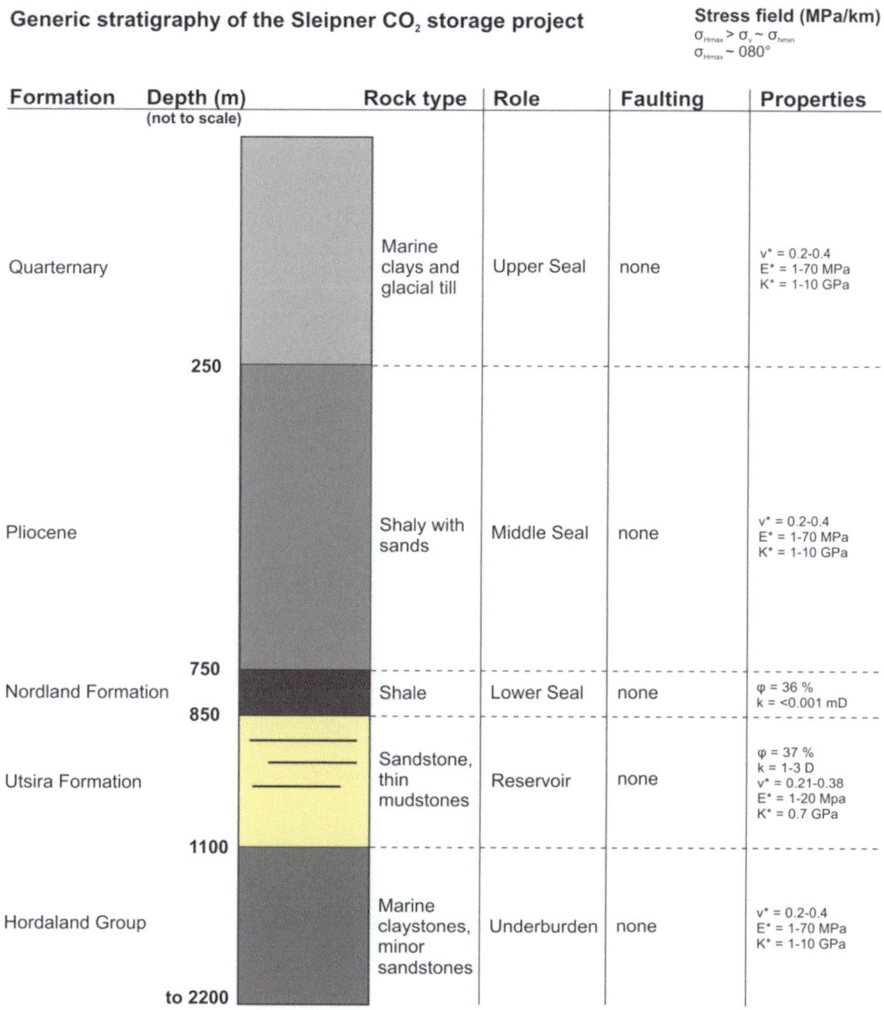

Fig. 9.11 Generic stratigraphy of the Sleipner CO_2 storage project. *Asterisk* indicates literature values instead of in-situ data. *v* Poisson's ratio, *E* Young's Modulus, *K* Bulk modulus, *φ* porosity, *k* permeability

The reservoir model and field operation data from the Sleipner CO_2 storage project exhibits a very close first order predictive match with the geomechanical facies model, as summarized in Table 9.8.

Table 9.8 Summary of the match between the geomechanical facies storage potential prediction and the actual Sleipner field data

Sleipner	Basin architecture	Sedimentary stratigraphy	Reservoir potential	Caprock extend	Faults	Preservation potential
Geomechanical facies prediction—terrestrial rift basin	Long narrow geometry	Alluvial fans, lakes and marine deposits	Thick channelized sands	Thick caprocks	Single boundary normal faults	Good preservation potential
Field data	Elongated reservoir	Basin restricted marine deposits	Elongated sands, about 300 m thick	The caprock is 100 m's thick and variable	Sparsely faulted	Little evidence of tectonic modification
Model/field data match	Good	Good	Good	Good	Good	Good

9.5.4 *Snøhvit, Norway: CO_2 Storage Project*

9.5.4.1 Extensional Terrestrial Rift Basin

The Snøhvit gas field is located in the Barents Sea, offshore northern Norway. Three gas reservoirs, operated by Statoil, are producing gas which is processed into LNG. It contains approximately 5–8 % CO_2 which is separated before liquefaction. The CO_2 is reinjected and stored in the early Jurassic Tubåen Formation at about 2600 m depth (Chiaramonte et al. 2011). Injection started in 2008 and is planned to continue for 30 years with a rate of 2000 tons/day. A total storage of 23 Mt is planned. The reservoir is formed by delta plain dominated fluvial distributary sandstones with some marine-tidal influence with a thickness of ~110 m and is located on fault blocks. The faults are ENE-WSW trending and have a maximum throw of <150 m (Hansen et al. 2013). Open fractures, dominantly with a N–S strike azimuth, are thought to locally influence fluid flow, however the reservoir quality of the sandstones with permeabilities in the range of 10–800 mD and porosities of 7–20 % is very high anyway (Wennberg et al. 2008). The local caprock is formed by the Nordmela Formation which has a thickness of 60–100 m and contains several shaly layers which are thought to act as flow barriers. Regional caprocks are formed by the thick marine shales of the Fulgen and Hekkingen Formations (Rodrigues Duran et al. 2013). Figure 9.12 presents the generic stratigraphy for the Snøhvit CO_2 storage project.

9.5.4.2 Geomechanical Facies Model

The geomechanical facies model would predict this to be a moderate storage opportunity with long narrow basin architecture, thick but channelized reservoir and caprock architecture and few fractures. The basin will have low risk of overprint, low risk of orogenesis modification and good preservation potential.

The reservoir model and field operation data from the Snøhvit CO_2 storage project exhibits a good first order predictive match with the geomechanical facies model, as summarized in Table 9.9.

9.5.5 *Buracica, Brazil: CO_2 EOR*

9.5.5.1 Extensional Terrestrial Rift Basin

The Buracica field is located in the Reconcavo Basin, a late Jurassic to early Cretaceous rift related basin, in the north–east of Brazil. CO_2 injection into the oilfield started in 1991 and until 2005 about 600,000 tons of CO_2 were injected into the reservoir which is a 9 m thick late Jurassic aeolian sandstone of the Sergi

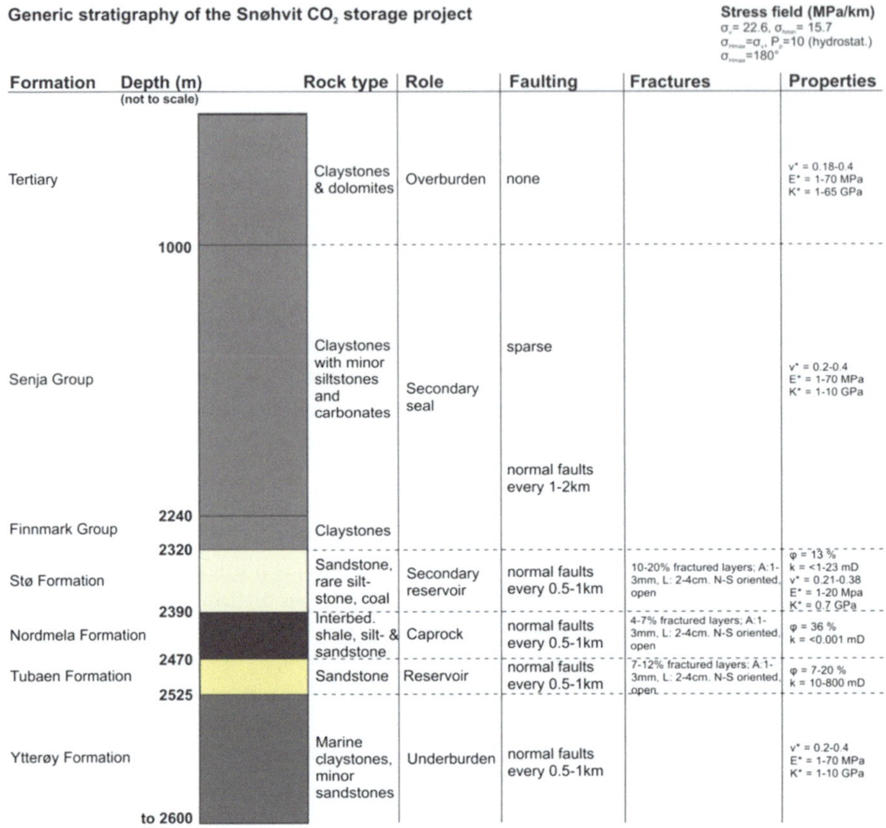

Fig. 9.12 Generic stratigraphy of the Snøhvit CO_2 storage project. *Asterisk* indicates literature values instead of in-situ data. *v* Poisson's ratio, *E* Young's Modulus, *K* Bulk modulus, φ porosity, *k* permeability

formation with an average porosity of 22 % and an average permeability of 570 mD (Estublier et al. 2011). Overall the reservoir formation is 200 m thick and consist of 14 aeolian and fluvial sandstone reservoirs that are separated by thin lacustrine deposits (Scherer et al. 2007). The reservoir is relatively shallow, with depths of 320–646 m below sea level (~500–850 m below surface), is dipping with ~6° south–eastward and has a lateral extend of 4.4 km in E–W direction and 3.2 km in N–S direction. At 470 m depth it has a temperature of 44 °C and an initial pressure of 5.5 MPa. The field is composed of three main tilted fault blocks with a maximum throw of about 150 m. The top seal is made of >150 m thick succession of early Cretaceous shales of the Itaparica and Taua formations (Hung Kiang et al. 1992; Rouchon et al. 2011). Figure 9.13 presents the generic stratigraphy for the Buracica CO_2 storage project.

Table 9.9 Summary of the match between the geomechanical facies storage potential prediction and the Snøhvit field data

Snøhvit	Basin architecture	Sedimentary stratigraphy	Reservoir potential	Caprock extend	Faults	Preservation potential
Geomechanical facies prediction—terrestrial rift basin	Long narrow geometry	Alluvial fans, lakes and marine deposits	Thick channelized sands	Thick caprocks	Single boundary normal faults	Good preservation potential
Field data	Elongated reservoir on horst structure	Fluvial and marine sandstones	Thick sands	Thick caprocks	Reactivated normal/reverse faults	Major phase of uplift in Pliocene/ Pleistocene
Model/field data match	Good	Good	Good	Good	Moderate	Moderate

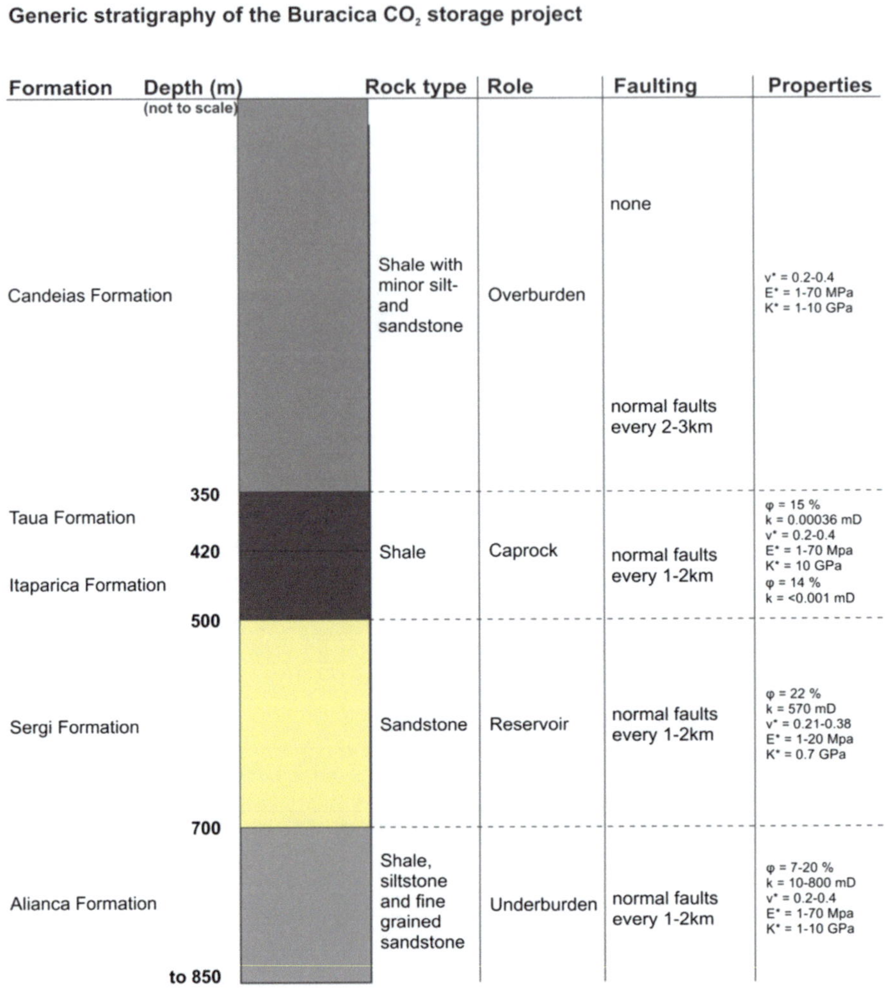

Fig. 9.13 Generic stratigraphy of the Buracica CO_2 storage project. *Asterisk* indicates literature values instead of in-situ data. v Poisson's ratio, E Young's Modulus, K Bulk modulus, φ porosity, k permeability

9.5.5.2 Geomechanical Facies Model

The geomechanical facies model would predict this to be a moderate storage opportunity with long narrow basin architecture, thick but channelized reservoir and caprock architecture and few fractures. The basin will have low risk of overprint, low risk of orogenesis modification and good preservation potential.

The reservoir model and field operation data from the Buracica CO_2 storage project exhibits a good first order predictive match with the geomechanical facies model, as summarized in Table 9.10.

Table 9.10 Summary of the match between the geomechanical facies storage potential prediction and the actual Buracica field data

Buracica	Basin architecture	Sedimentary stratigraphy	Reservoir potential	Caprock extend	Faults	Preservation potential
Geomechanical facies prediction—terrestrial rift basin	Long narrow geometry	Alluvial fans, lakes and marine deposits	Thick channelized sands	Thick caprocks	Single boundary normal faults	Good preservation potential
Field data	Elongate basin	Aeolian sandstone in alluvial environment, lacustrine shales	Overall thick sequence of fluvial and aeolian sands	Thick caprocks	Normal faults	Uplift in Tertiary
Model/field data match	Good	Good	Good	Good	Good	Moderate

9.5.6 *Miller Field, UK North Sea: Natural CO_2 Reservoir*

9.5.6.1 Extensional and Rotated Half-Graben Terrestrial Rift Basin

The Miller Field is located at the western margin of the north–south trending Viking Graben in the North Sea. It contains 28 mol% CO_2 and the reservoir and caprock have been exposed to these high concentrations of CO_2 since its emplacement and the CO_2 has been successfully stored for millions of years. The South Viking Graben is a half graben fault which is bounded against the west basement of the Fladen Ground Spur. Late Jurassic rifting and subsidence in the graben led to deposition of submarine fan systems which constitute the reservoirs in the Miller field (Eiken et al. 2011). It covers an area of 45 km^2 and shows limited faulting, faults in the Miller Field have NW–SE orientation (Rooksby 1991).

The Miller reservoir sandstones are composed of three main lithofacies. The first is clean, fine to medium-grained, well-sorted quartzose sandstone transported by, and deposited from sand-rich, high-density, low-efficiency turbidity currents. The second lithofacies is thinly bedded alternation of sandstone and mudstone, usually interbedded with the clean sandstones and deposits of the low-density turbidity currents. The third lithofacies is isolated mudstones locally interbedded with the main part of the reservoir, representing normal background sedimentation at the margins of the fan system, or during periods of non-deposition within the fan (Lu et al. 2009). During the period of highest sea level the Kimmeridge Clay Formation covered the Miller Field reservoir sands and formed a seal over the field with a thickness of several hundreds of meters. Figure 9.14 presents the generic stratigraphy for the Miller field natural CO_2 reservoir.

9.5.6.2 Geomechanical Facies Model

The geomechanical facies model would predict this to be a moderate storage opportunity with long narrow basin architecture and few fractures. The basin will have low risk of overprint, low risk of orogenesis modification and good preservation potential.

The reservoir model and field operation data from the Miller natural CO_2 reservoir exhibits a close first order predictive match with the geomechanical facies model, as summarized in Table 9.11.

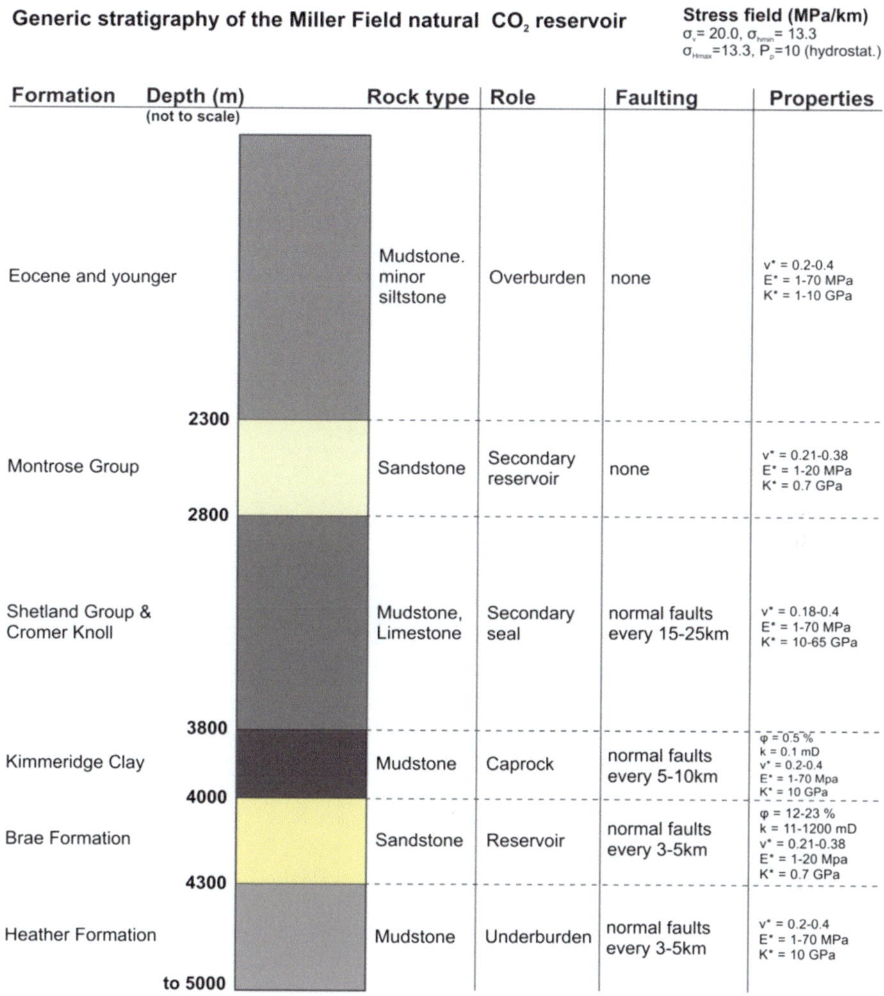

Fig. 9.14 Generic stratigraphy of the Miller field natural CO_2 reservoir. *Asterisk* indicates literature values instead of in-situ data. v Poisson's ratio, E Young's Modulus, K Bulk modulus, φ porosity, k permeability

9.5.7 St. Johns Dome, USA: Natural CO_2 Reservoir

9.5.7.1 Possible Foreland Basin

The natural CO_2 field of St. Johns is located in North–Eastern Arizona at the southern end of the Permian Holbrook basin and the southern edge of the Colorado Plateau (Blakey 1990; Rauzi 1999). The mechanism of basin subsidence of non-yoked Permian basins on the Colorado Plateau and Southern Rocky Mountains

Table 9.11 Summary of the match between the geomechanical facies storage potential prediction and the actual Miller field data

Miller	Basin architecture	Sedimentary stratigraphy	Reservoir potential	Caprock extend	Faults	Preservation potential
Geomechanical facies prediction—terrestrial rift basin	Long narrow geometry	Alluvial fans, lakes and marine deposits	Thick channelized sands	Thick caprocks	Single boundary normal faults	Good preservation potential
Field data	Small basin, covers an area of 45 km^2	Turbidite deposits	Thick sands followed by thick interbedded sands/muds	Thick caprock deposits	Limited faulting	Moderate preservation
Model/field data match	Good	Moderate	Good	Good	Good	Good

is hard to determine as the Late Paleozoic tectonics were subtle (Blakey 2008). The reservoir rocks of the Permian Supai Formation are dominantly siltstones, fine grained sandstones and carbonate layers and are in depth between 300 and 700 m below surface and have an approximate thickness of ~400 m. Continuous, thin anhydrite layers form seals within the reservoir complex and the top seal is formed by shales of the upper Supai formation, Permian limestones of the San Andres Formation and Triassic shales of the Moenkopi and Chinle Formations which have a total thickness of ~300 m. The CO_2 is mantle sourced and is probably related to the nearby Springerville volcanic complex (Gilfillan et al. 2011). CO_2 has been encountered in fractured Precambrian Granite which forms the basement (Embid 2009). Structurally the reservoir is located in a faulted anticline. It's spatial extent is ~60 × 35 km. The St. Johns Dome area is well known for its extensive travertine deposits which record a history of CO_2 leakage from the reservoir over the last 400 ka (Priewisch et al. 2014). Figure 9.15 presents the generic stratigraphy of the St. Johns Dome natural CO_2 reservoir.

9.5.7.2 Geomechanical Facies Model

The geomechanical facies model would predict this to be a good storage opportunity with a long and wide basin architecture, with thick reservoir sands and extensive caprocks with few fractures. The basin will have a cool geothermal gradient, a moderate risk of overprint or destruction, low risk of orogenesis modification and a good preservation potential.

The reservoir model and field operation data from the St. Johns Dome natural CO_2 reservoir exhibits a moderate-poor first order predictive match with the geomechanical facies model, as summarized in Table 9.12. This might indicate that the basin type has not correctly been identified.

9.5.8 Fizzy Field, UK Southern North Sea: Natural CO_2 Reservoir

9.5.8.1 Post Extensional Terrestrial Rift Sag Basin

The Fizzy field is located in block 50/26b of the UK sector of the southern North Sea and is part of the Southern Permian Basin (SPB) complex. The SPB is a major east–west striking sedimentary basin that has hosts the majority of gas and oil fields of the UK, the Netherlands, Denmark and Germany (Glennie 1998). The reservoir consists of Permian aeolian Rotliegend sandstones with a thickness of 100 m and good reservoir quality (18 % porosity and 260 mD permeability) and holds a gas column of 50 % CO_2, 41 % CH_4 and 9 % N_2 (Underhill et al. 2009). About 550 m of Permian evaporites (mainly anhydrite, minor salt and carbonates) and Triassic

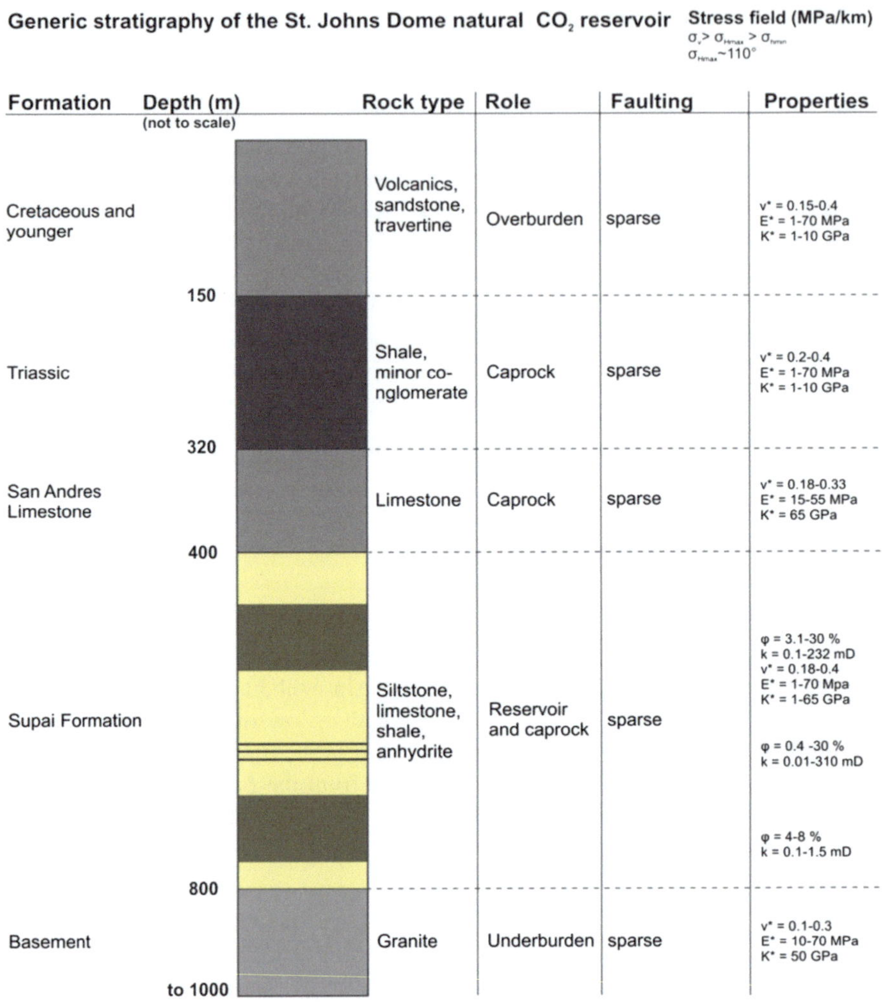

Fig. 9.15 Generic stratigraphy of the St. Johns Dome natural CO_2 reservoir. *Asterisk* indicates literature values instead of in-situ data. *v* Poisson's ratio, *E* Young's Modulus, *K* Bulk modulus, *φ* porosity, *k* permeability

shales form the top seal. The reservoir is located on a horst structure and is fault bound, with the bounding fault having a maximum throw of 500 m (Yielding et al. 2011). The fault has been reactivated and inverted during the Late Cretaceous, todays stress field is probably a strike-slip faulting regime. Figure 9.16 presents the generic stratigraphy of the Fizzy Field natural CO_2 reservoir.

Table 9.12 Summary of the match between the geomechanical facies storage potential prediction and the actual St. Johns Dome field data

St. Johns	Basin architecture	Sedimentary stratigraphy	Reservoir potential	Caprock extend	Faults	Preservation potential
Geomechanical facies prediction—foreland basin	Wide to very long basin with varying profile	Filled with mountain belt material, alluvia, fluvial and lake deposits	Thick and extensive sands	Thick and extensive caprocks	Predominantly thrusts rather than deep faults	Good preservation potential
Field data	Big reservoir (10 s of km)	Mainly redbeds with possible marine carbonates	Thick siltstones and fine sands	Thick caprocks	Deep seated faults	Extensive uplift and overprinting
Model/field data match	Good	Good	Moderate	Good	Poor	Poor

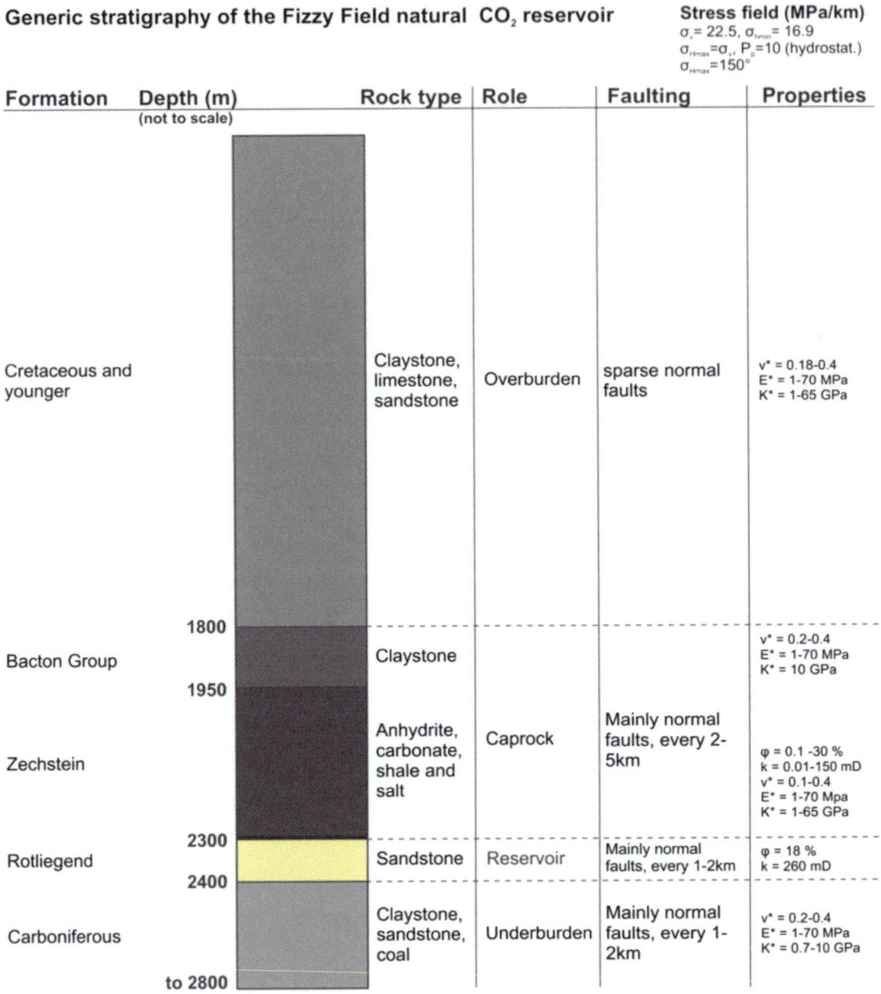

Fig. 9.16 Generic stratigraphy of the Fizzy Field natural CO_2 reservoir. *Asterisk* indicates literature values instead of in-situ data. *v* Poisson's ratio, *E* Young's Modulus, *K* Bulk modulus, φ porosity, *k* permeability

9.5.8.2 Geomechanical Facies Model

The geomechanical facies model would predict this to be a moderate storage opportunity with long narrow basin architecture, thick but channelized reservoir and caprock architecture and few fractures. The basin will have low risk of overprint, low risk of orogenesis modification and good preservation potential.

The reservoir model and field operation data from the Fizzy Field natural CO_2 reservoir exhibits a close first order predictive match with the geomechanical facies model, as summarized in Table 9.13.

Table 9.13 Summary of the match between the geomechanical facies storage potential prediction and the actual Fizzy field data

Fizzy	Basin architecture	Sedimentary stratigraphy	Reservoir potential	Caprock extend	Faults	Preservation potential
Geomechanical facies prediction–terrestrial rift basin	Long narrow geometry	Alluvial fans, lakes and marine deposits	Thick channelized sands	Thick caprocks	Single boundary normal faults	Good preservation potential
Field data	Elongated reservoir	Aeolian sandstones, lake and marine deposits	Thick dune sands	Thick caprocks	Reactivated normal faults	Moderate preservation
Model/field data match	Good	Good	Moderate	Good	Moderate	Moderate

9.6 Conclusions

A total of 61 natural analogue CO_2 storage sites where investigated and the key geological and physical controls on ensuring the longer term retention of CO_2 were identified. These controls were then related to different characteristics of the geomechanical facies within the storage system. The geomechanical facies are identified as combinations of geological facies operating in a specific engineering way, e.g. storage, retention or overburden.

The characteristics of the different geomechanical facies are controlled by the tectonic settings they were deposited in. This directly influences sediment stratigraphy, thickness and the distribution of the caprock and reservoir sediments. The tectonic setting also determines basin architecture, stress state, mechanical properties, fracture characteristics, burial depths, geothermal gradient, structural stability and preservation potential, all crucial inputs into assessing CO_2 storage site suitability.

Using the geomechanical facies approach, the geomechanical facies inputs crucial to the primary CO_2 storage requirements of storage volume and storage security can be evaluated and graded as good, moderate and poor based on an assessment of their net contribution towards providing ideal CO_2 storage conditions.

The results show that foreland basins and passive continental margin basins are very suitable basins for CO_2 storage. Strike-slip basins, terrestrial rift basins and back arc basins are also suitable for CO_2 storage with oceanic basins and fore-arc basins being moderately suitable and trench basins unsuitable tectonic settings for CO_2 storage.

The geomechanical facies approach was then used to evaluate a number of current anthropogenic CO_2 storage projects and natural CO_2 storage analogues. Generic conceptual profiles are presented for the sites and the main hydraulic and mechanical parameters for the different geomechanical facies in each site is presented. The sites include Otway-Australia, In Salah-Algeria, Buracica-Brazil, Sleipner-Norway, Snøhvit-Norway, Miller Field-UK North Sea, Fizzy Field, UK Southern North Sea and St. Johns Dome-USA.

Using the geomechanical facies framework to evaluate the storage potential correctly predicts that most of the sites would be suitable CO_2 storage opportunities. Only in cases where there has been significant overprint of the original basin architecture (e.g. Snohvit) or where the basin type is not well known (St. Johns) the approach has limitations.

References

Bachu S (2003) Screening and ranking of sedimentary basins for sequestration of CO_2 in geological media in response to climate change. Environ Geol 44:277–289

Baines SJ, Worden RH (2004) The long-term fate of CO_2 in the subsurface: natural analogues for CO_2 storage. Geol Soc Lond Spec Publ 233:59–85

Barton CA, Zoback MD, Moos D (1995) Fluid flow along potentially active faults in crystalline rock. Geology 23:683–686

Bissell RC, Vasco DW, Atbi M, Hamdani M, Okwelegbe M, Goldwater MH (2011) A full field simulation of the in Salah gas production and CO_2 storage project using a coupled geo-mechanical and thermal fluid flow simulator. Energy Proced 4:3290–3297

Blakey RC (2008) Chapter 7 Pennsylvanian–Jurassic sedimentary basins of the Colorado Plateau and Southern Rocky Mountains. In: Miall AD (ed) Sedimentary basins of the world. Elsevier, Amsterdam, pp 245–296

Blakey RC (1990) Stratigraphy and geologic history of Pennsylvanian and Permian rocks, Mogollon Rim region, central Arizona and vicinity. Geol Soc Am Bull 102:1189–1217

Busch A, Amann-Hildenbrand A, Bertier P, Waschbuesch M, Krooss BM (2010) The significance of caprock sealing integrity for CO_2 storage. Soc Pet Eng. doi:10.2118/139588-MS

Chadwick A, Arts R, Bernstone C, May F, Thibeau S, Zweigel P (2008) Best practice for the storage of CO_2 in saline aquifers: observations and guidelines from the SACS and CO2STORE projects. British Geological Survey, Nottingham

Chadwick RA, Williams GA, Williams JDO, Noy DJ (2012) Measuring pressure performance of a large saline aquifer during industrial-scale CO_2 injection: the Utsira Sand, Norwegian North Sea. Int J Greenh Gas Control 10:374–388

Chadwick RA, Zweigel P, Gregersen U, Kirby GA, Holloway S, Johannessen PN (2004) Geological reservoir characterization of a CO_2 storage site: the Utsira Sand, Sleipner, northern North Sea. Energy 29:1371–1381

Chiaramonte L, Johnson S, White JA (2011) Preliminary geomechanical analysis of CO_2 injection at Snøhvit, Norway. In: Presented at the 45th US rock mechanics/geomechanics symposium, American Rock Mechanics Association

Chiquet P, Broseta D, Thibeau S (2007) Wettability alteration of caprock minerals by carbon dioxide. Geofluids 7:112–122

Dai J, Yang S, Chen H, Shen X (2005) Geochemistry and occurrence of inorganic gas accumulations in Chinese sedimentary basins. Org Geochem 36:1664–1688

Dance T, Spencer L, Xu J-Q (2009) Geological characterisation of the Otway project pilot site: what a difference a well makes. Energy Proced 1:2871–2878

Delprat-Jannaud F, Korre A, Shi JO, McConnell B, Arvanitis A, Boavida D, Car M, Gastine M, Bateman K, Poulsen N, Sinayuc C, Vähäkuopus T, Vercelli S, Wojcicki A (2013) State of the art review of CO_2 storage site selection and characterisation methods (CGS Europe Report No. D3.3)

Edlmann K, Edwards MA, Qiao XJ, Haszeldine RS, McDermott CI (2015) Appraisal of global CO_2 storage opportunities using the geomechanical facies approach. Environ Earth Sci 73:8075–8096

Edlmann K, Haszeldine S, McDermott CI (2013) Experimental investigation into the sealing capability of naturally fractured shale caprocks to supercritical carbon dioxide flow. Environ Earth Sci 70:3393–3409

Eiken O, Ringrose P, Hermanrud C, Nazarian B, Torp TA, Høier L (2011) Lessons learned from 14 years of CCS operations: Sleipner, In Salah and Snøhvit. In: Energy procedia, 10th international conference on greenhouse gas control technologies, vol 4. pp 5541–5548

Embid EH (2009) U-series dating, geochemistry, and geomorphic studies of travertines and springs of the Springerville area, east-central Arizona, and tectonic implications [MS thesis]: Albuquerque, University of New Mexico

Estublier A, Dino R, Schinelli MC, Barroux C, Beltran AM (2011) CO_2 injection in Buracica: long-term performance assessment. Energy Proced 4:4028–4035

Faulkner DR, Jackson CAL, Lunn RJ, Schlische RW, Shipton ZK, Wibberley CAJ, Withjack MO (2010) A review of recent developments concerning the structure, mechanics and fluid flow properties of fault zones. J Struct Geol 32:1557–1575

Faulkner DR, Lewis AC, Rutter EH (2003) On the internal structure and mechanics of large strike-slip fault zones: field observations of the carboneras fault in southeastern Spain. Tectonophysics 367:235–251

Finkbeiner T, Zoback M, Flemings P, Stump B (2001) Stress, pore pressure, and dynamically constrained hydrocarbon columns in the South Eugene Island 330 field, northern Gulf of Mexico. AAPG Bull 85:1007–1031

Gilfillan SMV, Wilkinson M, Haszeldine RS, Shipton ZK, Nelson ST, Poreda RJ (2011) He and Ne as tracers of natural CO_2 migration up a fault from a deep reservoir. Int J Greenh Gas Control 5:1507–1516

Glennie KW (1998) Lower Permian—Rotliegend. In: Honoraryessor KWG (ed) Petroleum geology of the North Sea. Blackwell, Hoboken, pp 137–173

Hallam A (1981) Facies interpretation and the stratigraphic record. Freeman, Oxford

Hanks CL, Lorenz J, Teufel L, Krumhardt AP (1997) Lithologic and structural controls on natural fracture distribution and behavior within the Lisburne Group, northeastern Brooks Range and North Slope subsurface, Alaska. AAPG Bull 81:1700–1720

Hansen O, Gilding D, Nazarian B, Osdal B, Ringrose P, Kristoffersen J-B, Eiken O, Hansen H (2013) Snøhvit: the history of injecting and storing 1 Mt CO_2 in the Fluvial Tubåen Fm. Energy Proced 37:3565–3573

Harrington JF, Noy DJ, Horseman ST, Birchall DJ, Chadwick RA (2009) Laboratory study of gas and water flow in the Nordland Shale, Sleipner, North Sea. ResearchGate 59:521–543

Hillis RR (2003) Pore pressure/stress coupling and its implications for rock failure. Geol Soc Lond Spec Publ 216:359–368

Holloway S, Pearce JM, Ohsumi T, Hards VL (2005) A review of natural CO_2 occurrences and their relevance to CO_2 storage (No. CR/05/104 117). In: British Geological Survey External Report

Hung Kiang C, Kowsmann RO, Figueiredo AMF, Bender A (1992) Tectonics and stratigraphy of the East Brazil Rift system: an overview. Tectonophysics 213:97–138

IEAGHG (2009) CCS site characterisation criteria. IEA Greenhouse Gas R&D Programme

Jenkins CR, Cook PJ, Ennis-King J, Undershultz J, Boreham C, Dance T, de Caritat P, Etheridge DM, Freifeld BM, Hortle A, Kirste D, Paterson L, Pevzner R, Schacht U, Sharma S, Stalker L, Urosevic M (2012) Safe storage and effective monitoring of CO_2 in depleted gas fields. Proc Natl Acad Sci 109:E35–E41

Ladeira FL, Price NJ (1981) Relationship between fracture spacing and bed thickness. J Struct Geol 3:179–183

Lu J, Wilkinson M, Haszeldine RS, Fallick AE (2009) Long-term performance of a mudrock seal in natural CO_2 storage. Geology 37:35–38

Lyon G, Giggenbach WF, Sano Y (1996) Variations in the chemical and isotopic composition of Taranaki gases and their possible causes. In: Presented at the New Zealand petroleum conference, pp 171–178

Manzocchi T, Childs C, Walsh JJ (2010) Faults and fault properties in hydrocarbon flow models. Geofluids 10:94–113

McDermott CI, Edlmann K, Haszeldine RS (2013) Predicting hydraulic tensile fracture spacing in strata-bound systems. Int J Rock Mech Min Sci 63:39–49

McDermott CI, Lodemann M, Ghergut I, Tenzer H, Sauter M, Kolditz O (2006) Investigation of coupled hydraulic–geomechanical processes at the KTB site: pressure-dependent characteristics of a long-term pump test and elastic interpretation using a geomechanical facies model. Geofluids 6:67–81

Miocic JM, Gilfillan SMV, McDermott C, Haszeldine RS (2013) Mechanisms for CO_2 leakage prevention: a global dataset of natural analogues. Energy Proced 40:320–328

Miocic JM, Gilfillan SMV, Roberts JJ, Edlmann K, McDermott CI, Haszeldine RS (2016) Controls on CO_2 storage security in natural reservoirs and implications for CO_2 storage site selection. Int J Greenh Gas Control 51:118–125. http://www.sciencedirect.com/science/article/pii/S1750583616302626

Moss B, Barson D, Rakhit K, Dennis H, Swarbrick RE (2003) Formation pore pressures and formation waters. The millennium atlas: petroleum geology of the central and northern North Sea. The Geological Society of London, London, pp 317–329

Nara Y, Meredith PG, Yoneda T, Kaneko K (2011) Influence of macro-fractures and micro-fractures on permeability and elastic wave velocities in basalt at elevated pressure. Tectonophysics 503:52–59

NETL (2010) Site screening, selection, and initial characterisation for storage of CO_2 in deep geologic formations (No. DOE/NETL-401/090808). National Energy Technology Laboratory

Onuma T, Ohkawa S (2009) Detection of surface deformation related with CO_2 injection by DInSAR at In Salah, Algeria. Energy Proced 1:2177–2184

Pearce JM, Holloway S, Wacker H, Nelis MK, Rochelle C, Bateman K (1996) Natural occurrences as analogues for the geological disposal of carbon dioxide. Energy Convers Manag 37:1123–1128

Pearce J, Czernichowski-Lauriol I, Lombardi S, Brune S, Nador A, Baker J, Pauwels H, Hatziyannis G, Beaubien S, Faber E (2004) A review of natural CO_2 accumulations in Europe as analogues for geological sequestration. Geol Soc Lond Spec Publ 233:29–41

Priewisch A, Crossey LJ, Karlstrom KE, Polyak VJ, Asmerom Y, Nereson A, Ricketts JW (2014) U-series geochronology of large-volume Quaternary travertine deposits of the southeastern Colorado Plateau: Evaluating episodicity and tectonic and paleohydrologic controls. Geosphere 10:401–423

Rauzi SL (1999) Carbon dioxide in the St. Johns-Springerville area, Apache County, Arizona. Ariz Geol Surv 99

Reading HG (1978) Sedimentary environments and facies, 1st edn. Blackwell Scientific, Oxford

Rinaldi AP, Rutqvist J (2013) Modeling of deep fracture zone opening and transient ground surface uplift at KB-502 CO_2 injection well, In Salah, Algeria. Int J Greenh Gas Control 12:155–167

Ringrose PS, Roberts DM, Gibson-Poole CM, Bond C, Wightman R, Taylor M, Raikes S, Iding M, Østmo S (2011) Characterisation of the Krechba CO_2 storage site: critical elements controlling injection performance. Energy Proced 4:4672–4679

Roberts JJ (2012) Natural CO_2 fluids in Italy: implications for the leakage of geologically stored CO_2 (PhD). The University of Edinburgh, Edinburgh

Roberts JJ, Wood RA, Haszeldine RS (2011) Assessing the health risks of natural CO_2 seeps in Italy. Proc Natl Acad Sci 108:16545–16548

Roberts JJ, Wood RA, Wilkinson M, Haszeldine S (2015) Surface controls on the characteristics of natural CO_2 seeps: implications for engineered CO_2 stores. Geofluids 15:453–463

Rodrigues Duran E, di Primio R, Anka Z, Stoddart D, Horsfield B (2013) Petroleum system analysis of the Hammerfest Basin (southwestern Barents Sea): comparison of basin modelling and geochemical data. Org Geochem 63:105–121

Rooksby SK (1991) The Miller field, Blocks 16/7B, 16/8B, UK North Sea. Geol Soc Lond Mem 14:159–164

Rouchon V, Magnier C, Miller D, Bandeira C, Gonçalves R, Dino R (2011) The relationship between CO_2 flux and gas composition in soils above an EOR-CO_2 oil field (Brazil): a guideline for the surveillance of CO_2 storage sites. Energy Proced 4:3354–3362

Scherer CMS, Lavina ELC, Dias Filho DC, Oliveira FM, Bongiolo DE, Aguiar ES (2007) Stratigraphy and facies architecture of the fluvial–aeolian–lacustrine Sergi Formation (Upper Jurassic), Recôncavo Basin, Brazil. Sediment Geol 194:169–193

Shipton ZK, Evans JP, Kirschner D, Kolesar PT, Williams AP, Heath J (2004) Analysis of CO_2 leakage through "low-permeability" faults from natural reservoirs in the Colorado Plateau, east-central Utah. Geol Soc Lond Spec Publ 233:43–58

Shukla R, Ranjith P, Haque A, Choi X (2010) A review of studies on CO_2 sequestration and caprock integrity. Fuel 89:2651–2664. doi:10.1016/j.fuel.2010.05.012

Smith M, Campbell D, Mackay E, Polson D (2011) CO_2 aquifer storage site evaluation and monitoring. Heriot-Watt University, Edinburgh

Stevens SH, Pearce JM, Rigg AAJ (2001) Natural analogues for geologic storage of CO_2: an integrated global research program. Presented at the proceedings of first national conference carbon sequestration. Washington, DC

Torp TA, Gale J (2004) Demonstrating storage of CO_2 in geological reservoirs: the Sleipner and SACS projects. Energy 29:1361–1369

Underhill JR, Lykakis N, Shafique S (2009) Turning exploration risk into a carbon storage opportunity in the UK Southern North Sea. Pet Geosci 15:291–304

Veritas DN (2010) Guidelines for selection and qualification of sites and projects for geological storage of CO_2 (No. DNV No. 2009-1425)

Vidal-Gilbert S, Tenthorey E, Dewhurst D, Ennis-King J, Van Ruth P, Hillis R (2010) Geomechanical analysis of the Naylor Field, Otway Basin, Australia: implications for CO_2 injection and storage. Int J Greenh Gas Control 4:827–839

Wennberg OP, Malm O, Needham T, Edwards E, Ottesen S, Karlsen F, Rennan L, Knipe R (2008) On the occurrence and formation of open fractures in the Jurassic reservoir sandstones of the Snøhvit Field, SW Barents Sea. Pet Geosci 14:139–150

White JA, Chiaramonte L, Ezzedine S, Foxall W, Hao Y, Ramirez AL (2014) Geomechanical behavior of the reservoir and caprock system at the In Salah CO_2 storage project. Proc Natl Acad Sci USA 111(24):8747–8752

Wycherley H, Fleet A, Shaw H (1999) Some observations on the origins of large volumes of carbon dioxide accumulations in sedimentary basins. Mar Pet Geol 16:489–494

Yielding G, Lykakis N, Underhill JR (2011) The role of stratigraphic juxtaposition for seal integrity in proven CO_2 fault-bound traps of the Southern North Sea. Pet Geosci 17:193–203

Zweigel P, Arts R, Lothe AE, Lindeberg EBG (2004) Reservoir geology of the Utsira Formation at the first industrial-scale underground CO_2 storage site (Sleipner area, North Sea). Geol Soc Lond Spec Publ 233:165–180

Chapter 10
Risk Management for CO_2 Geological Storage Projects

Yvi Le Guen, Stéphanie Dias, Olivier Poupard, Katriona Edlmann and Christopher Ian McDermott

Abstract A number of key challenges relating to potential CO_2 reservoir capacity, injectivity and confinement need to be overcome when validating the performance of a storage system for its lifecycle. In the case of a failure of a storage operation, the environment, investments, and human health and safety, may be at risk. It is therefore important to use risk management methods to ensure that the project will meet its objectives in all aspects. The aims of risk management are both to identify and evaluate all the risks that could impact the project objectives, and to establish treatment, monitoring actions and plans to reduce the impact of risks thereby ensuring the project performance. This Chapter discusses the implementation of risk management for a CO_2 geological storage project.

10.1 Introduction

Deep CO_2 geological storage is one of the most promising solutions to reduce the CO_2 emissions to the atmosphere, and minimize the impact of greenhouse gas effects. Nevertheless, some key challenges relating to capacity, injectivity, and containment need to be addressed in order to ensure the performance of the storage

Y. Le Guen · S. Dias · O. Poupard (✉)
OXAND SA, 49 av Franklin Roosevelt, 77210 Avon, France
e-mail: olivier.poupard@oxand.com

Y. Le Guen
e-mail: yvi.leguen@oxand.com

S. Dias
e-mail: stephanie.dias@oxand.com

K. Edlmann · C.I. McDermott
School of Geosciences, University of Edinburgh, King's Buildings,
West Mains Road, Edinburgh EH9 3JW, UK
e-mail: katriona.edlmann@ed.ac.uk

C.I. McDermott
e-mail: cmcdermo@staffmail.ed.ac.uk

A. Niemi et al. (eds.), *Geological Storage of CO_2 in Deep Saline Formations*,
Theory and Applications of Transport in Porous Media 29,
DOI 10.1007/978-94-024-0996-3_10

system during its lifecycle, for a duration that can range from some years to hundreds of years. This means that we need to understand the risks associated with this technology and devise strategies to manage these risks, which include technological risks to the environment and human health and safety, social risks, policy risks, legal risks and the like.

Risk management for the geological storage of CO_2 is complex due to the

- wide variety of physical phenomena that need to be taken into account (including hydrodynamic, geochemical, geomechanical phenomena) and their inter-dependence,
- limited access to geological data and the associated uncertainties,
- variability between the sites and the time scales involved, ranging from very short term to centuries or a few millennia.

The safety management of the storage projects is one of the key obstacles to the large-scale deployment of the technology, not only for technical reasons, but also for societal reasons, including acceptance of industrial projects of a new type and growing awareness of people. Ensuring the safety of a geological storage project requires mitigating the main technological obstacle, namely that of reliably predicting the behaviour of the storage performance over its lifecycle. This goal is fraught with the inherent uncertainties related to deep geological reservoirs.

Currently, there is no recognized and by the international scientific community commonly accepted standard methodology, for the analysis and management of the risks related to the geological storage of CO_2. The activity of geological storage of CO_2 is still under development. Research over the past ten years proposed approaches for the identification of risk scenarios or tools for their representation (Pawar et al. 2015). However, these studies have not yet produced a generic, comprehensive and commonly accepted methodology to evaluate, in a quantitative way, the risks posed by the geological storage of CO_2.

Risk studies have been carried out in for various sites, such as:

- In-Salah, Algeria (Dodds et al. 2011; Paulley et al. 2011; Oldenburg et al. 2011),
- Weyburn, US (Stenhouse et al. 2005),
- Illinois Basin, US (Hnottavange-Telleen et al. 2009, 2011),
- Fort Nelson Carbon Capture, US (Sorensen et al. 2014),
- The North Sea, Captain Sandstone Aquifer, UK (SCCS 2015).

The aforementioned studies were conducted as part of research projects and had no regulatory framework for the definition of procedures and/or standards (Oldenburg et al. 2008; US EPA 2008). Examples of initiatives in Europe include those by Quintessa, who has developed a database on related FEPs (Features, Events and Processes) (Maul et al. 2005), by DNV (Det Norske Veritas 2009) and by TNO (Wildenborg et al. 2004; Yavuz et al. 2009). The project ANR CRISCO2 (Bouc et al. 2010) proposes a method to determine qualitative and quantitative criteria to ensure the safety of CO_2 storage. This study focuses on aquifer storage. The approach was developed on the basis of the Paris Basin case study and relies on reasonably conservative assumptions and simplified models for quick evaluation of the safety. CO_2-PENS software platform has been developed by the Los Alamos Laboratory (Pawar et al. 2006; Stauffer et al. 2009, 2011) and tests on a few storage cases within National Risk Assessment Partnership (NRAP) projects (Cugini et al. 2010) have been performed.

National Energy Technology Laboratory in the US has in a Best Practices Manual summarized the concepts of risk analysis (risk assessment) and numerical simulation by describing the experience gained by the DOE Regional Carbon Sequestration Partnerships as they implemented multiple field projects (NETL 2011). This manual focuses on the risks arising from unplanned migration of injected CO_2 from the reservoir and on the ways in which numerical codes have been used to model specific processes related to the behavior of injected CO_2 in the subsurface.

Guidelines and standards are needed to delineate best practices for conducting studies of long term predictive failure analysis for elements of geological storage of CO_2, namely:

- reservoirs and geological barriers,
- geological features (faults, fractures),
- wells (former ones or those constructed for the purpose of the storage project).

Potential impact studies concerning the effect of CO_2 geological storage on both surface targets, such as drinking water aquifers and soils as well as subsurface targets and on human health, fauna and flora are also needed. This is needed to provide objective demonstration of the performance of such projects elements. Risk management is the process that aims at identifying all the potential risks related to a project, organizing them in order to define which are the critical ones and outlining actions that may be taken to lower these risks. The different steps of risk management process are the following (Fig. 10.1):

- Communication and consultation;
- Establishment of the context;
- Risk assessment or risk analysis, that includes risk identification, estimation and evaluation;
- Risk treatment;
- Risk monitoring and review.

Management and communication will ensure that the policy is understood, implemented and maintained at all levels of the CO_2 project.

Risk management is an iterative process to be applied over the project lifecycle (Fig. 10.2), from site selection to abandonment stages and should be viewed as an essential component for any CO_2 storage project. Its main principles are to:

- Contribute to the achievement of project objectives regarding, for example, health and safety, environment and investments as well as the improvement of project performance;

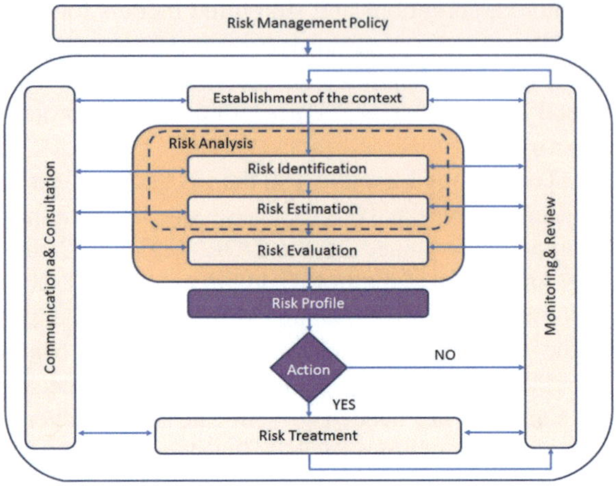

Fig. 10.1 Risk management process according to ISO 31000 (2009)

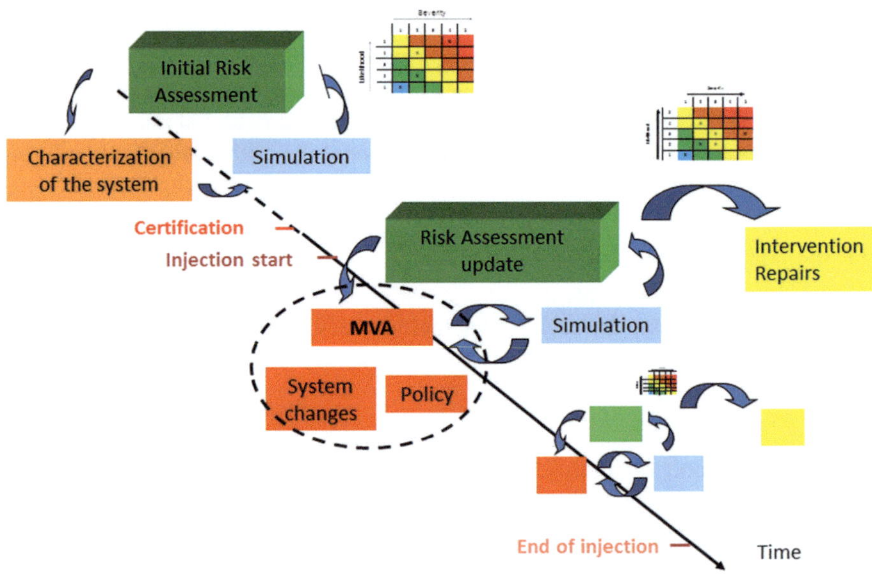

Fig. 10.2 Risk management workflow over CO_2 geological storage project life cycle

- Support the decision making process for risk treatment and definition of any Monitoring, Verification and Accounting (MVA) program, including prioritizing actions and justifying the choices;
- Provide management and/or authorities demonstration of effective and comprehensive management of risks;
- Provide consistent, comparable and reliable results of risk evaluation as a product of a transparent and structured methodology.

10.2 Risk Management Policy

According to the reference process for risk management (ISO 31000 2009), the risk management policy should:

- Clarify the project objectives and commitment for risk management;
- Specify the link between the risk management policy and the project objectives, and rationale for managing risk;
- Specify the processes, methods and tools to be used for managing the risks;
- Identify the roles and responsibilities in the project team for managing risks;
- Describe the way in which risk management performance will be measured and reported;
- Establish the project commitment to the periodic review and verification of the risk management policy.

10.3 Establishment of the Context

The definition of the context of the CO_2 project supports the risk management process, as it defines the contours of the risk management and the elements to be considered in the process. The context of this process will vary according to the needs of the project. It can involve, but is not limited to:

- Defining the scope, as well as the extent of the risk management activities to be carried out, including specific inclusions and exclusions;
- Defining the activity, process, function associated to the project in terms of time and location as well as their goals and objectives;
- Defining the way performance is evaluated in the risk management;
- Identifying and specifying the decision process (who, when, for which purpose and what).

The definition of the context of a project must include the elements listed below, which are described in more details in what follows:

- The scope of the project: environment of the project, storage characteristics and timeline.

- The internal and external entities involved in risk management, including the operators, stakeholders and contractors.
- The risk criteria to evaluate the significance of the risk: the probability and severity levels and the criticality matrix.

It is also important to point out that good quality information is essential in any risk management process. The data collection step is one of the most important steps in the Risk Management process, and must be started early in the process.

10.3.1 Scope of the Study

The first step in the definition of the scope of the risk assessment is the identification of the system under consideration and its breakdown into subsystems. It is also necessary to define the type of risks that will be considered. These risks must be included in the different risk families defined previously in the field of application of the risk management policy.

To define the scope of the CO_2 project, all the elements and interactors of the project must be defined. The following paragraphs describe what these elements and interactors could be. Here, we focus only on technical aspects. Note that the elements and the interactors depend on the system or subsystem under consideration.

For any CO_2 storage project, the system elements could be included into two main groups, namely (1) the natural system, i.e. the geology, which defines the target reservoir, the caprock above it, the overburden, the fresh water aquifers, and (2) the anthropogenic influence, i.e. the wells located in the storage area, including any injection wells, oil and gas production wells, water disposal wells, monitoring wells, shutoff wells, plugged and abandoned wells.

10.3.2 Internal and External Entities

The second step for establishing the context is to define the internal and external entities of the CO_2 project. The relationship between the CO_2 project and its environment (both internal and external) must be taken into account when identifying and assessing risks. The identification and definition of internal and external entities will support the identification of stakeholders. To ensure that all stakeholders are taken into account, both internal and external contexts are defined and linked to the subsystems by means on a functional analysis.

There is a wide variety of stakeholders that must be considered in the context of a project. The stakeholders can be considered either internal or external. Internal stakeholders would be departments or teams that interact with the project (several disciplines and groups will be involved, such as geologists, drilling engineers, geophysicists, to mention some), while external stakeholders would consist of entities that interact with the operator and may affect or have an impact on the project, such as regulators, other oil and gas companies, local communities and the

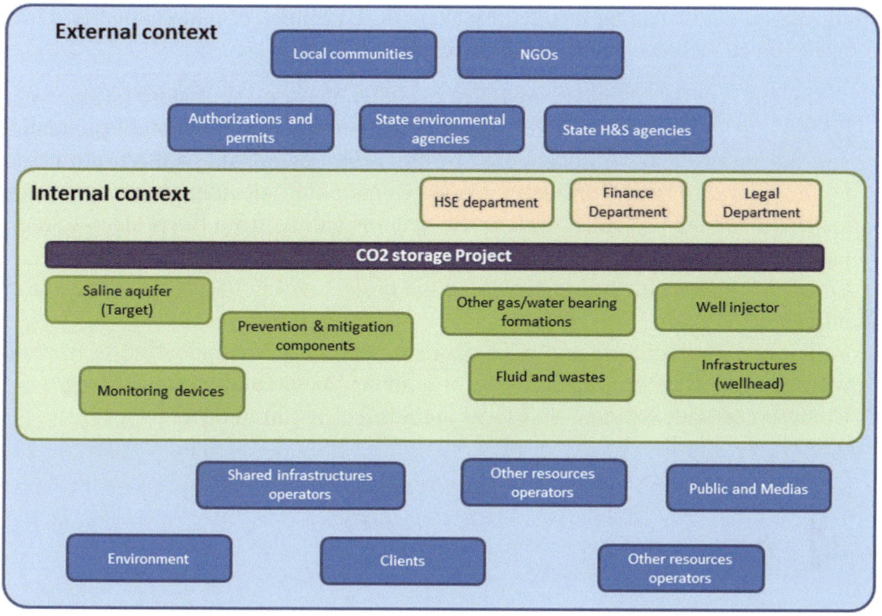

Fig. 10.3 Key actors/activities within in a CO_2 geological storage project

like. Figure 10.3 illustrates a representation of key stakeholders to be considered in a CO_2 geological storage project. The understanding of the internal context can include, for example:

- The capabilities, understood in terms of resources and knowledge (e.g. time, people, processes, systems and technologies);
- Information systems, information flows, and decision making processes (both formal and informal);
- Internal stakeholders;
- Standards and reference models adopted within the project.

External context is the external environment in which the project seeks support to achieve its objectives. The external context can include, but is not limited to:

- The cultural, political, legal, regulatory, financial, technological, economic, natural and competitive environment, whether international, national, regional or local;
- Key drivers and trends having impact on the objectives of the project.

10.3.3 Constraints

The definition of constraints is one of the important steps in the establishment of the project context. The project team needs to identify the activities, the media and the organizations which can be affected by any risk. Examples of constraints for a CO_2 storage project include, but is not limited to:

- Technical aspects, such as capability to safely transport and store CO_2;
- Health and safety aspects, to ensure that the personnel and the local population will not in any way be endangered by the activities related to the project;
- Public outreach and confidence aspects, including demonstrating a reliable leadership of the project as well as the benefits gained from the project, to reach public confidence;
- Financial aspects, including achieving the project within the frame of the agreed budget;
- Policy and strategic aspects: CO_2 storage project is an innovative technology which is involved in global and local climate change and energy strategies;
- Compliance with national and local authorities' requirements.

10.3.4 Risk Criteria

This step uses the project objectives to identify key performance indicators that will be used to estimate, evaluate and treat risks. In this step, the risk criteria will be defined and used to evaluate the importance of risk. The criteria must reflect the project values and objectives and be continuously reviewed. Risk criteria could be defined according to expert opinions, interviews with stakeholders and actors, or by expert elicitation with the project team (Edlmann et al. 2016).

10.3.4.1 Probability Grid

The definition of the probability grid is defined with the knowledge of the project team and eventually based on other expert opinions. When possible, a quantitative estimation has to be used, such as probability of a CO_2 leakage, probability of a mechanical failure for a well component, to mention some. This quantitative estimation is then converted to a probability level on the basis of the probability grid used for the project. An example of such a grid is presented in Table 10.1.

Table 10.1 Example of a probability grid

Description	Probability over CO_2 storage period[a]	Level
Very unlikely: very rare	<0.001 %	A
Unlikely: rare	(0.001 %; 0.01 %)	B
Possible: can be observed, feared	(0.01 %; 0.1 %)	C
Likely: already observed, will probably occur	(0.1 %; 10 %)	D
Very likely: expected to occur (almost certain)	>10 %	E

[a]Provided as an example

Table 10.2 Example of severity grid

Description	Level
Minor	1
Low	2
Medium	3
High	4
Very high	5

10.3.4.2 Severity Grid

The severity levels indicate the magnitude of the impact if an unwanted event, or failure, occurs. The definition of the severity grid is also defined with the knowledge of the project team and with subject matter expert opinions. It is the preliminary step of the consequence grid elaboration. Table 10.2 is an example of a severity grid with 5 levels.

10.3.4.3 Consequence Grid

The consequence grid provides a description of the different severity levels for each project objective identified. The objectives are expressed in using performance indicators to illustrate the different level of impact (severity levels) on the objectives of the project (see Sect. 10.3.3).

This grid is the link between the project objectives impacted and resulting severity level. It must be developed closely with the project stakeholders, and eventually with expert opinions. When the objectives of the project and the key indicators (severity levels) are defined, each project stakeholder defines the minimum and maximum severity levels regarding each objective and then proceeds to complete the intermediate levels.

10.3.4.4 Risk Matrix

The risk matrix shown in Table 10.3 is also named a criticality matrix. The criticality "C" represents a mathematical relation between the severity and probability level. Mostly, criticality is a function of the severity level and the probability level:

$$\mathrm{C\,(Criticality) = F\,(S(Severity), P\,(Probability))}$$

The higher the criticality level, the higher the risk.

Table 10.3 Illustration of a risk matrix

Probability	E	Medium	High	High	Extreme	Extreme
	D	Medium	Medium	High	High	Extreme
	C	Medium low	Medium	Medium	High	High
	B	Medium low	Medium low	Medium	Medium	High
	A	Low	Medium low	Medium low	Medium	Medium
		1	2	3	4	5
		Severity				

10.3.4.5 Level of Acceptability

The level of acceptability is the level used to decide if treatment actions or monitoring actions are required. The definition of the level of acceptability depends on each objective identified by the project team. This level is defined according to the technical, financial, legal, social and other criteria. It is defined within the project and reflects the appetence or aversion of risks by the company.

Acceptability level of risk delimits 2 zones (Table 10.3): a zone where risks are critical (not acceptable) and a zone where risks are acceptable. For example, if the acceptability level has a criticality level of 6, the critical risks are those located in the orange and red zones in Table 10.3. This enables the definition of an action plan to mitigate risks.

10.4 Risk Assessment

A risk cannot be managed unless it is first identified and its impact assessed. Consequently, after risk management policy and context have been completed, the first process in the iterative Risk Management process aims to identify all the knowable risks to CO_2 project objectives. Risk assessment is the process of systematically and continuously identifying, categorizing, and assessing the initial significance of risks associated with a CO_2 project. Risk identification determines risks that might affect the project and registration of their characteristics.

The assessment should be performed on a regular basis throughout the project timeline. The purpose is to identify risks to the maximum extent that is practicable. The fact that some risks are unknowable or emergent requires the 'identify risk' process to be iterative, repeating the 'identify risks' process to find new risks which have become knowable since the previous iteration of the process. During the progress of the project through its lifecycle, new risks may appear. The project team should be involved in this process so that they can develop and maintain a sense of ownership of, and responsibility for, the risks and associated risk response actions.

In the following section, we will focus the risk assessment on technological issues for a CO_2 geological storage project to illustrate the approach.

10.4.1 Risk Identification

A comprehensive identification based on a well-structured and systematic process is essential to ensure that all significant risks are considered. Different methodologies can be used for this: FMEA (Failure Mode and Effects Analysis), Fault tree analysis, Event tree analysis, or Features, Events and Processes (FEP) analysis (Wildenborg et al. 2004; Pawar et al. 2006; Oldenburg et al. 2011; Paulley et al. 2011). We propose the use of FMEA. This is a systematic approach that focuses on the function to be fulfilled by the subsystems and components. In this process, the list of risks is based on the failure modes that might prevent, degrade or delay the achievement or performance of the objectives of the CO_2 project; it uses the results of the establishment of the context. For each subsystem and component (Fig. 10.4), the failure modes, their causes and their potential impacts (or consequences) on the objectives are defined (Fig. 10.5). The outcome of the risk identification step is a comprehensive list of risks related to the project compiled in a risk register (Table 10.4). This risk register is the input for the risk estimation step.

An example of the risk identification is given in Fig. 10.6. In the figure a structured and comprehensive inventory of leakage pathways and leakage impact factors (risks) through the caprock, which could contribute to CO_2 storage performance, was generated based on information from the literature (e.g. Oldenburg et al. 2011; Vilarrasa et al. 2011; Al-Bazali et al. 2005; Li et al. 2006; Shukla et al. 2008, 2010; Bildstein et al. 2009; Class et al. 2009; Wollenweber et al. 2010; Busch et al. 2010; Amman et al. 2011; Le Guen et al. 2008, 2010; Viswanathan et al. 2008). This is used as the basis for subsequent risk estimation.

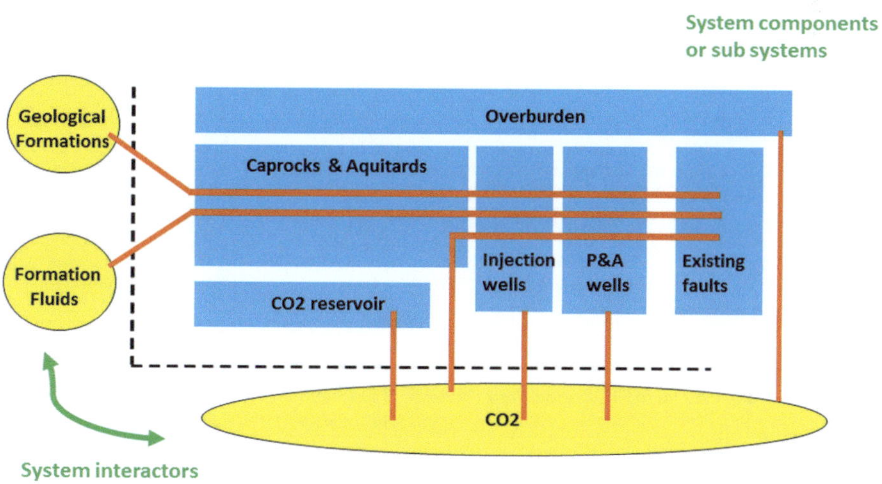

Fig. 10.4 Example of a schematic of a CO_2 geological storage system and possible interactions

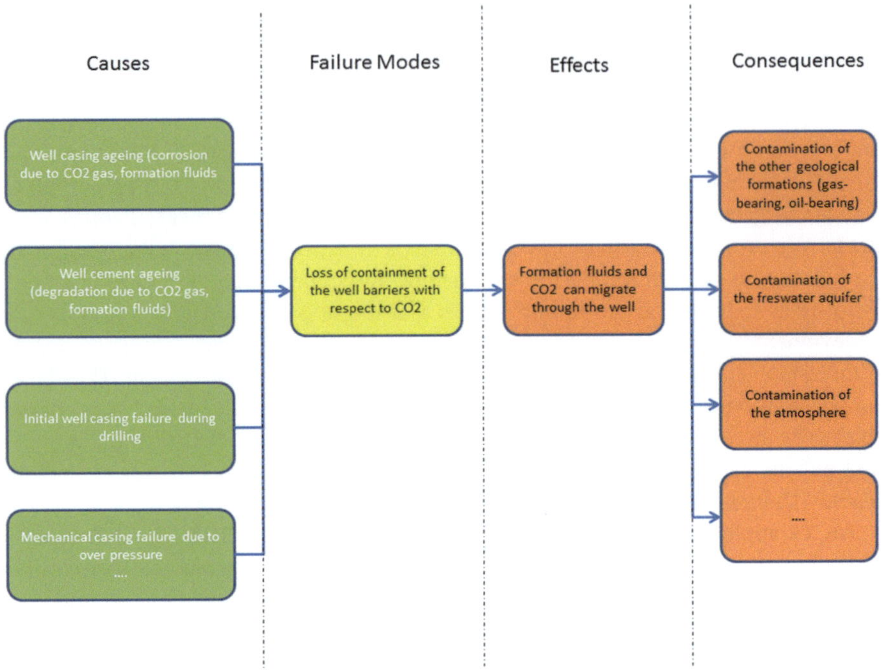

Fig. 10.5 Generic bow-tie illustrating central event 'loss of containment for an injecting well with respect to CO_2' and its associated causes and effects

Table 10.4 Example of the components and functions breakdown for a CO_2 geological storage project

Component	Sub-functions	Interactors	Failure mode
CO_2 reservoir	To ensure injectivity into the reservoir	CO_2	Loss of injectivity
	To ensure storage capacity	CO_2	Loss of storage capacity
Caprock	To resist to the formation fluids pressure	Formation fluids	Deformation of the caprock, cracks
	To ensure the sealing with respect to the formation fluids	Formation fluids	Loss of confinement
	To resist to the injected gas pressure	CO_2	Loss of mechanical resistance
	To ensure the sealing with respect to injected gas	CO_2	Loss of confinement
	To resist to geological formations deformation pressure	Geological formation	Deformation of the caprock, cracks
Wells (all types)	To resist to the formation fluid pressure	Formation fluids	Deformation of the caprock, cracks
	To ensure the sealing with respect to the formation fluids	Formation fluids	Loss of confinement
	To resist to the injected gas pressure	CO_2	Loss of mechanical resistance

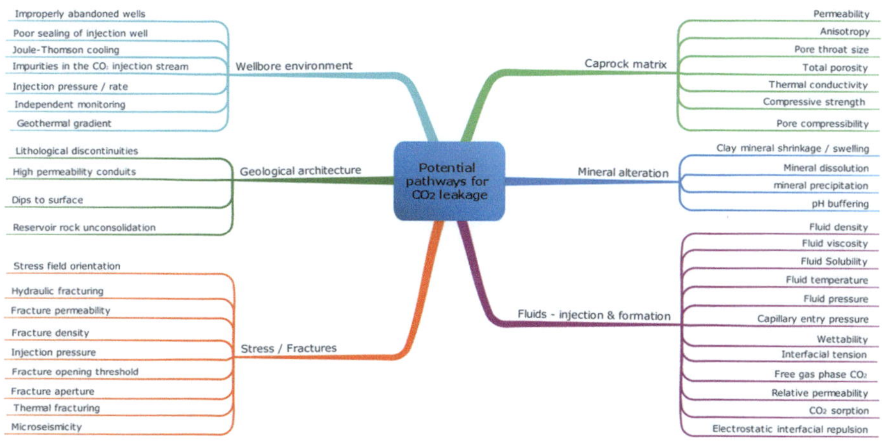

Fig. 10.6 The potential CO_2 leakage pathways and leakage impact factors (risks) influencing caprock leakage grouped by primary category

10.4.2 Risk Estimation

Risk estimation is the second step of risk assessment (Fig. 10.1) where the risk levels are estimated. The input data for this step is the list of risks established by the risk identification process.

The risk level, i.e. criticality, is a combination of:

- A severity level: the magnitude of the impact of a failure mode on the identified objectives. The definition of the different severity levels is established by defining the consequence grid;
- A probability level: the occurrence of the failure mode. The the different probability levels is established by defining the probability grid.

Estimation can be qualitative or preferentially quantitative estimation of the failure mode and its impacts on the associated performance indicators:

- Quantitatively, using statistical analysis, modeling and simulations;
- Qualitatively, on the basis of past records, experience, subject matter expert's opinions or literature review.

A failure mode can have multiple consequences and can impact various objectives, thus each risk must be estimated for every threatened objective. The resultant outcome of the "risk estimation" step is risk mapping, where each risk is plotted by means of its criticality value, represented by the severity level (y-axis in Fig. 10.7) versus probability level (x-axis in Fig. 10.7) of a failure mode.

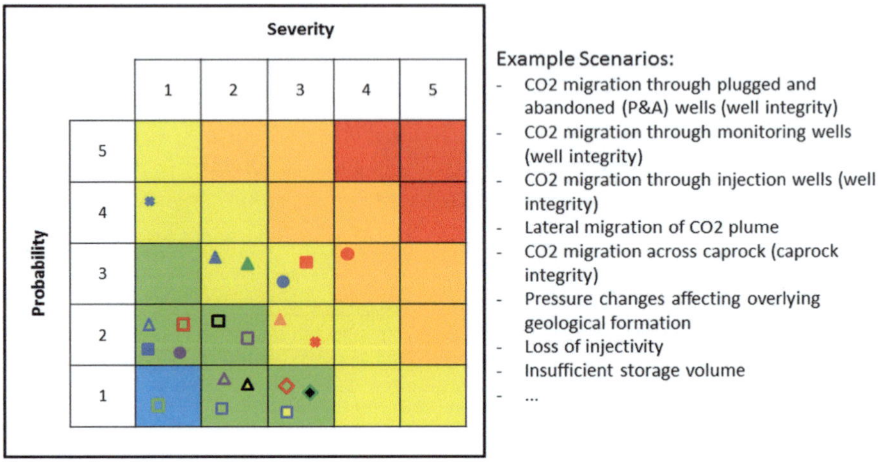

Fig. 10.7 Risk matrix example for a specific CO_2 geological storage project. *Dots* represent various scenarios identified with some examples given in the figure *right* hand panel

10.4.3 Expert Elicitation

Expert elicitation is an approach whereby the reasoned and subjective judgment of experts can be synthesised where there is uncertainty due to insufficient data, making explicit the inherent knowledge based on experience and expertise (Slottje et al. 2008). It is particularly useful in risk assessment when there is very limited "hard" input data. For example, at the beginning of an injection project, early decisions need to be made with regards to data collection, field operation and monitoring strategies. Then, an expert elicitation can be undertaken with the aim to identify, assess and rank potential leakage scenarios, in order to support the assessment and decision making process. The elicitation can be undertaken by a questionnaire where experts are asked for their best estimation of the semi-quantitative scenario uncertainty assessment criteria of severity (i.e. how extensive the leakage could be) and immediacy (i.e. what is the likely time frame of the leakage). Simple mathematical aggregation giving equal weight to all experts can be used for the data analysis and the severity and immediacy be plotted in a probability and impact matrix, assigning an impact rating from low, medium low, medium, high to very high and conclusions made based on the outcome. Such an exercise was carried out for the Heletz pilot injection project (Edlmann et al. 2016), as an demonstration example and the findings were in agreement with more conventional risk assessment studies at existing pilot CO_2 injections sites (Deel et al. 2007; Oldenburg et al. 2011; Watson 2014; Jewel and Senior 2012). Prudent expert elicitation can provide useful insight and guidance and make a valuable contribution to decision making.

10.4.4 Risk Evaluation

Risk evaluation is the third step of risk assessment. The input data for this step are the outcome of the risk analysis, i.e. the list of the risks with a level of severity, probability and criticality.

The purpose of risk evaluation, based on the outcomes of risk analysis, is to make decisions about which risks require treatment and to define priorities between treatment actions.

The final outcome of a risk evaluation is a prioritized risk register recommending further action. In addition, the risk evaluation yields a risk matrix in which risks are ranked (Fig. 10.7). Risks with the higher criticality levels should be treated with priority.

A detail of the map can be drawn for specific scenarios. An example for caprock integrity is detailed below (Fig. 10.8).

10.4.5 Description of Risk Treatment Process

Risk treatment defines the processes of selection and implementation of measures to modify the risk. Risk treatment is based on the outcomes of the risk evaluation which ranked the risks that have to be treated by priority.

Risk treatment involves:

- Identifying the key parameters driving the critical risks;
- Listing the range of options for treating risk, including (1) selecting a short list of actions among treatment options and applying those options to critical risks and (2) assessing the options;
- Defining the actions in terms of cost, nature, and duration.

Before a risk can be effectively treated, it is necessary to understand its cause, in order to identify and select the appropriate actions. Possible risk treatment actions are defined by the project team during review meetings and the selection of treatment options is made by project managers or those delegated by the project manager. Treatment options can include the following:

- *Avoid the risk* by deciding not to start or to stop any activity that contributes to the risk, in other words terminate the risk.
- *Change the nature and magnitude of probability of a risk* by prevention and/or monitoring, thereby lowering the probability of the risk occurring.
- *Decrease the severity of a risk* by protection and/or mitigation actions, thereby lowering the consequence(s) of the risk.

Other options can also be chosen, such as to tolerate the risk by deciding to start or to stop any activity that contributes to the risk.

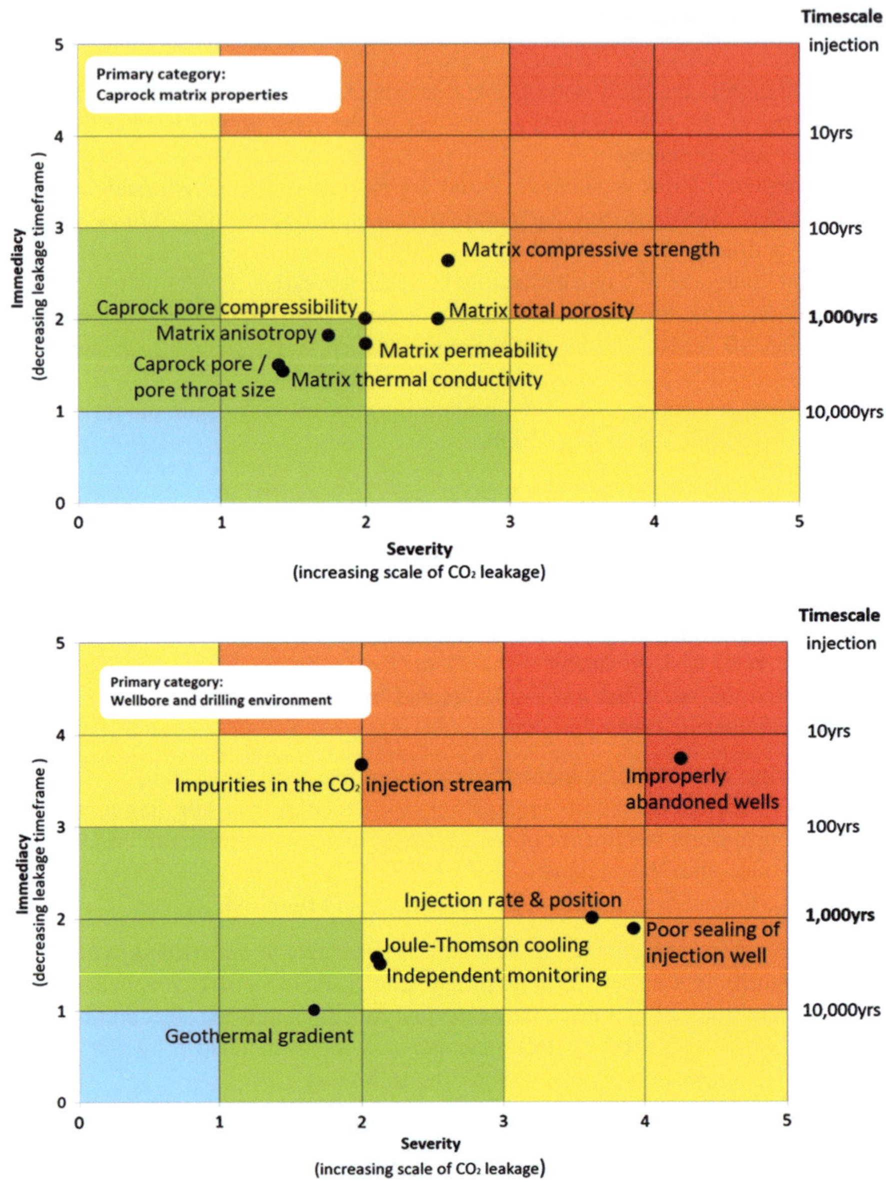

Fig. 10.8 Risk matrix plot for CO_2 leakage scenario. Example for poorly defined input parameters when assessing caprock integrity

The high-criticality risks evaluated in the risk assessment are analyzed and broken down into causes, failure modes and consequences. This process results in a list of parameters that may provide direct or indirect indication of the occurrence of the risks, thus requiring monitoring. This list will drive the ensuing selection of applicable MVA technologies. Relevant MVA techniques are proposed to deal with the parameters previously identified. The technical applicability of each solution for the CCS project has to be discussed, as well as its cost level. This results in a list of relevant MVA technologies recommended for the monitoring of the deep subsurface. Frequencies of data acquisition are then defined for each MVA technology, as well as the action plan stemming from the detection of any deviation from the base case scenario. Two types of MVA plan are recommended in CO_2 geological storage projects:

- **A "regular" monitoring plan**: continuous monitoring shows the system behaviour in accordance with the models. Regular additional surveys are performed to confirm this "normal" behaviour;
- **An "in case" monitoring plan**: continuous monitoring detects a system deviation. A risk-based decision-tree will have to be implemented. This decision tree will define the response plan for a proper mitigation of the risk as early as possible through complementary measurements or modification of the injection strategy.

10.4.6 Preparing and Implementing Risk Treatment Plans

In implementing a risk treatment plan, a strategy is formulated using for example: the description of the proposed actions to treat the risk, benefit expected to be gained (i.e. estimate the residual risk after action), cost and man resource required, and schedule.

The outcomes of the risk treatment are:

- A list of treatment actions associated to each risk, including the actions, resources and planning, in other words, a risk treatment plan;
- A new risk matrix, which takes into account the effect of the treatment actions. In this risk matrix the risks are plotted after the treatment actions (for actions which needed to be treated) with new severity and/or probability levels.

The inclusion of new data in a second-round assessment will support the development of a project-specific MVA plan to monitor the critical risks in the most efficient and cost-effective manner.

10.5 Risk Monitoring, Review and Reporting

10.5.1 Objectives

Risk monitoring allows risk evolution to be tracked over time. In an operational way, risk monitoring is focused on processes and causes of the risks. The purpose is to ensure that risk is known and controlled. Monitoring will also ensure that risk treatment actions are effective. Monitoring, review and reporting is an essential and integral step in the risk management process and it takes place throughout the risk management process (see Fig. 10.1).

10.5.2 Risk Monitoring

Risks need to be monitored to ensure that changing circumstances are recorded and duly reported and analysed. Monitoring actions must be continuous and need to be reinforced during particular actions or phenomena. The periodicity of risk monitoring has to be defined in the risk management policy. Very few risks will remain static. Therefore, the risk management process needs to be regularly repeated so that identified risks are up-to-date and the new risks are captured in the process.

10.5.3 Risk Review

Risk review establishes continuity and improvement of the whole risk management process. This stage helps to identify possible deviation from the objectives defined by the risk management policy (i.e., change of injection conditions). It also evaluates the benefits of the risk treatment actions implemented. Periodical re-assessment of the risks must be performed to control risk changes and residual risk levels. After treatment actions, the risks must be re-assessed to identify if the objectives of the treatment action have been achieved.

10.5.4 Risk Reporting

Risk reporting constitutes a necessary support for monitoring and reviewing risks. It relies on a functional risk management tool that provides risk reports. Risk reporting is important to ensure an efficient communication and traceability between all persons involved in the CO_2 project.

10.6 Conclusion

Geologic CO_2 storage project risk management provides a more accurate understanding of the relevant, project-specific technical risks while establishing a robust framework designed to mitigate subsurface risk through the life cycle of the project. By identifying knowledge gaps in current data, risk assessment activities can provide direction for future studies and characterization work. Additionally, geologic storage risk assessment supports the development of a project-specific, risk-based MVA plan.

As a project progresses, the risks can change. The risks that were once estimated to be high can diminish, becoming negligible, and conversely, risks that were once not relevant can become critical. As a result, the risks must be monitored to ensure they are successfully controlled throughout the lifetime of the project. Because risk management is an iterative process, as the details of a project change, the risk management plan may need to evolve to fit its needs. The risk management plan can be reviewed to ensure that it is still effectively controlling the risks for the project and can be modified if necessary.

The successful application of a risk management framework to CCS feasibility projects provide a step forward for the development of CCS. It supports the idea that a risk management framework, including technical risk assessment, can be effectively implemented for large-scale CCS projects. The risk management framework also provides an invaluable decision-making and communication tool that can validate project planning, educate stakeholders, and demonstrate project safety and reliability—all essential for the success of CCS.

References

Al-Bazali TM, Zhang J, Chenevert ME, Sharma MM (2005) Measurement of the sealing capacity of Shale cap-rocks. In: SPE 96100, presented at the SPE annual technical conference and exhibition held in Dallas U.S.A., October 09–12

Amann A, Waschbüsch M, Bertier P, Busch A, Kroossa BM, Littke R (2011) Sealing rock characteristics under the influence of CO_2. Energy Procedia 4:5170–5177

Bildstein O, Jullien M, Crédoz A, Garnier J (2009) Integrated modelling and experimental approach for caprock integrity, risk analysis, and long term safety assessment. Energy Procedia 1:3237–3244

Bouc O, Bellenfant G, Dubois D, Guyonnet D, Rohmer J, Wertz F, Gastine M, Jacquemet N, Vong CQ, Grataloup S, Picot-Colbeaux G, Fabriol H (2011) Safety criteria for CO_2 geological storage: determination workflow and application in the context of the Paris Basin. Energy Procedia 4:4020–4027

Busch A, Amann A, Bertier P, Waschbusch M, Kroos BM (2010) The significance of caprock sealing integrity for CO_2 storage. In: SPE 139588

Class H, Ebigbo A, Helmig R, Dahle H, Nordbotten J, Celia M, Audigane P, Darcis M, Ennis-King J, Fan Y, Flemisch B, Gasda S, Jin M, Krug S, Labregere D, Naderi Beni A, Pawar R, Sbai A, Thomas S, Trenty L, Wei L (2009) A benchmark study on problems related to CO_2 storage in geologic formations. Comput Geosci Geosci 13(4):409–434

Cugini A, Birkholzer J, Bromhal G, Brown C, Carroll S, Chiaramonte L, Daley T, DePaolo D, Fox M, Friedmann J, McGrail P, Pawar R, Strazisar B, Virden J, Guthrie G (2010) US-DOE's national risk assessment program: bridging the gap to provide the science base to ensure successful CO_2 storage. In: Proceedings of GHGT-10 conference, Amsterdam

Deel D, Mahajan K, Mahoney CR, McIlvried HG, Srivastava RD (2007) Risk Assessment and management for long-term storage of CO_2 in geologic formations—United States Department of Energy R&D. Syst Cybern Inf 5(1):79–84

Dodds K, Watson M, Wright I (2011) Evaluation of risk assessment methodologies using the In Salah CO_2 project as a case history. Energy Procedia 4:4162–4169

Edlmann K, Bensabat J, Niemi A, Haszeldine RS, McDermott CI (2016) Lessons learned from using expert elicitation to identify, asses and rank the potential leakage scenarios at the Heletz pilot injection site. Int J Greenh Gas Control 49

Hnottavange-Telleen K, Krapac I, Vivalda C (2009) Illinois Basin—Decatur Project: initial risk-assessment results and framework for evaluating site performance. Energy Procedia 1:2431–2438

Hnottavange-Telleen K, Chabora E, Finley RJ, Greenberg SE, Marsteller S (2011) Risk management in a large-scale CO_2 sequestration pilot project, Illinois, USA. Energy Procedia 4:4044–4051

ISO/FDIS 31000 (2009) Risk management, principles and guidelines, International Standard, ISO / FDIS31000: 2009 (E) Geneva : ISO

Jewel S, Senior B (2012) CO_2 storage liabilities in the North Sea: an assessment of risks and financial consequences, Summary report for DECC

Le Guen Y, Le Gouevec J, Chammas R, Gerard B, Poupard O, Van Der Beken A, Jammes L (2008) CO_2 storage—managing the risk associated with well leakage over long timescales. In: SPE-116424-MS

Le Guen Y, Huot M, Loizzo M, Poupard O (2010) Well integrity risk assessment of Ketzin injection well (ktzi-201) over a prolonged sequestration period. Energy Procedia 4:4076–4083

Li Z, Dong M, Li S, Huang S (2006) CO_2 sequestration in depleted oil and gas reservoirs—caprock characterization and storage capacity. Energy Convers Manag 47:1372–1382

Maul P, Savage D, Benbow S, Walke R, Bruin R (2005) Development of a FEP (features, events, processes) database for the geological storageof carbon dioxide. In Rubin ES, Keith DW, Gilboy CF (eds) Greenhouse gas control technologies, vol 1. Elsevier Ltd, pp. 701–709

NETL (2011) Risk analysis and simulation for geologic storage of CO_2, DOE/NETL-2011/1459. www.netl.doe.gov

Oldenburg CM (2008) Screening and ranking framework for geologic CO_2 storage site selection on the basis of health, safety, and environmental risk. Environ Geol 54:1687–1694

Oldenburg CM, Preston DJ, Nicot JP, Mazzoldi A, Gupta AK, Bryant SL (2011) Leakage risk assessment of the In Salah CO_2 storage project: applying the certification framework in a dynamic context. Energy Procedia 4:4154–4161

Paulley A, Metcalfe R, Limer L (2011) Systematic FEP and scenario analysis to provide a framework for assessing long-term performance of the Krechba CO_2 storage system at In Salah. Energy Procedia 4:4185–4192

Pawar RJ, Carey J, Chipera S, Fessenden J, Kaszuba J, Keating G, Lichtner P, Olsen S, Stauffer P, Viswanathan H, Ziock H, Guthrie G (2006) Development of a framework for long-term performance assessment of geologic CO_2 sequestration sites. In: Proceedings of the 8th international conference on greenhouse gas control technologies, Trondheim, Norway June 19–22, pp 52–54

Pawar RJ, Bromhal GS, Carey J, Foxall W, Korre A, Ringrose PS, Tucker O, Watson MN, White JA (2015) Recent advances in risk assessment and risk management of geologic CO_2 storage. Int J Greenhouse Gas Control 40:292–311

SCCS (2015) Optimising CO_2 storage in geological formations; a case study offshore Scotland, CO_2 MultiStore project, 15th Sept 2015. http://www.sccs.org.uk/expertise/reports/co2multistore-joint-industry-project

Shukla R, Ranjith PG, van Ruth P, Choi SK, Haque A (2008) Some aspects of caprock failure in geo-sequestration of carbon dioxide: a review. In: 12th International conference of international association for computer methods and advances in geomechanics (IACMAG), Goa, pp 2197–2203

Shukla R, Ranjith P, Haque A, Choi X (2010) A review of studies on CO_2 sequestration and caprock integrity. Fuel 89:2651–2664

Slottje P, Sluijs JP, Knol AB (2008) Deliverable "Expert Elicitation Protocol" of RIVM/SOR project IQARUS: identification quantitative assessment and reduction of uncertainties in burden of disease estimates for environmental Stressors (S/630004). RIVM letter report 630004001/2008

Sorensen JA, Botnen LS, Smith SA, Gorecki CD, Steadman EN, Harju JA (2014) Application of Canadian Standards Association guidelines for geologic storage of CO_2 toward the development of a monitoring, verification, and accounting plan for a potential CCS project at Fort Nelson, British Columbia, Canada. Energy Procedia 63:5959–5970

Stauffer P, Viswanathan H, Pawar RJ, Guthrie G (2009) A system model for geologic sequestration of carbon dioxide. Environ Sci Technol 43(3):565–570

Stauffer PH, Pawar RJ, Surdam RC, Jiao Z, Deng H, Lettelier BC, Viswanathan HS, Sanzo DL, Keatin GN (2011) Application of the CO_2-PENS risk analysis tool to the Rock Springs Uplift, Wyoming. Energy Procedia 4:4084–4091

Stenhouse SMJ, Zhou W, Savage D, Benbow S (2005) Framework methodology for long-term assessment of the fate of CO_2 in the Weyburn Field. In: Thomas DC, Benson SM (eds) Carbon dioxide capture for storage in deep geologic formations—results from the CO_2 Capture Project, (2), pp 1251–1262

U.S. Environmental Protection (2008) Agency vulnerability evaluation framework for geologic sequestration of carbon dioxide EPA430-R-08-009

Veritas DN (2009) CO2QUALSTORE: guideline for selection, characterization and qualification of sites and projects for geological storage of CO_2, report DNV 2009-1425

Vilarrasa V, Olivella S, Carrera J (2011) Geomechanical stability of the caprock during CO_2 sequestration in deep saline aquifers. Energy Procedia 4:5306–5313

Viswanathan HS, Pawar RJ, Stauffer PH, Kaszuba JP, Carey JW, Olsen SC, Keating GN, Kavestski D, Guthrie GD (2008) Development of a hybrid process and system model for the assessment of wellbore leakage at a geologic CO_2 sequestration site. Environ Sci Technol 42:7280–7286

Watson M (2014) Containment risk assessment. In: Cook P (ed) Geological storing carbon: learning from the Otway project experience. CSIRO Publishing, Melbourne

Wildenborg T, Leijnse T, Kreft E (2004) Long-term safety assessment of CO_2 storage: the scenario approach. In: Seventh international conference on greenhouse gas control technologies GHGT-7

Wollenweber J, Alles A, Busch A, Krooss BM, Stanjek H, Littke R (2010) Experimental investigation of the CO_2 sealing efficiency of caprocks. Int J Greenhouse Gas Control 4: 231–241

Yavuz F, van Tilburg T, David P, Spruijt M, Wildenborg T (2009) Second generation CO_2 FEP analysis: CASSIF—Carbon Storage Scenario Identification Framework. Energy Procedia 1 (1):2479–2485

Index

A. Niemi et al. (eds.), *Geological Storage of CO_2 in Deep Saline Formations*,
Theory and Applications of Transport in Porous Media 29,
DOI 10.1007/978-94-024-0996-3

E

N

O

T

MIX
Papier aus verantwortungsvollen Quellen
Paper from responsible sources
FSC® C105338

If you have any concerns about our products,
you can contact us on
ProductSafety@springernature.com

In case Publisher is established outside the EU,
the EU authorized representative is:
Springer Nature Customer Service Center GmbH
Europaplatz 3, 69115 Heidelberg, Germany

Printed by Libri Plureos GmbH
in Hamburg, Germany